# Furniture Design

Jerzy Smardzewski

# Furniture Design

 Springer

Jerzy Smardzewski
Poznan University of Life Sciences
Poznan
Poland

ISBN 978-3-319-19532-2        ISBN 978-3-319-19533-9   (eBook)
DOI 10.1007/978-3-319-19533-9

Library of Congress Control Number: 2015941136

Springer Cham Heidelberg New York Dordrecht London

Printed on acid-free paper

Springer International Publishing AG Switzerland is part of Springer Science+Business Media (www.springer.com)

# Preface

Wood is an excellent construction material, which has been used by people for thousands of years for the production of building constructions, machinery, tools, interior design, including furniture, accessories, and even jewellery. In particular, pieces of furniture made from wood are structures, which sometimes have not changed their form and technical solutions for several millennia. Many excellent furniture models were the work of outstanding designers, who have mastered all the details with reverence. To this day, they are appreciated and recognised among connoisseurs and collectors of works of art.

Today, many countries have recognised design as a priority direction for the development of education and economy, seeing in it the quintessence of innovation and an opportunity to modernise European economy. However, it should be noted that modern furniture is not only the fruit of the work of individual architects and artists. Creating an attractive, functional, ergonomic and safe piece of furniture requires an effort of many people working in interdisciplinary teams, and to them, among others, this book is addressed.

The aim of the book is to present the principles of designing furniture as wooden structures. It discusses issues related to the history of the furniture structure, classification and characteristics in terms of the most important features essential during designing, ergonomic approach to anthropometric requirements and safety of use. It presents methods and errors of designing, characteristics of the materials, components, joints and structures, and rules of developing design documentation. It also raises the issue of calculating the stiffness and endurance of parts, joints and whole structures, including the questions of the loss of furniture stability and of resulting threats to health and even life of the user.

Jerzy Smardzewski

# Contents

# Chapter 1
# The History of Furniture Construction

## 1.1 Introduction

At the dawn of human civilisation, when the concept of furniture was not yet known, man, driven only by the need to make life easier, in a natural way used various objects made spontaneously by nature. A trunk of a tree felled by the wind or rock served as a place to sit (Fig. 1.1), a flat stone block served as a base for performing a variety of common work, and soft moss or woollen skins served as a bed. Over the years, as a result of the creative activity of humans, artefacts began to be made which replaced the spontaneously made objects mentioned earlier. Over the centuries, due to the preferences of societies that lived in a given age, their forms changed. New types of furniture were created that fulfilled specific functions: to sit, lie down, for work, for dining, storage and others.

Generally, the remaining furniture constructions from the first dynasty of ancient Egypt are accepted as the beginning of the history of furniture (the years around 3100–2890 B.C.) (Setkowicz 1969). Meanwhile, there is much evidence to suggest that furniture was manufactured and used by humans in the late Palaeolithic and early Neolithic period.

Historically, the most commonly used material for manufacturing furniture was wood. Archaeological finds, however, indicate that in steppe and permafrost terrains, stone, metal and animal bones, especially mammoth bones, were also used. Despite the fact that the reconstruction of prehistoric homes with their equipment is not possible, there is not the slightest doubt that they housed furniture. Archaeological discoveries in the pool of the middle River Don (among others, in the region of the village of Kostienki near Voronezh 16) indicate that in this area, groups of mammoth hunters continued a semi-sedentary lifestyle, also in the period after maximum glaciation (the second pleniglacial, i.e. after 18 thousand–17 thousand years ago) (Kozłowski 1986; Escutenaire et al. 1999; Svoboda 2004). Such a lifestyle was conducive to the creation of innovations, which preceded the civilisation achievements of the first sedentary farmers and breeders. These

© Springer International Publishing Switzerland 2015
J. Smardzewski, *Furniture Design*,
DOI 10.1007/978-3-319-19533-9_1

**Fig. 1.1** Stone shaped by nature as one of the first objects that functioned as a piece of furniture

innovations include many areas related not only with settlement and economic strategies, symbolic culture and religion, but also with material culture. The building type, for both residential ground buildings and half dugouts, was characterised by unprecedented soundness and stability. Due to a lack of wood, caused by the gradual disappearance of trees in the periglacial steppe environment, mammoth bones were used to build foundations, and also the construction of walls and ceilings of structures.

The first houses built almost only from mammoth bones we know from Moravia (e.g. Milovice), as well as from eastern Europe: from Kostienki, approximately 18 thousand–17 thousand years B.C. (Anosovka II site, cultural layer Ia), and established 16 thousand–14 thousand years ago in the River Dnieper basin (sites—Mezhirich, Mezine, Dobranichevka in Ukraine) (Svoboda 2004).

The main theme that is presented in Palaeolithic arts is a woman's figure with strongly marked gender features, occurring during the period 30 thousand–20 thousand years ago throughout all of Europe from the Atlantic Ocean to the River Don, known as the so-called Palaeolithic Venus (Soffer et al. 2000). The woman presented in Fig. 1.2, on the contrary to many other figures of Venus, is sitting on a seat especially designed for this function. This indicates the fact that, regardless of the lack of reconstructions of complete homes from that period, people of the Palaeolithic Era not only made tools that were necessary to acquire food and its processing, but also made usable objects, including furniture.

**Fig. 1.2** Venus figure from Gagarino approx. 21000 B.C. (Gorodnjanski V.). Gagariono is located on a loess terrace on the north lip of a ravine on the right bank of the Don River about 5 km north of the junction of the Sosna, a tributary stream. It is north of the well-known Kostienki sites

Therefore, it seems that upper Palaeolithic mammoth hunters of the periglacial steppe, due to their semi-sedentary lifestyle, over 10 thousand years earlier achieved a standard of development that came close to the Neolithic societies in the Middle East (Kozłowski 1986). This thesis is all the more likely because in the Middle East many innovations, such as the appearance of monumental architecture and stone sculpture, attributed to the peoples of the pre-ceramic Neolithic Era, was in reality, as recent discoveries in eastern Anatol show, the work of settled hunters and gatherers (today defined as "sedentary foragers") in the eleventh–tenth century B.C. This indicates that many civilisation achievements did not depend directly on the production of food, but it was primarily the result of the sedentary lifestyle regardless of the type of farming.

We know much more about the creative accomplishments of peoples living in Europe in the Neolithic Era. Near the most famous stone circle of Stonehenge, British archaeologists unearthed a huge settlement dating back to before 4.5 thousand years B.C. This discovery sheds new light on the role of Stonehenge. According to anthropologists, the people who built it also created a similar wooden construction.

**Fig. 1.3** Interior of one of the households of the Neolithic village at Skara Brae: **a** stone cupboard, **b** stone beds (*Photograph* Nick Lee)

Durrington Walls is the largest British wooden henge, and this site is the remains of the largest known wooden British settlements from the Neolithic Era (from before 6 thousand–4 thousand years B.C.). It was created exactly at the same time when the first boulders in Stonehenge were begun to be erected, that is about 4.5 thousand years ago. Today, there is no doubt that the settlement was inhabited by hundreds of people. In September 2006, archaeologists unearthed inside the ring a floor of eight houses. They found imprinted traces of beds and other furniture, the remains of fires placed in the middle of the house, and various household rubbish. The next two houses were found in the western part of Durrington Walls. They were surrounded by their own palisades and ditches.

Another interesting find, presenting the art of making furniture in the ancient times, is the Neolithic village of Skara Brae located on the western coast of Orkney in Great Britain, from 3200 B.C. In the reconstructed rooms of the old one-room households one can find: wardrobe, beds and cupboards made of stone (Fig. 1.3). The residents used this material mainly because of low forest cover on the island, therefore minimal resources of wood, which was used rather as a fuel than a raw material for making furniture.

The use of furniture in the Neolithic Era is also shown by the stone figurines of sleeping or seated figures of women (Fig. 1.4).

Prehistoric designers, like modern designers, paid particular attention to a comprehensive approach to designing, meeting both the requirements concerning a satisfactory appearance, as well as the necessary functions. In fact forms of furniture

**Fig. 1.4** Mother Goddess from Çatal Hüyük, Turkey. Neolithic Era approx. 6000–5500 B.C. (Museum of Anatolian Civilizations in Ankara)

constituted small architectures with legs designed like columns, in other cases they were parts of anthropomorphic forms of animal origin. Furniture design and their form changed from simple to intricate, depending on the period in which they were made. Some of the oldest, well preserved and described objects in museum collections come from the region of the former Mesopotamia, richly gilded furniture constituting the furnishing of palace interiors. We learn of many of them thanks to the good condition of parts of ancient furniture that survived to modern times, such as original parts of gilded Egyptian furniture buried together with the mummies of Pharaohs in the hot sands of the desert. The style and form of furniture evolved much more quickly than other forms of architecture, thus reflecting new ideas and innovative solutions of past designers. In many cases, however, one can conclude that the functionality of those structures still remains impeccable and timeless. Tables and chairs used by Egyptian workers in 2800 B.C. looked and were used identically as chairs of workers in the year 1800, of the Finnish peasants from the region of Yamsakoski (Fig. 1.5). Also Dutch painters of the seventeenth century and American painters of the early nineteenth century presented interiors of rural huts identically.

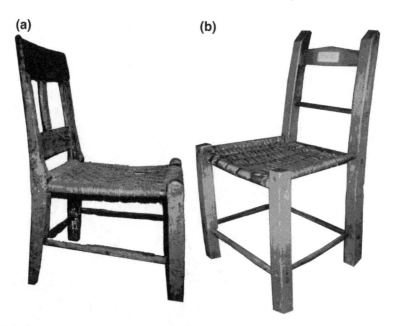

**Fig. 1.5** Similarity of furniture structure: **a** chair with a plaiting, Egypt around 2800 B.C. (British Museum), **b** chair with a plaiting, Finland around 1930 (Heritage Park in Jämsänkoski)

## 1.2   Antique Furniture

### 1.2.1   Furniture of Ancient Egypt

We mainly learn about the form and construction of furniture made in ancient Egypt from the perfectly preserved finds, reliefs and paintings that decorate the walls of the tombs of Pharaohs. It was found that many design solutions used by the contemporary artisans are also used today. All retained museum exhibits of furniture of ancient Egypt prove that the Egyptians used many techniques for decorating furniture. Gold plating and ivory incrustation were common methods of finishing the surfaces of furniture. These methods, as well as making legs in the shape of animal paws, became the common practice of carpenters from much later periods. To make valuable furniture, Egyptian carpenters used the wood of the ebony, cedar, yew, acacia, olive, oak, fig, lime and sycamore tree, often by importing this raw material from Asia Minor, Abyssinia or Namibia. Exterior elements of furniture were finished off not only with ebony wood combined with ivory, but also metal (brass, silver, gold), mother of pearl, lacquer, colourful faience and semi-precious stones (Gostwicka 1986). Ancient Egypt was familiar with bone glue, which was used in techniques of incrustation and veneering. The Egyptians used mortise and tenon joints and dowelled joints to combine most structural elements of furniture. In sarcophagi, chests and dressing tables they also used dovetail joints or bevelled joints (Setkowicz 1969).

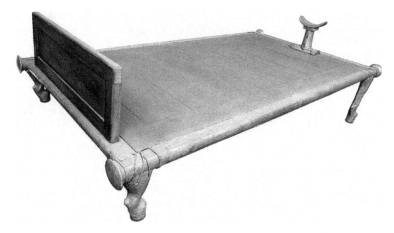

**Fig. 1.6** Bed of Queen Hetepheres IV, from the dynasty of Snefru, around 2575–2551 B.C. (reconstruction of the original from the Museum of Fine Arts, Boston)

The oldest known Egyptian bed, more or less from the times of the first dynasty, is a design consisting of a horizontal wooden frame, resting on four thick and massive bull legs carved in ivory. The legs were joined to the frame usually with mortise and tenon joints, while on the frame of the bed, belts made from leather or other kind of plaiting were stretched over.

Beds with higher peaks on the side of the head were made with footrests. In some bed designs, the connection of the legs with the frame of the bed system was strengthened by bindings from leather strap stretched across drilled holes (Fig. 1.6). The beds in ancient Egypt did not have headrests, but wooden boards were placed at the side of the feet. The board was fixed by two bolts coated with a copper sheet, which also matched the sockets in the frame that were lined with copper. The board at the ends of the legs was the only ornate part of such a bed. The legs in the shape of lion paws were usually directed towards the head.

It is also known that the Egyptians used litters already during the first dynasty. The litter of Queen Hetepheres has survived to our times. Elements of the litter are also joined by leather straps or with a tongue and groove joint. On the front side of the seat back, at the height of the armrests, was an ebony strip inscribed by gold hieroglyphics.

Sitting furniture in Ancient Egypt had a wide variety of forms and an extremely large number of design varieties. A chair, table, bed and corner settee from the fourth dynasty (2600 B.C.), preserved in the tomb of Queen Hetepheres, had legs in the shape of animal paws usually turned to the front and always parallel to each other. They were not set directly on the ground, but on rounded bricks or spheres. If the armchair had a backboard and side rests, they were filled with papyrus, bas-relief or an openwork crate with figural, human and animal motifs, or symbols consisting of hieroglyphics.

**Fig. 1.7** Stela of Amenemhat, the end of the Middle Kingdom of Egypt (middle of the second millennium B.C., Cairo Museum in Egypt)

Figure 1.7 shows a limestone depicting a scene of a funeral banquet, in which the whole family participates. The father, mother and son named Intef sit on a long bench with legs in the shape of lion feet, with two low backboards on both its sides. Next to the bench is a lower, columnar table set with the sacrifice of various kinds of meat and vegetables.

Tutankhamun's throne is extremely impressive, richly decorated with gold, silver, semi-precious stones and a coloured glass paste (Fig. 1.8). Glass paste is a glassy mass consisting of silicates, fused in refractory forms, mixed with crystal and dyed with metal oxides. A special property of glass paste was its susceptibility to plastic working. The legs of the throne have the shape of lion paws, the armrests are two-winged cobras in double crowns of Upper and Lower Egypt, spreading the wings over the cartouches of the king. The front edge of the seat is decorated with two lion heads. The backrest's decoration is a scene depicting Tutankhamun sitting on a soft upholstered throne with a pillow, holding one hand on the rest, and supporting his feet on a footrest. Ankhesenamun is standing in front of him and with his right hand he is rubbing Tutankhamun's arm with an ointment from a dish in his left hand. This drawing illustrates that not only beds, but also seats and backrests of chairs were spread over with soft pillows filled with feathers or wool.

The folding chair found in the annex of Tutankhamun's tomb has a very interesting design, which was transformed from a stool by adding a backrest (Fig. 1.9). It is made from ebony wood with irregular incrustations of ivory, which imitate the skin of a leopard, while the legs have the shape of duck heads. The backrest is also made from ebony wood incrusted with ivory, decorated with semi-precious stones, glass paste and a gold sheet.

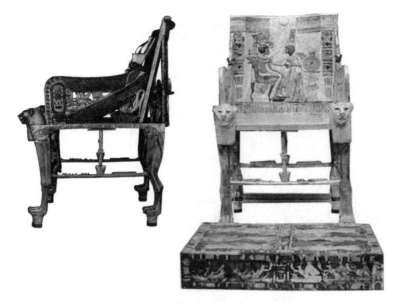

**Fig. 1.8** Tutankhamun's throne. West Thebes, Valley of Kings, Tutankhamun's tomb, 18th dynasty, around 1325 B.C. (Cairo Museum in Egypt)

When chairs appeared, the Egyptians also began to build tables. They were similar to each other in construction. The legs were usually made of thin stiles inclined at an angle to the ground and joined together in the central part with a connector, and in the upper part with a case. During funeral banquets, tables were erected on one, it seems, turned column (Fig. 1.10). This is an observation that is extremely intriguing, because so far traces of tools and equipment for turning, which the Egyptians could have used, have not been found. However, it seems highly unlikely that with such a developed technique, for those times, this technology was unknown to them. As reliefs show (Fig. 1.10), the seats of chairs were usually a bit higher than the working boards of tables, on which loaves of bread or incense burners lay.

Aside from chests for storing clothes, the ancient Egyptians made sarcophagi, coffins, dressing tables for storing toiletries and jewellery, as well as trunks. Many of them were made from wooden boards joined at the width, and in the corners dowelled, bevel and multi-dovetail joints were used.

The beautiful jewellery chest presented in Fig. 1.11 was found in the tomb of Yuya and Tjuyu. The chest is supported on long, slender legs adorned with a geometric ornament made from faience and ivory dyed pink. The same continuous pattern runs along the edge of the chest and its arc-shaped top. The sides are divided by a geometric frieze into two equal parts, the upper is adorned with a hieroglyphic inscription made in gilded wood, containing the cartouches of Amenhotep III and his wife Tiye, who was the daughter of Yuya and Tjuyu. The lower part is filled by a recurring motif consisting of symbols expressing the wish that the owner of the

**Fig. 1.9** Tutankhamun's ceremonial throne, around 1325 B.C. (Cairo Museum in Egypt)

chest enjoys life and fortune. The lid is decorated with two symmetrical panels of gilded wood and faience.

## 1.2.2   Furniture of Ancient Assyria and Persia

We learn about the material culture of the Assyrians mainly from archaeological excavations. Assyrian furniture was usually cast from bronze and shelving, much like in Egypt, placed on supports in the form of lion paws. Also, many carvings and sculptures take the form of animal heads, their full figures, as well as numerous

**Fig. 1.10**  Relief, Horemheb
at a dining table. Sakkara,
Horemheb's tomb, 18th
dynasty, around 1325 B.C.
(British Museum)

**Fig. 1.11**  Jewellery box.
Valley of Kings, Tjuyu's
tomb, 18th dynasty,
Amenhotep III's rule, the
years 1387–1350 B.C.
(Cairo Museum in Egypt)

**Fig. 1.12** Palace in Nimrud, relief presenting king Ashurnasirpal II on the throne, next to which a footrest is standing, around 865–860 B.C. (British Museum)

human figures. Characteristics of the Assyrian culture were also platforms that constituted small construction forms resembling skeletal structures, upon which the actual piece of furniture was set. Thanks to such a solution, the Persians obtained the impression that a seated person has an advantage over a standing one, which was to arouse due respect to the ruler. Structurally, this effect was achieved by applying numerous turned spheres and spirals on which the legs of chairs, armchairs and platforms were set (Fig. 1.12).

Chairs and armchairs were richly lined with colourful textile materials. This particular feature distinguished Assyrian furniture from many others, but these achievements were commonly used by both Greeks and Romans.

Assyrian beds were much higher than Egyptian ones, as the Persians received guests and celebrated at lavishly set tables in reclining positions. Assyrian furniture was extremely massive, heavy, oversized and impractical. Their main task was to emphasise social position of the user, rather than to provide him with maximum comfort of use.

### 1.2.3   Furniture of Ancient Greece

Although no good-quality museum exhibits were preserved, the constructions of Greek furniture can be fairly accurately recognised based on frescoes, paintings on pottery, bas-reliefs, as well as on the basis of numerous written messages. The earliest forms of furniture works in Greece clearly take advantage of Egyptian design, but their further evolution was directed in its own original forms, reaching peak development in the fifth century B.C. Greek artisans perfectly mastered the technology of bending and turning wood, gluing and veneering, as well as finishing wood surfaces with varnish and polychrome. They also expertly used the technique of joining elements using different connectors and glue. They knew of mortise and tenon joints, dowelled and dovetail joints, as well as frame constructions and weaved seats. They commonly used box wood, yew wood, walnut wood and ebony wood (Gostwicka 1986). In Greece, many new, original structural forms of furniture were created, especially chairs and stools. They were ascribed separate, distinctive names, which include: diphros, i.e. a small and lightweight stool, diphros okladias a type of folding stool consisting of a large number of elements under the seat, klismos, klinter or klisja is a lightweight chair with a backrest and legs that are characteristically bent forward and backward, intended mainly for women, kathedra or thronos represented by a heavy chair with a backrest, designed for men, having insignia showing their authority and family or social position.

A diphros stool was relatively low, without a backrest, with straight, turned legs in the shape of a mace, usually placed perpendicular to the ground (Fig. 1.13).

**Fig. 1.13** Diphros-type stool. Relief from Athens, around 430–420 B.C. (British Museum)

The seats of these stools were most often made of a strap weave, leather or fabric. A folding stool with crossed legs was called a diphros okladias, i.e. a folding diphros. Due to the mobility of the construction, the piece of furniture was made of metal or wood.

A klismos-type chair was made from wooden curved slats obtained by both plastic working and bending wood, as well as cutting elements from trees of a large, naturally shaped curve. There are also facts known that the Greeks joined the boughs and branches of trees, fixing them to the ground with anchors, in this way forcing the desired curvature for the planned construction of furniture. The bending of legs of klisja was also important for construction, since it increased the sections of the legs in the case part, enabling technical bonding using mortise and tenon joints (Fig. 1.14).

The throne was usually richly decorated with animal or human figures. In many cases, the Greek throne was similar to the ones manufactured in Egypt and Assyria.

**Fig. 1.14** Klismos-type chair with bent backrest. Relief from Athens, around 430–420 B.C. (British Museum)

**Fig. 1.15** A thronos chair made from rolled elements. The relief shows the seated Gaius Popillius, around 50 B.C. (Archaeological Museum in Thessaloniki)

In particular, the structures of footrests and armrests are characteristic. However, they are distinguished by the impressive height of the backrest and rolling technology used in shaping almost every part of the piece of furniture (Fig. 1.15). The thrones were placed in public places, temples, stadiums and theatres; they were made of marble and always richly decorated by Greek ornamentation.

The ancient Greeks, like the Persians, did not have tables of a function and purpose as we know today. Commonly only feast tables were used, set by the beds and reaching the height of the bed (Fig. 1.16). An exception was sacrificial, column tables, with a wide worktop and height adjusted to a standing person.

Greek artisans paid particular attention to furniture for lying down on. They were made from wood decorated with precious veneers, inlay, incrusted with precious metals or ivory. Wealthy Greeks had beds lined with soft leathers, cloths and woollen fabrics, on which linen sheets were placed. Pillows and mattresses filled with wool or feathers were used.

**Fig. 1.16** Beds and feast table. Chalkidiki, around 380 B.C. (Archaeological Museum in Thessaloniki)

## 1.2.4   Furniture of Ancient Rome

In Roman, furniture practicals ruled, which consisted in adapting all useful aesthetic, structural and organisational solutions of the peoples of the conquered countries. The civilisation predecessors of Rome were the Etruscans, who through their contacts with the Greeks, brought many interesting solutions into their territory. This is why in appearance, the Etruscan and Roman beds resemble Greek furniture (Fig. 1.17).

In Rome, furniture designed for sitting was known under the generic name sella; only a chair with armrests was called a kathedra (Setkowicz 1969). Furniture that emphasised power also included the stool kurulne (sella curtulis). This is a structure in the form of a stool, without a backrest, with legs bent and crossed, with a pillow placed on the seat. Initially, chairs and stools were made of wood, as well as metal incrusted with ivory, in later times they were also decorated with gold (Fig. 1.18).

Kathedra-type chairs, similar to the chair of the Greek klismos, had the legs bent outwards, but they were characterised by much greater sizes of cross sections of individual elements. The Solium was a representative chair, the seat of honour of the man of the house, as well as the throne of the head of the State and other dignitaries of the administration. At one of such piece of furniture, the seat is supported by two Sphinxes, which wings that are raised up high form armrests (Fig. 1.19).

The Romans used three types of beds: for resting—lectus cubicularis, for feasting—lectus tricliniaris, for work—lectus lucubratorius (Setkowicz 1969). The bed construction was mostly a frame supported on four legs. The supporting and

**Fig. 1.17** Roman bed with soft pillows, sarcophagus from Side, around 30 B.C. (Museum in Side, Turkey)

**Fig. 1.18** Metal stool like
sella, around 100 B.C.
(British Museum)

springy layer was made from brown tapes woven over one another, which formed a kind of chess board. Roman sleeping beds were usually built as lightweight and portable, in whole or in part from metal, with an S-shaped headboard, usually on both sides of the bed. Some beds of wealthy Romans were lined with mattresses

**Fig. 1.19** Representative
marble throne Solium (Louvre
Museum, Paris)

filled with wool and goose or swan feathers. The Romans were the first also to
make furniture woven from cane. Among the different types of wood, Roman
artisans mainly used walnut, ash, beech, ebony, and as auxiliary types palm, lemon
and thuja.

Roman dining tables were similar to Greek ones. Lightweight, movable,
three-legged or one-legged, low tables, reaching only the height of the bed, exhibited
elegance. Three-legged and four-legged tables had a structure made of cross-linked
bars making it possible to fold the frame after removing the work top. In contrast to
this design, in homes surrounding the ancient Agora, stone tables were erected with a
wide work top and massive stem in the form of a stone column. The tops of these
tables in the form of an elongated rectangle were either marble or wooden.

The Romans also knew the principle of building wooden cabinets, called ar-
maria, filled with shelves for storage. In the atrium of the Roman house stood a
chest, which stored the household valuables. The chest was made of wood and fitted
ornamentally with iron and bronze tapes, which protected it against an intruder.

## 1.3   Furniture of the Middle Ages

A characteristic feature of furniture in the early Middle Ages was the widespread use of wood turning techniques, including for legs, backrests, cases and connectors. Many of these elements referred to the products of Roman artisans in form and style. In the Byzantine period, furniture was incrusted with ivory and lined with rich and soft fabrics. Like the Greeks and Romans, Byzantine carpenters used support legs between which they fitted openwork or board backrests. A symbol of the significant influence of the antiquity on Byzantine work is also the throne of King Dagobert (Fig. 1.20).

In the Romanesque period, the technique of manufacturing furniture deteriorated significantly and came down to simple carpenter works, while for decorating purposes polychromy and coloured paintings on a chalk base were commonly used. At that time, in the north of Europe, furniture was produced mainly from oak wood, while in the south, from coniferous wood. The structures were massive, oversized with large sections of elements, often turned and carved (Fig. 1.21). Next to the chairs with a rectangular seat supported on four legs, three-legged frame structures with triangular seat boards were also made.

**Fig. 1.20**  The throne of king Dagobert made from gilded bronze, around the seventh century (Department of coins, medals and antiquities, France)

**Fig. 1.21** Romanesque armchairs built from massive, turned legs, around the twelfth–thirteenth century

We know about the constructions of Romanesque beds only from iconography. Based on them, it can be concluded that the furniture in the early Middle Ages differed significantly from antique furniture (Gostwicka 1981, 1987). The main element of the design of the bed is turned or cuboid corner posts tied with cases (Fig. 1.22).

For storage, mainly chests were used which were massive in structure made of thick fir-tree wood or oak boards that formed the side walls and bottom of the chest, the walls were joined using vertical posts. Cabinets, as a substitute for chests, were extremely rare. Their earliest designs come from the thirteenth and fourteenth century. Romanesque cabinets were characterised by smooth, although thick exterior walls, made of wooden, well carved boards.

The Gothic period began a clear breakthrough in European furniture making. In place of the massive and heavy furniture, light frame panel constructions appeared. This resulted in chests, which were hitherto widely used in the Romanesque period, being ousted by cupboards that were innovative in form and functionality, in which the frame elements were joined by mortise and tenon joints. Mostly, they served for storage, and pantries also for displaying expensive dishes and table settings, which testified to the luxury and wealth of the household members.

In the fifteenth century, multi-door cabinets started to be built, which were equipped with numerous lockers, drawers and flaps. These furnishings were placed on a high base with a frame construction with corner supports. Panel elements were usually veneered with precious veneers made of ash wood or maple wood, and in the Alpine countries, the wood of pines, fir trees, spruces, larches, cedar and

**Fig. 1.22**  Romanesque bed from the fourteenth century

sometimes walnut was used. Thanks to the special design of carving tools, a new ornament was also introduced, commonly used in the Gothic period—it was flat in the shape of scrolls of parchment paper. However, it was difficult to do in coniferous wood; therefore, the plant ornament was used interchangeably. Gothic carpenters used chisels and planes, which made the quality of wood carving, which decorated practically all furniture, more attractive. In addition, artisans used inlay, polychromy and gilding.

Some furniture for storage, in particular chests, which in the Gothic period was encased with backrests and armrests, also began to serve as seats (Fig. 1.23). Examples of such furniture have been preserved mainly in churches in the form of typical stalla.

Among the much distinctive furniture for sitting from this period, the most important is Italian constructions with folding frames, while typical pieces of furniture for lying on were beds of a chest structure with a canopy. Alcove beds were known then, walls of a frame panel structure built up on three sides, as well as curtain beds covered with a textile curtain on four sides. A breakthrough was also enclosing the bottom part of the bed with a chest construction designed for storing bed linen, curtains and heavy canopies.

Significant progress took place also in the group of furniture for work and dining. Tables lost their massive and immutable character to structures with divided and sliding panels, which easily provided a new usable surface. Furniture with

**Fig. 1.23** Gothic sitting and
storage furniture from around
the fifteenth century
(Avignon, France)

folding panels, fitted on hinges, also appeared. The tables of wealthy owners were
distinguished by carved coats of arms incorporating other ornaments that made the
piece of furniture more beautiful (Fig. 1.24).

## 1.4  Modern Furniture

### 1.4.1  Renaissance Furniture

When in Italy a new direction in artistic creativity appeared, which directed its
inventiveness towards potential users, furniture achieved previously unencountered
aesthetic and functional values. They ceased to be exclusively objects of common
use and transformed into works of applied art just as attractive as sculptures or
paintings. During this period, artists drew from the rich heritage of ancient culture,
using woodworking techniques similar to those used by sculptors working with
stone. Chairs with a cross-structure and richly decorated stools came to be widely
used (Fig. 1.25).

During the Renaissance, the masters in furniture production were the Italians
from Florence, Sienna, Lombardy and Venice. They mainly used the wood of the
walnut tree, which they decorated by dyeing and finishing with wax The seats and

**Fig. 1.24**  Late Gothic table from the fifteenth–sixteenth century

**Fig. 1.25**  Stools (Museum of the Radziwiłł family in Kiejdany)

backrests of chairs were made as lattice, panel and frame panel structures, on which pillows or mattresses were placed, which were filled with hair and covered in leather, damask, velvet or gobelin fabric.

The most representative piece of furniture for lying on during the early Italian Renaissance period was a bed without a canopy, surrounded by steps in the form of chests. Later, the steps were completely integrated with the cases of the bed.

In some designs, the legs were massive columns protruding slightly over the surface of the bed, while others provide high support for canopies.

**Fig. 1.26** Oak Renaissance chest, around 1520 (Victoria & Albert Museum, London)

A distinctive piece of furniture for storage was the chest. In the fifteenth century, finishing touches in the form of cornices, pilasters and plinths became widely used. The side walls of the chest were decorated with pilasters or flattened consoles. The top of the chest was shut by a flat top decorated with a protruding cornice. In a later period, it was begun to divide the chests into individual fields by using side skirts, figural reliefs, cartouches, medallions, trophies and other similar elements (Fig. 1.26).

As it can be seen, the place for flat painting decorations and inlays used thus far is taken up by a deep relief sculpture, which with the exception of the top of the chest, covers all the visible walls completely. The Renaissance chest, like in the Gothic period, was used as furniture for sitting.

Under the influence of the Italian Renaissance, furniture workshops all over Europe were creatively developing. An original accomplishment of the French Renaissance was the armchair with a high backrest and armrests with a seat in the shape of a trapezium (Fig. 1.27). The chair was light, and hence easy to transfer and move. Of similar design were chairs intended for women and made like an armchair that was deprived of armrests.

Furniture manufactured in Spain stood out with the wealth of its fittings and ornamental pins, which were used to fix leathers or upholstery fabrics. A little different in character were the products made by the Netherlands artisans. They were distinguished by legs that were turned in two stages, with a larger diameter towards the connector and a smaller one from the connector downwards. The chairs of the German Renaissance are typical cross-structures made from two pairs of bent beam elements and crossed in the front plane.

**Fig. 1.27** Caquetoire chair in
France, sixteenth century
(Gostwicka J.)

## 1.4.2 Baroque Furniture

The most characteristic pieces of furniture of the Baroque are those of Louis XIV, the absolutist ruler, who gave the majesty of the authority, which he represented, an appropriate exterior form. Therefore, the products manufactured in Manufacture Royale des Meubles de la Couronne founded in 1667 had an adequate aesthetic form. Their characteristic feature is, among others, using flat decorations in case furniture, called marquetry, while in skeletal furniture (such as chairs, armchairs, tables) wood carving decorations. The legs of the furniture mentioned were originally in the form of vertical, four-side beam elements tapered towards the bottom, stabilised at the bottom by connectors made of curved boards. In such cases, the wood was primed and gilded to emphasise the rank and wealth of the user. However, wood carvers were not the most valued craftsmen—the most respected were artists who worked with ebony wood, marqueters and bronzers (Setkowicz 1969; Gostwicka 1981, 1986, 1987).

The most important piece of furniture for sitting was a heavy, padded armchair with a high backrest (Fig. 1.28). Aside from the seat, it was distinguished by an obliquely set backrest, which usually reached above the head of a seated person, as well as an s-shaped bent armrest. The seat tapestry was made of velvet, embroidered fabrics, gobelin, extruded and gilded leather.

During the reign of Louis XIV, new types of furniture appeared, including the chest of drawers, desk, dining table, console tables, table clocks, cabinet clocks, base tables for vases. The chest of drawers replaced the chest and often

**Fig. 1.28** Baroque armchair
in Louis XIV style

**Fig. 1.29** Baroque armchair
in Louis XIV style

complemented tall cupboards in function (Fig. 1.29). Desks were an innovative
variation to the escritoire and cabinet from the period of the late Renaissance.

In the bedrooms, the bed was placed in an alcove separated by a balustrade or
columns from the rest of the room (Fig. 1.30). The bed was covered over the top
with a canopy decorated on the corners with feathers or finials. Also differently than
in the Renaissance, the wooden frame of the bed was covered with fabrics
completely.

English Baroque furniture was made of walnut wood and gilded, rich wood
carving ornaments. Embroidered fabrics, drapery and velvets were also used with
rich trimming near the fittings of beds, lining armchairs and sofas. Tables are
characterised by a strong resemblance to Dutch tables with screw turned legs.

**Fig. 1.30** Louis XIV's royal bed

The high backrests of chairs were usually carved and decorated with ornamentation in the form of sashes or acanthus leaves. Case furniture came in the forms of simple bodies. At this time, English furniture used profiling very sparingly and almost completely avoided any cornices. Characteristics of ornamentation of English furniture are marquetery decoration, the use of black and red lacquers according to Chinese designs, and the use of s-shaped bent supports in place of existing banister and volute supports (Setkowicz 1969). The most important piece of box furniture of this period in English furniture was the chest of drawers, which was represented by a chest fitted with drawers. However, there were no large clothes cupboards in England, because clothes during this time were still stored in drawers.

In the seventeenth century in the Netherlands, mostly oak and walnut wood was used, although laminated boards from exotic types of wood were also used: ebony, rosewood and others. Sometimes, the surfaces of furniture were also lined with ivory tiles placed in starry designs or with colourful veneers from exotic trees. An essential type of a Dutch wardrobe during this period was the dual four-door wardrobe. Chairs had the form known already from the late Renaissance, but the legs, like in tables, had the form of screw lines. Dutch Baroque furniture was mainly ordered and made in China. Hence, this also had a significant impact on their further shape, form and purpose. In particular, the impact of Chinese bent frames and bases gave impetus to the transition from the serious and raw forms of Baroque furniture to more casual and mobile Rococo forms.

At the turn of the seventeenth and eighteenth centuries, German wardrobes developed a kind of final shape and constituted a distinctive and recognisable piece of furniture of that period. In German literature, they were called Hamburg

**Fig. 1.31** Replica of Gdańsk
furniture from the seventieth–
eighteenth century—dresser
and table (manufacturer of
Gdańsk furniture Gorlikowski
Company, www.gorlikowski.
eu/)

wardrobes, while in Poland—Gdańsk wardrobes, as in both of these cities their most representative exponents were produced and collected (Fig. 1.31).

Gdańsk wardrobes from the second half of the seventeenth century had a two-storey body embedded on a massive base, while the columns, supporting the expansive entablature took over pilasters that were covered in ornaments. All the surfaces of the wardrobe were devoid of ornamental and sculptural decorations. The horizontal division was indicated by profiled cornices: over the base, between the levels and the crowning cornice. The architectural composition of the wardrobes was emphasised by three columns on each level. Gdańsk wardrobes were made of oak or coniferous wood, veneered with walnut wood, enriching the colour with inserts of ebony or dark-brown oak wood. At the end of the seventeenth century, there was a breakthrough in the form and function of wardrobes. The division into two levels disappeared, and the interior above the massive base was not divided. Such a design made it possible to hang clothes on fixed hangers. A refined piece of furniture in the high hallways of middle-class tenement buildings was a hallway wardrobe (Fig. 1.32). It was a massive structure having rich wood carving decorations and exposed cornices. The one-level, large-sized body was placed on a low base with drawers (Setkowicz 1969; Gostwicka 1981, 1986, 1987; Swaczyna 1992).

### 1.4.3   Rococo Furniture

French artists and craftsmen perfected the technique of finishing furniture by discovering a terrific black and red furniture lacquer, which imitated Chinese lacquer ideally. Designers put a lot of effort and creative passion into the aesthetics and

**Fig. 1.32** Hallway cupboard, beginning of the eighteenth century

comfort of furniture for sitting: chairs, armchairs, sofas, chaise longues, etc. Finely curved legs of furniture without connectors, rounded corners and smooth lines corresponded harmoniously with delicate tapestry fabrics and wall coverings. During this period, French furniture did not have flat surfaces and straight lines. Thanks to this, the cubic case furniture was enlivened: by profiles narrowing downwards, curved legs and surfaces of hatches, doors and side walls (Fig. 1.33).

Instead of dark ebony wood, light-coloured veneers of exotic trees began to be commonly used, such as amaranth, lemon, rosewood, olive, thuja and many others. Furniture surfaces, like in China, were covered using varnishes to such an extent that it almost completely covered the natural drawings of wood with paintings resembling flowery meadows. In particular, these treatments were done to: chests of drawers, night tables, escritoires and screens. Chests of drawers were usually equipped with two deep drawers, while all the visible surfaces were shaped in the form of arcs. Together with the change of tastes of users, the table also changed—it was started to form the legs in an S-shape.

The Rococo chair, not only in shape, but also in dimensions was adjusted to the needs of the user. In particular due to the prevailing fashion of wide women's dresses, the seat of the armchair was widened significantly, the low armrests spread outwards increased the surface of the seat, which was lined with soft materials. Usually, it was equipped with small cushions called manchettes. The skeletons of these pieces of furniture were made mainly of painted wood or gilded wood on a chalk glue base.

During the period of the French Rococo, an impressive amount of furniture for sitting was created, rich in features, satisfying the specific needs of their owners. Therefore, the following was created: bergere-type armchairs with semicircular armrests upholstered with the backrest, two-person marquises with an elongated

**Fig. 1.33**  Louis XV's davenport, 1760–1769

chaise-longue-like seat, duchesse-type lounges with two full backrests, the higher of which was placed at the side of the head, sofas, ottomans, i.e. small couches with an oval seat, stools, benches and beds without columns. In structure, beds did not differ significantly among themselves in the form of the canopy or the height of case; however, in relation to the beds from the previous period, they were characterised by a visible lightness, as well as a clear differentiation in the men's and women's bedroom design.

During the period of Louis XV, for the first time the concept of sets of furniture started to be differentiated. Common in the royal court was the use of a suit, i.e. a set of furniture designed for one room, having a specific function and matching the architecture and colour of the selected interior.

English furniture in the Rococo period was distinguished by enriched ornamental wood carving in the form of the acanthus leaf and human silhouettes and figural animal motifs, including lion heads, masks of satyrs, etc. Around 1725, the backrest form in furniture for sitting changes, which final shape was provided by Chippendale. This transformation was in an openwork forming of the backrest panel to the form of a ribbon ornament, which gave the piece of furniture a lightness that was a fundamental value in Rococo art. The characteristic of this period forms of curved legs and lacquer decorations are credited to the clear Chinese influence on English furniture (Fig. 1.34).

**Fig. 1.34** Chair made in
Chinese style (cherry),
1760–1800 (Victoria & Albert
Museum, London)

The oldest Chippendale chairs from the front have curved legs, finished with animal claws embracing a sphere. In the upper part, at the joint with cases, they are finished in a straight manner, usually in a baton form or with additional side volutes. The back legs are usually straight or slightly curved. The backrest is openwork, in a rectangular frame, with straight side posts and slightly curved upper joints, which is a characteristic solution for Chippendale-type chairs and what distinguishes them from similar chairs from the previous period (Fig. 1.35).

Chippendale designed and made almost the majority of structural types of furniture, including chairs, benches, sofas, clocks, beds, cabinets, dressers, bookcases, escritoires, desks and chests of drawers (Fig. 1.36).

During the same period, there was a clear impact of Chinese and Japanese design on the form of furniture produced in London's manufactures. To finish these pieces of furniture Chinese and Japanese varnishes were used, which were begun to be imported in the seventeenth century.

## 1.4.4   Classical Furniture

In France, under the influence of archaeological discoveries in Pompeii and Herculaneum, a significant turn occurred towards antique forms. Furniture was deprived of Baroque features, structures were significantly simplified, replacing curved lines with straight elements. Chairs, though almost unchanged, compared to the previous era, were characterised mostly by grooved legs. The connection of cases and legs was strongly emphasised by the presence of a cuboid cube usually

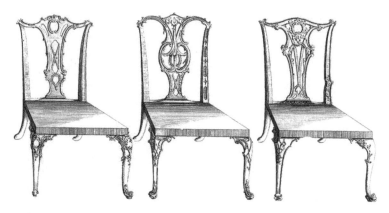

**Fig. 1.35** Drawings of chairs by Thomas Chippendale (1753) (Thomas Chippendale, The Gentleman and Cabinet Maker's Director, London 1754. The Chippendale Society)

**Fig. 1.36** Chest of drawers made by Thomas Chippendale, around 1760

decorated with a sculpted rosette. The backrests, usually concave, were in the shape of a medallion or openwork lyres. The skeletons were painted with light-coloured paints, gilded or finished with waxes. The seats were finished with gobelin fabrics (Fig. 1.37).

Chests of drawers, desks, escritoires had smooth surfaces, cornices, side frames and straight legs. At that time, the following appeared: buffet—commodes and shelf-commodes, chiffoniers—small cabinets with elongated proportions and fitted at the front with drawers, glazed bookcase cabinets, the desk in the form of an ordinary table. Furniture for writing was produced with special care for their

**Fig. 1.37** Armchair in the style of Louis XVI (Galeria Glorious Antiques Sp. z o.o.)

appropriate dimensions ensuring comfort of office work. Also a comfortable, large, extendable dining table appeared, known from the Renaissance period.

In England, Robert Adam, George Hepplewhite and Thomas Sheraton had a significant impact on the evolution of furniture design. In particular, the last two are considered the creators of native directions of the English furniture industry. Hepplewhite's chairs have straight legs, tapered towards the bottom, with a round or square section, while the backrests have oval forms, shield-shaped, heart-shaped and in the shape of a camel's hump, etc. (Fig. 1.38).

Sheraton developed new backrest shapes in the form of a rectangle with vertical lines (Fig. 1.39). The most commonly used motif, however, was three overlapping, sharply finished ovals or three palm leaves set into the oval. In addition, he also designed sofas and case furniture.

Among the case furniture, two- and three-chambered designs were produced for storing clothing and/or books. A number of chests of drawers could also be encountered: with three or more drawers at the bottom, with two door wings in the central part, with drawers placed on the sides, as well as many other forms in between designed on this basis (Fig. 1.40).

Among tables, round, oval and rectangular structures prevailed, with the function of folding or sliding one under the other. There were also dresser tables, toiletry tables, tables for writing in the form of an escritoire with a variety of compartments (Fig. 1.41).

During the period of English classicism, wooden elements of furniture were finished with a wax varnish, leaving its natural colour and texture of the base. The

**Fig. 1.38** George Hepplewhite chairs (drawings from George Hepplewhite, The Cabinet Maker and Upholsterer's Guide, 1794)

**Fig. 1.39** Drawings of chairs form the work by Thomas Sheraton, The Cabinet Maker and Upholsterer's Drawing Book (vol. 1), plates 33 and 35 (vol. 2), 1793

most commonly used wood was mahogany, which was finished by marquetry with light-coloured, speckled and veined veneers made of lemon or satin wood.

### 1.4.5 Empire Furniture

The empire style in France evolved as a symbol of Napoleon I's Empire, based on turbulent political and social changes. Empire furniture has a monumental, compact, heavy and massive form. Straight lines and mostly flat surfaces prevail here.

Case furniture was equipped with a wide, massive base that emphasised grandeur and lack of mobility. Usually, they were veneered with dark-red mahogany, which was contrasted with gilded browns and wood carved ornaments, inlay was

**Fig. 1.40** Chest of drawers
according to George
Hepplewhite's design

rarely encountered. The French empire caused the creation of new types of cabinets
and bookcases, as well as new constructions which were sometimes glass-cased on
three sides for china. Separate case furniture constituted chests of drawers and
chiffoniers, which served mainly to store underwear and other trinkets, as well as

**Fig. 1.41** Dressing table according to George Hepplewhite's design

**Fig. 1.42** Rectangular table with legs directly on the floor

escritoires which were used for writing and storing. Empire desks almost entirely preserved their style from the previous era, and only minor details and decorations are the result of interactions with contemporary trends.

Tables were made in different variations and sizes, including dining room tables, tables for flowers, coffee and tea tables, toiletry tables, tables for women's knitting, as well as tables with wash basins. The worktops of these constructions were characterised by a round, rectangular or polygonal shape (Fig. 1.42).

The legs of the Empire table do not always rest directly on the floor, but on supports in the shape of a board or multi-pointed stars. They usually have the form of columns, human and animal figures or a structure based on Roman models, including the winged female figure emerging from a lion leg.

Chairs, armchairs and other furniture for sitting differ slightly from those produced in the period of Louis XVI, they are heavier and more massive. It can be noticed that usually the front legs differ from the rear legs. They form a straight line and come right up to the armrest supported by a shape of an Egyptian woman's head, woman's bust, winged lions or swans (Fig. 1.43). Large carved figures of women mostly fulfilled the function of a table's legs, while the figures of Sphinxes were supported for the armrests of chairs and armchairs. Many armchairs also had spheres by the armrests.

Sofas were decorated with the silhouettes of swans with high outspread wings. In representative furniture, the seat and backrest were covered with silk fabrics, carpets and embroidered gobelin fabrics threaded with a gold tinsel.

Luxury beds were mostly built on a podium along with a magnificent canopy, from which drapes trimmed with gold dropped (Fig. 1.44). The connectors and posts were finished off with classic urns and vases made of bronze or gilded wood. Other beds usually did not have a canopy.

**Fig. 1.43** Armchair made of
beech wood, around 1830
(Louvre Museum, Paris)

### 1.4.6   Biedermeier Furniture

Furniture from this period is characterised by a solid built, choice of appropriate raw material and a design that meets the requirements of users. For their production, the following wood was mostly used: mahogany, walnut, birch, cherry, pear, elm, poplar, ash and oak.

Biedermeier case furniture is characterised by a similar, but greatly simplified form in relation to the analogous furniture manufactured in England at the end of the eighteenth century (Fig. 1.45). Most of the panel elements of furniture were joined using an oblique dovetail joint. The bottoms of drawers were inserted in a groove, while the face was situated in one plane with the rest of the top surface of the piece of furniture. Frame panel or, less often, board back walls were joined with the body using tongue and groove joints. A popular piece of furniture among townspeople of that period was the escritoire. Its essential element was the lifting flap on hinges, behind which there were numerous drawers and compartments (Sienicki 1954).

Glass cases were also made, glazed in the front and on the sides, which served to store china, small objects, miniatures and other memorabilia. Interchangeably with glass cases, lightweight shelves or glass dressers functioned.

Tables were constructed from a round or oval board supported on a frame, which was usually a single column or legs in the shape of a lyre. In the case of the table, there were one or two drawers. Worktops were covered with a mahogany veneer inlayed with narrow highlights of light-coloured veneers (string inlay) (Fig. 1.46).

**Fig. 1.44** Napoleon's bedroom in Fontainebleau

**Fig. 1.45** Biedermeier salon inside the Loewenfeld mansion (The Irena and Mieczysław Mazaraki Museum, Chrzanów, Poland)

The most significant structural changes occurred, however, in the group of skeletal furniture. Next to frame and rail systems came also rack constructions. In the case of rack construction, the side frames were strengthened by joints with a front and rear rails. To join them, usually mortise and tenon joints, double mortise

**Fig. 1.46** Table structure
made in Biedermeier style

and tenon joints as well as mortise and tenon joints with offset were used. The use
of double mortise and tenon joints as well as mortise and tenon joints with offset
increased the usable surface of contact of the joined elements, thus the strength and
rigidity of the joint. This was the next step in the development of skeletal furniture
designs. Backrests and support legs were joined in a similar way. Because during
this time pine wood was begun to be used, artisans started to use veneering of
elements of skeletal furniture. In contrast to the hardware made in classical style,
skeletal furniture is characterised by universality of use of curved supports, backrests and legs (Fig. 1.47).

The seats of the chairs were usually lined with light-coloured, floral or striped
cretonnes, baize fabrics, as well as silk, damask or embroidered fabrics.

**Fig. 1.47** Frame structure of
a piece of furniture from the
Biedermeier period

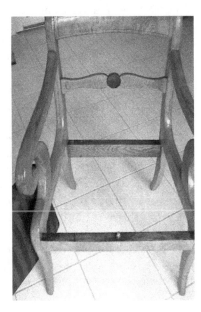

Large, wide sofas, usually in a heavy and massive form, were also commonly produced. The main decorations of wooden elements of the furniture include swans with outstretched necks, cornucopia, griffins, plant twines and similar motifs, most often gilded.

Also during this period, for the first time in the spring layer specially designed springs were used. Thanks to this, furniture was more attractive and the comfort of use improved significantly.

## 1.4.7   Eclectic Furniture

During the eclectic period, homage was paid to short-lived trends that were based on former historical styles. This wave of sentiments sparked interior designs in Turkish, Persian, Indian, Chinese and Japanese style. All of these stylish trends put together gradually led to the internal disintegration of Biedermeier style. In addition, they resulted in the collapse of an established furniture art, shaped on the healthy centuries' old craft traditions. The progressing industrialisation contributed to ousting craft workshops that constituted forges of talents and a place of collaboration for artists with craftsmen. Nevertheless, the industry contributed to the creation of a separate branch of furniture making—furniture manufactured in series and packed in boxes. This group of furniture constituted curved furniture.

In 1830, the German carpenter Michael Thonet (born in 1796 in Boppard on the Rhine, the son of a German tanner Franz Anton Thonet), wanted to avoid the costs associated with wood milling, he began experimenting with wood bending techniques. Initially, he immersed properly prepared strips of veneer in tubs of boiling glue, which after taking them out, he would bend and glue using presses in special moulds. Unfortunately, the bent glued elements obtained from this, due to insufficient properties of adhesive and wood, could not be formed into shapes that the creator expected from the future structural elements of furniture. After carefully observing the behaviour of wood fibres, Thonet perfected his technology and became the first producer in the world of curved wooden elements. In 1841, Thonet demonstrated for the first time his own-designed and manufactured furniture made of curved wood at an exhibition organised by Koblenz Art Association. There, he also met the Austrian Prince Metternich, who invited him to Vienna, where in 1849 he began mass production of furniture using the technique of bending wooden rods. For the production of furniture, Thonet used mainly beech wood, which in industrial conditions was easily subject to this technology. An advantage of this technology was also that it did not require from workers any special talents or skills. Thonet was ahead of his time and discovered a new market of affordable furniture produced in long series and intended for both the bourgeoisie and offices or cafes. Thonet's factories mainly manufactured chairs (Fig. 1.48) (chair model no. 14 was manufactured in more than 45 million copies with a 40 year period), armchairs, beds and rocking chairs (Fig. 1.49), living room sets, dining room sets and bedroom sets.

**Fig. 1.48**  Chair model 14,
designed by Michael Thonet,
1859–1860

**Fig. 1.49**  Rocking chair,
designed by Michael Thonet

The seats of the armchairs and chairs were carried out in three variants:

- from plywood, with a burnt out or pressed design, or with holes drilled in a decorative arrangement,
- weaved with Spanish reed,
- upholstered (leather or fabric).

The elements of furniture, especially skeletal furniture, were joined using mortise and tenon joints with a threaded stud and screws, which made it possible to disassemble the furniture for transport. Thonet's furniture is lightweight and durable, which they owe to the elimination of certain joints from the skeletal structure, so that one element fulfils a few functions. Characteristic of them is the openwork lightness of structure and the effect of transparency, and their style remained independent of the art of past periods.

After 1929, Thonet's factories begin mass production of curved furniture also made of steel tubes. Thonet's furniture is lightweight, often with the possibility of disassembling for transport, and also quite durable. They were always labelled with stickers with the name: Thonet, Thonet Brothers—Vienna (from 1853) or Thonet—Mundus (from 1923). A significant part of the production of curved furniture has also been located in Poland: in Radomsko, Jasienica, Jaworze and Buczkowice.

### 1.4.8   Art Nouveau Furniture

When the Art Nouveau period appeared in Europe, the shape of furniture was simplified. Due to the use of natural wood and perfectly matched components, furniture was considered to be products of exceptionally high quality. Arthur Heygate Mackmurdo's (1851–1942) furniture appeared at this time, an architect and designer from the circle of arts and crafts and the founder of the association of artists and craftsmen established in 1882 under the name Century Guild.

Mackmurdo furniture designed in the 1970s combined arts and crafts trends with the style referring to Japanese art. Characteristic of this period was searching for new, cheap and technological products, especially two technologies hot-curving wood and veneering, which exerted a significant influence on the development of design. The former was based directly on methods developed by Michael Thonet, gradually improved, but essentially the same. The latter perfectly met the requirements of Art Nouveau design, in practice providing the benefits of a twofold kind, cheap and aesthetically made products (Fig. 1.50).

### 1.4.9   Art Deco Furniture

Art Deco furniture, intended mainly for selected wealthy buyers, often resembled a casket made of a valuable, rare material, which making was an achievement of furniture art deriving from the best traditions of eighteenth century French Ébénistes. During this time, the richness of patterns and decorations were used which were available thanks to obtaining new, numerous kinds of wood from overseas. The most frequently used were ebony from Sulawesi, Brazilian rosewood, mahogany, amaranth, sycamore and various sets of these veneers with solid ash or maple. A specific contrast between the compact structure of some and the uneven surface of others enables an infinite number of combinations.

**Fig. 1.50**  Dresser from the Art Nouveau period

## *1.4.10  Early Twentieth-Century Furniture*

In the 1920s, the fame of the Weimar school of design, architecture and Bauhaus applied arts grew at a rapid pace, which was founded by Walter Gropius (1883–1970) in 1919 and moved to Dessau in 1925. Above all, the use of metal in the production of series furniture was a revolutionary move in the industry.

Metal met the criteria of modernity perfectly: it was a material that derived from industrial production, it had a distinctive appearance and uniform structure, and it could also be used in furniture of any sizes.

Innovation, however, did not consist in replacing one material with another. As Marcel Breuer (1902–1981) wrote, a Professor at Bauhaus in Dessau, metal marked the complete revision of hitherto concept of interior design: *Metal furniture is one of the ingredients of a modern facility. It has no style, because no one expects that it will express anything other than its function and structure. The modern interior should not be a self-portrait of the architect, who designed it, nor bear any features of the inhabitants. The piece of furniture must allow air to pass through, it cannot impede movements or obscure the view of the room. I chose metal exactly for these requirements of modern spatial planning. Despite its low mass, steel, and aluminum even more so, is able to support great pressure. The lightweight structure increases the flexibility of the piece of furniture. All models are based on standardized components, which one can freely move and exchange. Metal furniture is to act as a useful device in everyday life.* Modernity, understood as conforming a

**Fig. 1.51** Chair designed by
Ludwig Mies van der Rohe

piece of furniture to the requirements of functionality, the lack of ornamentation, the use of new technologies and materials, as well as mechanised manufacturing processes, has become the overriding aim of Bauhaus designers. For example, the designs of Gropius or Josef Albers (1888–1976), precursory in their idea of the freedom of disassembly, confirm this thesis.

However, the most significant architect creating within the circle of the school was Ludwig Mies van der Rohe (1886–1972), a supporter of modernity based on pure form, functionality and sophisticated aesthetics. It was not until 1930 that he arrived at Dessau, though his architectural concepts coincide with the school's programme already much earlier.

In 1927, Ludwig Mies van der Rohe created a series of his most interesting and most beautiful steel furniture, today considered classic. We owe the famous freely bending pipe chair to him (Fig. 1.51). It consisted of one segment of bent steel pipes, a wooden backrest and seat.

The famous Barcelona armchair (Fig. 1.52) should be mentioned here, built from a cubic skeleton of chrome-plated steel, which finish required precise manual work and leather cushions. With this project, van der Rohe demonstrated that modern

**Fig. 1.52** Barcelona
armchair, designed by
Ludwig Mies van der Rohe

aesthetics, despite rigorous simplicity, allows one to create objects of undeniable beauty. Using them in today's design (Knoll International) proves that their beauty is everlasting.

# References

Escutenaire C, Kozłowski JK, Sitlivy V, Sobczyk K (1999) Les chasseurs de mammouths dans la vallée de la Vistule., Musées Royaux d'Art et d'Histoire, Bruxelles

Gostwicka J (1981) Dawne stoły. Krajowa Agencja Wydawnicza, Warsaw

Gostwicka J (1986) Dawne krzesła. Krajowa Agencja Wydawnicza, Warsaw

Gostwicka J (1987) Dawne łóżka. Krajowa Agencja Wydawnicza, Warsaw

Kozłowski JK (1986) The Gravettian in Central and Eastern Europe. Advances in World Archaeology 5:131–200

Sienicki S (1954) Historia architektury wnętrz mieszkalnych. Budownictwo i Architektura, Warsaw

Setkowicz J (1969) Zarys historii mebla od czasów starożytnych do końca XIX wieku, Warsaw

Soffer O, Adovasio JM, Illingworth JS, Amirkhanov HA, Praslov ND, Street M (2000) Palaeolithic perishables made permanent. Antiquity 74(286):812–821

Svoboda J (2004) Afterwords: the Pavlovian as the part of the Gravettian. In: The Gravettian along the Danube, Brno, p 283–294

Swaczyna I (1992) Wybrane cechy konstrukcji jako kryterium identyfikacji mebla zabytkowego. Wydawnictwo SGGW, Warsaw

# Chapter 2
# Classification and Characteristics of Furniture

## 2.1 Characteristics of Furniture

Furniture is objects of applied arts intended for mobile and permanent furnishing of residential interiors. Among other things, it serves for storage, work, eating, sitting, lying down, sleeping and relaxing. Furniture can be used individually, in suites or sets.

A furniture suite (Fig. 2.1) is a collection of articles, often of different features, but with a similar purpose, having identical or very similar aesthetic form. They are made through the implementation of a specifically determined design work, in which goal might be, for example, furniture for the dining room: in a flat, residence or hotel. A characteristic feature of a suite is that individual pieces of furniture can be combined according to different, but logical rules. The following criteria for completing suites are most frequently adopted: type of material, wood species, type of surface finish, place of use of the furniture, and the historical period in which the furniture was made or what period it refers to stylistically. A lounge suite can consist of two or three armchairs, two double sofas or two corner reclining sofas. A suite is also three armchairs, pouffe and reclining double sofa. Another suite can be a corner reclining sofa and an armchair with a container. A suite for storage can consist of a clothes cupboard, a library bookcase, a bar and glass case, as well as a dresser, chest of drawers, glass case and cabinet. A suite for the dining room can include a dresser, cabinet, dining table, chairs and side table. A kitchen suite usually consists of upper and lower cabinets or built-in cabinets, but may be supplemented with a table and chairs, buffet or bar. A suite of office furniture can consist of a series of filing cabinets, shelves for files, cabinets with sliding shutters, work tables with chairs and dividing walls. A suite of study furniture can include a desk, side table, armchair, wardrobe, library bookcase and table with chairs.

A furniture set (Fig. 2.2) can contain both individual furniture pieces and furniture suites. Furniture constituting a set, unlike furniture included in suites, may have a different purposes and different aesthetic and structural forms.

© Springer International Publishing Switzerland 2015
J. Smardzewski, *Furniture Design*,
DOI 10.1007/978-3-319-19533-9_2

**Fig. 2.1** A furniture suite for the dining room (Furniture Collection Klose, Juvena, designed by Zenon Bączyk)

**Fig. 2.2** Furniture set (Furniture Collection Klose, Juvena, designed by Zenon Bączyk)

However, they can be grouped together in sufficiently harmonious collections. A characteristic feature of a set is that the individual furniture pieces or suites were created independently of each other and were not the product of a coherent idea of one designer or team of designers.

## 2.2   Classification of Furniture

Furniture belongs to the group of objects of applied arts, and many of them have similar structural, technological, functional, operational and aesthetic features. For these reasons, making a distinctive and obvious division of furniture is difficult and to a large extent depends on the experience and intuition of the author of such a division. The main difficulties which may arise in the future, when creating new divisions of furniture, result primarily from:

- the development of new technologies of production and use of new materials,
- the use of identical furniture in various places and in different conditions of exploitation,
- coincidence, i.e., that various furniture pieces have similar functions in similar places of use and
- blurring of boundaries of clear criteria for the division of furniture.

Taking into account the presented concerns associated with the division of furniture, as well as taking into account the need to systematise names for existing and new designs of furniture, it is necessary to build a more or less orderly and logical classification. Classification is the arrangement of objects, including furniture, depending on the classes, sorts, types, forms and general features. By building a useful classification of furniture, it can be divided according to the following criteria:

- purpose—according to the place of use,
- functionality—according to the nature of human activity associated with
- this or other type of furniture piece,
- form and construction—defining the form and technical solutions of the furniture piece, their mutual influence on each other and on the surrounding environment,
- technology—determining the type of materials used, type of treatment, the method of manufacture of the product and the methods of finishing the surface and
- quality—characterising the most important requirements in the processes of design, construction, manufacture and exploitation of the furniture.

## 2.2.1 Groups of Furniture According to Their Purpose

In terms of purpose, i.e. the conditions and nature of use, furniture can be divided into three distinct groups. For furnishing:

- offices and public buildings (office furniture, school furniture, dorm furniture, hotel furniture, cinema furniture, hospital furniture, canteen furniture, common room furniture, etc.),
- residential rooms in multi-family and free-standing buildings (flat furniture, kitchen furniture, bathroom furniture, garden furniture) and
- transport (ship furniture, train furniture, aircraft furniture).

This division is extremely important, especially when shaping the technical assumptions for a new product. The requirements and conditions of use included in the design and manufacturing process are different for ship furniture, different for office furniture and different for hospital or school furniture.

In the group of furniture for offices and public buildings, there is another subgroup related to specific human activities:

- furniture for administration,
- furniture for offices and studies and
- furniture for workers.

The nature of work and method of use of rooms in office buildings and public buildings requires designing furniture intended for managers, group leaders, assistants and secretaries, employees working in groups and individually, serving internal and external clients. Within this group, the separate subgroups constitute hospital furniture, school furniture, furniture for waiting areas at train stations, airports, as well as restaurant and cafe furniture. The nature of these furniture pieces should correspond to the specific requirements of many different and often anonymous users. School and office furniture should be well suited to the anthropometric parameters of individual groups of users. Hospital furniture should be conducive to rehabilitation and should minimise the negative phenomenon of prolonged pressure of the human body on a mattress or seat. Furniture intended for use in waiting rooms is required to ensure high durability and functionality, adapted to the nature of travel of prospective users.

Furniture for residential rooms in multi-family and free-standing buildings should comply with the requirements of individuals and families, living together in a house or flat, as well as be able to perfectly incorporate into the room and make it possible to perform everyday activities in these rooms. The furniture should meet all the functional needs of the following zones: relaxation and lounging, sleep, work, learning, preparing and eating meals, physiological needs and maintaining personal hygiene, and storage.

Marine, vehicular (car and train) and aircraft transportation have very high demands in terms of quality of material used in the manufacture of furniture, quality of make and safety of use of furniture built into the body of the transport units carrying people.

## 2.2.2  *Groups of Furniture According to Their Functionality*

In terms of functionality, furniture can be divided into the following groups:

- for sitting and lounging,
- for reclining,
- for working and eating meals,
- for learning,
- for storage,
- multifunctional furniture and
- complementary furniture.

Each of the given groups is characterised by specific properties and requirements:

- of exploitation, that is the character of the performed task,
- anthropotechnical, that is adjusting the user's anthropotechnical characteristics to the technical features of the operational object,
- sanitary and hygienic,
- pedagogical and
- construction.

A group of furniture for sitting and lounging comprises typical chairs, tabourets, stools, pouffes and bar stools, which do not or only partially provide support for the user's back (Fig. 2.3a–e), as well as armchairs, sofas, chaise lounges and corner sofas, supporting the whole body or its major part (Fig. 2.3f–i).

Furniture for reclining should ensure comfortable and continuous support of the human body in a reclining position. There are, however, structures that not only meet this basic function, but also provide support for the body in a sitting position. For this reason, furniture for reclining can be divided into two subgroups (Fig. 2.4):

- only with reclining function, such as beds, couches and mattresses (Fig. 2.4a–c) and
- with a reclining and sitting function, such as folding sofas, sofas and corner sofas (Fig. 2.4d–f). In this subgroup of furniture, the change of function can be achieved by using fittings and accessories that enable to transform the piece of

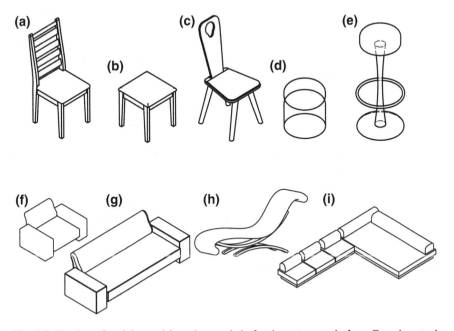

**Fig. 2.3** Furniture for sitting and lounging: **a** chair, **b** tabouret, **c** stool, **d** pouffe, **e** bar stool, **f** armchair, **g** sofa, **h** chaise lounge, **i** corner sofa

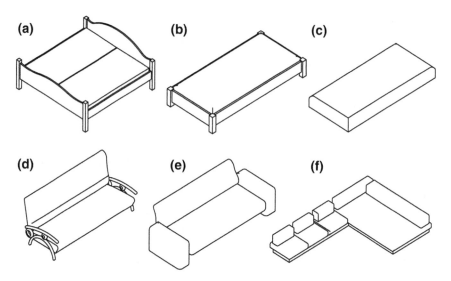

**Fig. 2.4** Furniture for reclining: **a** bed, **b** couch, **c** mattress, **d** folding sofa, **e** sofa, **f** corner sofa

furniture and unfold or fold the reclining surface. However, if the dimensions of the seat are significant, the reclining function can be provided without the need for transforming the geometry of the piece.

The group of furniture for working and eating meals mainly consists of tables, table add-ons, desks, side tables, buffets and reception bays (Fig. 2.5). Tables can be used to work, study, prepare and consume meals, games, as well as bases for apparatus, instruments, flowers or lighting. Here, we distinguish tables for the dining room, kitchen, conference rooms, construction offices, trade offices, editorials of magazines, schools, kindergartens, etc. The tables for dining rooms can usually change the geometry of the work surface, increasing its length and at the same time, area for future users. Tables and desks for offices usually have a fixed geometry of the work surface, but they have step or stepless adjustment of its position height. The change of the geometry of the work surface of office furniture is provided by applying side tables and add-ons. These furniture pieces are designed by particularly considering the arrangement of devices and objects that are the basic equipment of the workplace, including the computer, telephone, notepad, writing supplies, binders and other office accessories requiring compartments, drawers, boxes, slides, hangers and top extensions. A problem at the design stage of the integration of form, function and structure of the furniture piece is providing an exit and hiding cables delivering certain media to electrical devices.

Reception bays are a unique type of furniture designed to work, for they are an obvious flagship of an institution—they stand in the lobby or the hallway, where clients are welcomed. They should not only enable the performance of precision work in a sitting position, but also hard work in a standing position.

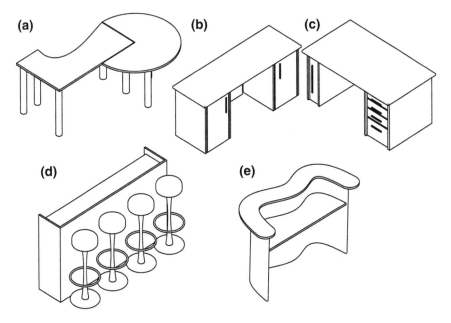

**Fig. 2.5** Furniture for work and dining: **a** table and table add-on, **b** side table, **c** desk, **d** buffet, **e** reception bayt

Furniture for learning is primarily benches, pupil tables, drafting tables and davenports (Fig. 2.6). When designing furniture belonging to this group, in particular school furniture, one must anticipate a different than normative structure load and guarantee the furniture's adequate stiffness, strength and stability. One must also remember to use such structural components for which the designer has certificates of their complete non-toxicity.

The group of storage furniture represents cabinets, bookcases, shelves, dressers, chests of drawers, cabinets, containers, dressing tables and library bookcases (Fig. 2.7). This is most numerous and diverse group of furniture, with a wide variety of forms and dimensions depending on the type, shape and size of the stored items.

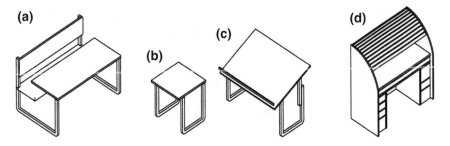

**Fig. 2.6** Furniture for learning: **a** bench, **b** pupil's table, **c** drafting table, **d** davenport

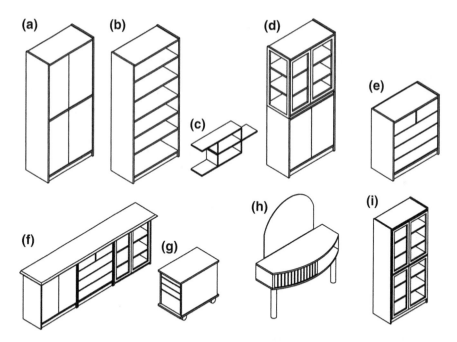

**Fig. 2.7** Furniture for storage: **a** wardrobe, **b** bookcase, **c** shelf, **d** dresser, **e** chest of drawers, **f** buffet, **g** container, **h** dressing table, **i** library bookcase

Depending on the degree of connection with the room, furniture from this group can be divided into mobile (not connected to the construction elements of the room, e.g. containers, chests of drawers, cabinets, buffets and chests) and stationary (connected impermanently with the construction elements of the room, e.g. wall cupboards, shelves, partitions or tall standing cupboards).

Multifunctional furniture originated from the need to meet the many different needs of users. These needs appeared for a number of reasons. The first were the dreams of wealthy clients of having products that were unique in form, with surprising technical solutions, and enabling users to meet several practical administrative needs. Modern multifunctional furniture pieces are often adapted by necessity to minimum living space, the nature of work and financial possibilities of future owners. Usually, multifunctional furniture can be found among sofas with a reclining function, couch beds, couch shelves, escritoires for work and storage (Fig. 2.8).

The group of complementary furniture constitutes flower beds, covers, partition walls and side tables. Mostly, they are manufactured for individual needs of clients who are furnishing their rooms completely and expecting a uniform form, construction and technology of manufacture in order to maintain consistency of the interior's look and aesthetics.

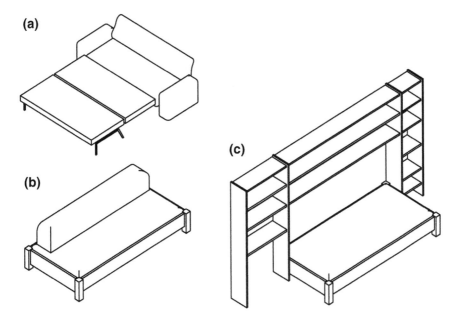

**Fig. 2.8**  Multifunctional furniture: **a** sofa with reclining function, **b** couch bed, **c** couch shelf

## 2.2.3  Groups of Furniture According to Their Form and Construction

The characteristics of form and construction of the furniture piece are determined on the basis of spatial organisation of form, interconnection of main structural components and architectural structure of the product. Depending on the spatial organisation of the form, i.e. from the spatial distribution of individual elements of the furniture piece, three basic schemes of furniture can be distinguished (Fig. 2.9):

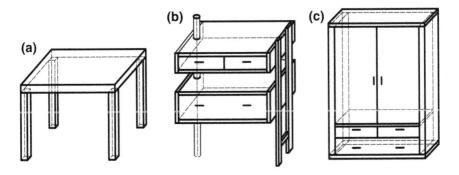

**Fig. 2.9**  Spatial organisation of the form of the furniture piece: **a** with an open spatial structure, **b** with a partially open spatial structure, **c** with a volume spatial structure

- with an open spatial structure, in which linear and surface elements dominate,
- with a partially open spatial structure, in which linear, surface and volume elements can dominate and
- with a volume spatial structure, in which volume elements dominate with small participation of linear and surface elements.

In furniture with a volume spatial structure, an additional feature can be distinguished—modularity. This feature of furniture increases functionality, many variants of use, as well as quality of the designed and furnished interiors. Modular furniture is designed on the basis of a body with a universal structure of elements, which maintains unified and repetitive closed dimensions in universal templates, with the possibility of any completion and decompletion of the system. Completion of systems can be done vertically, horizontally and matrixwise by using a system of simple connections between the elementary solids of furniture. Depending on the design, modular furniture can be divided into (Fig. 2.10):

- single-bodied,
- multi-bodied,
- universal for completion,
- on a frame and
- for hanging.

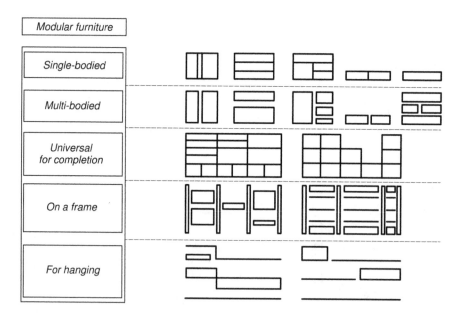

**Fig. 2.10** Examples of modular furniture

Due to the method of binding certain structural components, subassemblages and assemblages, furniture can be divided into:

- non-disassembling, produced in the form of compact blocks, which makes their disassembly impossible,
- disassembling, produced in the form of solids that provide the possibility of repeatable disassembly and reassembly and
- for individual assembly, sold in packages containing elements for repeatable assembly and disassembly.

Taking into account the characteristics arising from the design of the furniture, two main groups can be distinguished:

- case and
- skeletal.

Case furniture is built mainly from panel elements, in which thickness is several times smaller than the other dimensions. Typically, these elements are located relative to each other in such a way that closes space from five or six sides. Depending on the relative positions of the external elements of the furniture piece's body, two subgroups of case furniture have been distinguished (Fig. 2.11):

- flange, in which the top and bottom of the furniture piece (called flanges) are placed from the top and the bottom on the side walls. In this case, the upper element is called the top flange and the bottom element, the bottom flange. Such a piece of furniture can be supported on legs, a frame or a socle,
- rack, which are characterised by the fact that the top and bottom of the furniture piece can be found between the side walls. In this case, the structural solutions for the base part are as follows: the furniture piece can be supported on legs, a frame, socle, or on extensions of the side walls to the bottom, which are called racks.

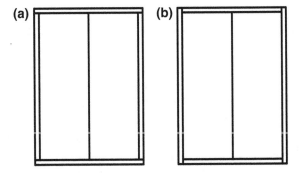

**Fig. 2.11** The division of case furniture depending on the relative positions of the sides, and top and bottom of the furniture piece: **a** flange, **b** rack

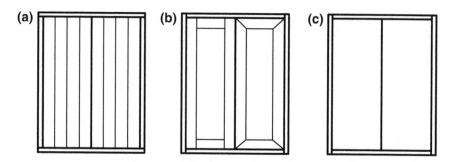

**Fig. 2.12** The division of case furniture depending on the design of visible parts: **a** slat, **b** frame panel, **c** board

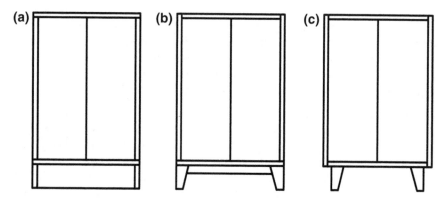

**Fig. 2.13** Body of furniture placed on **a** socle, **b** frame, **c** legs

Due to the nature of the structure, visible parts or external elements, case furniture is divided into (Fig. 2.12):

- slat,
- frame panel and
- board.

Given the solutions of designing support construction of case furniture, bases are distinguished in the form of a socle, a frame and legs (Fig. 2.13).

Depending on the construction and the technological characteristics, the following furniture is distinguished:

- segment, in which structure, external dimensions and functionality allow their vertical or horizontal configuration and
- compact, built from elements of unified dimensions and construction solutions that enable folding any suites of different purposes and dimensions.

Case furniture can be single- or multifunctional, which enables to use them in different ways, for example:

- a wardrobe, a tall case furniture piece designed primarily to store clothing, bedlinen, office binders, folders, drawings, etc.,
- library bookcase, a cupboard with shelves and mostly glass doors designed to store books,
- bookcase, a furniture piece consisting of a few or a dozen or so shelves hung on the side walls, with or without a rear wall, open or partially closed,
- glass case, glass cupboard or top part of a dresser, for storing decorative ceramics, dishes and jewellery,
- dresser, serves to store dishes, table linen, cutlery, etc.,
- buffet, a type of dresser, intended to be laid with occasional meals,
- chest of drawers, a low cupboard with drawers for storing bedlinen and clothing,
- overhead cupboard, cupboard at the ceiling used for storing rarely used items,
- wall unit, a set of furniture that fulfils all storage functions and at the same time serves to divide rooms and
- built-in set, wall set of furniture, usually closed, used for storage.

Skeletal furniture is made of elongated elements with small cross sections, in the shape of a square, rectangle, triangle, circle, oval, etc., not closing space within it, e.g. chairs, armchairs, tables and flower beds. The shape of the cross sections of these elements determines the next division of frame furniture into:

- beam furniture, in which the cross section of elements is a polygon and
- rod furniture, which is built using elements of circular, ellipse, oval or similar to circular cross sections.

Depending on how the seat or worktop is supported, we can distinguish furniture of the following designs (Swaczyna and Swaczyna 1993) (Fig. 2.14):

- board,
- cross,
- with rails,
- without rails,
- frame,
- column,
- rack and
- one piece.

Furniture of a board structure is made mainly from elements cut from timber (ready part) or by gluing friezes in the direction of their width and then cutting out the desired shape. The base of the furniture piece is made up of board elements, the width of which is several times greater than the thickness.

Furniture of cross-structures is made of beam or rod elements joined by hinges and slides. Usually, this construction enables folding and unfolding the piece of furniture both in the direction of the width of the seat and in the direction of its depth. This lets the seat be made of textile or leather materials. In this type of

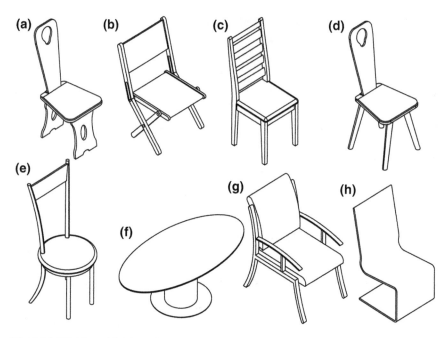

**Fig. 2.14** Division of furniture depending on seat or worktop support: **a** board, **b** cross, **c** with rails, **d** without rails, **e** frame, **f** column, **g** rack, **h** one piece

construction, only axial and transverse forces are transferred through structural nodes; therefore, high strength of wood against bending and torsion is not used.

Rail structure, as opposed to cross-construction, in the vast majority of cases, is inseparable. Thanks to permanent joining of legs with rails, a spatial frame is formed, which transfers axial and transverse forces, as well as bending and torsion moments. Thanks to these geometric features, the stiffness and strength of furniture of rail structure is much greater than with a different structures.

Structure without rail is distinguished by the fact that the base of the furniture piece is legs attached directly to the seat panel or worktop of a table. Cantilever mounting of the leg and usually deviating its axis from the vertical line causes that very large bending moments work on joints, which is the cause of damages of connections.

Furniture of a frame structure under the seat or worktop has fixed frames, to which the legs are attached. Usually, the frame is made of bent-glued elements, and friezes connected in the direction of their length or metal sections. The legs are usually connected with the frame using screws. However, it should be noted that this method of connecting construction components does not provide permanent stiffness and high strength of the base of the furniture piece. Bolt joints belong to the semi-stiff group of joints. Therefore, they have better mechanical properties than hinged joints, but at the same time, worse than adhesive joints belonging to stiff joints. With the passage of time, the stiffness of bolt joints decreases both due to the deformation of the hole by the hard core of the bolt or screw and due to the minimal changes in the moisture of the wood.

In furniture of a rack structure, the seat or worktop is attached on subassemblages that resemble racks. Usually, racks are made in the form of frames or rails connected to the seat with screws. It also happens that below the seat or worktop of a table, the side frames are connected with each other using front and rear rails. If bolt joints are not used, then the stiffness and strength of furniture of rack structures corresponds to the stiffness and strength of furniture of structure with rails.

The column structure of a skeletal furniture piece is distinguished by supporting the seat or table using one or two columns, which have been made in the form of turned wooden trunks, supported on legs or a plate, oval columns or multi-element supporters made of rods, strips or boards. Usually, the column enables to assemble in it a mechanism that allows one to change the height of the seat or worktop, as well as feeding cable lines connecting media to electrical devices installed on the table.

Furniture of one-piece structure is created mainly based on technologies of injecting plastic or natural fibrous materials. In furniture made of wood and wood-based materials, the technology of bending and simultaneous bonding of multiple thin layers of veneers is most commonly used.

## 2.2.4   Groups of Furniture According to Technology

By creating a division of furniture according to the technological characteristics, one should keep in mind both the variety of used machining processes and the types of materials used, as well as methods of finishing visible surfaces.

Technologies of manufacturing furniture can be brought together into four main groups: machine cutting, machine bending, weaving, and the technology of cutting and sewing cover materials. In terms of these technologies, furniture is divided into the following groups (Fig. 2.15):

- carpentry,
- bent,
- woven and
- upholstered.

Carpentry furniture is made of wood or wood-based materials by way of the classical machine cutting. These include the following:

- turned furniture,
- carved furniture and
- typical carpentry furniture.

Bent furniture is commonly manufactured using the hydrothermal treatment and wood bending or by bending together with simultaneous bonding of thin veneers of wood and wood-based materials. The modern technology of manufacturing

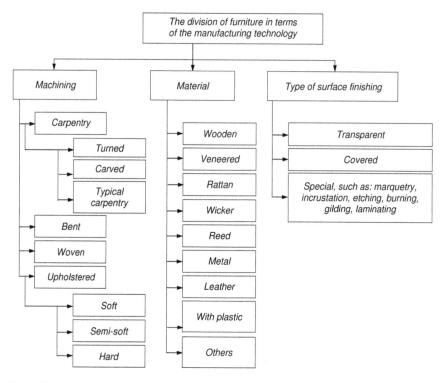

**Fig. 2.15** The division of furniture in terms of the manufacturing technology

components for bent furniture production was proposed by the Danish company
Compwood Machines Ltd. (www.compwood.dk). This technology consists in
softening the wood point by steaming in autoclaves or heating in a high-frequency
electric field and then compressing the wood in one axis along the fibres to up to
about 80 % of the initial length of the element. After such a process, the element is
suitable for free bending or bending using tapes and bending limiters.

Woven furniture is formed primarily by using thin and flexible materials, which
can be used to shape full or openwork surfaces of seats, backs and sides of chairs,
armchairs and sofas. They also constitute the decorative element of door panels or
sides of bodies of case furniture. Usually, for weaving, wicker, rattan, peddig,
bamboo, water hyacinth, raffia and loom are used.

Upholstered furniture has a complex construction and various ways of making
the elastic section and upholstered section. They are usually formed by stretching
over and fitting of a cut and sewn cover on a frame with an elastic layer. In terms of
the measurable criterion of the softness of the mattress or seat, upholstered furniture
can be divided into:

- soft,
- semi-soft and
- hard.

This division is quite subjective and in many cases depends on the individual feeling of softness of the seat by the designer or manufacturer. In order to objectify and quantify this division, a value of acceptable pressure on the human body in contact with the mattress should be assumed as the criterion of evaluating softness. Krutul's studies (2004a, b) show that the acceptable level of pressure on the human body, not hampering blood circulation in arteries, veins and capillaries, should not be higher than 32 mmHg = 4.26 kPa. On the basis of the results of the works carried out by Smardzewski et al. (2007, 2008), it was also shown that the value of contact pressure depends on the construction of the mattress and the kind of elastic materials used. The lowest pressure on the human body, with a value of up to 28 mmHg = 3.73 kPa, is obtained on water mattresses. For foam and hybrid systems of foam–spring sets, the value of pressure ranges from 45 to 120 mmHg (from 5.99 to 15.99 kPa). Taking this into consideration, the following numeric indexes can be suggested for the previously proposed division of furniture:

- soft, acceptable pressure on the human body no greater than 5 kPa,
- semi-soft, acceptable pressure on the human body ranging from 5.1 to 9 kPa and
- hard, acceptable pressure on the human body greater than 9.1 kPa.

Due to the type of material used, we divide furniture into:

- wooden, in which at least the basic construction consists of elements made of wood or wood-based materials,
- veneered, in which surfaces are veneered with a thin layer of precious wood— veneer, thin films or decorative papers,
- rattan, in which the basic bearing construction of the furniture piece consists of elements made of rattan bars (rotang) usually with a diameter of 35 mm,
- wicker, in which the entire construction or only the woven back or seat is made of wicker,
- reed, in which the entire construction or only the woven back or seat is made of reed,
- metal, in which the construction is made of metal, while wooden or wooden-based elements are only a supplement,
- with plastic, in which the construction is made of plastic,
- leather, usually upholstered furniture, which has all external surfaces or only use surfaces covered with natural leather and
- others.

Taking into account the method of finishing the surface of a ready product, furniture can be divided into:

- transparent, with a visible structure and drawing of the base,
- covered, with an invisible structure and drawing of the base and
- special, such as marquetry, incrustation, etching, burning, gilding and laminating.

## *2.2.5 Groups of Furniture According to Their Quality*

The concept of quality first appeared in the philosophy of Plato, who called it poiotes. Cicero, when creating Latin philosophical vocabulary, for the Greek term, created the Latin qualitas. Hence, qualitas infiltrated some of the Romance and Germanic languages, e.g. as qualita in Italian, qualite in French, quality in English and qualität in German. According to Aristotle, any quality boils down ultimately to two types. Quality in the first meaning is that what constitutes the diversification of the object in itself. The second type of quality constitutes the properties of variable things, i.e. properties in terms of which changes are distinguished (warm and cold, white and black, heavy and light). He also attributes to quality every advantage and defect (Iwaśkiewicz 1999, Łańcucki 2003).

In ancient Egypt, proof of the commitment of builders to quality was the construction of the pyramids, which to this day are testimony to the great craftsmanship of builders. In the Code of Hammurabi (from 1754 BC), it is concluded, among others, that if a mason built a house for a citizen and did not check his work and the wall leaned, this mason will strengthen this wall for his own silver.

The dictionary of the English language defines quality as a property, type, kind, value and set of features constituting that a given object is that object and no other, whereas in the colloquial language, quality usually means an evaluation of the degree in which a given object or service conforms to the requirements of the evaluator. Such an assessment may include all or some of the features of the evaluated object: weight, colour, shape, structure, chemical composition, physical characteristics, impact on the environment, efficiency in the performance of specific functions, etc.

In the literature on the subject (Iwaśkiewicz 1999; Łańcucki 2003), definitions of quality are presented in which, despite the large variety of wordings, there is one fundamental idea that quality means fulfilling the customer's requirements.

The evaluation of the quality of furniture comes down to a detailed analysis of the features of design, construction, functional, ergonomic, safety for the environment and directly for the usage.

Each piece of furniture, as an object of applied arts, in a long life cycle usually passes over two stages: the stage of development, i.e. design and manufacture, and the stage of use, i.e. fulfilling needs. In the first stage, the designer and the manufacturer can only in theory or on a small sample evaluate the quality of the modelled and manufactured products. The real assessment of the quality of the furniture piece takes place only during everyday use by dozens or hundreds of users. Therefore, furniture should be evaluated from two points of view (Fig. 2.16):

- manufacturing and
- requirements of the user.

Among the manufacturing characteristics, construction, technological and economical values are distinguished, which affect both the aesthetics and the safety of using furniture and the technical cost (as low as possible) of manufacturing.

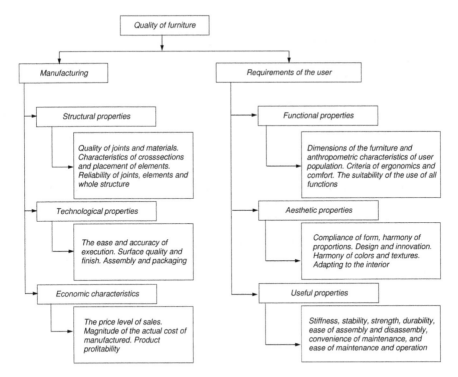

**Fig. 2.16**  The division of furniture in terms of quality

Among the features associated with the requirements of the user, the following should be named functional, aesthetic and user-friendly.

## 2.3    Characteristic of Case Furniture

Case furniture is the goods of a volume structure, in which the surface elements limit and close a given space. Among the large group of elements of a case furniture piece, we distinguish:

- the main element, without which the furniture piece loses its functional nature and the construction becomes a mechanism,
- a supplemental element, without which the construction can be maintained, but the furniture piece does not perform the intended functions and
- compensation element, without which the function and the construction can be preserved; however, its use significantly improves the stiffness, strength, durability and reliability.

Each of these elements can be made as:

- a panel element, from a chipboard, carpentry board, MDF, HDF, cellular, composite with the addition of lignocellulosic particles, plywood, etc., and
- an element of solid wood.

Usually, a case furniture piece is made up of a body, front build, interior build, read wall and frame. Each of these parts has distinctive surfaces, the position and visibility of which, in terms of the user, allow the diversification of finishing and build quality. Here, we distinguish the following surfaces (Fig. 2.17):

- front, visible in furniture piece from the standard position of use, such as the visible surfaces of doors, flaps of bars, fronts of drawers, slides, as well as surfaces of recesses (also glazed) and
- external, visible in the furniture piece at a non-standard position of use.

These include external surfaces of side walls, lids, top flanges located at a height not exceeding 1.8 m, bottoms, bottom flanges, shelves and compartments situated at a height no less than 0.9 m from the floor,

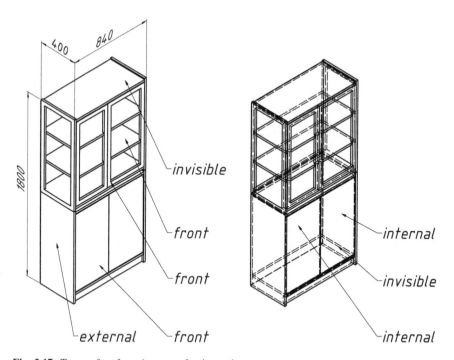

**Fig. 2.17** Types of surfaces in a case furniture piece

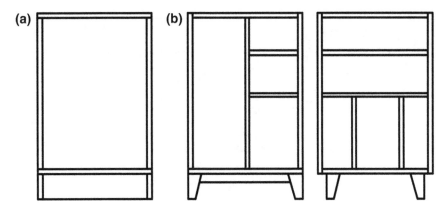

**Fig. 2.18** Body of the furniture piece: **a** single-chamber, **b** multi-chamber

- internal, visible in the open furniture piece at a normal position of use. These include the interior surfaces of cabinets, couch chests, drawers and other containers,
- invisible, invisible in the furniture piece at a normal position of use. These also include external surfaces of tops situated at a height not exceeding 1.8 m and bottoms situated at a height no less than 0.9 m from the floor.

Usually, the body of the case furniture piece consists of two side panels, a bottom, a top, partitions and a rear wall. Depending on the number of partitions dividing the space into functional usable areas, we divide the bodies into (Fig. 2.18):

- single-chamber and
- multi-chamber.

The body of the furniture piece can be supported on a frame, stand on rests being an extension of the side walls and can be raised on a socle or rest directly on the floor. Each of these types of structures significantly determines the strength and durability of the furniture piece. This also affects the essential design qualities of the product, harmony of shapes and proportion of dimensions. Usually, furniture mounted on slender, towering frames has a lightweight construction, elegance of form and freedom of use. However, if the body rests on a minute frame, massive socle or directly on the floor, it is usually associated with a heavy form and limited functionality.

The legs of case furniture are made of wood, plastic and metals (Fig. 2.19). Usually, they are attached directly to the underside, perpendicular to its surface. Such fitting provides the legs an axial load and high shearing strength. A disadvantage of this kind of support of the body is the small resistance to lateral forces, causing destructible bending moments at their base. Therefore, in order to obtain an interesting aesthetic effect, they are often designed as cast, welded or moulded elements.

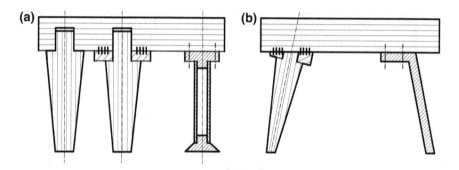

**Fig. 2.19** Legs supporting the body of the case furniture piece: **a** perpendicular fitted wooden or metal legs, **b** wooden, metal or plastic legs situated obliquely to the underside

The frames of case furniture most often have two or more legs. They are made of wood, metal, plastic and wood-based materials. Due to the structural bonding of the legs in a uniform frame, from the point of view of durability, the frame constitutes a far preferable solution than single legs (Fig. 2.20).

The base of the furniture in the form of a socle is a mechanically very effective bearing structure. Depending on the number of elements, the following socles have been distinguished (Fig. 2.21):

- three-element,
- four-element and
- five-element.

Three- and four-element socles not only serve an important supporting function, but also improve the stiffness of the body of the furniture piece usually by about 5 %. A five-element socle has an additional lateral element which improves the stiffness

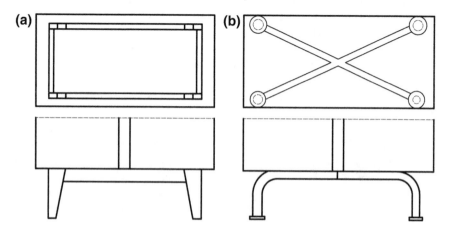

**Fig. 2.20** Frames supporting the body of the case furniture piece: **a** wooden, **b** metal or plastic

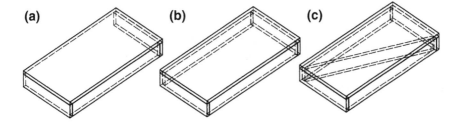

**Fig. 2.21** Types of socles: **a** three-element, **b** four-element, **c** five-element

of the body by even up to 100 %. This is due to the fact that most of the elements of the furniture piece's body during natural exploitation are subject to torsion, while the lateral element is subject to bending, which is why the indicator for resistance to bending is used, not for torsion.

Storage in case furniture is enabled by an appropriate number of compartments, shelves, cases and drawers. As elements fixed permanently to the body of the furniture piece, compartments have an impact on its stiffness and strength. The increase in the number of horizontal or vertical compartments clearly leads to a reduction in the thickness of the main and complementary elements. The shelves are movable elements, not permanently bonded with the body of the furniture piece, and in the calculation of the construction's stiffness, it is not taken into account in total equations. However, due to the fact that they carry significant loads, their number indirectly affects the stiffness of the piece of furniture. Indeed, at a specific constant stiffness of the construction, the increase of load causes a linear increase of the displacement of corner nodes of the construction.

Cases and drawers, like shelves, are movable parts (subassemblages) of box furniture and are not included in the calculations of stiffness. Safety of use of furniture for storage requires that all movable elements are protected against ejecting. To this end, for shelves, special supports are applied that prevent movement in the horizontal plane, and for cases and drawers, there are stoppers halting the container after it slides out of the inside of the body to a certain point. For aesthetic reasons, the fronts of cases and drawers can be entered into the body, harmonised with narrow planes of side walls and partitions or placed on these surfaces (Fig. 2.22). For the same reasons, they are done in the same way as other

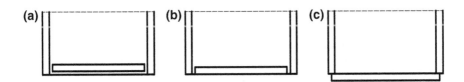

**Fig. 2.22** Ways of inserting the fronts of drawers and doors to the body of the furniture piece: **a** inserting, **b** harmonising, **c** applying on

elements of the front, that is doors, flaps or covers, referring to their design, structural form and technology of finishing.

Like the fronts of drawers, doors can be entered into the body, harmonised with narrow planes of side walls and partitions or placed on these surfaces. Access to the interior of the body is possible by turning or sliding the door. Usually, the rotation of the door, made in the form of boards or panels, takes place on the horizontal or vertical axis, while sliding takes place vertically or horizontally along appropriate straight runners. An interesting variation of a door is blinds, which are made of thin wooden strips glued to a canvas, plastic (PVC ABS, polypropylene) lamellas or aluminium lamellas joined by hinges. Blinds can run in horizontal and vertical runners, in which movement track can be either a straight line or any curve composed of many gentle arcs.

The construction of a case furniture piece is a kind of bonding of its volume, surface and linear elements in a way that is appropriate and enables safe use. It consists of elements, subassemblages and assemblages.

An element of a case furniture piece is a single part made from one material or a few kinds of materials permanently bonded with one another, e.g. top or bottom, side wall or front of drawer veneered with natural veneer.

A subassemblage is an assembled set of elements, which constitute a separate whole during the phase of assembly of the furniture piece, e.g. the socle, body and panel-structured door.

An assemblage is a set of assembled subassemblages or subassemblages and elements, constituting a separate structural unit, fulfilling most commonly a specific function in the piece of furniture, e.g. a top extension on a dresser, a top shelf on a cabinet, a cabinet which is a constituent part of a cabinet and buffet.

Designing furniture, in engineering activities, comes down to the careful design of construction documentation in paper or virtual form. In both of these cases, it is necessary to index and name individual elements, subassemblages and assemblages of furniture, in accordance with the nomenclature used in the industry. It should be noted that along with the introduction of EN-PN norms in Poland, definitions arising as calques of English translations have significantly distorted traditional Polish names and descriptions of constituent elements in furniture. Therefore, this book uses the nomenclature contained in the standard BN-87/7140-01—Furniture, Terminology.

In case furniture, the following elements are distinguished (Figs. 2.23, 2.24, 2.25 and 2.26):

- the front on the drawer, an element which is the front wall of the drawer,
- bottom, an element which is the fixed bottom limitation of the subassemblage or assemblage in the furniture piece, e.g. the bottom of a drawer and the bottom of a chest for bedlinen,
- laminated board, an element which is the complement of panel elements or structures, mainly for securing or decorating,
- bar, an element which is most often the functional equipment of the furniture piece, e.g. closet bar,

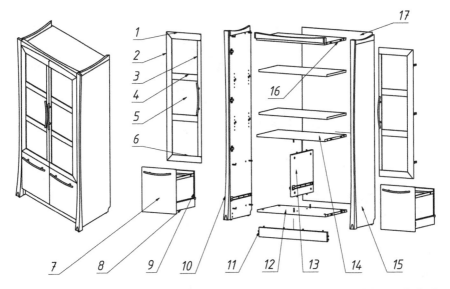

**Fig. 2.23** Names of elements of case furniture—library: *1* top transverse rail, *2* left longitudinal rail, *3* right longitudinal rail, *4* central transverse rail, *5* panel, *6* bottom transverse rail, *7* front of the drawer, *8* side wall of the drawer, *9* rear side of the drawer, *10* laminated board, *11* socle skirt, *12* bottom, *13* vertical partition, *14* horizontal partition, *15* side wall, *16* top, *17* rear wall

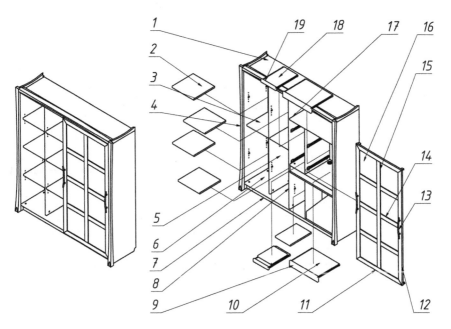

**Fig. 2.24** Names of elements of case furniture—wardrobe: *1* top, *2* shelf, *3* horizontal partition, *4* laminated board, *5* vertical partition, *6* bottom, *7* laminated bottom board, *8* vertical partition, *9* front of the slide, *10* slide, *11* bottom transverse rail, *12* right longitudinal rail, *13* handle, *14* central transverse rail, *15* central longitudinal rail, *16* panel, *17* horizontal partition, *18* slide, *19* front of the slide

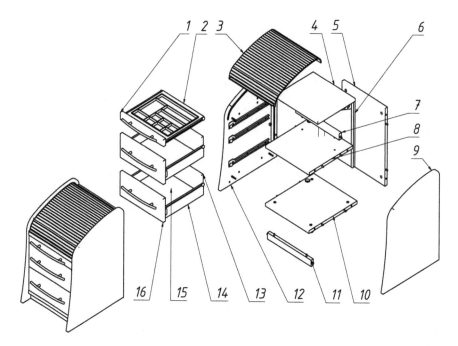

**Fig. 2.25** Names of elements of case furniture—container: *1* front of the case, *2* case, *3* blinds, *4* top, *5* rear wall, *6* partition wall, *7* skirt, *8* horizontal partition, *9* right side wall, *10* bottom, *11* socle skirt, *12* left side wall, *13* rear wall of the drawer, *14* side wall of the drawer, *15* bottom of the drawer, *16* front of the drawer

- door, a movable element with a vertical axis of rotation, which closes the furniture piece. Depending on the direction of opening, distinguished are right doors—opened to the right—and left doors—opened to the left, and in special cases, it may be a subassemblage,
- sliding doors, a movable element shifted horizontally, which closes furniture piece,
- flap door, door lowered down, in other words, doors with a horizontal axis of rotation,
- strip, the element that complements basic construction, acting as a slider, supporter, resistant, connector, etc., depending on the function it might be a sliding strip, skirting, ring beam strip, thickening strip, closing trim, bearing strip, supporting strip, etc.,
- leg, the bearing element of the base of furniture, depending on the location in the unit it may be a front or rear leg,
- panel, the plate element filling the space between the rails of the frame,
- worktop, the element or subassemblage that constitutes the usable, top plane of the furniture,
- bottom gantry and top gantry, the element that constitutes a slider of drawers in case and skeletal furniture,

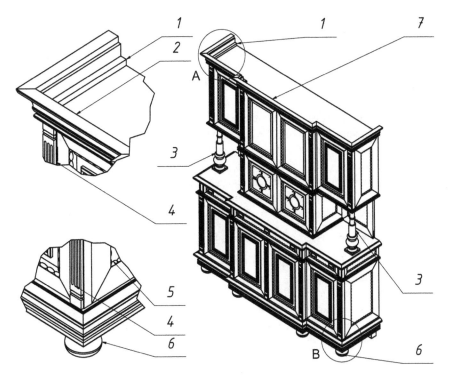

**Fig. 2.26** Names of elements of case furniture—dresser: *1* transverse rail, *2* longitudinal rail, *3* post, *4* decorative strip, *5* quarter round, *6* leg, *7* longitudinal rail

- thickener, the complementing element, mainly of compensatory, reinforcing and decorative character,
- shelf, loosely arranged horizontal board element for placing various objects,
- runner, the strip element, sometimes specially profiled to slide drawers, projectors, etc., and depending on the location there are top, lateral and bottom runners,
- vertical partition, an element which constitutes a fixed vertical division of parts of space in case furniture,
- horizontal partition, an element which constitutes a fixed horizontal division of space in case furniture,
- rail, the component element of the frame, and depending on the location, there are longitudinal, transverse, central rails, etc.,
- post, the element that constitutes vertical reinforcement or filling in the base or the body of the unit,
- side wall, an element which is the fixed side exterior limitation of the case furniture piece, and in special cases, it may be a subassemblage,
- partition wall, an element which is a fixed vertical separation of the entire space in the case furniture piece,

- front wall, an element forming a fixed front outer limitation of the case furniture piece,
- rear wall, an element forming a fixed rear outer limitation of the case furniture piece,
- wall, an element forming a fixed outer limitation of the assemblage or subassemblage acting as a container, e.g. a drawer or container for bedlinen,
- depending on the position, there are side, front, rear walls, etc.,
- seal, a compensatory element made of wood, rubber or other material,
- bottom (bottom flange for flange structures), an element which is the fixed outer bottom limitation of the body of the case furniture piece,
- top (top flange for flange structures), an element which is the fixed outer top limitation of the body of the case furniture piece and
- slide, a movable element which provides an additional worktop when pulled out.

Subassemblages and assemblages include (Figs. 2.27 and 2.28):

- side, subassemblage constituting a side limitation,
- socle, subassemblage consisting of a bottom and at least 3 interconnected socle skirts, which is a type of base in the case furniture,
- body, subassemblage consisting mostly of structurally bonded side walls, bottom, top and rear wall, constituting the bearing structure of the furniture or assemblage of case structure,

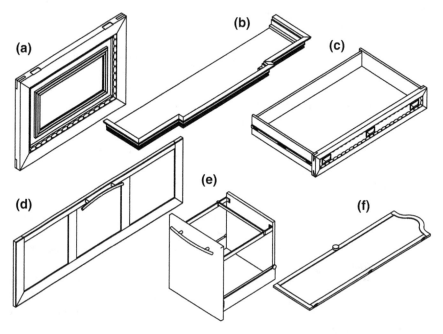

**Fig. 2.27** Names of subassemblages of case furniture: a, d, **f** frame panel door, **b** top, c, **e** drawer

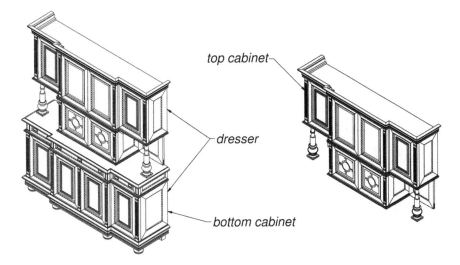

**Fig. 2.28**  Names of assemblages of box furniture

- base, an assemblage or subassemblage that has a bearing function for the furniture piece or assemblage,
- frame, subassemblage consisting of connected rails, e.g. door frame,
- chest, assemblage fulfilling the function of a container,
- base frame, subassemblage of a skeletal structure which is a kind of base,
- cabinet, an assemblage fulfilling the function of a container, e.g. desk cabinet and davenport cabinet,
- drawer, subassemblage which is a movable container, open at the top and
- blinds, subassemblage for closing a case furniture piece by drawing, flexible (bendy) in the direction of motion.

## 2.4   Characteristic of Skeletal Furniture

Skeletal furniture is the products of linear and surface structure, in which elements do not close a space. Like in the group of case furniture, we distinguish:

- main element, forming the function and construction of the furniture piece (leg, backrest board, seat frame, top panel—usually the worktop of the table),
- complementary element, providing only the function of furniture piece (insert of worktop, armrest) and
- compensatory element, which can improve satisfactory stiffness, strength, durability and reliability of construction (support, bar).

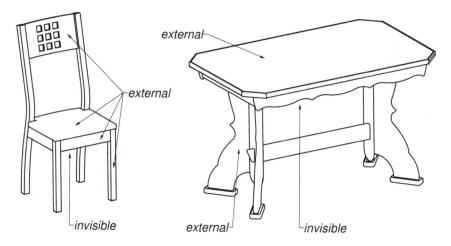

externa

external

external

invisible          external          invisible

**Fig. 2.29** Types of surfaces in a skeletal furniture piece

Each of these elements can be made as:

- beam or rod element derived from wood, or any profile made of metal or plastic,
- element in the form of a board made of wood or fitting made of metal or plastic and
- panel element, obtained from an MDF board, HDF board, composite board with the addition of lignocellulosic particles, plywood, etc.

Usually, skeletal furniture consists of a frame (i.e. a skeletal bearing structure), the seat and the backrest (in the case of chairs), as well as of the top panel (in the case of a table, it is usually a worktop). Characteristic surfaces of these parts significantly affect diversity of finish and workmanship. Distinguished here are (Fig. 2.29):

- external surfaces, visible in the furniture piece at a position of use. This includes all surfaces of beam, rod, board and panel elements and
- invisible surfaces, those that are invisible in the furniture piece during its use in accordance with the intended purpose. This includes bottom surfaces of seats, worktops, invisible surfaces of rails, frames and runners.

The most characteristic representative of skeletal furniture is the chair. Usually, frame elements (of bearing structure) of chairs are connected using inseparable connections, which provide greater stiffness and strength of the construction compared to connections using metal joints (bolts and screws for wood) (Fig. 2.30).

The seats of chairs usually constitute a separate subassemblage, but they may well be a constituent element of the frame. Depending on the purpose, the seats of chairs are made in upholstered form, from plywood and plastic mouldings, flat or profiled boards, leather, wicker or rattan weaves (Fig. 2.31).

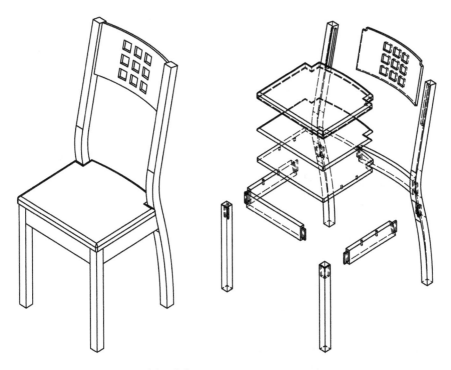

**Fig. 2.30** Bearing structure of the chair

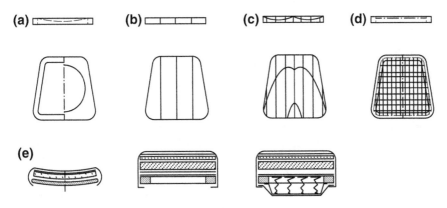

**Fig. 2.31** Examples of constructions of seats: **a** plywood, **b** straight wood panel, **c** profiled wood panel, **d** woven, **e** upholstered

The backrest of a chair, similarly to the backrest of an armchair, ensures additional transfer of the human body's weight. It usually constitutes the form of connection of backrest legs. There are also backrests that are an independent element or subassemblage fitted to the seat (in stools), frames (in bent furniture) or to

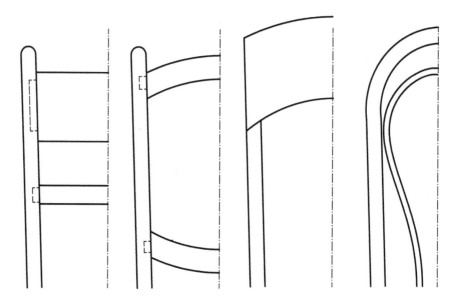

**Fig. 2.32** Examples of backrests of chairs

the rails (in carpentry furniture) (Fig. 2.32). As in the seat, depending on the purpose of the chair, the filling between the vertical elements of the backrest are made in upholstered form, made of plywood and plastic mouldings, flat or profiles boards, leather, wicker or rattan weaves.

Tabourets and benches are a kind of skeletal furniture mostly used in kitchens and public buildings. These are usually constructions without a backrest; however, due to ergonomic reasons, for benches for waiting rooms in public offices, backrests are designed which support the entire back or only a portion at the height of the loin (Fig. 2.33). Like with chairs, the seats of tabourets and benches can be hard or upholstered.

The shape and detailed features of tables depend on the place of use. For each specific group of users in kindergarten, school, office, laboratory, computer laboratory, dining room, kitchen, restaurant, etc., often personalised parameters are defined, which decide on the shapes, dimensions and construction solutions. In tables intended for flats, the height of the top surface is adapted to users sitting on chairs. In places of recreation and leisure, low tables are designed, called occasional tables, tailored to people sitting in armchairs.

Bearing structures of high tables occur mainly in rail forms (Fig. 2.34), with the legs of the following cross sections: polygonal, rectangular, triangular, circular, oval and irregular. In longitudinal cross sections, the legs have rectangular or trapezoidal shape, usually converging downwards. For transport reasons, it is recommended that the legs are attached to the rails in a disjoint manner.

Depending on the purpose and functions of the designed table, the top boards can have a regular form of squares, rectangles, ellipses, ovals and circles, or in the

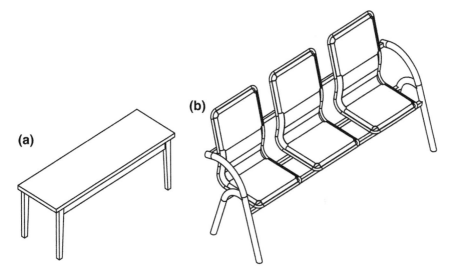

Fig. 2.33 Examples of constructions of benches: **a** from wood, **b** from metal

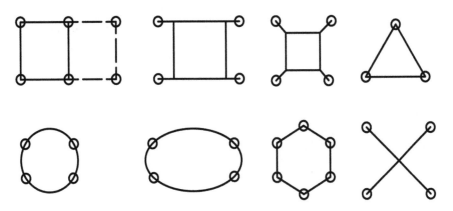

Fig. 2.34 Examples of the shapes of rails

form of irregular surfaces limited by a polygon, splain or polyline. In employee table for offices, designers search for shapes of top surfaces which allow their connection into new usable forms (Fig. 2.35).

Each of the shapes of the top surface should enable its pivoting, rotating, pulling out and disassembling. In addition, in school and office furniture, it is important to provide the possibility of step or stepless height adjustment of their position.

Top surfaces of tables can be made entirely from solid wood, chipboard, carpentry boards, cellular and MDF boards, boards veneered with natural veneer, decorative paper, HPL, CPL laminate, as well as metal and plastics. There are also boards made of plastic, marble, glass or a few materials simultaneously, e.g. metal

**Fig. 2.35** Examples of the shapes of top surfaces of tables and possibilities of their connection

and glass, wood and ceramic tiles, wood and glass, etc. Depending on the purpose, worktop surfaces should be finished with materials of varying durability and resistance to the impact of fluids, temperatures, UV radiation, mechanical damage, etc.

From the point of view of use, tables of variable surface of the worktop are favoured. According to Swaczyna and Swaczyna (1993), they can be divided into three groups:

- with an increased working surface,
- with a lifted working surface (board) and
- unfolding or connected tables.

Among tables with an increased worktop, Swaczyna and Swaczyna (1993) distinguish tables with:

- moving boards supported after being unfolded on pulled out legs,
- moving boards supported on extended rails,
- sideboards supported on additional supports and
- boards enlarged using auxiliary boards (inserts).

Tables with movable boards supported after extended on pulled out legs are characterised by the mobility of the structure of the frame, generally thanks to fitting wheels. Legs with a pull-out function are attached to the rails by hinges. Worktops of these tables can be divided into two or three parts and also joined by hinges from the bottom surface. The worktop divided into two parts leans on a turntable set on the rails. In order to increase the usable area, the legs must be pulled out (usually at an angle of less than 90°), the worktop turned by 90°, and one wing of the worktop spread out (Fig. 2.36a). If the worktop is divided into three parts, then its vertical parts should be raised to a horizontal position and the legs pulled

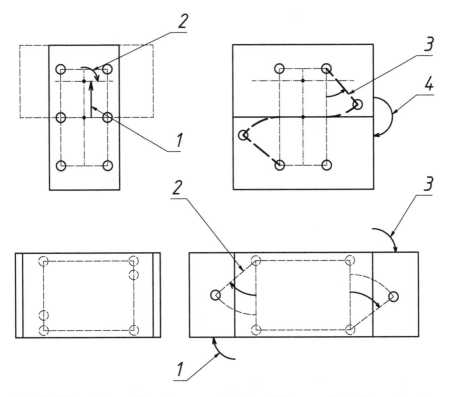

**Fig. 2.36** Tables with movable worktops supported after unfolding on pulled out legs: **a** two-piece board, **b** three-piece board; *1, 2, 3, 4*—sequence of actions when unfolding the table

out (Fig. 2.36b). These types of constructions are applied to kitchen tables and small dining rooms.

Tables with movable boards supported on extended rails can have divided worktops and ones connected with hinges, as well as inserts. A characteristic feature of this group of furniture is the extended construction of the rails.

In divided worktops, they rest directly on rails or on the turntable fitted to the rails. Tables with inserts are distinguished by two worktops secured to the extended rails. In order to increase the usable area of the table with an unfolded worktop, the frame must be extended and the board spread out by turning the top element 180° (Fig. 2.37a) or by turning and spreading out the upper wing (Fig. 2.37b). If the worktop of the table is assembled of two parts and fixed to the rails, user should spread out of the worktop, take out inserts and the stabilising elements from a pocket and slide worktop back again. (Fig. 2.37c). These types of constructions are applied to kitchen furniture and dining rooms.

Tables with sideboards supported on additional supports are characterised by external invariability of the geometry of the frame. Increasing the usable surface of the worktop usually occurs through the elevation of its vertical parts to the

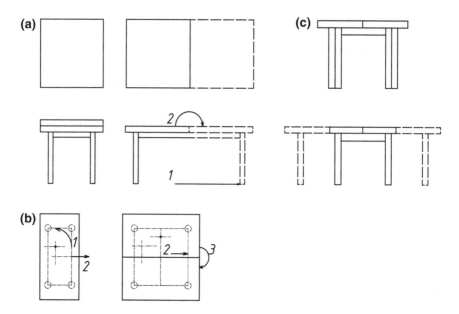

**Fig. 2.37** Tables with moving boards supported on extended rails, **a** two-piece board, **b** two-piece board—turning, **c** two-piece board with inserts; *1, 2, 3*—the sequence of actions when unfolding the table

horizontal position and extending the vertical supports being movable elements of the rails (Fig. 2.38a) or extending additional worktops resting under the surface of the main board (Fig. 2.38b). With this construction solution, auxiliary worktops are fixed to supports which are also runners enabling to hide and pull them out.

Tables with worktops enlarged by inserts are distinguished by: two equal sliding worktops and usually geometrically unchangeable frame. Enlarging the surface of the worktop takes place by sliding out worktop that is fixed to runners (wooden or metal) and inserting into the gap inserts in the shape of a rectangle. Inserts can be placed loosely in a magazine created in the space between the lower surfaces of the worktop and the transverse skirtings fastening the longitudinal rails (Fig. 2.39a). An interesting solution is also the eccentric fastening of the insert (Fig. 2.39b). After extending the worktops, the insert rotates 180° in the rails and again connects the

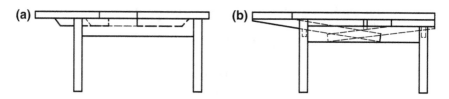

**Fig. 2.38** Tables with sideboards supported on additional supports: **a** two-piece board, **b** auxiliary worktops

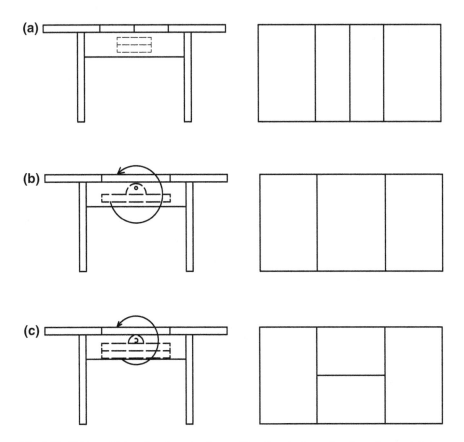

**Fig. 2.39** Tables with boards enlarged using auxiliary boards (*inserts*): **a** loose inserts, **b** inserts on the eccentric joint, **c** inserts on the eccentric joint and with a turntable

worktops. At the eccentric embedding of the insert, it is recommended also to install a turntable which enables to install an insert divided into two parts, thus significantly increasing the work surface (Fig. 2.39c). With this construction solution, after extending the boards, the insert should be rotated once 180° and then the upper element of the board unfolded 180°, after which the worktops should be connected together.

Tables with an unfolded worktop are rare today. Worktops of these tables are one piece and rotationally fixed to the column or one edge of the frame (Fig. 2.40).

Unfolding or connected tables are usually marked by a fixed geometry of worktops and frames, and they are distinguished by low weight and mobility by fixing wheels to the legs. Such a construction of tables enables the quick and efficient completion of furniture and adapting rooms (usually offices) to the user's needs.

**Fig. 2.40** Tables with an
unfolded worktop

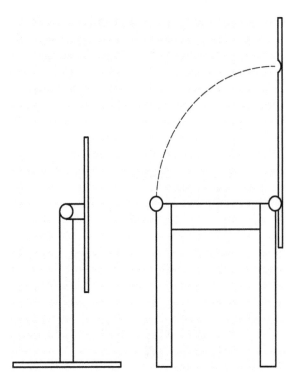

The following elements are distinguished in skeletal furniture, chairs and tables
(Fig. 2.41):

- backrest board, an element of skeletal furniture for sitting, constituting the
  backrest or base for upholstering in the backrest,
- seat board, an element of skeletal furniture for sitting, constituting the seat or
  base for upholstering in the seat,
- hanger head, a rolled or sharp-edged element, constituting the co-axial extension
  of the hanger post,
- hanger hook, an element, usually formed in the shape of the letter J or S,
  constituting the basic functional part of the hanger,
- block, an element which is the strengthening of the main structural nodes
- and fulfils functions, e.g. of slider and resistance,
- connector, a type of curve-shaped bar in bent furniture,
- bar, an element constituting additional strengthening of structural connections,
  and depending on the location in the furniture piece, the bar may be longitudinal, transverse, etc.,
- leg, bearing element of the base of the furniture piece, and depending on the
  location in the furniture, it can be the back or front leg,
- support leg, an element fulfilling the function of a back leg and support of the
  backrest simultaneously in skeletal furniture for sitting,

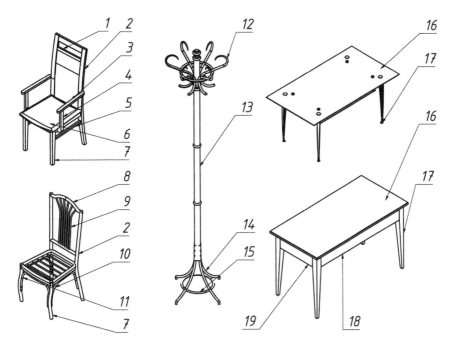

**Fig. 2.41** Elements of skeletal furniture: *1* backrest board, *2* support leg, *3* armrest, *4* armrest support, *5* bar, *6* seat board, *7* front leg, *8, 10, 11* arch, *9* filling, *12* hanger hook, *13* column, *14, 17* leg, *15* rim, *16* worktop, *18* longitudinal rail, *19* transverse rail

- support legs, an bent element in the shape of an inverted U, and these are two support legs connected from the top in bent furniture for sitting,
- rim, a bent element with a closed circumference, usually constituting additional strengthening of construction connections in the base of skeletal bent furniture,
- rail, an element constituting the primary horizontal structure of the base in skeletal furniture, and depending on the location in the furniture, the rail may be frontal, side, rear, longitudinal or transverse,
- arch, a bent element in one or a few planes of an open circumference, usually constituting strengthening of connections at the base of bent skeletal furniture,
- worktop, the element or subassemblage that constitutes the usable, top plane of the furniture,
- support, an element supporting the armrest in skeletal furniture for sitting,
- armrest, an element constituting the support of arms in skeletal and upholstered furniture for sitting,
- semi-rim, a bent element with an open circumference, in the shape of the letter U, usually constituting the strengthening of connections at the base of bent skeletal furniture,
- rail, the component element of the frame, and depending on the location, there are longitudinal, transverse, central rails etc.,
- column, a vertical bearing element, e.g. of a free-standing hanger,

- muntin, an element for filling a particular space in an openwork manner
- in skeletal furniture,
- insert, a movable element for increasing the dimension of the worktop (Fig. 2.39) and
- inset, an element that fills the space between the rails and other external limitations of the backrest, side, etc., in an openwork manner, and depending on the location, there are backrest insets, side insets, etc., and in special cases, it can be a subassemblage.

Subassemblages and assemblages of skeletal furniture include (Fig. 2.42):

- side, subassemblage constituting the side limitation, and depending on the type of furniture distinguished are chair sides and armchair sides,
- backrest, an assemblage or subassemblage, upholstered or not, serving as a backrest,
- rail module, an assemblage of several rails joined together,
- base, an assemblage or subassemblage that has a bearing function for the furniture piece or assemblage,
- armrest module, in skeletal furniture usually built from an armrest and support,

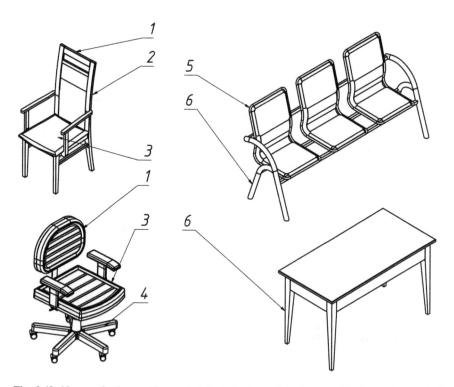

**Fig. 2.42** Names of subassemblages of skeletal furniture: *1* backrest, *2* side frame, *3* seat, *4*, *6* frame, *5* seat backrest subassemblage

- frame, subassemblage consisting of connected rails, fulfilling the bearing function for the upholstery pillow, also in the base or body of the furniture, and in bent furniture, it can constitute a bent element with a closed circumference, and depending on the function, there are upholstery frames, door frames, etc.,
- seat frame, a bent element with a closed circumference or U-shaped, which constitutes the main structural connection of the seat,
- seat, an upholstered assemblage or subassemblage for sitting,
- frame, subcomponent of a skeletal structure which is a kind of base
- in case furniture and skeletal furniture,
- skeleton, an assemblage or subassemblage composed of structurally bound beam elements, rod elements, pipe elements, etc., which constitutes the bearing structure of the skeletal furniture or its part, e.g. upholstered.

## 2.5   Characteristic of Upholstered Furniture

Upholstered furniture belongs to a group of products of complex structure and a multifaceted manufacturing process. In the designed and used interior, furniture fulfils two main functions: usable, ensuring comfortable relaxation, recreation, sleep, etc., and aesthetics connected with enriching the decor of the room.

Upholstered furniture is products of linear, surface and volume structure. Among the large group of elements of an upholstered furniture piece, we distinguish:

- main element, without which the furniture piece cannot provide the necessary stiffness, strength and reliability (e.g. the rail of an upholstery frame),
- complementary element, without which the product does not meet the expected usability functions (e.g. the foam insert of the seat, the bottom of a container) and
- compensatory element, which can improve the quality of the bearing structure (e.g. an additional rail of the frame, additional supports and bars).
- The main elements can be made of:
- boards: chipboard, carpentry, MDF, HDF, cellular, composite with the addition of lignocellulosic particles, plywood, etc., and
- wood, metal and plastics.

Complementary elements are made of:

- natural raw materials (straw, seagrass, coconut fibre, wool, feathers, horsehair, etc.),
- elastomers (polyurethane, polypropylene), latex, and silicone and
- elastic units, and pneumatic and hydraulic units.

Compensatory elements, like main elements, can be made of:

- boards: chipboard, carpentry, MDF, HDF, cellular, composite with the addition of lignocellulosic particles and plywood, and
- wood, metal and plastics.

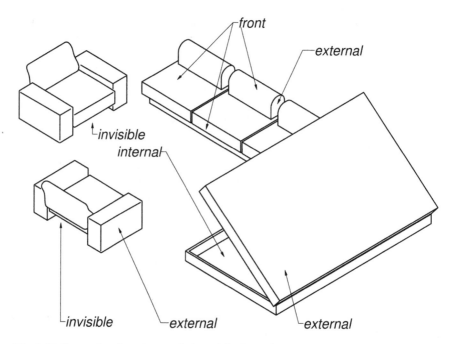

**Fig. 2.43** Types of surfaces in an upholstered furniture piece

As a rule, an upholstered furniture piece consists of a skeleton glued over with foam, a permanent (or movable) subassemblage of the seat, backrest or bed and sewn cover. Each of these parts has characteristic surfaces, which position and visibility, in terms of the user, enable to diversify the finish and construction quality. Here, we can distinguish the following surfaces (Fig. 2.43):

- front, visible in the furniture from the standard position of use, like the surfaces of a seat, backrest and sides, with which the user comes into contact directly, as well as wooden finished surfaces of the frame,
- external, visible in the furniture like in the previous usable position, however with the difference that the user does not come into contact with these surfaces directly, e.g. the back surface of the backrest, external surfaces of the sides and armrests, sides. These surfaces are finished with coordinates, i.e. materials of identical colour as the front surfaces, but of lower quality and lower price,
- internal, visible in the open furniture piece at a normal position of use. These include surfaces of couch cases and containers for bedlinen and
- invisible, invisible in the furniture piece at a normal position of use. This includes all surfaces of covered frames and partially upholstered frames.

The seat or bed is fixed permanently to the rack, which constitutes the subassemblage of an upholstered furniture piece; however, when these parts are movable, in the form of pillows, then they are called assemblages. Naturally, such assemblages,

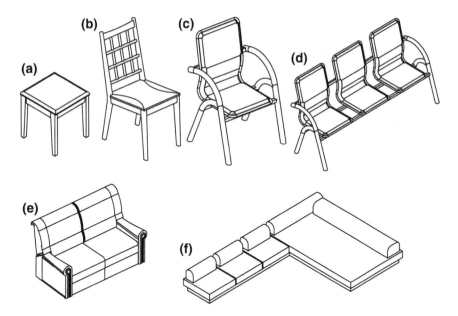

**Fig. 2.44** Upholstered furniture is designed for sitting: **a** tabouret, **b** chair, **c** armchair, **d** bench, **e** sofa, **f** corner sofa

from a commercial point of view, can be stand-alone ready products, intended for separate sale.

For functional reasons, upholstered furniture can be divided into three groups:

- for sitting: tabourets, chairs, armchairs, benches, sofas, couches and corner sofas (Fig. 2.44),
- for reclining: couches, beds, lounges, chaise lounges and mattresses (Fig. 2.45) and
- for sitting and reclining: sofa, couches, sofas with a reclining function and corner sofas with a reclining function (Fig. 2.46).

The tabouret is characterised by the simplest design. In contrast, the chair has a backrest, and an armchair also has a backrest and armrest. The bench can be a simple continuation of a tabouret structure with an increased width, as well as a richly developed form of an armchair, and also with an increased width of the seat. The construction of sofas differs from the structure of armchairs only by the width of the seat and the number of persons that can use the furniture at the same time.

The most important constituent of furniture for reclining is the bed often made as an assemblage in the form of a mattress. The simplest solution is distinguished as the recliner, which does not have a container for bedlinen and with rails fitted straight to the bed. In contrast to it, the couch has a container. A bed is usually a massive frame made from wood or wood-based materials and a mattress supported on the bearing frame. Beds usually do not have containers, which is why chests of

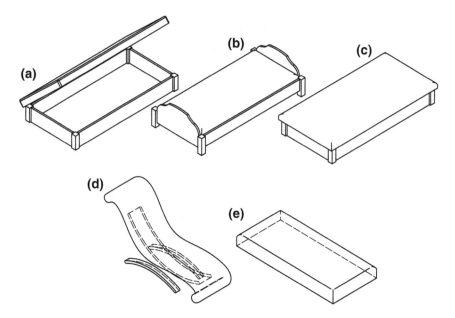

**Fig. 2.45** Upholstered furniture designed for reclining: **a** couch, **b** bed, **c** lounge, **d** chaise longue, **e** mattress

drawers are placed in the bedroom. The double function of the sofa–couch, called action "couch beds", is built from a seat and backrest joined with a container using a special lock.

In upholstered furniture, the following elements are distinguished (Figs. 2.47, 2.48 and 2.49):

- strip, the element that complements basic construction, acting as a slider, supporter, resistant, connector, etc., and depending on the function, it might be a sliding strip, thickening strip, bearing strip, supporting strip, etc.,
- connector, a type of curve-shaped bar in bent furniture,
- bar, an element constituting additional strengthening of a structural connection, and depending on its location in the furniture piece, a bar can be longitudinal, transverse, etc.,
- leg, bearing element of the base of the furniture piece, and depending on the location in the furniture, it can be the back or front leg,
- support leg, an element fulfilling the function of a back leg and support of the backrest simultaneously in skeletal furniture for sitting,
- support legs, a bent element in the shape of an inverted U, playing the role of two support legs connected from the top in bent furniture for sitting,
- armrest, a bent element with a closed circumference, usually constituting additional strengthening of construction connections in the base of skeletal bent furniture,

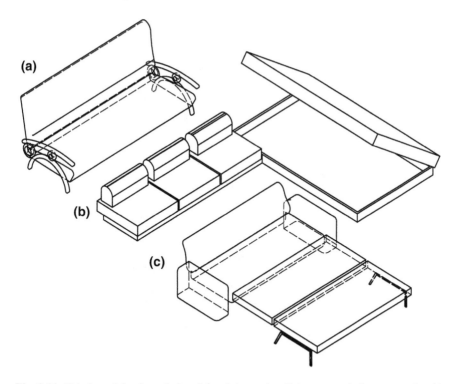

**Fig. 2.46** Upholstered furniture designed for sitting and reclining: **a** couch, **b** corner sofa with reclining function, **c** sofa with reclining function

- rail, an element constituting the primary horizontal structure of the base in skeletal furniture, and depending on the location in the furniture, the rail may be frontal, side, rear, longitudinal or transverse,
- arch, a bent element in one or a few planes of an open circumference, usually constituting strengthening of connections at the base of bent skeletal furniture,
- panel, the plate element filling the space between the rails of the frame,
- support, an element supporting the armrest in skeletal furniture for sitting,
- armrest, an element constituting the support of arms in skeletal and upholstered furniture for sitting,
- shelf, loosely arranged horizontal board element for placing various objects,
- semi-rim, a bent element with an open circumference, in the shape of the letter U, usually constituting the strengthening of connections at the base of bent skeletal furniture,
- rail, the component element of the frame, and depending on the location, there are longitudinal, transverse, central rails etc.,
- post, the element that constitutes vertical reinforcement or filling in the skeleton and
- muntin, an element for filling a particular space in an openwork manner in skeletal furniture.

**Fig. 2.47** Elements of
upholstered furniture—
armchair: *1* support leg,
*2* armrest, *3* armrest support,
*4* horizontal rail, *5* rail *6* front
leg

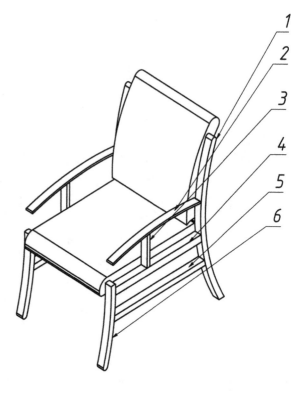

**Fig. 2.48** Elements of
upholstered furniture—couch:
*1* mattress, *2* bedside short,
*3* bottom of the container,
*4* bedside long

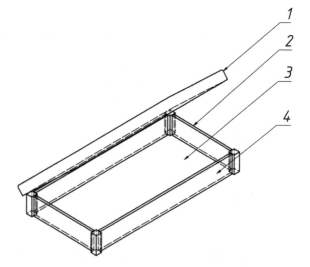

**Fig. 2.49** Elements of
upholstered furniture—bed:
*1*, *4* leg, *2* front head board,
*3* bedside long

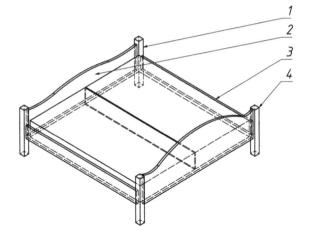

We distinguish the following subassemblages and assemblages (Figs. 2.50 and 2.51):

- side, subassemblage constituting the side limitation, often also the basis in upholstered furniture for sitting and reclining, and depending on the type of furniture, the following is distinguished: armchair side, couch side and bed side,
- bed, upholstered assemblage or subassemblage for reclining,
- mattress, type of cushion acting as a bed, the mattress can have one, two, or more parts
- backrest, an assemblage or subassemblage, upholstered or not, serving as a backrest,
- case module, an assemblage of several cases joined together,
- headrest, upholstered assemblage or subassemblage used to support the head
- in upholstered furniture for sitting and reclining,
- base, an assemblage or subassemblage that has a bearing function for the furniture piece or assemblage,
- cushion, upholstered subassemblage constituting a loosely inserted bed, seat, rest, etc.,
- armrest module, in skeletal furniture usually built from an armrest and support,
- frame, subassemblage consisting of connected rails, fulfilling the bearing function for the upholstery pillow or in the skeleton of the furniture, and in bent furniture, it can constitute a bent element with a closed circumference, and depending on the function, there are upholstery frames, side frames, door frames, etc.,
- sitting frame, a bent element with a closed circumference or U-shaped, which constitutes the main structural connection of the seat,
- seat, an upholstered assemblage or subassemblage for sitting,
- chest, assemblage fulfilling the function of a container or also the base in upholstered furniture, e.g. a chest for bedlinen,

**Fig. 2.50** Subassemblages of
upholstered furniture—
armchair: *1* backrest, *2* seat,
*3* side frame

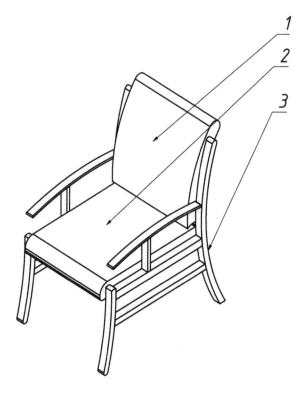

**Fig. 2.51** Assemblages of
upholstered furniture—bed:
*1* front head, *2* mattress, *3* rear
head

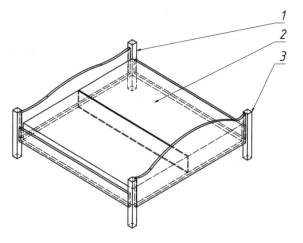

- frame, subcomponent of a skeletal structure which is a kind of base in box and skeletal furniture,
- front head, subassemblage constituting a fixed external constraint, most often also the base of an upholstered furniture for reclining from the side of the user's head,
- rear head, subassemblage constituting a fixed external constraint, most often also the base of an upholstered furniture for reclining—from the side of the user's legs and
- skeleton, an assemblage or subassemblage composed of structurally bound beam elements, rod elements, pipe elements, etc., which constitutes the bearing structure of the skeletal furniture or its part, e.g. upholstered.

# References

BN-87/7140-01—Furniture, terminology. http://compwood.dk. Accessed 05 Nov 2007
Iwaśkiewicz A (1999) Zarządzanie jakością. PWN, Warsaw
Krutul R (2004a) Odleżyny. Pielęgnacja w warunkach szpitalnych. Ogólnopolski Przegląd Medyczny 4:38–42
Krutul R (2004b) Odleżyna, profilaktyka i terapia. Revita
Łańcucki J (2003) Podstawy kompleksowego zarządzania jakością TQM. Wydawnictwo Akademii Ekonomicznej w Poznaniu, Poznań
Smardzewski J, Grbac I, Prekrat S (2007) Nonlinear elastic of hyper elastic furniture foams. In: 18th Medunarodno Zanstveno Svjetovanje Ambienta'07, University of Zabreb, p 77-84
Smardzewski J, Kabała A, Matwiej Ł, Wiaderek K, Idzikowska W, Papież D (2008) Antropotechniczne projektowanie mebli do leżenia i siedzenia. In: Raport końcowy projektu badawczego MNiSzW nr 2 PO6L 013 30, umowa nr 0998/P01/2006/30
Swaczyna I, Swaczyna M (1993) Konstrukcje mebli. Cz. 2. WSiP, Warsaw

# Chapter 3
# Ergonomics of Furniture

## 3.1 Introduction

By using furniture, the human being becomes a part of the system known as the anthropotechnic system. The elements of this system are as follows: the animate part, so the human body, and the inanimate part, that is the technical element (Winkler 2005). The anthropotechnic system is a result of the deliberate impact of the human being on a technical product, in this case, on a piece of furniture. The overall aim of initiating such systems is to improve the aesthetics of home interiors and raise their functionality, security and comfort of use. By using technical objects, we usually try to adjust them to the nature of work and one's own psychophysical abilities. Such activity aims to humanise work through the organisation of the system human–machine–environment conditions, so that it is done as lowest possible biological cost, yet highly effectively (Fig. 3.1).

The creator of the concept of ergonomics (from Greek *ergon*—work, *nomos*—principle, law) is the Polish scientist Professor Wojciech Bogumił Jastrzębowski (1799–1882), who was the first in 1857 to use the term ergonomics and identified the need to develop this science. In the periodical *Przyroda i przemysł*, in a series of articles entitled: *Outline of ergonomy, i.e. science of work, based on the laws drawn from the science of nature*, Jastrzębowski defined ergonomics as a science of using the powers and abilities given to man from his Creator.

The Englishman, Kenneth Frank Hywel Murrell (1908–1984) wrote in 1949—*Ergonomics: the study of the relationship between man and his working environment*.

According to the Polish Society of Ergonomics, ergonomics is an applied science aiming to optimally adjust tools, machinery, equipment, technology, organisation and the material working environment, as well as objects of common use to the physiological, psychological and social requirements and needs of the human being. Whereas according to International Ergonomics Association (IEA), ergonomics is the scientific discipline concerned with the understanding of interactions among humans and other elements of a system and the profession that applies

J. Smardzewski, *Furniture Design*,
DOI 10.1007/978-3-319-19533-9_3

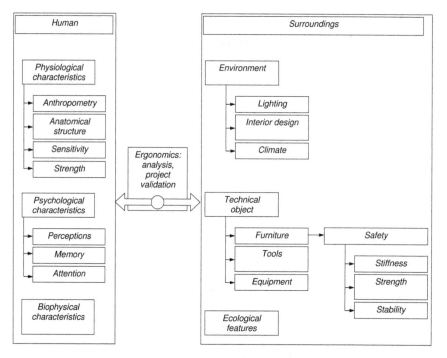

**Fig. 3.1** Ergonomics as a connection between human and his surroundings

theory, principles, data and methods to design in order to optimise human well-being and overall system performance.

The main area of interest of modern ergonomics, which grew from the teachings about work, is still the working and leisure environment of the human being. The ergonomic quality of this environment depends, among others, on the proper design and construction of the basic components of locations of work, study, dining, relaxation, sleep, etc.

Modern furniture, regardless of its intended purpose, is characterised by the domination of form over function and practical use values. However, it needs to be kept in mind that a furniture piece, as an object of applied art, should harmoniously combine all the requirements, including aesthetic, functional, use, constructional, technological, economic, and social.

## 3.2  Basic Design Requirements

Furniture is among the oldest of objects of applied arts, which the human uses. It should be noted here that these constructions differ little from those used today. Chairs, cabinets and even a folding bed found in Tutankhamen's tomb (Destroches-Noblecourt 1963)

**Fig. 3.2** Chair, relief from
Mastaba, 6th dynasty reign of
Pepi I, 2289-2255 BC (British
Museum)

show close similarities to the constructions of contemporary furniture (Fig. 3.2). And
some joints, e.g. mortise and tenon joints, have not changed their basic shape for more
than 3000 years.

In Poland, many objects of use constituting elements of the equipment of rooms
were initially defined as *equipment*. The term "furniture" is not encountered until
inventories from the second half of the eighteenth century. In city registries from
the end of the eighteenth century, we find the description: *today the term furniture,
which names equipment, is widespread, which serves comfort, benefit and decoration* (Sienicki 1954).

Throughout the history of the furniture industry, there have been two styles: folk
and courtly, as different from one another as different was the economic situation of
the citizens. An indicator of the quality of furnishing court interiors, aside from
relating to the currently prevailing fashion taste of the customer, was also the degree
of his wealth and skill level of the contractor. They were also furniture inherited
from their ancestors. Monarchs or other wealthy customers of most furniture maker
artists commissioned the interior design of their newly built palaces.

In the folk art, artisans and serfs, esthetics forms of furniture were modeled on
furniture from rich homes. Usually the folk carpenters completely transformed the
form of furniture by introduction new shapes and functions suitable for the local
materials and less perfect techniques (Sienicki 1954). An artisan, being both the
designer and producer, knew the customer perfectly, his tastes, dimensions and
design expectations and was able to freely adjust the technical object to the needs of
the future user. At the same time, he perfected the manufacturing method of furniture in the scope of the technique itself and methods of organising production.
The problem of the anonymous user did not appear until the nineteenth century, due
to the great economic and social changes, when industry began to dominate over the
craft in furniture, which was able to produce furniture for the broad masses of the
population. Then the division of furniture on folk and courtly, lost its raison d'être.

A new systematics of furniture was developed as a result of evolution of their destination. It was begun to distinguish home, office, school, hotel, ship, garden and many other types of furniture. The materials, fittings and additional accessories, especially in kitchen furniture, began to decide of the value of the furniture. Throughout the process of forming a new product, however, the subject for which the product was being made was not noted. At the turn of the nineteenth and twentieth centuries, the strong development of industrial production forced to strictly connect it with the need of ergonomic design. Furniture manufacturers began to realise the postulate of humanising technology. They also began to boldly use such design methods which in the designed product would exhibit a strong relation with the user, thus creating the anthropotechnic system. With this, it was made sure that humanocentric requirements dominated over other technical and technological criteria. As a result of the implementation of many research works, another division of furniture was made, which takes into account the age of the user, as well as anthropometric characteristics. This division differentiates furniture constructions designed for adults, for youth and for children. In standardisation progress, many constructional, dimensional and safety use requirements of furniture were also specified.

In addition to these requirements, the contemporary laws of the market impose increasingly higher and often contradictory expectations towards new products. Therefore, the task of today's designer–constructor is to identify these requirements, since the degree of satisfying them depends largely on his skills (Fig. 3.3).

Studies of individual market segments of home furniture (Pakarinen and Asikainen 2001; Smardzewski and Matwiej 2004), taking into account their breakdown by design features to solid wood furniture, furniture from boards and upholstered furniture, show that on a ten-point scale, the priority values for potential consumers when deciding whether to purchase a given product are quality and design. The second, in terms of importance, group of factors affecting a consumer's decision to purchase a given product is, respectively, price, services and material (Fig. 3.4).

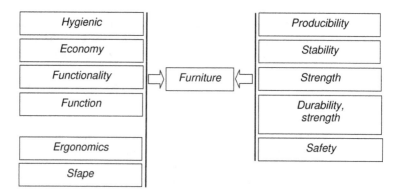

**Fig. 3.3** The requirements for furniture included in the design process

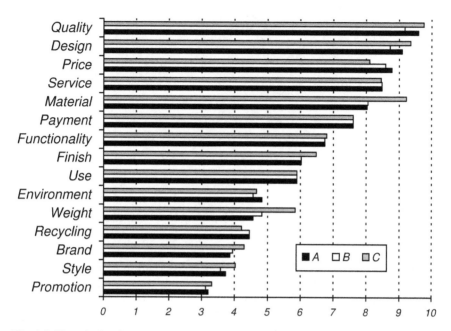

**Fig. 3.4** The criteria taken into account by customers when purchasing furniture: *A*—furniture from solid wood, *B*—furniture from boards, *C*—and upholstered furniture (own development based on: Pakarinen and Asikainen 2001; Smardzewski and Matwiej 2004)

Therefore, designing furniture, aside from the source of creative inspiration, requires support from many complex engineering processes, which consist of the knowledge of the following: markets of producers and consumers, purpose, function and features of furniture, properties of materials used, construction and manufacturing technology, method of securing a furniture piece during distribution, methods of conservation and renovation, as well as recycling used products. Therefore, mutual interaction of all known and unknown design parameters has to be taken into account in the process of designing the cooperation of competent partners.

The innovativeness of designers in creating and constructing avant-garde forms is most visible in skeletal and upholstered furniture. And case furniture presents a more traditional form. For the manufacture of furniture, today practically all available materials are used, including wood, wood-based materials, metal, glass, synthetic materials, rocks, fabrics, leathers, grass, shoots of bushes and others (Table 3.1).

Only the highest quality products having valid hygienic approval and durability certificates should be selected during the design process as constructional materials and those enhancing the surface of the furniture. They should ensure the designer and constructor of the harmlessness of components used, both for the user and for the natural environment.

**Table 3.1** Raw materials
used in designing furniture

| Type of material | Use in furniture (%) |
|---|---|
| Wood-based materials | 40 |
| Wood | 24 |
| Metal | 16 |
| Rattan | 7 |
| Bamboo | 4 |
| Fabric and leather | 4 |
| Glass | 2 |
| Plastic | 2 |
| Others | 1 |

Modern furniture gives rise to four principal issues related to ecology (Dzięgielewski 1996; Dzięgielewski and Smardzewski 1995; Dzięgielewski and Smardzewski 1997):

• the hygiene of the materials used in the construction of the furniture and their impact on man and the environment in which he resides,
• the impact of technologies on the natural environmental of applying modern auxiliary materials for the production of furniture, such as adhesives and painting–varnishing materials,
• disposal of waste occurring in the process of finishing and securing surfaces of furniture elements,
• the possibility to remove or secondary use of materials of worn furniture.
• Taking this into account, the following requirements concerning ecological furniture can be determined:
• designing furniture with a long life span in terms of its design and functionality,
• the use of ecological constructional and finishing materials,
• eco-friendly method of packing furniture,
• good possibility of using, processing or removing the materials of worn furniture.

In today's furniture constructions, the primary constructional materials, aside from wood, are wood-based materials such as chipboards, fibreboards, MDFs and plywood. Chipboards and MDFs constitute a minimum source of emission of carcinogenic free formaldehyde into the air. It is assumed that the emission of formaldehyde should not exceed 0.1 ppm (Łęcka et al. 1995). For the first time, the production of formaldehyde-free boards was undertaken in Germany and Ireland, through the use of adhesives based on isocyanate resins. However, the content of formaldehyde does not provide full information about formaldehyde emitted into the atmosphere from furniture.

Therefore, the assumption alone that we use ecological constructional materials in the design of furniture does not solve the issue of the "greenness" of the furniture. In designing and constructing ecological furniture, another important factor is also the application of organic adhesives and finishing materials (Proszyk and Bernaczyk 1994; Radliński 1996; Scheithauer and Aehlig 1996).

## 3.2.1 Aesthetic Requirements

Industrial design is a creative activity, which aims to define the external characteristics of objects produced industrially. These characteristics define the structural
and functional relations that make a given industrial design a coherent whole, both
for the manufacturer and the user. Industrial design aims to cover all of these
aspects of the human environment, which are either conditioned by industrial
production or are their direct result. Design is based on those elements from the
field of art, which express beauty in an expressive way—a positive aesthetic
property of existence resulting from maintaining proportions, harmony of colours,
sounds, appropriateness, moderation and usability, and perceived by the senses.

In ancient Greece, the concept of beauty was much wider than in later years.
This was related above all to the idea of goodness, spirituality, morality, thought
and reason, at the time it was identified with accuracy as a condition of beauty and
art at the highest standard. It was argued that beauty is mainly a result of maintaining proportions and an appropriate system. This view was considered most
relevant for centuries. It was professed, among others, by Pythagoreans, claiming
that beauty consists in an excellent structure, resulting from the proportion of parts,
their harmonious layout. They argued that it is an objective feature. According to
Aristotle, beauty is something that causes positive emotions. Plato believed that true
beauty is supersensual and is goodness as great as truth.

In the fifth century B.C., Sophists rejected the objectivity of perceiving this value
and limited its concept to that which is perceived as pleasant by the sense of sight
and hearing. Stoics, as well as some artists living in the Enlightenment, were also
close with this theory. Marcus Vitruvius Pollio—a Roman architect and an engineer
of war living in the first century B.C., in his work on architecture, wrote that: *the
value of beauty will be achieved in buildings through the symmetry and appropriate
ratio of the elements—height and width.* He believed that it is similar in sculpture,
painting and nature. This theory was supported, among others, by Leonardo da
Vinci, Michelangelo, St. Augustine, Albrecht Dürer and Nicolas Poussin. While
Plotinus preached that not only proportions and the right layout of parts determine
beauty, but above all, the soul.

During the Renaissance, the importance of the terms "perfection" and "beauty"
was considered, by analysing the works of certain artists. Petrarch and Giorgio
Vasari put the value of beauty as an objective value against other aesthetic concepts,
like grace, regarded by them as a subjective value.

These days, beauty and other aesthetic values are dealt with by the field of
philosophy—aesthetics (Gr. *aisthetikos*—literally *concerning sensory cognition,
but also sensitive*).

Modern industrial design is interdisciplinary knowledge, without which a
company that has the ambitions to achieve success on the market by introducing
innovative design and constructional solutions cannot function.

The importance of industrial design today is so strong, that the member states of
the European Union, including Poland, are investing significant resources in the

creation, development and promotion of design centres. The results of companies, which have already achieved significant successes on the market, may be convincing, in which the financial outlays on B+R activities (*Business+Research*) in departments dealing with researching, developing and designing a product amount to on average 8 % of annual turnover of these companies. For these reasons, the purpose of industrial design should not only be giving products, including furniture, the most attractive form, but also to build a strategy associated with this form of identity of the good on the market and creating the image of the manufacturer producing a given product. Therefore, the designer is obliged to actively participate in creating, shaping and developing new consumer needs. Thus, he should exert influence on creating a new and better material environment, in which the human being—the future user of the product—will function. It is this task that well-known designers consider to be most important and requiring the consolidation of efforts of many university-based and research and development centres.

In the evaluation of furniture design as applied arts, the basis of reactions and aesthetic experience is eye contact and the intellectual interpretation of the impression. An addition to this assessment is also the criticism of:

- usage,
- functionality,
- construction,
- technological values, including materials and finishing method.

The beauty of an industrial product is also affected by colour, which is an integral component of beauty. In many cases, colour determines both of the psychophysical condition of the user and of the success of a given product on the market (Dzięgielewski and Smardzewski 1995).

When discussing such a significant impact of colour on the market success of products, it also needs to be mentioned about the subjective perception of colours by every human being. Some colours usually excite, stimulate and activate, while others alienate, calm, soothe, inspire concentration and set a nostalgic or melancholic mood. The colour of a product and the colours of the surroundings have a significant impact on the well-being and organism of the user. Usually, the principle is adopted that cold colours soothe and warm colours stimulate:

- yellow is the colour of the Sun, it has a beneficial effect on the nervous system, and it stimulates to work, reduces fatigue and sets a mood of focus and perseverance. It is the colour of science, intellect, optimism and joy,
- orange strengthens the psyche, symbolises vitality and increases optimism. According to psychologists, the best colour for calm children is specifically orange,
- red is associated with the colour of fire, and it acts as a stimulant, raises blood pressure, increases physical activity and whets the appetite, inspires and improves the mood and sets an atmosphere for movement and rapture,
- purple has a beneficial effect on the nervous system, it stimulates, however, for people who are more sensitive, and it can also cause a gloomy and melancholic

mood. This colour is chosen at times of economic downturn of the market and investment limitations,

- blue calms and inspires, it is associated with space, it creates an atmosphere of harmony, being in blue spaces slows down the pulse and reduces blood pressure, and a blue environment also has a good effects on health of people who have a fever,
- green is the colour of peace, balance and mental focus; it regenerates one's strength, has a relaxing and soothing effect, regulates the work of the cardiovascular system, relieves symptoms of stress and gives optimism; therefore, it is excellent in counteracting depression. Dark green is associated with a typically masculine space, so this colour often adorns the walls, as well as fabrics of upholsteries of study furniture, libraries, billiard rooms, smoking areas and law firms. Bright green is a colour that symbolises hope and youth, and so it often appears in rooms of teenagers. According to psychologists, the best colour for children with a great temperament is specifically green,
- brown is a colour that gives a feeling of warmth, security and stability, and it is associated with the colour of items that have been known to people for thousands of years now.

Without a doubt, colour has a significant impact on realisation of the design of a specific product. It is first colour which affects the observer and causes certain reactions. Therefore, very good knowledge of the issues related to colours in industrial design is extremely important.

A designed object should arouse interest in the beauty of its proportions. Therefore, it is necessary that all its elements form a complete whole, composed in a logical and harmonious way (Dzięgielewski and Smardzewski 1995).

### 3.2.2 Functional Requirements

The expected function of a product has an essential influence on its form and constructional solutions. Every piece of furniture should meet a specific function, strictly connecting with the method, character and place of its use. In many cases, the function of a product is the factor which is the most inspiring to seek new shapes and artistic expression. For millennia, the human being has been refining the functionality of furniture, thus affecting its structural form and technology of production. By derogating from formerly existing conventions, we now sit on a chair with relaxed muscles, which is why the form of its seat is different. The distance of the seat from the base (floor), as well as the angle of this seat and backrest, has also changed. In this way, dimensional proportions changed, as did the distribution of fixed and moving parts of a furniture piece and the place of its use. Mostly, furniture with one well designed and technically solved function was thought of. Multifunctional furniture appeared along with the deterioration of living conditions and the necessity to maximise the use of housing space. These were products,

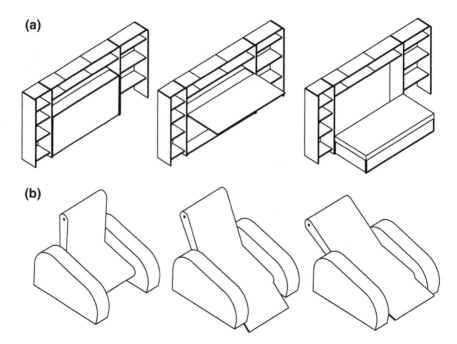

**Fig. 3.5** Multifunctional furniture: **a** negative interaction between functions, **b** creative overlapping of functions

which after transformation of certain elements enabled the following: work, dining, storage, sleep and rest. As a rule, they were distinguished by simple form and unsophisticated style, heavily dominated by technology ensuring the fulfilment of a number of user requirements at the same time. However, it should be pointed out that providing comfort for the user through comprehensively fulfilling many very different functions in one furniture piece is not possible, as these functions will mutually limit themselves greatly. In a classic shelf-couch, none of the functions of the furniture piece are adequately ensured (Fig. 3.5a). The number of compartments for storage is limited by the dimensions and location of the mattress, which is folded vertically. The construction of the mattress depends on the depth of the shelf, which limits the number of elastic layers affecting comfort of sleep and rest. The surface of worktops, which are usually lifted boards of the bottom of the couch's container, depends on the dimensions of the mattress. Therefore, the user of a hybrid furniture piece (multifunctional) cannot count on comfortable rest, pleasant work at a table and free access to many of the stored items. In furniture of this type, however, the constructional and technological solutions which ensure the synthesis of many different functions and closing them in one compact block of a furniture piece are inspiring.

   Modern multifunctionality of furniture, however, has a second, more beneficial side. Designers and constructors work on the forms of furniture that stand out with many highly integrated and mutually dependent functions. This group can include kitchen furniture with functions of storage, supporting silent opening and closing, lighting and memorising the number and type of stored products. Another example can be upholstered furniture designed for sitting and leisure with the function of unfolding the seat, footrest extension, massage, heating, assisting sitting down and standing up. As it is demonstrated, none of these functions limit others and itself (Fig. 3.5b). Harmonious cooperation between the form and functions of a furniture piece, as well as innovative design solutions, modern technology and original ecological materials, ensures that a product designed in such a way will be considered a high-end product of excellent quality. In this sense, the multifunctionality of furniture, and so the ability to adapt it to the growing needs of customers, is a huge advantage, which builds a strong brand of the product and distance products that are less developed.

Therefore, the modern concept of the function of a furniture piece has acquired a new meaning and a new quality. A piece of furniture with a function is mostly a product imbued with electronics and automation, improving the quality, comfort and safety of use, presented in modern proportions and design lines. Therefore, it is also worth noting that multifunctionality, in contrast to previous periods, is increasingly accompanying luxurious furniture of high quality. Therefore, the innovative constructional solutions introduced must work efficiently, reliably and safely.

In addition to the importance of the functions of a product, in industrial design, its functionality also plays a key role, i.e. adjusting the functions of a furniture piece to both physical and psychological requirements of the user (Dzięgielewski and Smardzewski 1995). The functionality of a furniture piece is primarily determined by the direct contact of the object or its part with the body of the user. The actual expression of this functionality is the dimensions, shape and quality of realisation, as well as in some cases, the type of material used.

## 3.2.3  Construction and Technology Requirements

The constructional requirements of the designed furniture are as follows: simplicity of the concept, rational selection of materials, satisfactory stiffness, stability and strength of the system, proper realisation of joints and technology of machining. The simplicity of the concept of a furniture piece, and its individual elements, joints and mechanisms affect the performance characteristics of the product and the technical and economic indicators of production. The choice of the type of materials, including their physical and mechanical parameters, depends on both the durability and reliability of the product, as well as the material absorbence of the production process. The designer–constructor, in the process of developing a new project, should not only focus on integrating the aesthetic, functional and technological requirements, but must also determine the usable strength of the product by choosing relevant cross sections of the elements, the dimensions of joints, types of connections and methods of connecting various materials, which will ensure failure-free work throughout the entire period of exploitation.

At the design and construction stage of a furniture piece, an analytical or numerical examination is also necessary for the strength of the construction under static and dynamic load, at a changing position of operational forces and at different support schemes of the tested system. The conditions of work of multifunctional, mobile and transformed furniture should especially be simulated in theoretical calculations. On the basis of these calculations, the durability of movable parts of furniture must be evaluated, such as doors, inserts, drawers, containers, flaps, as well as the permissible carrying capacity for shelves, partitions, frames, racks, hanger rods, as well as the global stiffness of the product and its individual elements, which permanent deformation may worsen the aesthetics of the piece of furniture.

A numerical optimisation of the construction may also turn out to be useful. By building functions of the aim related to achieving the minimum own cost value of production, at the permitted usable strength of the product, satisfactory dimensions of transverse cross sections of elements and joints are quickly found, which guarantee minimal wearing of materials and production forces at a maximum reliability of the construction.

The technology requirements are mainly related to the nature of production (unit, discrete, serial and mass) and a technological process that is possible to complete. In preparing a product for mass or serial production, requirements related to maximum use of existing resources of the company should be taken into account. In particular, the dimensions of elements are optimised and unified, adjusting them to maximum performance during the cutting of boards, veneering, sawing, sanding, etc. Operations are standardised in order to reduce the number of changes and prolong production cycles. Furniture for discrete production is designed in a different way. Production plans are created only for registered orders of customers; therefore, their number is known and highly atomised in terms of constructional and technological diversity. In the case of furniture produced in such a way, most attention during designing is devoted to finding a rational compromise between form, function, construction and technology. Therefore, it should be noted that even with a large diversity of products, elements can be unified which are not significant for the client, but important structurally or technologically. Only products designed and manufactured for individual customer orders, in micro- and small enterprises, provide designers with considerable freedom and constructors and technologists with greater flexibility to meet these visions, without any damages to the durability of the construction or safety of the user.

### 3.2.4  Technical and Economy Requirements

Technical and economic requirements are characterised by their cost-effectiveness: design methods, structures of the product (material specifications), manufacturing technologies, method of validation, distribution, warranty and post-warranty services, marketing campaigns, etc. By starting to design a new product, all the criteria should be defined which enable making all possible savings, both during the design or manufacture of the product and during the exploitation by the customer. Guided by the purpose and place of use of the product, the designer should limit the selection of costly and durable materials only for elements and subassemblages, which have usable significance, with which the user will be in direct contact with or to those which may determine the strength of the furniture piece. Depending on the type of market group of customers, the cost of production of furniture can be minimised by using cheaper fittings, material substitutes or alternative technologies.

In a market economy, all activities are oriented to optimise the quality of the manufacturing processes, in particular the reduction of the costs of a full design–construction–manufacturing cycle. In design and construction offices of furniture factories, expectations related to form usable function and structural solutions are usually maximised, forgetting about the strong connection of technological innovation with the final cost of technical manufacture of the product, which is the essence of market success and prosperity both of the products, and their manufacturers.

Furniture design is a process of selecting geometric, material and dynamic constructional characteristics including the aesthetic, technical, technological and economic existence of a product in terms of appropriate criteria selected based on the principles of construction. The aim of a scientific study of the costs of construction is to search for logically organised relations, addressing the impact of structural properties on costs, which will enable to formulate the principles of construction.

The processes of designing and constructing furniture have a layered structure with the gradual transition from abstraction to specifics at different levels (functional, physical impacts and processes, as well as a description of the material form); hence, the most difficult are the initial phases of formulating the task for assessing the minimum costs of planned production. A relatively small number of known structural characteristics of uncertain and complex impact on the cost of the finished product make it difficult to reach a compromise of technical and economic criteria.

In design methods with early recognition of costs, designing should be combined with construction catalogues, catalogues of costs and an analysis of values. In this way, for each of the constituent functions, the best economical form of realisation of an element or subassemblage can be chosen, e.g. type of material, type of connection and technology of implementation (Branowski et al. 1985; Snijders et al. 1995; Stockert 1989).

Calculating the cost of future products must be closely linked with the methods of their calculation commonly used in the industry, carried over in environments of integrated computer-aided design (CAD)–enterprise resource planning (ERP) systems, supporting the design and manufacture of furniture.

The construction of furniture is a concept activity that seeks to meet the set demands in the best way possible at the given moment. The laws of the market require that today's products meet customers' wishes in the most optimal manner possible. The degree of satisfying the numerous and partly contradictory demands and expectations to a great extent depends on the constructor (Fig. 3.6).

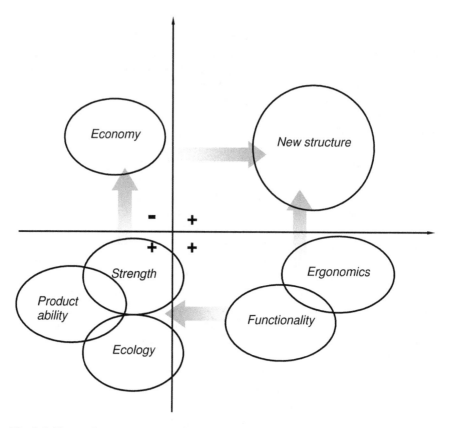

**Fig. 3.6** The requirements accompanying the creation of a new design of a furniture piece

Meeting the marketing challenge of long-term competitiveness forces to minimise the relation of outlays to the effects, which for furniture products can be reduced to the principle of satisfactory quality at minimal own costs.

The analysis of value is a planned process aimed to achieve the required functionality of a product at the lowest costs possible, without reducing the quality, reliability, efficiency and worsening the conditions of exploitation and delivery. This is a proven method in practice, especially useful when designing new technical and organisational solutions concerning marketing, organisation, construction, technology, quality, inventory management and sales.

## 3.3  Form and Construction of Furniture

Almost every day we encounter objects that captivate us by their form, function, colour and craftsmanship. Sometimes, these are objects designed by famous designers. We wonder then why a given object looks so good and why we like it.

**Fig. 3.7** Division of the section into two unequal parts

Other times, we see furniture manufactured in past periods and with astonishment we ask ourselves the question of how "they" did it? Usually, the answer to this question is to understand the principle of proportional distribution.

The oldest examples of the occurrence of the principle of proportional distribution are the works of nature, including the shape of galaxies, flowers, plants, snail shells and pine cones. The Egyptians used the rule of the golden ratio 2700 years before Christ. We know that the Pythagoreans knew the rule of golden ratio as early as 500 B.C. However, mathematically, this rule was described for the first time by Euclid in his most important work *Elements* (*Book 6, Proportions*, 300 B.C.). He stated that there is such an excellent division of a section, in which the ratio of the length of the smaller part to the length of the larger part is identical to the ratio of the larger part to the whole, while the value of this coefficient shows the proportion of 0.61803389...: 1. The obtainment of this value can be explained by Fig. 3.7 and the following equations. First, we divide any section into two unequal parts, indicating the longer section by $x$, the shorter by $(1 - x)$ and the ratio $x/(1 - x)$ as $\varphi$ (phi).

Then, let us create the proportion in the form of:

$$\frac{1}{\varphi} = \frac{1 - x}{x} = \frac{x}{1} \tag{3.1}$$

By multiplying both sides by $x$ and regrouping the expressions, the above equality is reduced to the form of the overall square equation:

$$x^2 + x - 1 = 0. \tag{3.2}$$

It has two real solutions:

$$x_{1,2} = \frac{-1 \pm \sqrt{5}}{2}, \tag{3.3}$$

of which one is positive:

$$x = \frac{-1 + \sqrt{5}}{2} = 0,61803389\ldots \tag{3.4}$$

Hence, the number of the golden ratio (phi) is:

$$\varphi = \frac{x}{x-1} = 1,6180342\ldots \tag{3.5}$$

The number $\varphi$ is sometimes also referred to as the golden number. Further, approximations of the value of this number can be obtained by calculating the quotients of adjacent Fibonacci numbers: 0, 1, 1, 2, 3, 5, 8, 13, 21, 34, 55, 89, 144, 233, which gives subsequently: 0, 1, 2/1, 3/2, 5/3, 8/5, 13/8 21/13, 34/21, 55/34, 89/55... $\rightarrow \varphi$. Already, the last of the fractions listed here gives an approximation of the golden number to the nearest 0.001.

At the beginning of the twentieth century in Saggara (Egypt), archaeologists opened a crypt, which contained the remains of the Egyptian architect Khesi-Ra (Shmelev 1993). The historical text on the crypt indicated that Khesi-Ra lived in the period of Imhotep, that is during the reign of Pharaoh Djoser (2700 B.C.). The figure of the architect was presented surrounded by different figures in different proportions. Initially, Egyptologists acknowledged the descriptions on the crypt as unreadable and incomprehensible. Only in the 1960s, after conducting many tedious analyses and comparisons, did Egyptologists agree that on the crypt there is an inscription of the principle of the golden ratio (Fig. 3.8).

It threw new light on the technical culture of ancient Egypt. And since the inscription concerns the period of the reign of Pharaoh Djoser, it is very likely that the pyramid of Djoser was the first experimental construction designed under the leadership of Khesi-Ra. The later pyramid of Pharaoh Khufu from 2560 B.C. is also an example applying the rule of the golden ratio by builders. Putting together the individual dimensions of the structure (Fig. 3.9), it can be noticed that (FG)/ EG = 230.36/(2 * 186.47) = 0.61768664. This gives the value of the coefficient $\varphi = 1.6189439$. This result differs from the value of the golden ratio number only by 0.005622 %.

In 1202, the Italian mathematician Leonardo Fibonacci completed his work entitled *Liber abaci*. In the third part of this book, he studied the problem of a series of specific numbers, which along with the increase of their value provide an identical quotient as the coefficient value of the golden ratio. That is why Fibonacci described the solution of a certain problem that occurs in nature, namely a man put into a box with four enclosed walls a pair of rabbits and asked the question, how many pairs of rabbits will one pair provide during a year, if presumably every month each pair breeds a new pair, which in the second month becomes fertile? By explaining the population problem of rabbits, Fibonacci arranged a string of numbers fulfilling the following pattern:

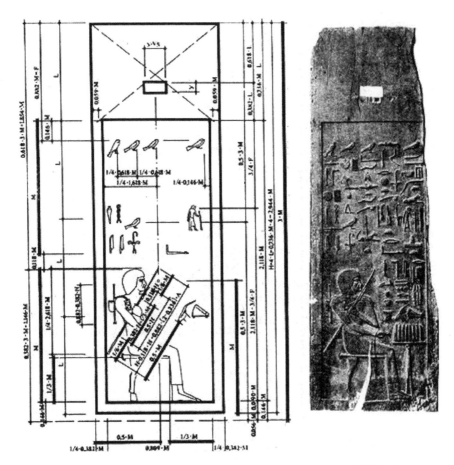

**Fig. 3.8** The tombstone crypt of the architect Khesi-Ra (Saggara, Egypt 2700 B.C.). *Source* Shmelev (1993): Phenomenon of the Ancient Egypt. Lotaz, Minsk)

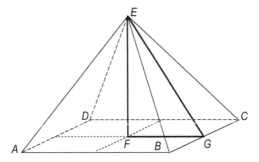

**Fig. 3.9** The dimensions of the pyramid of Khufu 2560 B.C.: *EF* height to theoretical tip = 146.73 m, *EG* average height of the side wall = 186.47 m and *2FG* average length of side = 230.36 m

$$1 = 1 + 0,$$
$$2 = 1 + 1,$$
$$3 = 2 + 1,$$
$$5 = 3 + 2,$$
$$8 = 5 + 3,$$
$$13 = 8 + 5,$$
$$21 = 13 + 8,$$
$$34 = 21 + 13,$$

$$\ldots$$

By dividing any number from the series by the number directly preceding it, Fibonacci obtained a value of the quotient close to the number of the golden ratio. These dependencies can also be presented in the form of spirals of the golden ratio (Fig. 3.10).

The rules of reproduction of the herd of rabbits adopted by Fibonacci can also refer to other objects of nature: trees, flowers, shells, fruit, as well as systems of stars and galaxies. The drawing (Fig. 3.11) illustrates a tree, which grows similarly to how rabbits reproduce. Each branch during the first year only grows and in each following year releases one young branch.

The number of petals of many flowers, including the popular daisies, is generally Fibonacci's number and amounts to 3 or 5, or 8, or 13. An even more surprising result is given by observations of the distribution of leaves on twigs and branches on the boughs of oak. It can be observed that not all the leaves lie one above the other, but twigs similarly. On the contrary, they arrange along a straight line, rather than along the spiral (helix), which orbits the stalk. The cycle of this curve is called the distance of the leaves embedded exactly one above the other, along the branches

**Fig. 3.10** Golden spiral (helix)

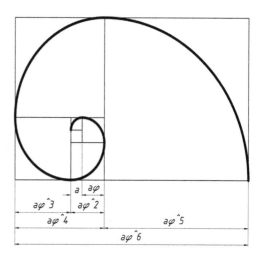

**Fig. 3.11** An example of tree growth according to Fibonacci's model

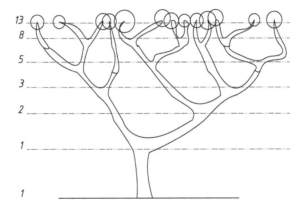

**Fig. 3.12** The spiral (helix) layout of the common pine cone

or stalks. The helix of a given plant can be characterised by two numbers: the number of rotations of the helix cycle around the branch or stalk and the number of intervals between subsequent leaves lying over one another. It turns out that for very many plants, these two numbers are Fibonacci numbers. For example, the tree of the beech has a cycle consisting of three leaves and performs one rotation, and the American willow has a cycle consisting of 13 leaves and performs 5 rotations. Another example of the occurrence of Fibonacci numbers in nature, well known to people working with trees, is a layout of scales in pine cones. The drawing (Fig. 3.12) illustrates a pine cone, on which spirals can be observed and created by its scales. These spirals can be dextrorotatory or laevorotatory. The pine cones of even the same pine species do not always have the same number of spirals, and also, laevorotatory or dextrorotatory do not always prevail. But in most cases, they are arranged along the spirals, in which number parameters strictly correspond to consecutive Fibonacci numbers.

So how can the rule of the golden ratio be used in furniture design? In furniture for the living room, bedroom, dining room etc., the only limitations are the ana-tomic dimensions of the selected population constituting future users of this fur-niture. For this group of furniture, one can use one's imagination in creating forms and choosing proportions. The process of designing furniture, which should be integrated in the environment of technical objects, such as dishwashers, refrigera-tors, sinks and dryers, is somewhat different. So, let us take a look at the following

**Fig. 3.13** The design of the system of fronts of drawers in the attempt to unify their dimensions

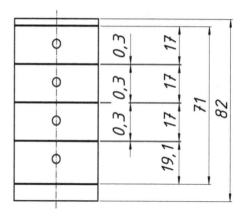

two examples of designing proceedings, which aim to propose a system and dimensions of the fronts of drawers (with a total height of 71 cm) in a kitchen cabinet with the dimensions: height 82 cm and width 42 cm.

For the manufacturer of furniture, the most convenient would be an equal division of the height of 71 cm in 4 equal parts. For calculating the height of one drawer front, we would then use the formula:

$$4h + 3s = 71, \tag{3.6}$$

where

$h$  height of one drawer front,
$s$  width of the gap between the fronts of the drawers, e.g. 0.3 cm,

which gives $h = 17.525$ cm.

Obtaining such a dimension in the production process is not possible; therefore, the manufacturer shall unify the dimensions of the first three drawers to a height equal to 17 cm and increase the height of the bottom drawer to 19.1 cm (Fig. 3.13).

By applying the golden ratio principle, the fronts of the drawers can be planned out using the following formula:

$$h_1\varphi^3 + h_1\varphi^2 + h_1\varphi + h_1 + 3s = 71, \tag{3.7}$$

which gives

| $h_1 = 7.40$ cm | $\rightarrow$ | 8.1 cm |
|---|---|---|
| $h_2 = 11.97$ cm | $\rightarrow$ | 12 cm |
| $h_3 = 19.37$ cm | $\rightarrow$ | 19 cm |
| $h_4 = 31.34$ cm | $\rightarrow$ | 31 cm |

By bringing the obtained dimensions to the technology requirements, the heights of the fronts of drawers should correspond to the markings provided in Fig. 3.14.

**Fig. 3.14** The design of the system of fronts of drawers taking into account the golden ratio principle

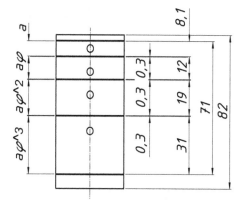

**Fig. 3.15** Construction of Fibonacci's compass

$AF=AH=340\ mm$
$BG=210\ mm$
$AB=CD=BE=CE=130\ mm$
$EG=80\ mm$

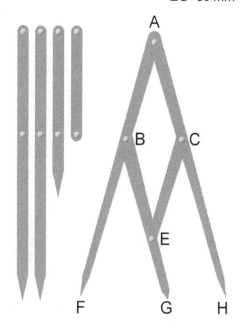

From a practical point of view, the designer–constructor of furniture should be equipped with IT tools (CAD system) supporting the design practice of the golden ratio methodology. In studios where computer technologies are not used, Fibonacci's compass will be useful, which facilitates the choice of lengths of sections in accordance with the golden ratio principle. Dimensional proportions and the structure of such a compass are shown in Fig. 3.15.

## 3.4   Anthropotechnical Designing

Designing mainly refers to specific objects. In the field of designing furniture, it will always be a technical artefact of usable character: a chair, a table, a wardrobe, a bed, etc. Like science, designing is an intellectual plan, realised using tools suitable for a particular discipline, within the framework of which this designing is done. According to Łopaczewska (2003), the following types of design are distinguished:

- technical design—usually associated with the manufacturer, it is sometimes treated as a fragment of so-called technical preparation of production (cf. Chap. 4),
- ergonomic design—also known as humanist design—is designing objects in accordance with the requirements of ergonomics.

The term ergonomic design appeared in the early 1970s, when a new type of technical design was clearly emphasised, oriented at the so-called human factor. In literature, ergonomic design was called designing control systems and it constitutes an important part of comprehensive technical design, which in this particular way ensures compliance of the designed object with ergonomic requirements. According to Słowikowski (2000): *The specificity of ergonomic design consists in the dualism of subject of designing. This is the biotechnical system (specifically, anthropotechnic), the parts of which have extremely different characteristics. One of the parts—the human being, constitutes an invariant, which features have been determined by nature, which is why the designer is left with adjusting the second: the machine, to the features of the first. This is a pragmatic interpretation of the anthropocentric principle, referred to the process of designing technical objects creating the system human-technical object.*

Therefore, ergonomic designing leads to defining the system human–technical object from the point of view of the principles of ergonomics (Kroemer et al. 1994). The systematics of the validity of aims in ergonomic works can be considered at three levels:

- developing acceptable conditions, which expose the user of furniture to the loss of health or life,
- creating widely endorsed (due to the current state of knowledge) social, technical and organisational conditions,
- defining and developing conditions of physical, mental and social comfort, adaptable by future users in the scope of personal qualities, abilities and expectations.

In accordance with current trends, the designer designs the system human–technical object, but not the technical structure of the object—the piece of furniture. Ergonomics, therefore, fulfils a dual role (Słowikowski 2000):

- It sets goals of design, demanding from the designer an anthropocentric approach, i.e. shifting the characteristics and needs of the user over constructional and technical requirements,

- enriches methodological design, since it is an integral component of the art of engineering or broader, the art of design.

The engineering technical design procedure constitutes the basis for ergonomic design. The mutual relation of technical and ergonomic design is based on the principle resulting from the characteristic for anthropocentric ergonomics recognition of the priority of features and needs of the human being in shaping the technical structure (Łopaczewska 2003). Technical structure is, of course, furniture, while their most demanding representative is directly the group of usable furniture, that is those that during use are in direct contact with the human body (furniture for lying down, sitting, resting). A person who has direct contact with a piece of furniture becomes part of a system, called the anthropotechnic system, consisting of an animate part (the human body) and inanimate part (technical element—the furniture piece) (Winkler 2005).

Therefore, in the design process, the priority is to increase the certainty of the product's functioning by taking into account all interactions together that occur between the object and the user in terms of vision, hearing and tactile stimuli. Therefore, when designing a new piece of furniture, the form, construction, technology of implementation, functionality and ergonomics of similar products must be critically assessed, because the object of ergonomic design is precisely the system human–technical object.

### 3.4.1 Anthropometric Measures of the Human

Anthropometry (Gr. *anthropos*—human, *metreo*—measure), according to the definition in the encyclopaedia PWN, is a group of technologies of making measurements of the body or skeleton of a modern and fossil human being, and it enables an accurate and comparable study of diversity and variability of measured characteristics of the human in his personal and evolutionary development.

Knowledge of the human's dimensions, called anthropometric dimensions, constitutes the basic component of the knowledge base necessary to shape workstations in terms of the user's comfort and functionality of the designed products. The application of these dimensions enables one to carefully design a usable space, choose the appropriate size of furniture and their component parts and propose the most favourable arrangement of this furniture in relation of each other and the user.

The frequency distribution of anthropometric characteristics generally takes the form of a Gaussian distribution. Therefore, when there is no possibility of designing furniture for 100 % of the population, in literature, it is recommended to adopt border values of characteristics corresponding to the lower centile (5C), upper centile (95C) and average value (50C). The threshold values determine what number of a given population is located in a given range. Hence, the lowest dimension defined by centile 5C is not achieved by only 5 % of the population, while 95 % of the population is found below centile 95C. Centile 50C (average)

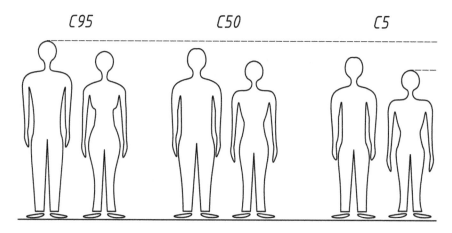

**Fig. 3.16**  An illustration of restrictive measures in the design of furniture

symmetrically divides the population of users into ones achieving a specific dimension and those who do not achieve this dimension (Batogowska and Malinowski 1997). According to the principle of restrictive measures, a properly designed piece of furniture should, therefore, take into account the dimensions adjusted to the dimensions of the users, at least 90 % of the population, that is those, whose dimensions fall between the values of the 5th and 95th centile (Fig. 3.16). Thus, the percentage of people, for whom a usable space or piece of furniture will not be adjusted, will amount to 5 and 10 %, respectively.

Usually, it is assumed that the dimensions that correspond to the characteristics

- of the 5th centile are applied to determine dimensions, borders of zones of reach, positions of important handles, fittings and locks,
- of the 95th centile are used for internal dimensions, spaces for the lower limbs, trunk, etc.

For these reasons, furniture for lying down should be designed for users from the population represented by the 95th centile and furniture for work and dining according to the characteristics of values of the 50th centile, as well as furniture for storage according to 5th or 50th centile. In order to unify the application of this rule, anthropometric data are provided in the system (for men and women) of the 5th, 50th and 95th centile. Figure 3.17 and Table 3.2 summarise the basic anthropometric features for a person in a seated position, and Fig. 3.18 and Table 3.3 illustrate the anthropometric features for a person in a standing position.

In the practice of design, there is a fundamental difficulty, due to the variation in the dimensions of individual members of the population. The reason for this differentiation is ethnic origin, gender, height, development, age and social and professional class. In order to facilitate the selection of these dimensions, anthropometric data have been provided below for the adults of selected countries

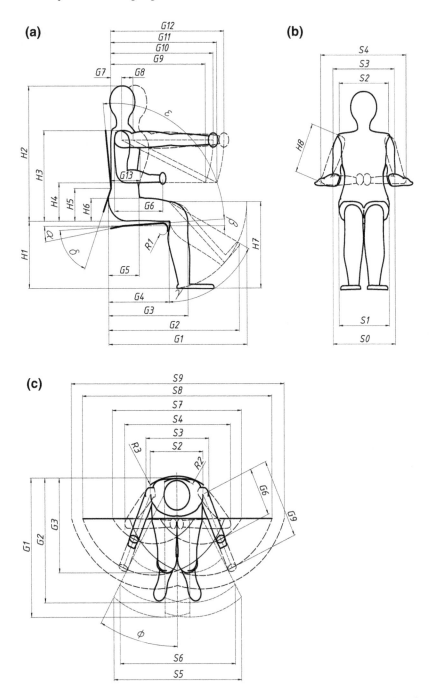

**Fig. 3.17** More important anthropometric measures of persons in the seated position: **a** side view, **b** front view, **c** top view

**Table 3.2** Dimensions of a person in the seated position for the 5th, 50th and 95th centile (own development based on Jarosz 2003)

| Name of the feature | No. of the feature acc. to PN-EN ISO 7250-1:2010 | Parameter | Unit | 5th centile | | 50th centile | | 95th centile | |
|---|---|---|---|---|---|---|---|---|---|
| | | | | Woman | Man | Woman | Man | Woman | Man |
| Reach of straight leg | | G1 | mm | 912 | 980 | 1003 | 1080 | 1119 | 1194 |
| Reach of shin at knee bend | | G2 | mm | 774 | 913 | 851 | 1007 | 952 | 1112 |
| Depth of the seat from the knee | 4.4.7 | G3 | mm | 532 | 554 | 578 | 601 | 624 | 646 |
| Depth of the seat | 4.4.6 | G4 | mm | 441 | 480 | 486 | 536 | 552 | 588 |
| 5/6 of the depth of the seat | | 5/6 G4 | mm | 367 | 400 | 405 | 446 | 460 | 490 |
| Width of space for knees | | G3_G4 | mm | 91 | 74 | 92 | 65 | 72 | 58 |
| Depth of buttocks–abdomen | 4.2.17 | G5 | mm | 197 | 230 | 256 | 284 | 343 | 344 |
| Maximum protrusion of a table worktop | | G3_G5 | mm | 335 | 324 | 322 | 317 | 281 | 302 |
| Reach of forearm at elbow bend | | G6 | mm | 274 | 294 | 315 | 335 | 326 | 376 |
| The distance of the acromion point (a) from the line of the back | | G7 | mm | 57 | 53 | 60 | 60 | 80 | 70 |
| The distance of the acromion points (a) and (a') when the body is bent | | G8 | mm | 193 | 214 | 235 | 241 | 227 | 297 |
| Frontal hold reach to the table worktop | | G9 | mm | 601 | 637 | 692 | 746 | 758 | 801 |
| Frontal hold reach | 4.4.2 | G10 | mm | 687 | 736 | 772 | 836 | 836 | 898 |
| Hold reach from the acromion point (a') to the table worktop when the body is bent | | G11 | mm | 816 | 877 | 949 | 1010 | 1070 | 1126 |
| Frontal hold reach when the body is bent | | G12 | mm | 880 | 950 | 1007 | 1077 | 1125 | 1195 |
| Depth of chest | 4.2.16 | G13 | mm | 228 | 214 | 263 | 242 | 321 | 280 |
| Popliteal height | 4.2.12 | H1 | mm | 361 | 388 | 402 | 428 | 448 | 488 |
| Seating height | 4.2.1 | H2 | mm | 792 | 833 | 861 | 907 | 916 | 980 |
| Shoulder seating height | 4.2.4 | H3 | mm | 515 | 547 | 566 | 586 | 625 | 688 |

(continued)

**Table 3.2** (continued)

| Name of the feature | No. of the feature acc. to PN-EN ISO 7250-1:2010 | Parameter | Unit | 5th centile | | 50th centile | | 95th centile | |
|---|---|---|---|---|---|---|---|---|---|
| | | | | Woman | Man | Woman | Man | Woman | Man |
| Elbow seating height | 4.2.5 | H4 | mm | 198 | 194 | 240 | 225 | 292 | 301 |
| Height of loin support | | H5 | mm | 170 | 220 | 170 | 220 | 170 | 220 |
| Thigh gap | 4.2.13 | H6 | mm | 115 | 115 | 139 | 146 | 169 | 171 |
| Thigh height | | H7 | mm | 471 | 500 | 517 | 544 | 567 | 606 |
| Arm length from acromion (a) to radiale (r) | | H8 | mm | 317 | 353 | 326 | 361 | 333 | 387 |
| Worktop height | | HP | mm | 254 | 253 | 305 | 291 | 367 | 381 |
| Knee height | | HK | mm | 112 | 112 | 116 | 116 | 118 | 118 |
| Thigh height | 4.2.14 | H1pHK | mm | 471 | 500 | 517 | 544 | 567 | 606 |
| Knee height | | H7_H1 | mm | 110 | 112 | 115 | 116 | 119 | 118 |
| Height to the rails | | H1pH6 | mm | 476 | 503 | 541 | 574 | 617 | 659 |
| Height to the worktop of the table | | H1pH4 | mm | 559 | 582 | 642 | 653 | 740 | 789 |
| Body height | 4.1.2 | HC1 | mm | 1536 | 1660 | 1634 | 1778 | 1740 | 1890 |
| Elbow seating width | | S0 | mm | 391 | 481 | 454 | 548 | 517 | 621 |
| Hip seating width | 4.1.11 | S1 | mm | 325 | 319 | 365 | 353 | 410 | 592 |
| Shoulder width | 4.2.8 | S2 | mm | 332 | 376 | 372 | 440 | 416 | 445 |
| Shoulder seating width | 4.2.9 | S3 | mm | 355 | 456 | 453 | 518 | 525 | 588 |
| Elbow width | | S4 | mm | 391 | 481 | 454 | 548 | 517 | 621 |
| Width of reach of straight legs | | S5 | mm | 770 | 1124 | 846 | 1238 | 944 | 1368 |
| Width of reach of legs bent at the knees | | S6 | mm | 654 | 1046 | 718 | 1154 | 804 | 1274 |
| Width of frontal hold reach at elbow bend | | S7 | mm | 880 | 964 | 1002 | 1110 | 1068 | 1197 |
| Width of frontal hold reach on the table worktop | | S8 | mm | 1420 | 1544 | 1636 | 1812 | 1772 | 1907 |

(continued)

**Table 3.2** (continued)

| Name of the feature | No. of the feature acc. to PN-EN ISO 7250-1:2010 | Parameter | Unit | 5th centile | | 50th centile | | 95th centile | |
|---|---|---|---|---|---|---|---|---|---|
| | | | | Woman | Man | Woman | Man | Woman | Man |
| Width of frontal hold reach on the table worktop when the body is bent | | S9 | mm | 1420 | 1544 | 1636 | 1812 | 1772 | 1907 |
| Radius of rounding of the front edge of the seat | | R1 | mm | 30 | 50 | 30 | 50 | 30 | 50 |
| Radius of rounding of the loin rest | | R2 | mm | 220 | no limits | 220 | no limits | 220 | no limits |
| Radius of rounding of the shoulder rest | | R3 | mm | 450 | no limits | 450 | no limits | 450 | no limits |
| Angle of inclination of the backrest part of the seat to the back | | $\alpha$ | deg | 1 | 4 | 1 | 4 | 1 | 4 |
| Angle of inclination of the seat to the back | | $\beta$ | deg | 5 | 7 | 5 | 7 | 5 | 7 |
| Angle of inclination of the shin from the vertical | | $\gamma$ | deg | 45 | 60 | 45 | 60 | 45 | 60 |
| Angle of inclination of the loin support | | $\delta$ | deg | 90 | 70 | 90 | 70 | 90 | 70 |
| Angle of inclination of the backrest to the back | | $\varepsilon$ | deg | 95 | 110 | 95 | 110 | 95 | 110 |
| Angle of inclination of the lower limb | | $\varphi$ | deg | 25 | 35 | 25 | 35 | 25 | 35 |
| Angle of inclination of the side edge of the backrest to the left | | $\rho$ | deg | 7 | 7 | 4 | 8 | 5 | 8 |

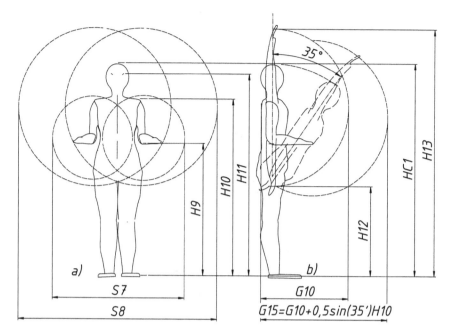

**Fig. 3.18** More important anthropometric measures of a person in a standing position: **a** front view, **b** side view

of the European Union for the purpose of designing (Jarosz 2003) (Tables 3.4, 3.5, 3.6, 3.7, 3.8 and 3.9).

Often, designs include the individual character of the workstation and place for sleep or rest. Then, the designer should both independently calculate the value of individual measures and specify the value of the centile representing the user, for whom the project is created (Dirken 2001; Haak and Leever-van der Burgh 1994; Molenbroek 1994; Panero and Zelnik 1979; Pheasant 2001). The method of calculating the dimension for centile C has been given below:

$$SC = SR \pm Z \cdot SD \qquad (3.8)$$

where

- SC—dimension calculated for centile C,
- SR—average measures established on the basis of anthropometric tables of a selected population,
- Z—coefficient established from Table 3.10. The coefficient can assume ± values. For dimensions above the average, the coefficient has the value (+), while for dimensions below the average value, the coefficient has the value (−),
- SD—standard deviation for the dimension SR determined on the basis of anthropometric tables of the selected population.

**Table 3.3** Dimensions of a person in the standing position for the 5th, 50th and 95th centile (own development based on Jarosz 2003)

| Name of the feature | No. of the feature acc. to PN-EN ISO 7250-1:2010 | Parameter | Unit | 5th centile | | 50th centile | | 95th centile | |
|---|---|---|---|---|---|---|---|---|---|
| | | | | Woman | Man | Woman | Man | Woman | Man |
| Elbow height | 4.1.5 | H9 | mm | 1012 | 1027 | 1017 | 1098 | 1124 | 1207 |
| Shoulder height | 4.1.4 | H10 | mm | 1329 | 1438 | 1397 | 1514 | 1465 | 1591 |
| Eye height | 4.1.3 | H11 | mm | 1492 | 1582 | 1561 | 1684 | 1630 | 1775 |
| Fist height | 4.4.4 | H12 | mm | 653 | 672 | 722 | 768 | 785 | 846 |
| Frontal hold reach | 4.4.2 | G10 | mm | 687 | 736 | 772 | 836 | 836 | 898 |
| Top hold reach | | H13 | mm | 1849 | 2053 | 2004 | 2127 | 2121 | 2304 |
| Body height | 4.1.2 | HC1 | mm | 1536 | 1660 | 1634 | 1778 | 1740 | 1890 |

**Table 3.4** Anthropometric data for the region: Northern Europe (Jarosz 2003)

| No. of the feature | Name of the feature | Men | | | Women | | |
|---|---|---|---|---|---|---|---|
| | | Centiles | | | | | |
| | | 5 | 50 | 95 | 5 | 50 | 95 |
| 4.1.1 | Body height when standing | 1710 | 1810 | 1910 | 1580 | 1690 | 1790 |
| 4.1.12 | Hip width when standing | 310 | 340 | 370 | 315 | 350 | 405 |
| 4.2.1 | Seating height | 900 | 950 | 1000 | 840 | 900 | 950 |
| 4.2.2 | Eye height when seating | 770 | 820 | 870 | 710 | 760 | 820 |
| 4.2.9 | Shoulder width (upper body width) | 425 | 460 | 500 | 365 | 400 | 430 |
| 4.2.11 | Hip width when seating | 320 | 350 | 390 | 325 | 375 | 630 |
| 4.2.12 | Shin length (popliteal height) | 415 | 455 | 505 | 370 | 410 | 450 |
| 4.2.14 | Thigh height | 505 | 550 | 600 | 460 | 500 | 550 |
| 4.3.1 | Arm length | 185 | 195 | 205 | 160 | 175 | 195 |
| 4.3.3 | Arm width | 80 | 90 | 95 | 70 | 80 | 85 |
| 4.3.7 | Foot length | 240 | 260 | 280 | 230 | 250 | 275 |
| 4.3.9 | Head length | 185 | 195 | 205 | 170 | 180 | 195 |
| 4.3.10 | Head width | 145 | 155 | 170 | 140 | 150 | 160 |
| 4.3.12 | Head circumference | 550 | 580 | 600 | 520 | 550 | 580 |
| 4.4.2 | Frontal reach (fingers straight) | 820 | 870 | 930 | 740 | 810 | 870 |
| 4.4.7 | Length buttock–knee | 580 | 630 | 670 | 540 | 590 | 450 |

**Table 3.5** Anthropometric data for the region: Central Europe (Jarosz 2003)

| No. of the feature | Name of the feature | Men | | | Women | | |
|---|---|---|---|---|---|---|---|
| | | Centiles | | | | | |
| | | 5 | 50 | 95 | 5 | 50 | 95 |
| 4.1.1 | Body height when standing | 1670 | 1770 | 1860 | 1550 | 1660 | 1750 |
| 4.1.12 | Hip width when standing | 310 | 350 | 375 | 315 | 360 | 410 |
| 4.2.1 | Seating height | 880 | 940 | 980 | 820 | 880 | 930 |
| 4.2.2 | Eye height when seating | 740 | 800 | 850 | 700 | 750 | 810 |
| 4.2.9 | Shoulder width (upper body width) | 420 | 460 | 490 | 365 | 420 | 455 |
| 4.2.11 | Hip width when seating | 310 | 365 | 390 | 330 | 400 | 440 |
| 4.2.12 | Shin length (popliteal height) | 420 | 465 | 500 | 390 | 425 | 460 |
| 4.2.14 | Thigh height | 495 | 550 | 595 | 460 | 500 | 540 |
| 4.3.1 | Arm length | 175 | 190 | 205 | 160 | 175 | 190 |
| 4.3.3 | Arm width | 80 | 90 | 95 | 70 | 75 | 85 |
| 4.3.7 | Foot length | 240 | 265 | 285 | 220 | 240 | 260 |
| 4.3.9 | Head length | 180 | 190 | 200 | 170 | 180 | 190 |
| 4.3.10 | Head width | 145 | 155 | 165 | 135 | 145 | 155 |
| 4.3.12 | Head circumference | 540 | 575 | 600 | 520 | 550 | 590 |
| 4.4.2 | Frontal reach (fingers straight) | 800 | 850 | 890 | 740 | 800 | 840 |
| 4.4.7 | Length buttock–knee | 550 | 610 | 660 | 530 | 580 | 630 |

**Table 3.6** Anthropometric data for the region: Eastern Europe (Jarosz 2003)

| No. of the feature | Name of the feature | Men | | | Women | | |
|---|---|---|---|---|---|---|---|
| | | Centiles | | | | | |
| | | 5 | 50 | 95 | 5 | 50 | 95 |
| 4.1.1 | Body height when standing | 1660 | 1750 | 1850 | 1540 | 1630 | 1720 |
| 4.1.1 | Hip width when standing | 305 | 345 | 385 | 315 | 360 | 405 |
| 4.2.1 | Seating height | 860 | 910 | 960 | 830 | 870 | 910 |
| 4.2.2 | Eye height when seating | 730 | 790 | 850 | 670 | 730 | 790 |
| 4.2.9 | Shoulder width (upper body width) | 410 | 450 | 490 | 370 | 410 | 450 |
| 4.2.11 | Hip width when seating | 310 | 360 | 400 | 325 | 380 | 435 |
| 4.2.12 | Shin length (popliteal height) | 395 | 445 | 490 | 375 | 405 | 430 |
| 4.2.14 | Thigh height | 490 | 550 | 590 | 445 | 510 | 540 |
| 4.3.1 | Arm length | 175 | 190 | 205 | 155 | 175 | 190 |
| 4.3.3 | Arm width | 80 | 90 | 100 | 75 | 80 | 85 |
| 4.3.7 | Foot length | 245 | 265 | 285 | 225 | 245 | 265 |
| 4.3.9 | Head length | 180 | 190 | 200 | 170 | 180 | 190 |
| 4.3.10 | Head width | 150 | 155 | 165 | 145 | 150 | 160 |
| 4.3.12 | Head circumference | 540 | 570 | 600 | 530 | 550 | 580 |
| 4.4.2 | Frontal reach (fingers straight) | 800 | 840 | 890 | 740 | 780 | 820 |
| 4.4.7 | Length buttock–knee | 550 | 600 | 650 | 520 | 570 | 610 |

**Table 3.7** Anthropometric data for the region: south-eastern Europe (Jarosz 2003)

| No. of the feature | Name of the feature | Men | | | Women | | |
|---|---|---|---|---|---|---|---|
| | | Centiles | | | | | |
| | | 5 | 50 | 95 | 5 | 50 | 95 |
| 4.1.1 | Body height when standing | 1640 | 1730 | 1830 | 1530 | 1620 | 1720 |
| 4.1.12 | Hip width when standing | 310 | 340 | 370 | 315 | 350 | 400 |
| 4.2.1 | Seating height | 860 | 900 | 960 | 800 | 860 | 900 |
| 4.2.2 | Eye height when seating | 740 | 790 | 840 | 680 | 730 | 780 |
| 4.2.9 | Shoulder width (upper body width) | 420 | 450 | 490 | 365 | 405 | 430 |
| 4.2.11 | Hip width when seating | 310 | 355 | 390 | 320 | 370 | 430 |
| 4.2.12 | Shin length (popliteal height) | 410 | 455 | 485 | 340 | 380 | 420 |
| 4.2.14 | Thigh height | 490 | 535 | 580 | 425 | 460 | 495 |
| 4.3.1 | Arm length | 175 | 190 | 205 | 160 | 175 | 190 |
| 4.3.3 | Arm width | 80 | 90 | 95 | 70 | 75 | 85 |
| 4.3.7 | Foot length | 245 | 265 | 285 | 220 | 240 | 260 |
| 4.3.9 | Head length | 175 | 190 | 205 | 160 | 175 | 190 |
| 4.3.10 | Head width | 145 | 155 | 165 | 140 | 150 | 160 |
| 4.3.12 | Head circumference | 550 | 570 | 590 | 530 | 550 | 570 |
| 4.4.2 | Frontal reach (fingers straight) | 790 | 830 | 880 | 740 | 780 | 830 |
| 4.4.7 | Length buttock–knee | 570 | 600 | 650 | 530 | 570 | 610 |

**Table 3.8** Anthropometric data for the region: France (Jarosz 2003)

| No. of the feature | Name of the feature | Men | | | Women | | |
|---|---|---|---|---|---|---|---|
| | | Centiles | | | | | |
| | | 5 | 50 | 95 | 5 | 50 | 95 |
| 4.1.1 | Body height when standing | 1660 | 1770 | 1890 | 1530 | 1630 | 1740 |
| 4.1.12 | Hip width when standing | 310 | 340 | 370 | 315 | 350 | 400 |
| 4.2.1 | Seating height | 870 | 930 | 980 | 820 | 860 | 910 |
| 4.2.2 | Eye height when seating | 750 | 800 | 850 | 690 | 730 | 780 |
| 4.2.9 | Shoulder width (upper body width) | 410 | 450 | 490 | 370 | 410 | 430 |
| 4.2.11 | Hip width when seating | 315 | 350 | 380 | 320 | 375 | 430 |
| 4.2.12 | Shin length (popliteal height) | 390 | 445 | 490 | 345 | 385 | 425 |
| 4.2.14 | Thigh height | 495 | 540 | 580 | 455 | 490 | 525 |
| 4.3.1 | Arm length | 180 | 195 | 210 | 160 | 170 | 180 |
| 4.3.3 | Arm width | 80 | 90 | 95 | 70 | 75 | 85 |
| 4.3.7 | Foot length | 245 | 265 | 285 | 220 | 235 | 255 |
| 4.3.9 | Head length | 180 | 195 | 205 | 170 | 180 | 190 |
| 4.3.10 | Head width | 145 | 155 | 165 | 135 | 140 | 150 |
| 4.3.12 | Head circumference | 540 | 570 | 600 | 520 | 550 | 570 |
| 4.4.2 | Frontal reach (fingers straight) | 800 | 850 | 910 | 730 | 780 | 830 |
| 4.4.7 | Length buttock–knee | 560 | 620 | 660 | 520 | 570 | 610 |

**Table 3.9** Anthropometric data for the region: Iberian Peninsula (Jarosz 2003)

| No. of the feature | Name of the feature | Men | | | Women | | |
|---|---|---|---|---|---|---|---|
| | | Centiles | | | | | |
| | | 5 | 50 | 95 | 5 | 50 | 95 |
| 4.1.1 | Body height when standing | 1580 | 1710 | 1830 | 1510 | 1600 | 1700 |
| 4.1.12 | Hip width when standing | 295 | 340 | 370 | 300 | 350 | 400 |
| 4.2.1 | Seating height | 830 | 890 | 950 | 800 | 850 | 900 |
| 4.2.2 | Eye height when seating | 720 | 790 | 850 | 680 | 740 | 810 |
| 4.2.9 | Shoulder width (upper body width) | 400 | 440 | 480 | 350 | 390 | 430 |
| 4.2.11 | Hip width when seating | 300 | 345 | 390 | 310 | 360 | 425 |
| 4.2.12 | Shin length (popliteal height) | 400 | 440 | 480 | 340 | 380 | 420 |
| 4.2.14 | Thigh height | 470 | 520 | 570 | 445 | 480 | 520 |
| 4.3.1 | Arm length | 170 | 185 | 215 | 155 | 175 | 200 |
| 4.3.3 | Arm width | 80 | 85 | 90 | 70 | 75 | 80 |
| 4.3.7 | Foot length | 240 | 270 | 300 | 215 | 245 | 280 |
| 4.3.9 | Head length | 175 | 185 | 200 | 165 | 180 | 190 |
| 4.3.10 | Head width | 145 | 155 | 165 | 140 | 150 | 160 |
| 4.3.12 | Head circumference | 520 | 565 | 600 | 505 | 535 | 565 |
| 4.4.2 | Frontal reach (fingers straight) | 760 | 820 | 880 | 720 | 770 | 820 |
| 4.4.7 | Length buttock–knee | 540 | 590 | 640 | 510 | 560 | 600 |

**Table 3.10** Values of the coefficient Z

| Z | | 0 | 0.01 | 0.02 | 0.03 | 0.04 | 0.05 | 0.06 | 0.07 | 0.08 | 0.09 |
|---|---|---|---|---|---|---|---|---|---|---|---|
| 0.0 | − | 50.00 | 49.60 | 49.20 | 48.80 | 48.40 | 48.01 | 47.61 | 47.21 | 46.81 | 46.41 |
| | + | 50.00 | 50.40 | 50.80 | 51.20 | 51.60 | 51.99 | 52.39 | 52.79 | 53.19 | 53.59 |
| 0.1 | − | 46.02 | 45.62 | 45.22 | 44.83 | 44.43 | 44.04 | 43.64 | 43.25 | 42.86 | 42.47 |
| | + | 53.98 | 54.38 | 54.78 | 55.17 | 55.57 | 55.96 | 56.36 | 56.75 | 57.14 | 57.53 |
| 0.2 | − | 42.07 | 41.68 | 41.29 | 40.90 | 40.52 | 40.13 | 39.74 | 39.36 | 38.97 | 38.59 |
| | + | 57.93 | 58.32 | 58.71 | 59.10 | 59.48 | 59.87 | 60.26 | 60.64 | 61.03 | 61.41 |
| 0.3 | − | 38.21 | 37.83 | 37.45 | 37.07 | 36.69 | 36.32 | 35.94 | 35.57 | 35.20 | 34.83 |
| | + | 61.79 | 62.17 | 62.55 | 62.93 | 63.31 | 63.68 | 64.06 | 84.43 | 64.80 | 65.17 |
| 0.4 | − | 34.46 | 34.09 | 33.72 | 33.36 | 33.00 | 32.64 | 32.28 | 31.92 | 31.56 | 31.21 |
| | + | 65.54 | 65.91 | 66.28 | 66.64 | 67.00 | 67.36 | 67.72 | 68.08 | 68.44 | 68.79 |
| 0.5 | − | 30.85 | 30.50 | 30.15 | 29.81 | 29.46 | 29.12 | 28.77 | 28.43 | 28.10 | 27.76 |
| | + | 69.15 | 69.50 | 69.85 | 70.19 | 70.54 | 70.88 | 71.23 | 71.57 | 71.90 | 72.24 |
| 0.6 | − | 27.43 | 27.09 | 26.76 | 26.43 | 26.11 | 25.78 | 25.46 | 25.14 | 24.83 | 24.51 |
| | + | 72.57 | 72.91 | 73.24 | 73.57 | 73.89 | 74.22 | 74.54 | 74.86 | 75.17 | 75.49 |
| 0.7 | − | 24.20 | 23.89 | 23.58 | 23.27 | 22.96 | 22.66 | 22.36 | 22.06 | 21.77 | 21.48 |
| | + | 75.80 | 76.11 | 76.42 | 76.73 | 77.04 | 77.34 | 77.64 | 77.94 | 78.23 | 78.52 |
| 0.8 | − | 21.19 | 20.90 | 20.61 | 20.33 | 20.05 | 19.77 | 19.49 | 19.22 | 18.94 | 18.67 |
| | + | 78.81 | 79.10 | 79.39 | 79.67 | 79.95 | 80.23 | 80.51 | 80.78 | 81.06 | 81.33 |
| 0.9 | − | 18.41 | 18.14 | 17.88 | 17.62 | 17.36 | 17.11 | 16.85 | 16.60 | 16.35 | 16.11 |
| | + | 81.59 | 81.86 | 82.12 | 82.38 | 82.64 | 82.89 | 83.15 | 83.40 | 83.65 | 83.89 |
| 1.0 | − | 15.87 | 15.62 | 15.39 | 15.15 | 14.92 | 14.69 | 14.46 | 14.23 | 14.01 | 13.79 |
| | + | 84.13 | 84.38 | 84.61 | 84.85 | 85.08 | 85.31 | 85.54 | 85.77 | 85.99 | 86.21 |
| 1.1 | − | 13.57 | 13.35 | 13.14 | 12.92 | 12.71 | 12.51 | 12.30 | 12.10 | 11.90 | 11.70 |
| | + | 86.43 | 86.65 | 86.86 | 87.08 | 87.29 | 87.49 | 87.70 | 87.90 | 88.10 | 88.30 |
| 1.2 | − | 11.51 | 11.31 | 11.12 | 10.93 | 10.75 | 10.56 | 10.38 | 10.20 | 10.03 | 9.85 |
| | + | 88.49 | 88.69 | 88.88 | 89.07 | 89.25 | 89.44 | 89.62 | 89.80 | 89.97 | 90.15 |
| 1.3 | − | 9.68 | 9.51 | 9.34 | 9.18 | 9.01 | 8.85 | 8.69 | 8.53 | 8.38 | 8.23 |
| | + | 90.32 | 90.49 | 90.66 | 90.82 | 90.99 | 91.15 | 91.31 | 91.47 | 91.62 | 91.77 |
| 1.4 | − | 8.08 | 7.93 | 7.78 | 7.64 | 7.49 | 7.35 | 7.21 | 7.08 | 6.94 | 6.81 |
| | + | 91.92 | 92.07 | 92.22 | 92.36 | 92.51 | 92.65 | 92.79 | 92.92 | 93.06 | 93.19 |
| 1.5 | − | 6.68 | 6.55 | 6.43 | 6.30 | 6.18 | 6.06 | 5.94 | 5.82 | 5.71 | 5.59 |
| | + | 93.32 | 93.45 | 93.57 | 93.70 | 93.82 | 93.94 | 94.06 | 94.18 | 94.39 | 94.41 |
| 1.6 | − | 5.48 | 5.37 | 5.26 | 5.16 | 5.05 | 4.95 | 4.85 | 4.75 | 4.65 | 4.55 |
| | + | 94.52 | 94.63 | 94.74 | 94.84 | 94.95 | 95.05 | 95.15 | 95.25 | 95.35 | 95.45 |
| 1.7 | − | 4.46 | 4.36 | 4.27 | 4.18 | 4.09 | 4.01 | 3.92 | 3.84 | 3.75 | 3.67 |
| | + | 95.54 | 95.64 | 95.73 | 95.82 | 95.91 | 95.99 | 96.08 | 96.16 | 96.25 | 96.33 |
| 1.8 | − | 3.59 | 3.51 | 3.44 | 3.36 | 3.29 | 3.22 | 3.14 | 3.07 | 3.01 | 2.94 |
| | + | 96.41 | 96.49 | 96.56 | 96.64 | 96.71 | 96.78 | 96.86 | 96.93 | 96.99 | 97.06 |
| 1.9 | − | 2.87 | 2.81 | 2.74 | 2.68 | 2.62 | 2.56 | 2.50 | 2.44 | 2.39 | 2.33 |
| | + | 97.13 | 97.19 | 97.26 | 97.32 | 97.38 | 97.44 | 97.50 | 97.56 | 97.61 | 97.67 |

(continued)

**Table 3.10**  (continued)

| Z | | 0 | 0.01 | 0.02 | 0.03 | 0.04 | 0.05 | 0.06 | 0.07 | 0.08 | 0.09 |
|---|---|---|---|---|---|---|---|---|---|---|---|
| 2.0 | − | 2.28 | 2.22 | 2.17 | 2.12 | 2.07 | 2.02 | 1.97 | 1.92 | 1.88 | 1.83 |
| | + | 97.72 | 97.78 | 97.83 | 97.88 | 97.93 | 97.98 | 98.03 | 98.08 | 98.12 | 98.17 |
| 2.1 | − | 1.79 | 1.74 | 1.70 | 1.66 | 1.62 | 1.58 | 1.54 | 1.50 | 1.46 | 1.43 |
| | + | 98.21 | 98.26 | 98.30 | 98.34 | 98.38 | 98.42 | 98.46 | 98.50 | 98.54 | 98.57 |
| 2.2 | − | 1.39 | 1.38 | 1.32 | 1.29 | 1.25 | 1.22 | 1.19 | 1.16 | 1.13 | 1.10 |
| | + | 98.61 | 98.64 | 98.68 | 98.71 | 98.75 | 98.78 | 98.81 | 98.84 | 98.87 | 98.90 |
| 2.3 | − | 1.07 | 1.04 | 1.02 | 0.99 | 0.96 | 0.94 | 0.91 | 0.89 | 0.87 | 0.84 |
| | + | 98.93 | 98.96 | 98.98 | 99.01 | 99.04 | 99.06 | 99.09 | 99.11 | 99.13 | 99.16 |
| 2.4 | − | 0.82 | 0.80 | 0.78 | 0.75 | 0.73 | 0.71 | 0.69 | 0.68 | 0.66 | 0.64 |
| | + | 99.18 | 99.20 | 99.22 | 99.25 | 99.27 | 99.29 | 99.31 | 99.32 | 99.34 | 99.36 |
| 2.5 | − | 0.62 | 0.60 | 0.59 | 0.57 | 0.55 | 0.54 | 0.52 | 0.51 | 0.49 | 0.48 |
| | + | 99.38 | 99.40 | 99.41 | 99.43 | 99.45 | 99.46 | 99.48 | 99.49 | 99.51 | 99.52 |
| 2.9 | − | 0.19 | 0.18 | 0.18 | 0.17 | 0.16 | 0.16 | 0.15 | 0.15 | 0.14 | 0.14 |
| | + | 99.81 | 99.82 | 99.83 | 99.83 | 99.84 | 99.84 | 99.85 | 99.85 | 99.86 | 99.86 |

*Example*

If the average height of an adult male amounts to SR = 1766 mm, the standard deviation of this measure amounts to SD = 76 mm, and then, in reading the value of coefficient Z (for the corresponding centile) from Table 3.10, we calculate the sought height value:

| for 5C | SC = 1766 − 1.64 · 76 = 1641 mm, |
|---|---|
| for 85C | SC = 1766 + 1.04 · 76 = 1845 mm, |
| for 99C | SC = 1766 + 2.33 · 76 = 1943 mm. |

When designing interiors, we refer the dimensions of many objects to the dimension of the dominating object or also the dominating dimension of the user. By knowing his main features, determining the functionality of the designed furniture piece or interior, as well as the average value and standard deviation of this dimension, the value of coefficient Z can be calculated, which will allow to design other dimensions of a given product or other objects:

$$Z = (SC-SR)/SD. \qquad (3.9)$$

*Examples*

A man who is 1920 mm tall represents the group of tall people, and 98.87 % of the population is shorter, because

$$Z = (1920-1706)/94 = +2.28,$$

and in Table 3.10, the coefficient Z = + 2.28 corresponds to the 98.87th centile.

In a kitchen furnished with cabinets with a height of 900 mm, the lowest usable plane is the bottom surface of the sink away from the floor by 750 mm. The centile corresponding to the first height specifies how many users will be taller:

$$Z = (750 - 766)/43 = -0.37.$$

Based on the calculated $Z$ value, it can be concluded that $100 \% - 35.57 \% = 64.43 \%$ of tall users will have to bend down while doing the dishes and straighten up again when moving on to work at the cabinets.

When designing, the dimension differences resulting from the age of users using the same furniture should also be taken into account. Compiling the average dimension for a man and woman of different ages can be conducted on the basis of the equation:

$$SR_{A+B} = \%A \cdot SR_A + \%B \cdot SR_B \tag{3.10}$$

and

$$SD^2_{A+B} = \%A \cdot SD^2_A + \%B \cdot SD^2_B + \%A \cdot \%B \cdot (SR_A - SR_B)^2 \tag{3.11}$$

When compiling the dimensions for any two groups, one needs to know the average dimensions and standard deviations of these dimensions for each of the groups. For example, at the age of 65, women dominate (58.65 %), while for the average population, the representation of gender is more uniform ($A = 50.27 \%$ of men and $B = 49.73 \%$ of women). Assuming the average height of the body of a man aged 18–65 years old equal to, e.g., $SR_A = 1766$ mm and standard deviation $SR_A = 75.46$ mm, as well as the dimensions of women corresponding to these parameters: $SR_B = 1646$ mm and $SR_B = 67.91$ mm, the average height of the human body between the ages of 18 to 65 years old should amount to:

$$SR_{A+B} = 0.5027 \cdot 1766 + 0.4973 \cdot 1646 = 1706.32 \text{ mm},$$
$$SD_{A+B^2} = 0.5027 \cdot 75.4596^2 + 0.4973 \cdot 67.9113^2 + 0.5027 \cdot 0.4973 \cdot (1766 - 1646)^2,$$
$$SD_{A+B^2} = 2862.45 + 2293.52 + 3599.90 = 8775.87,$$
$$SD_{A+B} = 93.6 \text{ mm}.$$

Adding and subtracting dimensions takes place according to the superposition principle:

$$SR_A \pm B = SR_A \pm SR_B, \tag{3.12}$$

$$SD^2_{A \pm B} = SD^2_A + SD^2_{B \pm 2} \cdot r \cdot SD_A \cdot SD_B. \tag{3.13}$$

Coefficient $r$ (Table 3.11) expresses relations between two dimensions of the body.

**Table 3.11** Coefficient value of the relations between the dimensions of the body

| Dimension | Height | Width | Depth |
|---|---|---|---|
| Height | 0.65 | | |
| Width | 0.30 | 0.65 | |
| Depth | 0.20 | 0.40 | 0.20 |

*Example*

In order to determine the height of the worktop of a table, the seat height should be set at $SR_A = 446$ mm and height from the seat to the surface of the worktop $SR_B = 244$ mm. For the adult population of users, the result of this compilation provides the average value:

$$SR_{A+B} = 446 + 244 = 690 \text{ mm},$$
$$SD^2 = 26^2 + 24^2 + 2 \cdot 0.65 \cdot 26 \cdot 24 = 2063.2,$$
$$SD = 45.42 \text{ mm}.$$

The average height between the floor and the worktop surface should therefore amount to 690 mm. In order to calculate the height taking into account the base under the feet (also the soles of footwear), an additional 30 mm should be considered. The new height of the worktop should therefore amount to 720 mm. Of course, a height that takes into account the needs of the population from 1 to 99 % would be satisfying, and therefore

$$1C = SR - 2.33 \cdot SD = 690 - 2.33 \cdot 45.42 = 584 \text{ mm} + 30 \quad (\text{sole}) = 610 \text{ mm},$$
$$99C = SR + 2.33 \cdot SD = 690 + 2.33 \cdot 45.42 = 789 \text{ mm} + 30 \quad (\text{sole}) = 820 \text{ mm}.$$

In the group of office furniture, chairs are often equipped with a lifting and tilting mechanism of the seat. Therefore, the height of the worktop should be 60 mm higher than at traditional chairs. According to the calculated values, the height of the worktop should range from 610 to 880 mm.

## 3.4.2 Requirements for Office Furniture

### 3.4.2.1 Dimensional Requirements

Designing office space requires attention to the proper arrangement of individual and group workstations. Firstly, the availability of the workstation must be taken into account, i.e. the possibility of free access of workers to workstations. Office space is a social and group space; therefore, it needs to be remembered that the adaptation of workstations in an office was conducive to proper cooperation between design teams and members of these teams (Janiga 2000). Depending on the purpose, office furniture is designed for individual office work, construction and designing works, the work of groups and problem teams, call centre, training tasks, conferences, individual and collective customer service. Examples of some of these types of furniture are shown in Figs. 3.19, 3.20, 3.21 and 3.22.

Due to different purposes, each of the given types of furniture is characterised by different design solutions which ensure mobility or stability, variability of geometry and dimensions, sufficient stiffness, stability and strength.

When designing office furniture, it is assumed that the workstation should be adjusted in terms of space to the dimensions of 90 % of the adult population of both women and men. The minimum dimensions assumed are those which are not achieved by 5 % of the population, while the maximum dimensions are those which 95 % of the population does not exceed. The key figures and anthropometric criteria used in designing were collected in many positions of the literature on the subject (Batogowska and Słowikowski 1989; Drożyńska 1997; Gedliczka 2001; Janow and Bielow 1971; Jarosz 2003; Juergens et al. 2005; Kamieńska-Żyła 1996; Konarska 2001; Niejmah and Smirnov 1984; Nowak 1993; Pakarinen and Asikainen 2001; Troussier et al. 1999; Tytyk 2001; Winkler 2005). Spatial parameters were also developed concerning the positions most commonly adopted by employees at computer stations (Corlett 1999; Drożyńska 1997; Kamieńska-Żyła 1996; Troussier et al. 1999).

**Fig. 3.19** Workstation: a individual, b team

**(a)**

**(b)**

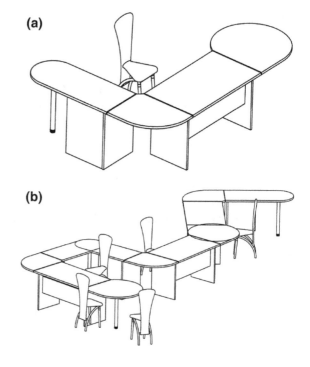

**Fig. 3.20** Call centre type of workstation

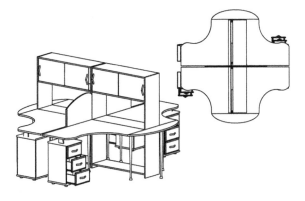

**Fig. 3.21** Workstation of a trainer and lecturer

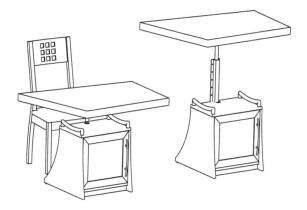

**Fig. 3.22** Dining room set: table and rattan chairs

Figure 3.23 shows the spatial structure of the workstation at a computer in relation to anthropometric data of the Polish population.

When working at a computer takes place in a standing position, separate regulation of the height of the worktop should be taken into account both under the keyboard and under the monitor. This will allow to use the furniture piece also for work in a seated and kneeling position. The ranges of regulations of parameters of a workstation in a standing, seated and seated–kneeling position are shown in Fig. 3.24.

It is worth mentioning that well designed and properly aligned furniture has a significant impact on the correct posture of employees. Office furniture should be dimensioned in accordance with the provisions of PN-EN 527-1:2004, PN-EN 1335-1:2004 (Fig. 3.25). Furthermore, furniture must also meet the basic constructional and functional requirements (Baranowski 2004).

In the design of seats, innovative solutions and mechanisms should be used, which are to ensure the dynamic position of the body. The key issue here is the ability to adjust the chair or armchair to the individual needs of users. Therefore, an office chair should be characterised by regulated seat height, inclination of both the seat and the backrest, height and spacing of armrests and rotation of armrests.

When designing seats, the constructional parameters of individual elements are specified more and more accurately in order to ensure a seated position, at which the spine can maintain its natural shape. It is also recommended that the height of

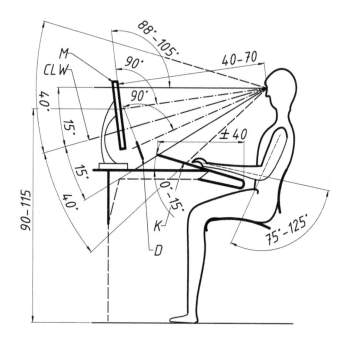

**Fig. 3.23** Spatial parameters of workstations at a computer (own development on the basis of a collective work 1990): *M* monitor, *CLW* central line of vision, *D* document, *K* keyboard (cm)

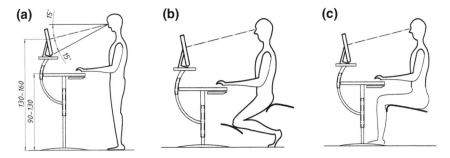

**Fig. 3.24** Spatial parameters of a workstation in the following position: **a** standing, **b** seated, **c** seated–kneeling (cm)

the backrest amount to 55–60 cm, and its profiling ensures the support of the spine in the lumbar region, and at this point it should be convex, while at the chest height slightly concave. It is also important that the depth of the seat amounts to 40–45 cm and its width allows the user to assume a comfortable seated and standing position and provides him freedom of movement in carrying out any activities. The construction of the chair must be above all static, stable and durable, which is also why the base of the seat should be fitted with at least a five-armed frame with swinging wheels, rollers or driving balls.

For tables, the recommended optimum surface of the worktop amounts to 0.96 m$^2$. Most commonly, this requirement is met by using rectangles of the dimensions 120 × 80 cm or 160 × 90 cm. When the table is equipped with drawers, their runners must have stops which secure them from falling out, as well as preventing the simultaneous opening of several drawers. An important feature of tables is the possibility of regulating the height of the worktop. Along with a proper chair, this regulation helps to configure a comfortable workstation. Currently, it is expected to allow regulation of this height from 68 cm to 90 cm.

When designing an ergonomic workstation, free space for the legs must be considered. This space is located below the surface of the worktop and should meet the standards of PN-EN 527-1:2004 (Fig. 3.25).

### 3.4.2.2 Requirements Concerning Safety of Use

Specific quality characteristics attributed to this group of furniture, which distinguish them from other types, are conditioned by the nature of use and certain external factors, which in the practice of approval tests is called exposures and operational loads of the furniture. In considering the movable element, e.g. a drawer of an office container, we shall notice what requirements it must meet. These include the following: capacity adapted to the overall dimensions of the folder or binder, the presence of staples for documents or dividers, the amplitude of

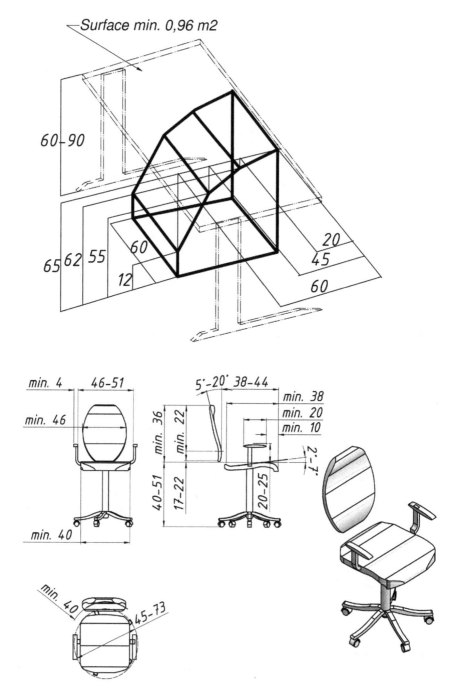

**Fig. 3.25** Dimensions of an office table and chair (cm)

extension much larger than the dimension of its depth, which is connected with the necessity of installing special types of runners with a large vertical load indicator and multiple extension range, an auto-lock system of the extension of other drawers in order to preserve the stability of the furniture piece and the installation of a handle with an internal lock. Among furniture of case construction of a significant height, with a large number of drawers, there could be a risk of the loss of stability after extending all the weight-bearing drawers, as a result of overload. Because such a danger must be absolutely excluded, especially in this group of furniture, the construction of cupboards and file cabinets should have a system of fittings that limit the user's ability to open all of the drawers simultaneously. In turn, double-blind doors are mounted together with a special block system with a tie.

Modern office furniture has a very complex design and is equipped with electromechanical devices that facilitate the use of the furniture and optimal use of office space. Mobility, that is a group of construction characteristics conducive to easy movement, is currently the key advantage of office furniture. Their different character is also shown by the modern architectural concept of the office's interior. In accordance with this concept, the floor of office space consists of a so-called technical floor, between its layers the complete cabling of office equipment, having output sockets, is mounted. Then, office equipment, such as computers, telephones or lighting, is connected to appropriate sockets in the floor, through special channels. Currently, a wide variety of cabling solutions can be encountered from the worktop of tables and desks to complex sockets in the floor. An interesting proposition in this regard is sleeves of a spinal construction. Such solutions are used also in work tables with regulated height. The system of hidden cabling ensures an aesthetic look of the office, because it eliminates the possibility of being caught up in loose cables. In addition to compliance with stiffness and strength standards, an office piece of furniture must also meet additional conditions related to use (Beszterda 2001; Dzięgielewski and Smardzewski 1995; Smardzewski 2004b, c; Smardzewski and Beszterda 2001; Smardzewski and Rogoziński 2001a, b). The construction of this furniture with internal channels must be completely safe to use. For this reason, some elements of the furniture piece should be made from appropriate insulators, ensuring electrostatic not allowing for electrical arching within the structure of the elements.

The emerging new aspects of functionality of a furniture piece, and therefore, new structural solutions, must, regardless of the complexity of the structure, meet the standardised safety rules of use, strength and stiffness. The methods of calculating stiffness, stability and strength of the construction of office furniture have been presented in the works of Dzięgielewski and Smardzewski (1995), Smardzewski and Rogoziński (2001a, b). In Poland, mandatory requirements relating to the quality of office furniture are determined by the norms: PN-EN 1335-1:2004, PN-EN 1335-2:2009, PN-EN 1335-3:2009, PN-EN 527-3:2004 and PN-EN 16139:2013-07 according to them:

- stability under vertical load aims to demonstrate adequate resistance to the construction falling over due to people using its worktops,
- stability with extended drawers aims to demonstrate adequate resistance to the construction falling over with drawers of a maximum load and completely extended,
- a strength test of vertical force with a value of 1000 N aims to test the strength of the worktop and construction of a desk/table affected by occasional, short-lasting loads,
- the test of resistance to horizontal force with a value of 450 N aims to test the strength of the construction of a desk/table affected by forces exerted by the user when moving the furniture piece,
- testing fatigue as the result of horizontal force with a value from 0 to 300 N aims to test the strength of the construction of a desk/table affected by small forces exerted cyclically, as well as the opening and closing of drawers and cautious moving, in order to ensure that during use, the construction can endure moving without any apparent deformations of its worktop.
- testing fatigue as a result of vertical force with a value of 400 N aims to verify the strength of the construction of a desk/table affected by forces directed downwards,
- test of falling aims to test the ability of tables to counteract falling over,
- testing the stiffness of the construction aims to determine the stiffness indicator of the construction of a desk/table determining its resistance to unacceptable deformations of the frame caused by horizontal force with a value of 200 N.

In addition, safety requirements have been defined in the scope of

- the shape and dimensions of corners, edges, in order to avoid slamming, pinching, cutting, physical injury or damage to the things of the person using it,
- the shape and dimensions of movable and regulated parts, in order to be able to avoid injuries and unintentional launching,
- the quality of joints exposing to excessive damage or loosening,
- protection against contaminating the body or clothing of the user,
- stability during stresses on the front edge of the seat, tilting of the user to the sides, resting on backrest or sitting on the front edge of the seat,
- resistance to rolling an unladen chair,
- the strength and durability in order to avoid injury by the user of the chair: sitting both in the middle and outside the centre of the seat, while moving forward and to the sides when sitting, along with leaning over the armrests, at pressing on the armrests when standing up from the chair.

It is worth noting that by acceding to designing office furniture and office space, one should be aware that individual elements of the equipment of a workstation should provide the possibility of free and easy arrangement of office space, so that the office does not limit employees and the company in realising business tasks (Smardzewski and Rogoziński 2001a, b).

### 3.4.3 Requirements for School Furniture

#### 3.4.3.1 Dimensional Requirements

In school children, there is a conflict between the natural tendency to unrestricted physical movement and the need to maintain a seated position for a longer period of time (Troussier et al. 1999). Headache, knee pain, back pain and attention deficit are among the most common adverse effects of prolonged sitting (Drożyńska 1997; Krutul 2004a; Snijders et al. 1995). In order to reduce these negative effects, education programmes in schools contain many forms of physical activation for students. The actions undertaken in this area only minimise the effects of erroneous furniture design. In order to completely solve problems connected with the functionality of school furniture, more effort should be put into creating innovative rules of furniture design. More attention must be devoted to the proper, ergonomic design of school furniture.

Diversity in designing should, therefore, take into account and promote solutions that not only improve the figure, but also ensure freedom and mobility of both children and furniture (Molenbroek 1994).

Due to the complexity design problems, innovative solutions for school furniture should arise in multidisciplinary teams composed of orthopaedic doctors, rehabilitation experts, teachers, psychologists, constructors, technologists, environmentalists, economists and of course designers (Fig. 3.26).

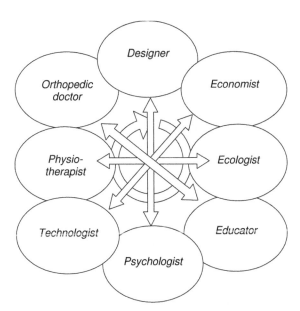

**Fig. 3.26** The composition of the multidisciplinary design team dealing with creating modern school furniture

Only such a team is able to develop the values of factors responsible for optimum shapes and constructional solutions of furniture. Figure 3.27 shows the most important parameters affecting optimal design of school furniture.

Differences in the height of Polish children (boys and girls) aged from 7 to 12 years (elementary school) amount to 30 cm on average, while in children between 13 and 15 years (middle school youth) only 12 cm. Therefore, due to this variation of height of children at a similar age, and living both in Poland as well as in other countries of the European Union (Drożyńska 1997; Jarosz 2003; Nowak 1993), when designing school furniture, it is necessary to answer the following questions: which measuring system best fulfils the needs of this population of users, for which project is it being done and which dimensions of the body should be measured in a population of children representing one class in order to find the best recommendations. The requirements in the scope of equipping school with furniture

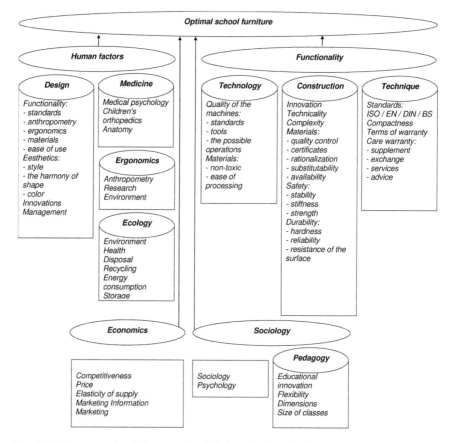

**Fig. 3.27** Parameters that influence optimal design of school furniture

intended for didactic rooms are specified in the norms PN-EN 1729-1:2007 and PN-90/F-06009. Tables and chairs should be marked with a label that specifies using a digit symbol or colour (Tables 3.12 and 3.13) and the furniture size. Marking furniture enables teachers to organise individual children easily to their proper benches and chairs. In this case, three important rules must be considered:

- Before inserting furniture to the teaching room, they need to be marked by a number of colours corresponding to the proper scope of a child's height,
- Pairs folded from benches (tables) and chairs should have an identical feature,
- In each room, there should be furniture having at least three-dimensional characteristics.

The younger the child, the less the developed bone structure, and it is more important to adopt the correct posture during school activities. This posture is significantly shaped by the chair and table; therefore, when designing school furniture, seven primary criteria must be taken into account, which are illustrated in Fig. 3.28.

1. The backrest of the chair should stiffly support the back at the height of the lumbar region, below the shoulder blades.
2. The height of the front edge of the table must correspond to the height of the bottom surface of the bend forearm in a position when the arm is placed vertically.
3. To guarantee freedom of movement, the distance between the edge of the table and pupil's body needs to be determined, as well as the distance between the lumbar support and the seat.
4. Distance between the lower surface of the table and the thigh must be ensured.
5. The part of the shank near the knee cannot put pressure on the front edge of the seat.
6. A clear space between the back part of the calf and front edge of the seat must be taken into account.
7. The foot must rest on the floor. The seat and table height should be adjusted to the thickness of the shoe sole, around 2 cm.

An example set of school furniture is shown in Fig. 3.29. It allows freedom of movement of the child both while listening (Fig. 3.29a), writing (Fig. 3.29b) and reading (Fig. 3.29c).

The results of anthropometric studies of children and adolescents at school age served to develop the optimal dimensions of school furniture, which are contained in the norms (EN 1729-1:2007; PN-ISO 5970:1994; PN-90/F-06009; PN-90/F-06010.01; PN-90/F-06010.05; PN-88/F-06010.02; PN-88/F-06010.03), and the principles of dimensioning have been illustrated in Fig. 3.30.

All these standards define the requirements for school tables and chairs, describing the dimensions, angles, furniture class, required identification markings and colours, adapted to the average values of height of European children.

**Table 3.12** The dimensions and numbers of chair sizes with a seat tilt of −5° +5° according to PN-EN 1729-1:2007 (mm)

| Number of the size | 0 | 1 | 2 | 3 | 4 | 5 | 6 | 7 |
|---|---|---|---|---|---|---|---|---|
| Colour code | white | orange | purple | yellow | red | green | blue | brown |
| Popliteal height (without shoes) | 200–250 | 250–280 | 280–315 | 315–355 | 355–405 | 405–435 | 435–485 | 485+ |
| Height (without shoes) | 800–950 | 930–1160 | 1080–1210 | 1090–1420 | 1330–1590 | 1460–1765 | 1590–1880 | 1740–2070 |
| Seat height (±10 mm) $h_8$ | 210 | 260 | 310 | 350 | 380 | 430 | 460 | 510 |
| Effective seat depth $t_4$ [±10 mm (0–2)], [±20 mm (3–7)] | 225 | 250 | 270 | 300 | 340 | 380 | 420 | 460 |
| Minimum seat width $b_3$ | 210 | 240 | 280 | 320 | 340 | 360 | 380 | 400 |
| Seat surface depth $t_7$ | $t_4 - 20$ | $t_4 - 20$ | $t_4 - 20$ | $t_4 - 30$ | $t_4 - 30$ | $t_4 - 30$ | $t_4 - 30$ | $t_4 - 30$ |
| Height up to the point of the loin support $h_6$ (−10 to +20 mm) | 140 | 150 | 160 | 180 | 190 | 200 | 210 | 220 |
| Backrest height | 100 | 100 | 100 | 100 | 100 | 100 | 100 | 100 |
| Minimum backrest width $b_4$ | – | 210 | 250 | 270 | 270 | 300 | 330 | 360 |
| Minimum horizontal backrest radius $r_2$ | – | 300 | 300 | 300 | 300 | 300 | 300 | 300 |
| Backseat inclination $\beta$ in degrees | – | 95–110 | 95–110 | 95–110 | 95–110 | 95–110 | 95–110 | 95–110 |

**Table 3.13** The dimensions and numbers of table sizes used with chairs with a seat tilt of −5°+5° according to PN-EN 1729-1:2007 (mm)

| Number of the size | 0 | 1 | 2 | 3 | 4 | 5 | 6 | 7 |
|---|---|---|---|---|---|---|---|---|
| Colour code | white | orange | purple | yellow | red | green | blue | brown |
| Poplitea height (without shoes) | 200–250 | 250–280 | 280–315 | 315–355 | 355–405 | 405–435 | 435–485 | 485+ |
| Height (without shoes) | 800–950 | 930–1160 | 1080–1210 | 1090–1420 | 1330–1590 | 1460–1765 | 1590–1880 | 1740–2070 |
| Worktop height (±10 mm) $h_1$ | 400 | 460 | 530 | 590 | 640 | 710 | 760 | 820 |
| Minimal worktop depth $t_1$ | – | 500[a] | 500[a] | 500[a] | 500 | 500 | 500 | 500 |
| Minimal worktop length per person | – | 600[b] | 600[b] | 600[b] | 600[b] | 600 | 600 | 600 |
| Minimal the horizontal distance between the front legs of the structure per person | – | 500[c] | 500[c] | 500[c] | 500[c] | 500 | 500 | 500 |

[a]For certain teaching conditions, it can be reduced to 400 mm
[b]For specific teaching conditions, it can be reduced to 550 mm
[c]For certain teaching conditions, it can be reduced to 450 mm

**Fig. 3.28** Preferential points to dimensioning school tables and chairs

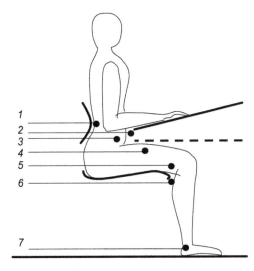

The requirements for seats are the following basic characteristics:

- The surface of the seat should be situated horizontally. It is recommended also to tilt it towards the backrest up to the value of −5°+5°,

**(a)**        **(b)**        **(c)**

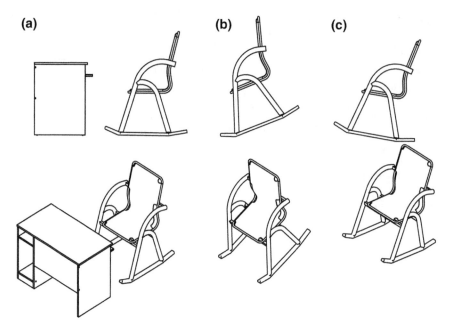

**Fig. 3.29** A set of school furniture allowing the child freedom of movement while **a** listening, **b** writing, **c** reading

- The surface of the seat cannot be flat. It is recommended to use a bend at the front and back edge, so that the radius of rounding the front edge was significantly greater than the radius of rounding the rear edge,
- The depth of the seat should be adjusted in such a way that the distance between the edge of the worktop and the body's trunk amounts from 5 to 10 cm, and
- the backrest supporting the back should have roundings and be tilted from the vertical by 5°–20°.

The requirements for tables are as follows:

- The worktop surface of the table for one pupil cannot be smaller than 60 × 50 cm,
- The worktop of the table should have the possibility of tilting from the horizontal by 0°–20°, and
- The space under the worktop is to ensure the ability to store schoolbags or other equipment of the pupil, not limiting freedom of movement in any way.

The standards do not specify the forms of design, construction, technology and quality of materials; however, this data must be specified by the manufacturer of furniture.

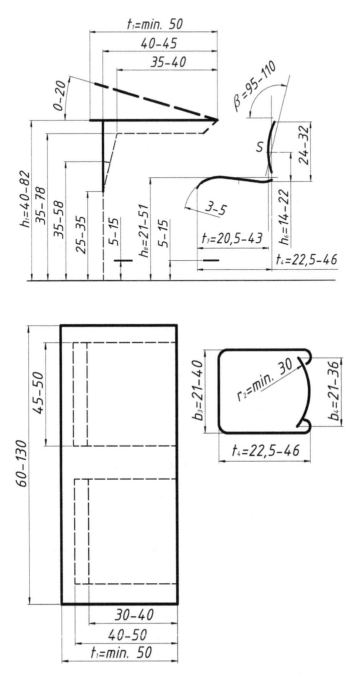

**Fig. 3.30** The principles of dimensioning school benches (cm)

### 3.4.3.2   Requirements for Safety of Use

The fulfilment of ergonomics requirements by school furniture still does not entitle
them to be placed in a classroom. It is important that the product meets the expected
usable requirements, which the specifics of schoolrooms and the nature of the pupil
environment are presented with. Therefore, during the design process, it needs to be
made sure that the surfaces of worktops of tables are finished with materials of high
resistance to high temperatures, abrasion, scratches, impacts and discolouration, and
also that they do not give off light reflections and enable easy maintenance of
hygiene. It is also recommended to pay attention to mobility, the possibility of
compiling and storing furniture in heaps even in small spaces. In order to ensure
pleasant contact of the pupil with school furniture, the use of cold materials should
be avoided, which lower the temperature of the human body locally (hands, fore-
arms, thighs, knees). It is also recommended to use wood and wood-based materials, in particular MDFs, chipboards, natural veneers and plywood.

Any incompatibility of school furniture with intended use presents a danger of
injury of the pupil. For this reason, during designing, the possibility of different
loads of construction from that which is standard should be foreseen and the furniture should ensure adequate stiffness, strength and stability (Dzięgielewski and
Smardzewski 1995). At this stage, it is important to eliminate the possibility of
injury to a child by rounding the edges and corners, adequately placing fittings and
elements, as well as carefully selecting materials. One must also remember to use
such structural components for which the designer has certificates of their complete
non-toxicity. Detailed requirements and testing principles of school furniture have
been provided in the norm PN-EN 1729-2:2007.

### 3.4.4   Requirements for Kitchen Furniture

### 3.4.4.1   Dimensional Requirements

The kitchen in a home should be organised as a sterile workshop, where all the tools
are at hand, and at the same time, they do not interfere in carrying out tasks. In order
for work in the kitchen not to require overcoming significant distances, household
appliances should be arranged in accordance with the order of work carried out
when preparing a meal. For this, a layout is created which contains three zones:

- the zone of storing products, to the right of it,
- the zone of processing products and dishwashing, behind it to the right, and
- the zone of preparing meals and their thermal treatment.

The functional zones set out in this manner create an ergonomic triangle
(Fig. 3.31a), the sides of which have a different length depending on the size and
shape of the kitchen. This triangle is overlapped by a technical triangle, which
vertices create three basic technical devices of the kitchen's equipment:

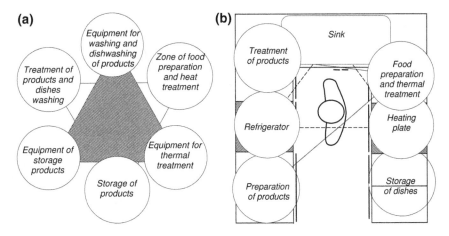

**Fig. 3.31** Areas **a** and functional equipment **b** in the kitchen

- equipment for storing products (refrigerator, cooler), to the right of it,
- equipment for washing products and washing dishes (sink, dishwasher, dryer), behind it on the right, and
- equipment for the thermal treatment of dishes (stove, oven, hob).

This layout is suitable for right handers. Left-handed people should arrange the zones in the order corresponding to a mirror reflection of the zones described above.

Between the elements of a kitchen's technical equipment, the following distances should be ensured:

- from the refrigerator to the sink 120–210 cm,
- from the sink to the stove 120–210 cm (in small spaces not less than 80 cm), and
- from the refrigerator to the stove 120–270 cm.

There should not be any door on any of the three sides of the triangles. However, next to one of the appliances, preferably in front of the stove or hob, the main workstation for preparing meals can be planned. In order to ensure convenience of movements, the sides of the triangles determined by the zones or appliances cannot be shorter than 3 m and longer than 6 m.

In the zone of processing products and dishwashing, a place for a drawer with a tray for cutting tools should be designed, as well as a container for waste, which should be located under the worktop of the workstation, and not—what is a common design error, under the sink. This way, after preparing a meal, the waste can be thrown down directly into the container.

The hob and oven are usually separate devices, of which the oven, like the microwave, is usually placed in a block of cupboards, at arm's height no less than 115 cm from the floor.

Figure 3.32 shows the preferred dimensions of kitchen furniture. It is recommended that the height of the worktop of bottom cupboards was adapted to the

**Fig. 3.32** Recommended
functional dimensions of
kitchen furniture (cm)

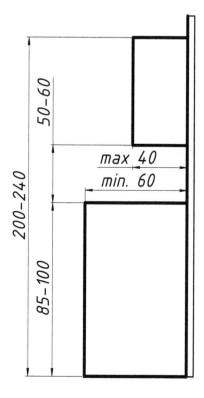

height of the user, enabling the convenient preparation of meals. This is usually from 70 to 90 cm. The optimal height of bottom cupboards differs, however, depending on the type of work done. The height of the hob should not be less than 70 cm. Worktops for placing pots are placed slightly higher—80 cm, and the boards intended for preparing meals is the highest, even 100 cm. If in the kitchen the working zones are not clearly separated and the worktop serves not only for putting dishes aside, but also for preparing meals, all surfaces should be designed at the same height. According to the norm PN-EN 1116:2005, the working height of bottom cupboards should be in accordance with one of the following standard heights: 85 (+5) cm, 90 (+5) cm and 100 (+5) cm. However, regardless of the provided heights of fixed work surfaces, lower surfaces extended at a height of 70–80 cm are additionally desirable (Janiga 2000; Janow and Bielow 1971).

In this norm PN-EN 1116:2005, the height of wall furniture has not been specified. However, it is recommended that the distance between the worktop of the bottom cupboards and the bottoms of wall cupboards amounts to 50–60 cm. Only the distance between the hob and cooker hood should be greater. For an electric cooker, it is recommended 65 cm, for a gas cooker—75 cm. Depending on the height of the user of the kitchen, the suspension height should be adjusted in such a way that enables him to reach the cupboard effortlessly. In practice, the height from

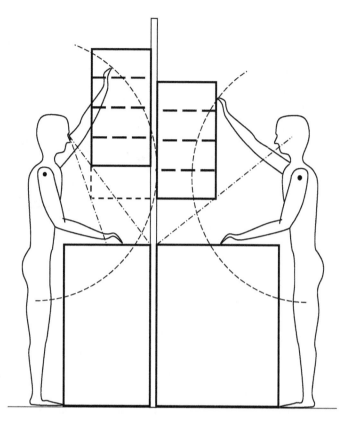

**Fig. 3.33** The availability of wall cupboards, depending on the height of their suspension and depth of bottom cupboards (cm)

the ground to the top edge of the wall cupboard should amount to 200–240 cm. Shelves, which we reach on a daily basis, are placed at a height of 170–190 cm, and others no higher than 230 cm. In order to ensure good visibility of worktop of bottom cupboards, the height of suspending wall cupboards should be increased. However, ensuring availability of higher shelves cannot be forgotten. In the event of increasing the depth of worktops of bottom cupboards, despite maintaining sufficient visibility of them, it significantly limits availability to the items in wall cupboards (Fig. 3.33).

The depth of hanging cupboards according to the recommendations of the standard should not exceed 40 cm, while the depth of the worktop, together with finishing narrow planes and wall-fixing elements, should not be less than 60 cm. This dimension is significantly affected by the depth and suspension height of wall cupboards. Figure 3.34 illustrates the availability of the worktop depending on the depth of the bottom and top cupboards.

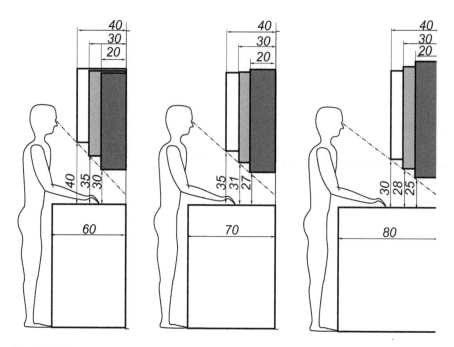

**Fig. 3.34** The availability of the worktop depending on the depth of the bottom cupboards (cm)

As it can be seen, reducing the depth of bottom cupboards causes that in order to ensure identical availability of the worktop, the height of suspending wall cupboards should be significantly increased. This is particularly important for wall cupboards with a maximum width of 40 cm. For this furniture, the suspension height increases more than cupboards with a depth of 20 cm.

The provided inconveniences can be corrected by adjusting the shape of the side walls of the cupboards to the requirements of functional kitchen furniture. Figure 3.35 shows an example of increasing availability of bottom cupboards, as well as the availability of the worktop. Through simple technological treatments, correcting the functionality of a set of kitchen furniture can be beneficial. The user can stand closer to the furniture and easily reach into a space that is further away, in which products or items are stored. Another option is to replace bottom cupboards with shelves, cupboards with many systems of drawers and containers.

### 3.4.4.2  Errors in Measuring Kitchen Areas

In contrast to designing home, school and office furniture, where operational space can be developed freely along with the migration of furniture, designing kitchen furniture requires taking into account its inseparable relation to the walls of the building. Therefore, when designing kitchen furniture all kinds of problems arising

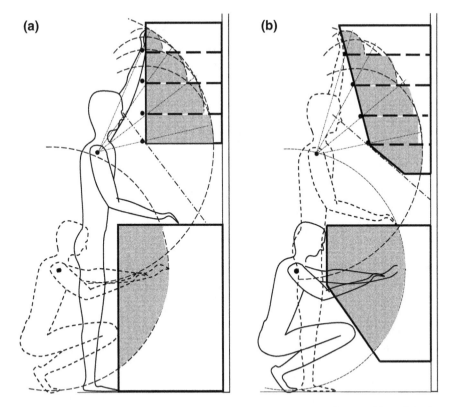

**Fig. 3.35** Availability of kitchen furniture cupboards: **a** classic layout, **b** proposition of shaping the side walls of bottom and wall cupboards

from the unevenness of walls must be solved. Their deviations from the vertical, horizontal and angles in relation to the values assumed in project. Underestimating this issue ultimately reduces ergonomic values, loss of functionality and the inability to mount furniture.

Despite the different systems of walls in rooms intended for kitchens (one and two walls situated at any angle in relation to one another, three walls oriented towards each other in the shape of the letter U or Z, as well as four and more walls situated freely in relation to one another), a general pattern of the layout of walls can be assumed to analyse design errors. On the basis of this pattern, Fig. 3.36 illustrates building construction errors that affect the dimensions of kitchen furniture. In other wall layouts, both building errors and their consequences in the dimensions of building a kitchen are analogous.

Figure 3.36 shows that not taking into consideration dimensional differences $dL$ and $dS$ will cause the following errors in producing furniture:

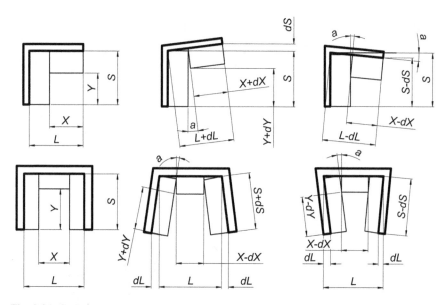

**Fig. 3.36** Geometric errors of buildings affecting the dimensions of kitchen furniture

- assuming the dimension of the built-up space $L + dL$ or $S + dS$, the designer will commission the manufacturer to build a set which will have larger width in relation to the primary dimension $L$ or $S$,
- assuming the dimension of the built-up space $L$ or $S$, after mounting the furniture, free spaces $dL$ and $dS$ will be left between the furniture pieces and the walls of the building, while the worktop will be too short.

A common mistake made by designers is the failure to take into account the height of the sill, as well as situating the sink in the light of the window, which makes it impossible to open the window wings. However, the omission of installations protruding from the building's walls will enable to mount cupboards manufactured in accordance with the previously adopted depth of the bodies.

The issue of the correctness of construction of a set of kitchen furniture does not end with properly carried out inventory of the room. In practice, it turns out that the dimension of the wall, as defined earlier, is not the only parameter on which the correctness of the construction of a set of kitchen furniture depends.

Failure to take into account in the design stage of the free space intended for distance strips between the bottom cupboards set in a corner results in the inability to open fronts, as there will be a collision of handles. Therefore, the designer should adjust the width of distance strips, which also fulfil the function of elements reducing the impact of hot air emitted when opening the oven on the fronts of kitchen cupboards.

### 3.4.4.3 Requirements for Safety of Use

In accordance with the Ordinance of the Minister of Economy of the 5th of July 2001, only safe furniture can be entered onto the market. This means that the manufactured furniture piece, in ordinary or other reasonably anticipated conditions of its use, does not pose any danger to the user. Hence, a safe furniture piece is one which is characterised by sufficient stability, strength, stiffness and hygiene. This means that the manufacturers of furniture, in wanting to introduce a new product onto the market, should be able to confirm the safety and high quality of the product issued by an independent research unit. Of course, kitchen furniture is also subjected to this regulation. Thereby, only a furniture piece that meets the requirements contained in the norm PN-EN 14749:2007 can be approved for product turnover. At the design stage, constructional hypotheses must be verified using numerical or mathematical calculating instruments (Dzięgielewski and Smardzewski 1995; Korolew 1973; Łajczak 2004; Smardzewski 2002a, b, c, d, 2004a, b, c).

In addition to the strength of kitchen furniture, the durability of use must also be carefully designed, by skilfully choosing materials that enhance the wooden surface or wood-based boards. Guidelines in this regard can be found in the appropriate norms: PN-F-06001-2:1994, DIN 68 861:1981 1985, PN-EN 438-1,2:1997 and also in the publications of Krzoska-Adamczak (1996, 2001), Krzoska-Adamczak and Nowaczyk (2005). Table 3.14 presents some of the most important requirements for furniture surfaces.

When using glass in kitchen furniture, one must find out whether it has obtained positive results in approval tests confirming that it is safe glass (PN-EN 14749:2007).

**Table 3.14** Classification of furniture areas according to the norm DIN 68,861

| Criterion | Class of use and surface resistance | | | | | |
|---|---|---|---|---|---|---|
|  | A | B | C | D | E | F |
| Time of working (min) | 16 h | 10 s–16 h | 2–10 | 2–10 | 2–10 | 2 |
| Number of fluids | 27 | 27 | 10 | 10 | 2 | 2 |
| Abrasion (rotations) | >650 | >350–650≤ | >150–350≤ | >50–150≤ | >25–50≤ | 25≤ |
| Scratches (N) | 4.0 | 2.0–4.0 | 1.5–2.0 | 1.0–1.5 | 0.5–1.0 | – |
| Cigarette heat (1–5) | 5 | 4 | 3 | 2 | 1 | – |
| Increased temperature in dry test (°C) | 180 | 140 | 100 | 75 | 50 | – |
| Increased temperature in wet test (°C) | 100 | 75 | 50 | – | – | – |

Where, A horizontal surfaces used in conditions of high risk of damage, B worktops of tables, desks, C horizontal surfaces intended for putting things aside, D front surfaces, E internal visible surfaces and F invisible internal surfaces

### 3.4.5  Requirements for Furniture for Sitting and Relaxing

#### 3.4.5.1  Dimensional Requirements

Scientific progress in the field of analysing the seated position and adjusting furniture to the physiological characteristics of a human being began from Staffel's claim (Izrael Abraham Staffel 1814–1884), that: *chairs are almost without exception constructed more for the eye, for the beautiful form, than for the back*. It was at that time when the principle was formed that the spine of a sitting person should be supported only in the lumbar region. The next step in formulating the principles of proper sitting was striving to ergonomically shape the upper part of the backrest, adjusted to the chest part of the spine (Stępowski 1973). Currently, the common belief dominates among designers and users of furniture that regardless of the subjective perception of comfort of a seat, chair seat, armchair and sofa seat should be designed in accordance with the physiology of the human.

Some furniture for sitting (chairs, armchairs, stools), together with tables, benches or desks, can be used to work, study and dine, while soft armchairs and sofas generally constitute equipment for places to relax and rest. Each of these groups of furniture, due to the clear differentiation of purpose, should be characterised by different requirements: dimensional, mechanical and functional.

In the furniture for sitting, used at tables, not only the dimensions of the chair, armchair or table must be dealt with, but also the appropriate dimensional relations between these furniture pieces should be selected.

The correct dimensional proportions for the chair have been illustrated in Fig. 3.37 and listed in Table 3.2. Considering the need of free movement of the lower limbs, Fig. 3.38 shows the space in which the user can move them once bent in the knee joint, while in Fig. 3.39, the lower limbs are straight at the knee joint.

When designing tables, it is worth noting the correct height of the worktop, the height of the rails (restricting the movement of the lower limbs), the width of the board guaranteeing operational freedom on the worktop surface and the depth of the board conditioning convenient reach in a comfortable seated position (Fig. 3.40).

The optimal dimension of the worktop of a workstation or for eating meals should also depend on the hold reach of the upper limbs, taking into account the zones of precise, accurate, general and inaccurate work (Figs. 3.41, 3.42, 3.43, 3.44 and 3.45). Composing models of a chair and table also enables to determine the most appropriate dimensions for a set and demarcating the ergonomic space for the free use of the chosen arrangement (Fig. 3.46).

By building virtual models of chairs and tables, which can adjust their dimensions to individual types of human phantoms, already during the designing stage, the product being made can be validated and show the weakness of operational or constructional solutions adopted. By composing a virtual model of an actual dining room table with the model of a chair for the user, for example, of characteristics of the 50th centile (Fig. 3.46), the basic ergonomic and functional errors of this system can be eliminated. For a full analysis of the functionality of the object, it is possible

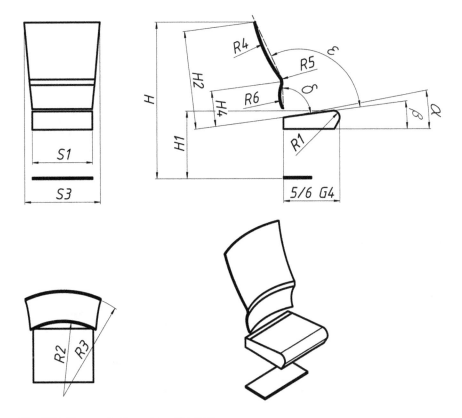

**Fig. 3.37**  Ergonomic dimensions of a chair

to slide out worktops and set up extra chairs (Fig. 3.47). A virtual synthesis of many probable functional systems of furniture enables to point out potential conflicts between the frame of the table and the comfort of the user.

Furniture intended for relaxation should be characterised by different mechanical and dimensional parameters than furniture for work and dining. Grandiean (1973, 1978) formulated the requirements in relation to the geometry of the seat and backrest of an armchair for healthy people and for people with spinal disorders (Fig. 3.47). The numerical values of these parameters have been provided in Table 3.15.

Preferred angles of seat and backrest inclination, depending on the purpose of the furniture, are also given in Table 3.16.

One of the basic dimensional parameters of furniture for sitting and resting is the height of the seat. It should be less than the distance between the popliteal bend and the base supporting a person sitting down, regardless of the adopted position during work, study or relaxation. In a seated position, the seat height should be 3–5 cm below the popliteal bend (Stępowski 1973). A seat that is situated too high (Fig. 3.48a) can cause pressure on the artery, thus burdening the circulatory system

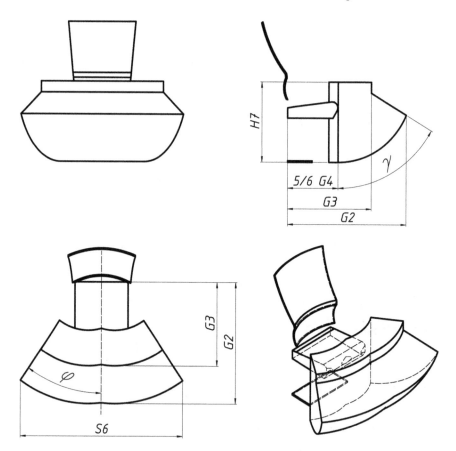

**Fig. 3.38** The dimensions of space for the free movement of the lower limbs bent in the knee joint

too much. And a seat that is too low (Fig. 3.48b) requires one to contract one's legs, which causes pressure on internal organs and increased pressure on sciatic protuberances. Incorrect load of the spine in the lumbar region also has a negative impact on the well-being of the user of furniture for sitting. Usually, it is caused by a seat that is too deep (Fig. 3.48c). For this reason, it is recommended that the popliteal part of the thigh protrudes from the seat by 1/3 (Stępowski 1973; Nowak 1993).

### 3.4.5.2   Requirements for Safety of Use

According to McCormick (1957), a person is more comfortable in a seated position if two basic conditions are met:

- The weight of the body is adequately distributed between the seat and the backrest and
- muscle tension necessary to maintain a seated position is as small as possible.

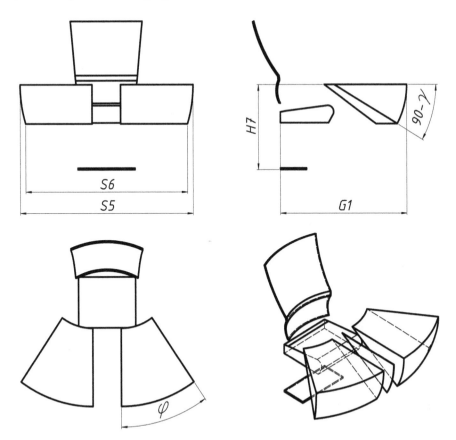

**Fig. 3.39** The dimensions of space for the free movement of the lower limbs straight at the knee joint

Identifying technical requirements for furniture for sitting and relaxation requires collecting data on:

- dimensional characteristics of the human body, including the dimensions and weight of the body and its individual parts (anthropometry),
- distributing the centres of gravity and the range of movements of parts of the human body (biomechanics),
- the construction of the body, that is tissue types, their distribution and mechanical properties (anatomy),
- reactions of the user's body to the impact of external factors, including the behaviour of the circulatory, nervous and bone system (physiology), and
- the construction and technology of making furniture for sitting, that is geometry and dimensional proportions of the seat construction and characteristics of upholstery systems and materials used (stiffness, strength, thermal insulation, air circulation, etc).

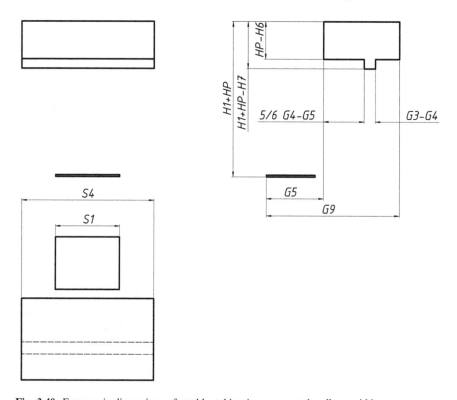

**Fig. 3.40** Ergonomic dimensions of a table, taking into account the elbow width

The factors that have the most important impact on the comfort of sitting are as follows:

- the level of contact pressures occurring at the junction of the seat and the user's body, and the value of stresses inside the soft tissues,
- body build, including weight, sizes and gender,
- time of sitting, and
- way of sitting.

Usually, the value of contact stresses between the human body and an upholstered system increases in proportion to the body weight of the user and strictly depends on the stiffness of the materials used (Smardzewski et al. 2008). A mattress obtains optimum stiffness when the spring layer is characterised by Young's modulus of a value similar to the elasticity of the human body's soft tissue (Smardzewski et al. 2006).

Depending on the geometry and constructional solution of the seat, contact stresses can develop within the muscle tissue stresses limiting the normal functioning of the circulatory system. By averaging the pressure in the arterial system,

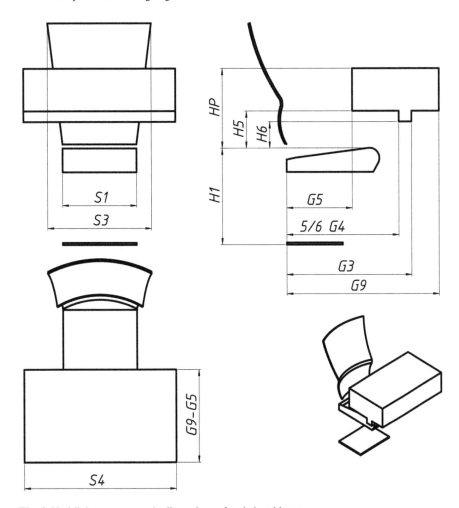

**Fig. 3.41** Minimum ergonomic dimensions of a chair–table set

we obtain a value of around 100 mmHg, in the capillaries around 25 mmHg, and in the final part of the venous system, this pressure amounts on average to 10 mmHg (Guzik 2001). The border value of surface pressure equal to 32 mmHg refers to the pressure reducing the light of capillaries (Krutul 2004b; Stinson 2003). Each pressure greater than this value may result in limiting or closing the lumen of veins, and arteries, which slows down the flow of blood or stops its circulation, causing local ischaemia. Time for which the stress is exerted on the human body plays an extremely important role. Large stresses within a short period of time can lead to deep damage to the muscle tissue. And stresses of small values, but lasting for a longer period of time, cause damage to soft tissue (Stinson 2003). Such a situation can take place in the event of long-term maintenance of a uniform seated position.

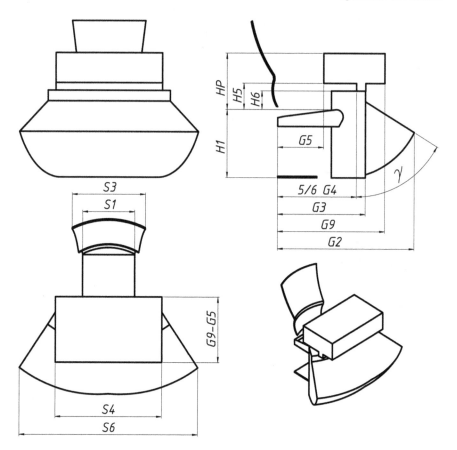

**Fig. 3.42** Dimensions of a chair–table set including space for the free movement of the lower limbs bent in the knee joint

The effect of this can be the compression of the spine muscle and gluteal muscle by the sacral vertebra and sciatica, which results in an increase of stiffness of the muscles where the bone meets the muscle. This is revealed by pressure pains at the height of the ischiatic bone or so-called pins and needles of the lower limbs caused by ischaemia. The studies of Gefen et al. (2005) showed a stiffening of muscle tissue during 35 min under external pressure 35 kPa (262.5 mmHg) and during 15 min under pressure of 70 kPa (525 mmHg).

By simulating the immobility of a person sitting on a hard ground, using the finite element method, these authors generated in the cross section of the stress compressing at a level of 35 kPa. In this way, they demonstrated that the intensity of the injury damaging cells grows within the first 30 min of sitting in stillness and is the cause of the formation of pressure pain.

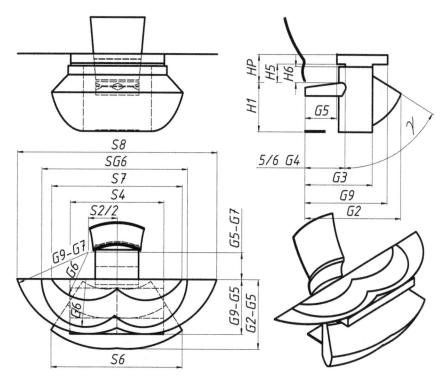

**Fig. 3.43** Dimensions of a chair–table set including frontal hold reach on the worktop surface and space for the free movement of the lower limbs bent in the knee joint

Defloor and Grypdonck (1999) identified the influence of the seated position on the distribution and value of contact stresses between the seat and the body of the user. The dimensions were carried out on 56 volunteers aged 19–46 years (average age 23.8 years, average weight 64.8 kg, BMI (body mass index 22.4 kg/m$^2$). The lowest stresses, at the level of 39.5 mmHg, have been assigned for the position adopted on a seat inclined backwards with a tilted backrest and support of legs on a footrest. The position with feet resting on the floor, without the possibility of resting the arms, forced the creation of surface stresses with an average value of 48.7 mmHg. However, the highest average value of surface stresses, equal to 55.4 mmHg, was obtained in a position when the body slides down on the seat by 200 mm from the upright position, and the line of the trunk forms an angle of 45° with the line of the seat surface.

By examining the impact of changes in the angle of inclination and the height of the backrest on pressure in the intervertebral disc, Nachemson (1976) demonstrated that the smallest stress and pressure in the disc occur when the angle of inclination is 120° and the support of the lumbar is located at a height of about 20 cm. The highest pressure occurred when the angle of inclination amounted to 90° (Fig. 3.49).

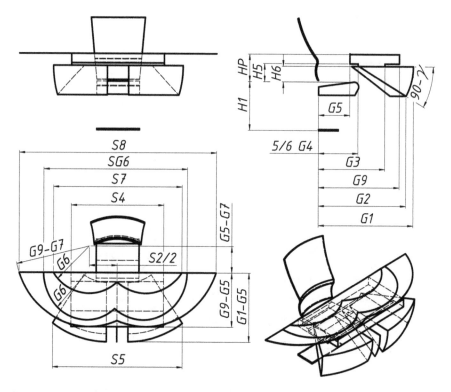

**Fig. 3.44** Dimensions of a chair–table set including frontal hold reach on the worktop surface and space for the free movement of the lower limbs straight at the knee joint

## 3.4.6  Requirements for Beds

### 3.4.6.1  Dimensional Requirements

Sleep and rest are an indispensable condition for the proper course of all life processes. While sleeping, the muscles and the skeletal system relax, as does the nervous system and many other processes relevant from the point of view of physiology. The proper conduct of all these processes is primarily dependent on comfortable, healthy sleep and therefore on individually suited, proper construction of the furniture intended for lying down, providing comfort of the bed in a place that gives a sense of security (Grbac 1985a, b, 1988; Grbac and Dalbelo-Bašić 1996; Grbac et al. 2000).

As opposed to furniture intended for sitting, furniture for lying down is characterised by much simpler anthropometric determinants expressed by dimensional dependencies (Fig. 3.50). This takes into account mainly the height of the person and a specific movable space, and when designing, the spring elements must be chosen in such a way that muscle stresses are as small as possible (Kapica 1991, 1993a, b, c, d, 1988, Kapica et al. 1999).

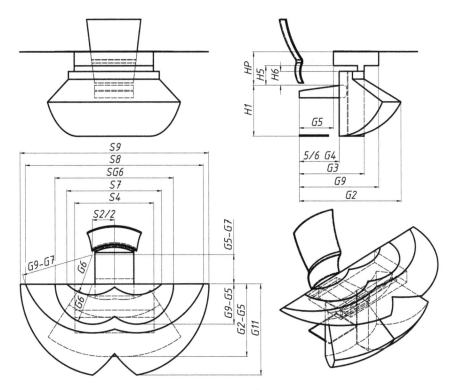

**Fig. 3.45** Dimensions of a chair–table set including frontal hold reach on the worktop surface when the user is bent and space for the free movement of the lower limbs bent in the knee joint

The bed is designed on the basis of the shape of a rectangular cushion of minimum dimensions:

- for one person: 1900 × 800 mm,
- for two persons: 1900 × 1200 mm, (PN-89/F-06027.04).

The minimum dimensions of length and width of the mattress should take into account the measures of people belonging to the population of the extreme upper position of statistical distribution of selected dimensions, resulting from anthropometric measurements (Smardzewski and Matwiej 2004). Furthermore, when determining the width, the place should be estimated which is taken up by a person lying down on the side with folded legs in a position when the thighs form an angle of 90° with the trunk (Kapica 1993a). Grbac (2006) suggests that the length of the bed was greater by at least 200 mm from the height of the user, while the width amounted to 1400 mm (for a single bed).

During sleep or rest, the Human body exerts pressure on the base and vice versa. Improper construction of mattress may cause: insomnia, sore parts of the body and/or the skeletal system, and even curvature of the spine. The largest loads occur

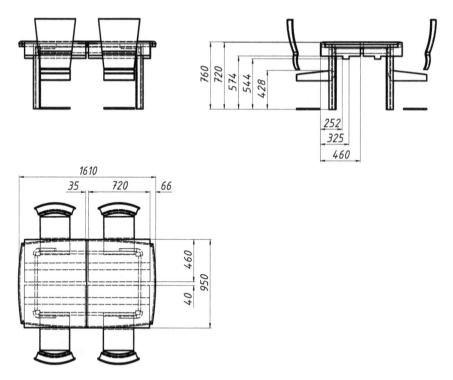

**Fig. 3.46** Ergonomic evaluation of a chair and oval table in the operational position S4 of a male of 50C (dimensions in mm)

in the hip area and the shoulder joints, which require from the construction of the mattress the largest area in contact with the body and the appropriate degree of deformation. The forces caused by the weight of the body should be spread over on the largest possible surface (Karpiński and Deszczyński 1997).

Mattresses of a simple construction and inappropriate filling do not always adjust to every body shape, providing support only to its convex parts. This part of the furniture, on which one statistically spends 1/3 of the day, should be appropriately adjusted to the user's body weight. While lying on one's back, the body's weight is transferred mostly to the area of the shoulder blades and buttocks, while the lumbar region which does not have support descends downwards (Fig. 3.51).

In Fig. 3.51, loads of the human's body weight have been marked by vectors, as well as the forces of the elasticity of the mattress balancing them. As it can be seen, some vectors are not coaxial, which indicates the presence of shear forces and compressive and bending forces, affecting the human skeletal system, nervous system and muscular system.

A similar situation occurs in the position of the body on the side (Fig. 3.52). Above all, the shoulder joint and hip joint are burdened here. The lumbar area has no support, which results in its lowering under the influence of body weight.

**Fig. 3.47** Dimensional characteristics of the cross section of armchair: *SH* seat height, *AH* armrest height, *SW* seat angle of inclination, *RW* backrest angle of inclination in the lumbar region, *KW* angle of backrest and *ST* the seat depth (Grandiean 1973)

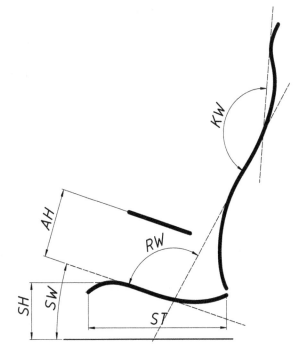

**Table 3.15** Dimensional characteristics of the cross section of an armchair intended for use by healthy people and people with spinal disorders (Grandiean 1973)

| Seat characteristics | Healthy persons in the position | | Persons with spinal disorders, relaxation position | Compromise position between reading and relaxing |
|---|---|---|---|---|
| | Reading | Relaxing | | |
| Seat angle of inclination *SW* (°) | 23 | 26 | 20 | 23 |
| Backrest angle of inclination in the lumbar region *RW* (°) | 103 | 107 | 106.5 | 107 |
| Seat height SH (mm) | 400 | 390 | 415 | 400 |
| Seat depth *ST* (mm) | 470 | 470 | 480 | 480 |
| Height of loin support (mm) | 140 | 140 | 90 | 80–140 |
| Armrest height *AH* (mm) | 260 | 260 | 260 | 260 |

Resting on a mattress that is too soft, which under the influence of body weight adopts a basin form, may cause deformation of the spine caused by hours of unfavourable concentration of stresses, within its range, intensifying reflexive stress of the muscles surrounding it. In addition, the inadequate support of the cervical

**Table 3.16** Angular dependencies for armchairs and sofas at non-production activities (Slavikova 1988)

| Type of inclination | Angular parameters for armchairs and sofas | | | |
|---|---|---|---|---|
| | For lecture halls | For audiences | For relaxing | For travellers |
| Seat inclination from the SW level | 4–8° | 7–11° | 10–15° | 15–20° |
| Backrest inclination in relation to the seat RW | 105° | 110° | 115° | 127° |

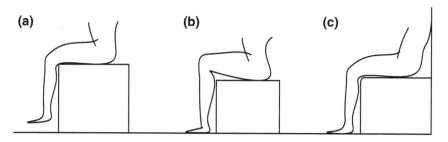

**Fig. 3.48** Errors in adjusting the human–seat system: **a** seat too high, **b** seat too low, **c** seat too deep (Nowak 1993)

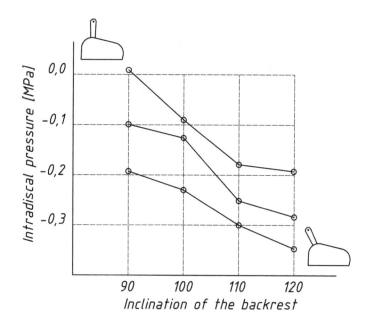

**Fig. 3.49** The impact of changes in the inclination and height of lifting the back when sitting on changes in pressure in the intervertebral disc (own development based on Nachemson 1976)

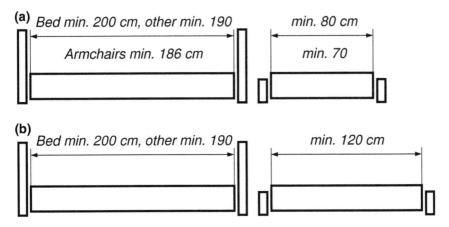

**Fig. 3.50** The basic functional dimensions of beds and unfolding chairs: **a** single bed, **b** double bed (cm)

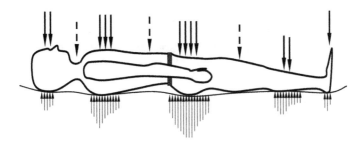

**Fig. 3.51** The impact of the mattress while lying on the back

spine is conducive to disorders of blood flow in the arteries, with all the pains, ailments, migraines and sleep disorders accompanying it.

All the previously presented effects of the sleeping space on a person in a lying down position are intensified by a very hard bed (Fig. 3.53).

Sleep on a hard base is very harmful, especially more uncomfortable in relation to sleeping on a mattress that is too soft. The influence of the base is more noticeable on the convex parts of the body, since they take on the entire load, while elastic forces of the base do not affect a sleeping person. A properly selected mattress, adjusted, variable elasticity of its surface ensures proper load distribution on the various parts of the body, reducing the harmful stresses arising around the spine and joints (Figs. 3.54 and 3.55).

The proper formation of the mattress enables to support the lumbar region correctly when lying on the back and on the side. This prevents this part of the body from "sliding down", thanks to which the physiological curvatures of the spine are maintained (Karpiński and Deszczyński 1997).

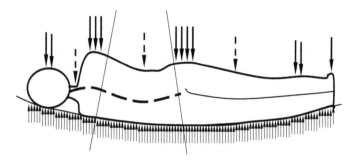

**Fig. 3.52**  The impact of a mattress that is too soft while lying on the side

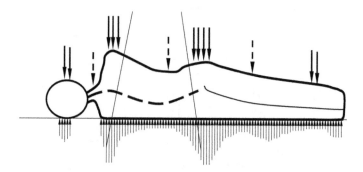

**Fig. 3.53**  The impact of a mattress that is too hard while lying on the side

Stresses occurring with lying down still for more than 2–3 h cause irreversible changes. Excessive point stresses caused by lying down on an inappropriate mattress can suppress and even stop the flow of blood in the skin above bone protuberances which leads to a lack of oxygen and nutrients supply of cells, and as a result, the death of the anoxemic cell. According to Krutul (2004b) in places where the bones push onto tissues harder, pressure increases and the lumen of blood vessels is reduced, which leads to the damage of skin tissue. The stresses of the external surface are 3–5 times smaller than internal forces that occur as a result. Therefore, the limit value of the pressure amounting to 32 mmHg (closing the lumen of capillaries) must be appropriately reduced and to the level from 6.4 to 10.6 mmHg. Each pressure greater than these values may result in closing the light of veins, then of arteries, which slows down the flow of blood or stops its circulation, causing local ischaemia and—in time—necrosis.

When designing furniture for lying down, a number of anthropometric and physiological rules should be taken into account, resulting from the conditions of their use.

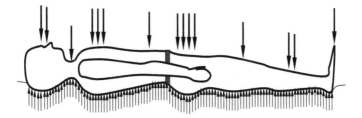

**Fig. 3.54** The impact of the orthopaedic mattress while lying on the back

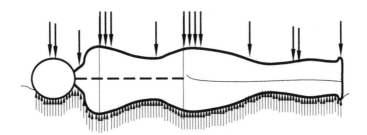

**Fig. 3.55** The impact of the orthopaedic mattress while lying on the side

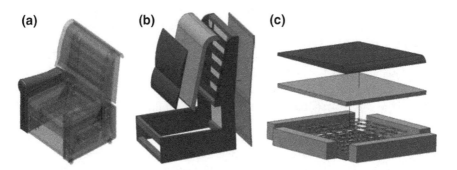

**Fig. 3.56** Construction of an upholstered armchair: **a** furniture, **b** backrest, **c** seat

In the case of sofas and armchairs, the seat and backrest must be designed appropriately, which differ both in terms of functional and structural requirements (Fig. 3.56).

Both the seat and the backrest in furniture for sitting should have a specified and varying degree of softness adapted to the size of the load attributed to a given part. The load caused by the weight of the user is distributed unevenly on the backrest and the seat of the furniture piece and on particular surfaces of these parts (Sienicki 1954; Smardzewski 1993a, b, 2002c; Smardzewski and Grbac 1998).

### 3.4.6.2   Quality of Sleep

Sleep, according to the definition provided by PWN Encyclopaedia (2004), is the functional state of the nervous system, opposed to the state of being awake, which essence is the temporary loss of consciousness. Both in the human and higher animals, it occurs in a daily rhythm, alternating with being awake; it is regulated by the brain centres of sleep and inducing centres. Sleep is essential for the regeneration of the organism and the central nervous system.

Questions about the essence of sleep, which absorbs about 1/3 of human life (Pietruczuk et al. 2003), have appeared since the dawn of time: approximately 2400 years ago, Hippocrates advised to sleep at night and to be active during the day, and he claimed that sleep is caused by the inhibition of blood flow and its direction to the internal organs (Nowicki 2002). The scientific analysis of sleep began to gain interest at the beginning of the twentieth century, when Berger applied an encephalographic study, providing objective criteria of brain function when sleeping, and Legendre and Pieron put forth the hypothesis that sleep is the result of increased secretion of substances called hypnotoxin during the state of being awake (Nowicki 2002). In 1937, Harvey, Loomis and Hobart made divisions of the depth of sleep, and Aserinsky and Kleitman in (1953) discovered the diversity of sleep in two separate states: NREM sleep (non-rapid eye movement) and REM sleep (rapid eye movement) (Nowicki 2002).

Somnology—a new discipline of science concerning the issues of sleep, associated with all of its aspects, is increasingly important also in designing ergonomic furniture products intended for sleeping. It provides a wealth of information on both the essence of sleep, as well as the risks resulting from disturbances of its course.

Users sleeping or trying to fall sleep on improperly designed beds often complain about morning lack of sleep, backache, headache and migraines; they are irritated and feel fatigue. In extreme cases, this can lead to insomnia. Sleep begins with NREM which lasts on average 80–100 min, followed by REM which lasts approximately 15 min (this is the phase in which dreams appear and the muscles totally relax). In adults, this cycle repeats itself during sleep 4–5 times (Fig. 3.57). Along with the length of sleep, the deepest phase NREM decreases, and the duration of the REM phase increases, which near the end of sleep usually lasts approximately 40 min. A dream is remembered when one wakes in this phase. Depriving people who are sleeping of going into the REM phase can entail a particularly high risk to health (especially mental). After a few days of not sleeping well, one may have delusions or collapse into a state of psychosis. Such a risk may result directly from a poorly designed, uncomfortable construction of the bed. In addition, sleep disorders are classified as a state of stress, and this can disrupt mutual communication between two systems: the immune system and the nervous system (Pietruczuk et al. 2003), which in turn may have an impact on overall health, decrease of immunity, etc.

Studies conducted on representatives of the Polish population show that the length of sleep at night is approximately 7.5 h (Pracka et al. 2003). Sleep function disorders and diseases associated with insomnia entail serious economic

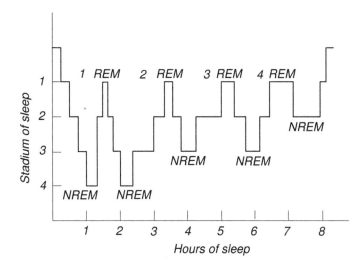

**Fig. 3.57** A graphic representation of the ideal course of an 8-h sleep of an adult (Nowicki 2002)

consequences. This is confirmed by American studies, which demonstrate that in 1990, in the USA, the costs of hospitalisation associated with sleep disorders amounted to 16 milliard USD (Grabowski and Jakitowicz 2006). Women complain about insomnia much more often, and the quality of sleep deteriorates with age (Nowicki 2002). In the elderly, it is also found that the NREM phase shortens and REM phase of sleep gets longer (Pietruczuk et al. 2003). According to Polish data, as many as approx. 24 % of Poles suffer from insomnia (Kiejna et al. 2003).

Furniture designers should, therefore, take into account the above arguments and focus not only on the form, but also on functional and structural solutions (Smardzewski et al. 2008). The mattress, which constitutes an assemblage, sub-assemblage or finished upholstered product designed for sleeping and/or resting, consists of many layers, and these can be made of materials which differ not only in shape and dimensions, but particularly in the physico-mechanical properties. Therefore, specifying the technical characteristics of mattresses, the quality of sleep should be examined, and the parameters specified, which have an impact on it, as well as take into account the requirements concerning the geometry of the product and characteristics of the materials.

### 3.4.6.3  Requirements for Safety of Use

The main factors that determine the comfort and safety of use of furniture for reclining are as follows: air circulation, temperature on the surface of the mattress, moisture circulation, softness of the mattress and durability (Smardzewski et al. 2008).

During the daily duties and activities, when the human is at the so-called wakefulness phase, appropriate physical conditions are required for the proper functioning of each organism. It is similar during the sleep. Optimal conditions allowing for sufficiently long, healing and relaxing sleep are characterised by the following values:

- atmospheric pressure: about 785 mmHg,
- relative humidity of air: in the winter about 50 % and in the summer about 60 %, and
- temperature of the air in the room: 14–18 °C (Grbac 2006).

**Air circulation**
The proper ventilation of the mattress is provided by adequate base and frame, which construction is of vital importance in shaping the form of the mattress—especially its overall dimensions. Mattresses based on spring systems usually provide comfortable air circulation inside their structure. Changing the geometry of the bed, its deformation under the weight of the user promotes air circulation and refreshing of the mattress. Foam materials usually have a porous structure and, as a general rule, enable air circulation in the mattress. The exception is latex foam. Made of vulcanised rubber, it does not provide air migration and forces shaping of artificial channels during its production.

**Temperature**
Maintaining a specific microclimate not only in the room in which the person is sleeping, but also in the immediate vicinity of the body (under the covers, blanket) is a very important thing for the organism.

Hyunja and Sejin (2006) have shown that a very important criterion for assessing the quality of mattresses is characteristics of the phases of sleep. The authors examined the impact of the quality of mattresses on the quality of sleep by measuring the temperature of the skin. Several different types of mattresses were chosen for the study, and the temperature was studied using polysomnogram. The polysomnogram used brainwaves recorded by electroencephalogram (EEG) and generated graphical video record of eye movements (EEA), chin movements (EMG), and pulse (ECG). 16 volunteers were examined, who tested the comfort of mattresses. The volunteers spent 6 days and nights in the laboratory. The study lasted for 7 h on each of 3 nights. On the basis of studies, it has been shown that the average temperature of the skin, deep sleep (phase III and IV of sleep), effective sleep, walking during sleep (WASO —wake after sleep onset), as well as phase I of sleep depends largely on the type of mattress. When comfortable mattresses were used, the average temperature of the skin of volunteers was significantly higher than when using uncomfortable mattresses, on which the share of effective sleep phase and deep sleep phase was much higher, and the percentage share of WASO stage and phase I was lower.

According to Krauchi (2007), sleepiness is associated with the body's thermo-regulation and begins to be felt in the event of a drop in its temperature by about 0.15 °C, with simultaneous increase in skin temperature by about 1.5 °C. A study carried out by Someren (2004) has confirmed that the high temperature of the

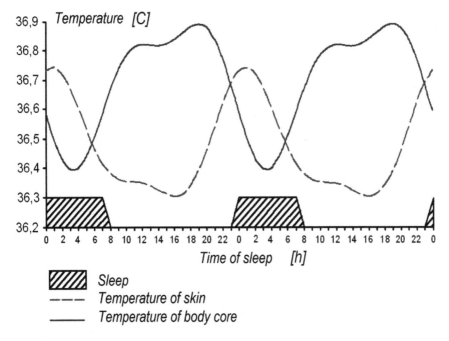

**Fig. 3.58** Distribution of temperature of the skin and core of the body during two-day alternate rhythm of awakeness and sleep phase (Someren 2004)

human body prevents from entering the drowsiness phase. Only neurons stimulating the skin to increase the temperature, and at the same time to exhaust the heat to the outside, cause excitation of the daily tendency to sleep (Fig. 3.58). Studies carried out by Lee and Park (2006) show that when you sleep on a comfortable mattress, the skin temperature is maintained at a higher level than on an uncomfortable mattress.

Temperature drop inside the body during sleep is a normal thing (Okamoto et al. 1998). The user's body temperature within 1.5 h drops by about 0.4 °C, and after about 8 h of sleep returns to the normal value of 36.6 °C. It should not be feared that in the room with air temperature of 14–18 °C, a sleeping person will feel the effect of the so-called cold surface of the bed. The same research confirmed that the bed, irrespectively of the temperature in the room, can have a temperature higher by about 2 °C and within 4 h rise to the temperature higher than the initial temperature by about 10 °C. The ability to give out heat by the user of the bed or to receive heat by the bed depends on the type of the used material contained in the lining and covering layer of the product, as well as on the bedding. The bedding should therefore provide adequate thermoinsulation conditions allowing for temperature migration so as to prevent the overheating of the body, which might cause decrease of drowsiness and at the same time affect the sleep process.

Okamoto et al. (1998) conducted a study, which purpose was to determine the impact of feather-covered mattresses on the quality of sleep and climate in bed.

Two mattresses were used for the study, one of which was covered with a feather lining with a thickness of 2.95 cm and the other with a futon-type woollen lining with a thickness of 4.80 cm. The temperature was maintained within the limits of 18–19 °C and humidity of air within 50–60 %. The quality of sleep was controlled using EEG, measurement of temperature in the anus and on the surface of the skin and air humidity. During the study, no significant differences in the quality of sleep in all its phases were observed. However, lower temperature in the anus was registered on the mattress with feather lining than on the futon type. There were no differences shown in humidity of air (climate).

**Moisture circulation**

During one night, human loses approximately 0.5–0.75 l of water (Grbac 2006). This results in a significant increase in water vapour content inside the mattress (Svennberg and Wadsö 2005). Cunningham's research (1999) indicates that the relative moisture of the mattress both in its lower and middle layer may rise even above 70 %.

Therefore, it is important that the materials that make up the spring, lining and cover layer, as well as bedding, have the ability of very good absorption and giving out of moisture (steam and sweat). Residual moisture at the appropriate temperature may well encourage the development of different kinds of fungi. According to Svennberg and Wadsö (2005), due to the ability of the mites to absorb water vapour from the air thanks to the salt compounds, relative moisture above 58 % at a temperature of 20 °C is dangerous. Therefore, it is recommended to use solutions enabling users to change the side of the mattress depending on the season, which is conditioned by the quantity and speed of the given out moisture. A study carried out by Svennberg and Wadsö (2005) confirms that with the increase in the thickness of the mattress (made with polyester foam and cotton cover), its quasi-insulation properties increase, and the combination of different materials significantly affects the microclimate in the immediate vicinity of the human body during sleep. Relative humidity of the air in the room, despite the migration of moisture within the mattress, does not undergo significant changes. Together with the loss of moisture, human gets rid of the old cuticle through decortication, which causes that the bed becomes over time an ideal habitat for mites, producing one of the strongest allergens—a protein enzyme. It is therefore necessary to use such materials that can prevent the settling of micro-organisms or at least reduce this phenomenon. An example might be a coconut mat or foam, which has antibacterial and antiallergic properties, or one of the newest materials—a special 3D, light and flexible polyester mesh DryMesh. This material has been designed so as to allow free airflow, reduce moisture condensation and thus reduce the likelihood of the formation of mildew. This mesh, due to its properties, is also used in yachts and boats.

**Softness of the mattress**

The human changes the position of the body about 40–50 times during sleep, assuming—depending on the gender, age, mobility and preferences—up to 21 positions (Grbac and Domljan 2007). This restless sleep is associated most often with the wrong degree of softness of the materials used in the upholstery layers.

After reaching a certain local border value of pressure, human feels uncomfortable and tries to fix this by changing the body position. According to Grbac (2006), the pressures on the body cannot exceed 28 mmHg. Krutul (2004b) and Defloor (2000) suggest that the pressure above 32 mmHg causes closing the lumen of the smallest blood vessels—capillaries and is so-called border value for the formation of compression ulcers. Beldon (2002), after Fronek and Zweifach (1975), reports that the limit of pressure for the human body amounts to 30–35 mmHg. The same author cites research of Bennett et al. (1981), which confirms that for ill people, compression ulcers already start occurring at the pressure of 22 mmHg. Usually, pressure below 45 mmHg can be treated as so-called relieving pressure, and pressures below 32 mmHg are called "reduced pressures". Presently—in terms of the value of pressures on the body, unequalled are waterbeds. Studies show that stresses occurring around the pelvis bones have the highest values that amount to 25 mmHg. For comparison, in an ordinary mattress, pressures in this point have a value of about 58 mmHg (Grbac 2006). The disadvantage of this type of solution is the need of using special disinfectants and relatively big weight. A mattress with the size of 2000 × 1600 mm can weigh even about 200 kg (Meier 1998).

During sleep, neuromuscular activity is at a low level and the major force acting on the body and partly on the spine is gravity. This force causes deformations of the soft tissue during the rest on the mattress. In the work of Normand et al. (2005), the method of measuring pressures that occur in the spine has been presented, with and without lumbar support, with the use of a pneumatic mattress in variable laboratory conditions. It was concluded that lumbar support causes more uniform support above the hips, in the lumbar and thoracic area, and causes lumbar lordosis when lying on the back (Table 3.17). Quality of support in six different experimental conditions was examined. These conditions were changing in such a way as to

**Table 3.17** The average pressure force (N) in different places of the body support (Normand et al. 2005)

| Area of support/kind of mattress | Air cushion | |
|---|---|---|
| | Without support for loins | With support for loins |
| *Loins* | | |
| Without mattress | 11.9 | 174.1 |
| Polyurethane foam | 26.7 | 41.6 |
| Spring mattress | 53.6 | 164.3 |
| *Hips* | | |
| Without mattress | 448.8 | 388.4 |
| Polyurethane foam | 79.7 | 71.7 |
| Spring mattress | 115.7 | 94.0 |
| *Breasts* | | |
| Without mattress | 106.9 | 60.8 |
| Polyurethane foam | 43.8 | 43.3 |
| Spring mattress | 107.7 | 117.2 |

reflect the body support on a bed without mattress, with foam and with mattress as well as when supporting loins with an air cushion or not. Pressures on the body were measured by the sensory compression mat Tekscan. On the basis of the studies, it has been shown that without supporting the loins with an air cushion, the pressures focus mainly in the area of the hips. And during a gradual filling up the cushion with air, a gradual, even increase of pressures in the lumbar area and alignment of pressures around the hips and breasts occurred. Such support reduces the pressures in the lumbar area of the spine.

Park et al. (2001) showed that critical assessment of a comfortable mattress during its use should include the assessment of the shape of the spine line, the size and the distribution of pressures on the human body and the manufacture class of the mattress (including the quality of the mattress). Deformations of the spine line were studied by comparing the spatial (3D) image of the shape of the spine of the user during sleep with the 3D shape of the spine of a standing person, with the method RMS (Root Mean Square). The pressures were studied using a sensometric mat, and the comfort of the mattress was evaluated on the basis of the survey responses of the mattresses users. Studies have shown that the mattress providing shape of the spine line as similar to the shape kept in a standing position as possible is the most comfortable. The larger the differences in the shape are, the less comfortable the mattress is. Properly constructed bed should therefore support the spine in a continuous manner, without unnecessary implementation of a shared sleeping surface. The function of a uniform distribution of reaction forces of the base on the body should be maintained for various positions, and when changing the position of the body protuberances, the deformation should occur on small radii of curvature with simultaneous support of the entire body (Kapica 1993a).

In the construction of an ergonomic mattress, it is therefore important not only to properly support the body in various positions, but also zonal differentiation of stiffness of the mattresses. This involves uneven distribution of pressures generated by the base. Studies have shown that the greatest pressures occur always around the shoulders, hips, and elbows, and the smallest pressure accumulates around the knees and ankles (Buckle and Fernandes 1998). It is also confirmed by a study carried out by Defloor and Schuijmer (2000).

Currently known construction solutions of mattresses require an individual approach to the issue of selection of mattress for the user, which is associated with subjective feeling of its softness or hardness, and softness is not the only criterion of assessing the quality of upholstered furniture. Such a determinant can also be the stiffness coefficient k, clearly determining the hardness of the examined system (Smardzewski 1993a).

**Durability**

Durability of the mattress depends on the degree of exploitation of the product, which has a direct connection with the properties of the materials used and the quantity of absorbed and retained moisture. The increase in the number of loading cycles (0–5000) decreases the stiffness coefficient by about 6 %, regardless of their construction (Kapica 1993b). Although methods of improving elastic properties of biconical springs are known, which have an impact on reducing permanent deformations (Kapica and Smardzewski 1994), durability of mattresses is estimated for about 10 years, but an exchange (for hygienic reasons) is recommended after about 5 years of use. In addition, by building the so-called layered systems of foams of various compressibility, one can adjust the pliability of the upholstery layer made of polyurethane foam (the degree of regulation of the pliability of the system depends on the diversification of compressibility of the component foams, and is the larger, the bigger this diversification is), and the characteristics of deformation of the systems depend on the array, their thickness and positioning in relation to one another (Kapica and Pechacz 1983).

**Dangers**
The extensive use of foam materials as the filling of elastic layers (ousting the use of springs) causes certain dangers, related to their chemical structure. Studies of Kozłowski et al. (1988) showed that during combustion of upholstery materials (wool, silk, flax, cotton and polyurethanes), very dangerous products in terms of toxicity are emitted, e.g. carbon dioxide, carbon monoxide, cyanates, isocyanates or hydrogen cyanide (polyurethanes), and depending on the material, the temperature of the glow amounts to 185–340 °C (Navratil and Osvald 1997). It is therefore necessary to choose solutions that will to the highest possible degree minimise the risk of sustaining of glow or fire. The covering layers must be absolutely made of materials with flame-resistance clearance.

# References

Aserinsky E, Kleitman N (1953) Regulary occuring periods of eye motility, and concomitant phenomena, during sleep. Science 118:273–274
Baranowski M (2004) Projekt stanowiska pracowniczego dla konstruktora-technologa do mieszkania typu M-2, M-3. Typescript Department of Furniture Design, Agricultural Academy of Poznań
Batogowska A, Malinowski A (1997) Ergonomia dla każdego. Sorus, Poznań
Batogowska A, Słowikowski J (1989) Atlas antropometryczny dorosłej ludności Polski dla potrzeb projektowania, bulletin 14. Prace i Materiały IWP, Warsaw
Beldon P (2002) Transfoam Visco™: evaluation of a viscoelastic foam mattress. Br J Nurs 11 (9):651–655
Bennet I, Kavner D, Lee BY, Trainor FS, Lewis JM (1981) Skin blood flow in seated geriatric patiens. Arch Phys Med Rehabil 62:392–398
Beszterda S (2001) Modelowanie ergonomiczne wybranych układów mebli biurowych. Typescript Department of Furniture Design Agricultural Academy of Poznań
Branowski M, Ciesielski K, Zalewski Z (1985) Analiza wartości. Poznań, Wydawnictwo Politechniki Poznańskiej, Poznań
Buckle P, Fernandes A (1998) Mattress evaluation—assessment of contact pressure, comfort and discomfort. Appl Ergon 29(1):35–39
Corlett EN (1999) Are you sitting comfortably. Int J Ind Ergon 24:7–12
Cunningham MJ (1999) Modelling of some dwelling internal microclimates. Building and Environment 34:523–536

Defloor T (2000) The effect of position and mattress on interface pressure. Appl Nurs Res 13(1):2–11

Defloor T, De Schuijmer J (2000) Preventing pressure ulcers: an evaluation of four operating-table mattresses. Appl Nurs Res 13(3):134–141

Defloor T, Grypdonck MH (1999) Sitting posture and prevention of pressure ulcers. Appl Nurs Res 12:136–142

Destroches-Noblecourt C (1963) Tutenkhamen. New York Graphic Society, New York

DIN 68 861:1981 (1985) Möbeloberflächen, Teil 1–8

Dirken H (2001) Productergonomie: ontwerpen voor gebruikers. Delftse Universitaire Pers, Delft

Drożyńska J (1997) Komputerowe stanowisko pracy. Warszawa

Dzięgielewski S (1996) Technologia mebli tapicerowanych. Produkcja przemysłowa. Wydawnictwa Szkolne i Pedagogiczne, Warsaw

Dzięgielewski S, Smardzewski J (1995) Meblarstwo. Projekt i konstrukcja. Państwowe Wydawnictwo Rolnicze i Leśne, Poznań

Dzięgielewski S, Smardzewski J (1997) Permanent deformations of upholstered furniture. In: Zbornik prednasok z odborneho seminara Calunnicke Dni, Technika a Technologie, Katedra Nabytku a Drevarskich Vyrobkov, Drevarska Fakulta Technickej Univerzity vo Zvolene, pp 18–19

Encyklopedia (2004) Nowa encyklopedia powszechna PWN. PWN, Warsaw

Fronek K, Zweifach BW (1975) Microvascular pressure distribution in skeletal muscle and the effect of vasodilation. Am J Physiol 1:791–796

Gedliczka A (2001) Atlas miar człowieka. Dane do projektowania i oceny ergonomicznej. Centralny Instytut Ochrony Pracy, Warsaw

Gefen A, Gefen N, Linder-Ganz E (2005) In vivo muscle stiffening under bone compression promotes deep pressure sores. J Biomech Eng 127:512–524

Grandiean E (1973) Wohnphysiologie. Grundladen gesunden Wohnens. Verlag fur Architektur Artemis, Zurich

Grandjean E (1978) Ergonomia mieszkania. Arkady, Warsaw

Grabowski K, Jakitowicz J (2006) Zaburzenia snu jako problem społeczny i ekonomiczny. Sen 6 (1):41–46

Grbac I (1985a) Research into the durability and elasticity of different mattress structures. Master's thesis, Faculty of Forestry, Zagreb

Grbac I (1985b) Rationalization of the bed construction. Bilten ZIDI 13(2):143–13

Grbac I (1988) Research into the quality of mattress and its structural upgrade. Ph.D dissertation, University of Zagreb, Faculty of Forestry, Zagreb

Grbac I (2006) Krevet i zdravlje. Sveciliste u Zagrebu. Sumarski Fakultet, Zagreb

Grbac I, Dalbelo-Bašić B (1996) Comparasion of thermo-physiological properties of diferent mattress structures. In: Proceedings of the 18th Internacional Conference on Information Technology Interfaces, Pula, pp 113–118

Grbac I, Domljan D (2007) Namjestaj i zdrav zivot. Sigurnost 49(3):263–279

Grbac I, Ivelić Ž, Markovac Ž (2000) Usage and abusage of terminology for furniture assigned for sleeping. In: International conference: ecological; biological and medical furniture—fact and misconceptions, Zagreb, October 13th 2000, University of Zagreb, Faculty of forestry Croatia, UFI-Paris, pp 83–92

Guzik P (2001) Ocena adaptacji układu krążenia do zmiany kąta pochylenia w przebiegu próby pionizacji. Ph.D dissertation, Medical University, Poznań

Haak AJH, Leever-van der Burgh D (1994) De menselijke maat: een studie over de relatie tussen gebruiksmaten en menselijke afmetingen, bewegingen en handelingen. Delftse Universitaire, Delft

Hyunja L, Sejin P (2006) Quantitative effects of mattresses types (comfortable vs. uncomfortable) on sleep quality through polysomnography and skin temperature. Int J Ind Ergon 36:943–949

Janiga J (2000) Podstawy fizjologii pracy i ergonomii. Legnica

Janow W, Bielow A (1971) Chudożestwiennoje konstruirowanie mebeli. Izdatielstwo Liesnaja promyszliennost, Moscow

References                                                                           181

Jarosz E (2003) Dane antropometryczne populacji osób dorosłych wybranych krajów Unii Europejskiej i Polski dla potrzeb projektowania. Instytut Wzornictwa Przemysłowego, Warsaw

Juergens HW, Aune IW, Piepe U (2005) Internationaler Anthropometrischer Datenatlas. http://www.ilo.org. Accessed 20 Jul 2005

Kamieńska-Żyła M (1996) Ergonomia stanowiska komputerowego. Wydawnictwa AGH, Kraków

Kapica L (1988) Badania porównawcze konstrukcji zespołów tapicerowanych mebli do leżenia. Przemysł Drzewny 1:2–5

Kapica L (1991) Przesłanki ergonomiczne kształtowania konstrukcji wyrobów meblarskich. Przemysł Drzewny 10:1–3

Kapica L (1993a) Wymagania projektowo-konstrukcyjne dla mebli przeznaczonych do leżenia ze względu na fizjologię. Przemysł Drzewny 5:1–4

Kapica L (1993b) Cechy konstrukcji mebli przeznaczonych do siedzenia ze względu na fizjologię. Przemysł Drzewny 3:1–5

Kapica L (1993c) Wpływ obciążeń zmęczeniowych na miękkość zespołów tapicerowanych mebli. Przemysł Drzewny 7:22–24

Kapica L (1993d) Wpływ konstrukcji warstwy wyściełającej na odkształcenia zespołów tapicerskich mebli. Przemysł Drzewny 7:1–3

Kapica L, Pechacz J (1983) Odkształcalność układów z miękkich pianek poliuretanowych stosowanych w meblarstwie. Przemysł Drzewny 11:13–16

Kapica L, Smardzewski J (1994) Wpływ geometrii sprężyn dwustożkowych stosowanych w formatkach bonnell na ich sztywność. Przemysł Drzewny 10:1–3

Kapica L, Smardzewski J, Grbac I (1999) Deformation of multiple-Layer sponges at spring constructions. In: Proceedings of International Conference Ambienta'99, University of Zagreb, pp 103–110

Karpiński J, Deszczyński J (1997) Aspekty biomechaniczne konstrukcji nowoczesnego materaca ortopedycznego. Magazyn Medyczny 7:50–52

Kiejna A, Wojtyniak B, Rymaszewska J, Stokwiszewski J (2003) Prevalence of insomnia in Poland—results of the national health interview survey. Acta Neuropyschiatrica 15:68–73

Konarska M (2001) Ergonomia pracy biurowej. CIOP, Warsaw

Korolew W (1973) Osnovy racionalnovo proiektirovania miebeli. Lesnaja Promyszliennost, Moscow

Kozłowski R, Wesołek D, Muzyczek M (1988) Toksyczne produkty rozkładu i spalania materiałów wyposażeniowych wnętrz mieszkalnych. In: Konferencja naukowo-techniczna Higieniczność materiałów stosowanych do produkcji mebli i wyposażenia wnętrz mieszkalnych, Poznań 11, pp 174–187

Krauchi K (2007) The human sleep—wake cycle reconsidered from a thermoregulatory point of view. Physiol Behav 90:236–245

Kroemer K, Kroemer H, Kroemer-Elbert K (1994) Ergonomics. How to design for ease and efficiency. Prentice Hall, New York

Krutul R (2004a) Odleżyny. Pielęgnacja w warunkach szpitalnych. Ogólnopolski Przegląd Medyczny 4:38–42

Krutul R (2004b) Odleżyna, profilaktyka i terapia. Revita

Krzoska-Adamczak Z (1996) Właściwości płyt MDF ze szczególnym uwzględnieniem ich higieniczności. In: Nowe technologie, obrabiarki, urządzenia, materiały i akcesoria dla meblarstwa. Wydawnictwo Instytutu Technologii Drewna, Poznań

Krzoska-Adamczak Z (2001) Materiały do uszlachetniania powierzchni mebli kuchennych. Przemysł Drzewny LII 9:10–13

Krzoska-Adamczak Z, Nowaczyk M (2005) Europejski projekt badawczy dotyczący metod oceny odporności powierzchni mebli. Przemysł Drzewny LIII 3:21–25

Lee H, Park S (2006) Quantitative effects of mattress types (comfortable vs. uncomfortable) on sleep quality through polysomnography and skin temperature. Int J Ind Ergon 36:943–949

Łajczak SZ (2004) Opracowanie parametrycznego modelu mebli kuchennych wraz z generatorem list materiałowych i kreatorem ceny łącznej produktu. Typescript Department of Furniture Design Agricultural Academy of Poznań

Łopaczewska K (2003) Wymagania i zalecenia ergonomiczne w procesie opracowania produktu. Instytut Wzornictwa Przemysłowego, Warsaw

Łęcka J, Morze Z, Lewandowski O (1995) Emisja formaldehydu z płytowych elementów meblowych. In: Jakość w meblarstwie oraz stan badań i normalizacji. SITLiD, Poznań

McCormic E (1957) Antropotechnika, przystosowanie konstrukcji maszyn i urządzeń do człowieka. WNT, Warsaw

Meier M (1998) Układy tapicerskie w meblach do spania i siedzenia. In: Konferencja Postęp w produkcji mebli tapicerowanych dla mieszkań i obiektów publicznych (hoteli, sal kinowych i teatralnych, szpitali, itd.), Poznań, Mai, pp 13–17

Molenbroek JFM (1994) Op maat gemaakt: menselijke maten voor het ontwerpen en beoordelen van gebruiksgoederen. Delftse Universitaire, Delft

Nachemson AL (1976) Towards a better understanding of low-back pain. A review of the mechanics of the lumbar disc. Rheumat Rehabil 14(75):129–143

Navratil W, Osvald A (1997) Priemyselna calunnicka vyroba a jej poziarne rizika. In: International Conference Calunnickie dni. Techika a technologie. Zvolen, Juni 18–19, pp 50–58

Niejmah AF, Smirnov SS (1984) Miebel dla administrativnykh zdanij. Liesnaja Promyshliennost, Moscow

Normand MC, Descarreaux M, Poulin C, Richer N, Mailhot D, Black P, Dugas C (2005) Biomechanical effects of a lumbar support in a mattress. J Can Chiropr Assoc 49(2):96–101

Nowak E (1993) Antropometria na potrzeby projektowania. Instytut Wzornictwa Przemysłowego, Warsaw

Nowicki Z (2002) Uwagi ogólne dotyczące problematyki snu. Sen 2(suppl. A):A1–A6

Okamoto K, Nakabayashi K, Mizuno K, Okudaira N (1998) Effects of truss mattress upon sleep and bed climate. Appl Human Sci 17(6):233–237

Park SJ, Lee HJ, Hong KH, Kim JT (2001) Evaluation of mattress for the Koreans. In: 45th Annual meeting of the human factors and ergonomics society, pp 727–730

Pakarinen TJ, Asikainen AT (2001) Consumer segments for wooden household furniture, Holz als Roh u. Werkstoff 59:217–227

Panero J, Zelnik M (1979) Human dimension and interior space: a source book of design reference standards. Architectural Press, London

Pheasant S (2001) Bodyspace: anthropometry, ergonomics and the design of work. Taylor & Francis, London

Pietruczuk K, Jakuszkowiak K, Nowicki Z, Witkowski JM (2003) Cytokiny w regulacji snu i jego zaburzeniach. Sen 3(4):127–133

PN-88/F-06010.02 School furniture. Pupil tables: laboratory tables for drafting and drawing, for learning foreign languages. Basic functional dimensions

PN-88/F-06010.03 School furniture. Demonstrative tables and stands for technical audio-visual teaching aids. Basic functional dimensions

PN-89/F-06027.04 Furniture. Furniture for reclining. Basic functional dimensions

PN-90/F-06009 School and kindergarten furniture. Requirements and testing

PN-90/F-06010.01 School and kindergarten furniture. Basic functional dimensions. General provisions

PN-90/F-06010.05 School and kindergarten furniture. Furniture for storage

PN-F-06001-2:1994 Home furniture. Quality classification

PN-EN 438-1,2:1997 High-pressure decorative laminates (HPL). Boards from thermosetting resins. 1—Requirements, 2—Determining properties

PN-EN 527-3:2004 Office furniture. Work tables and desks. Part 3: Methods of determining stability and strength of mechanical construction

PN-EN 527-1:2004 Office furniture. Work tables and desks. Part 1: Dimensions

PN-EN 1116:2005 Kitchen furniture. Co-ordinating sizes for kitchen furniture and kitchen appliances

PN-EN 1335-1:2004 Office furniture. Office chair for work. Part 1: Dimensions - determining dimensions

PN-EN 1335-2:2009 Office furniture. Office chair for work. Part 2: Safety requirements

PN-EN 1335-3:2009 Office furniture. Office chair for work. Part 3: Methods of testing safety

PN-EN 14749:2007 Domestic and kitchen storage units and worktops. Safety requirements and test methods

PN-EN 16139:2013-07 Furniture. Strength, durability and safety. Requirements for non-domestic seating

PN-EN 1729-1:2007 Furniture. Chairs and tables for educational institutions. Part 1: Functional dimensions

PN-EN 1729-2:2007 Furniture. Chairs and tables for educational institutions. Part 2: Safety requirements and methods of testing

PN-ISO 5970:1994 School and kindergarten furniture. Tables and chairs. Basic functional dimensions

PN-EN ISO 7250-1:2010 Basic human body measurements for technological design. Part 1: Body measurement definitions and landmarks

Pracka D, Pracki T, Nadolska M, Ciesielczyk K, Ziółkowska-Kochan M, Tafil-Klawe M, Jakitowicz J (2003) Epidemiologiczna ocena zmian jakości snu w wybranych grupach społecznych i wiekowych. Sen 3(4):139–144

Proszyk St, Bernaczyk Z (1994) Technologie postformingowego uszlachetniania powierzchni elementów płytowych laminatami HPL. In: Nowe technologie w zakresie uszlachetniania powierzchni materiałów płytowych i wyrobów z drewna. Warsaw, VIII Konferencja Naukowa SGGW

Radliński A (1996) Problemy usuwania odpadów lakierniczych w przemyśle meblarskim. Poznań, Nowe technologie, obrabiarki, urządzenia, materiały i akcesoria dla meblarstwa, Wydawnictwo Instytutu Technologii Drewna, Poznań

Scheithauer M, Aehlig K (1996) Lotne organiczne składniki (VOC) oraz zapach powierzchni lakierowanych—kryterium oceny cech ekologicznych nowoczesnych mebli. In: Wymagania Unii Europejskiej a stan higieniczności mebli i materiałów stosowanych do ich produkcji, Wydawnictwo Instytutu Technologii Drewna, Poznań

Shmelev IP (1993) Phenomenon of the Ancient Egypt. Wydawnictwo Lotaz, Minsk

Sienicki S (1954) Historia architektury wnętrz mieszkalnych. Budownictwo i Architektura, Warsaw

Slavikova M (1988) Zakladni rozmery pro kresla a pohovky urcene k sedeni. Drevo 10:286–290

Słowikowski J (2000) Metodologiczne problemy projektowania ergonomicznego w budowie maszyn. CIOP, Warsaw

Smardzewski J (1993a) Sztywność dwustożkowych sprężyn tapicerskich. Przemysł Drzewny 2:6–9

Smardzewski J (1993b) Model ergonomiczny formatki sprężynowej. Przemysł Drzewny 11:6–8

Smardzewski J (2002a) Strength of profile-adhesive joints. Wood Sci Technol 36:173–183

Smardzewski J (2002b) Technological heterogenity of adhesive bonds in wood joints. Wood Sci Technol 36:213–227

Smardzewski J (2002c) Jak optymalizować konstrukcję mebla tapicerowanego. Meble materiały akcesoria 9(30):40–42

Smardzewski J (2002d) Parametryczne modelowanie mebli kuchennych w środowisku AutoCAD. Roczniki Akademii Rolniczej w Poznaniu, CCCXLVI, Technologia Drewna 36:13–19

Smardzewski J (2004a) MebelCAD, Parametryczna nakładka dla systemu AutoCAD Meble materiały akcesoria 12(55):40–42

Smardzewski J (2004b) Stereomechanika połączeń mimośrodowych. Modelowanie półsztywnych węzłów konstrukcyjnych mebli. In: Branowski B, Pohl P (ed) Wydawnictwo Akademii Rolniczej im. A. Cieszkowskiego w Poznaniu

Smardzewski J (2004c) Modeling of semi–rigid joints of the confirmate type. Ann Warsaw Agric Univ Wood Technol 55:486–490

Smardzewski J, Beszterda S (2001) Ergonomic designing of office furniture in CAD systems. Electron J Polish Agric Univ Wood Technol 4(2)

Smardzewski J, Grbac I (1998) Numerical analysis of ergonomic function of upholstered furniture. In: Medunarodno Zanstveno Svjetovanje Ambienta'98, University of Zabreb, pp 61-68

Smardzewski J, Matwiej Ł (2004) Bezpieczeństwo użytkowania mebli tapicerowanych. Przemysł Drzewny 4:36–38

Smardzewski J, Rogoziński M (2001a) Problematyka badań wytrzymałościowych mebli o złożonej konstrukcji. Przemysł Drzewny 2:11–16

Smardzewski J, Rogoziński M (2001b) Jakość mebli biurowych w świetle wymagań norm europejskich. Przemysł Drzewny 6:3–9

Smardzewski J, Kabała A, Matwiej Ł, Wiaderek K, Idzikowska W, Papież D (2008) Antropotechniczne projektowanie mebli do leżenia i siedzenia. In: Raport końcowy projektu badawczego MNiSzW nr 2 PO6L 013 30, umowa nr 0998/P01/2006/30

Smardzewski J, Wiaderek K, Grbac I (2006) Numerical analysis of contact problems of human body and elastic mattress. In: Medunarodno Zanstveno Svjetovanje Ambienta'06, University of Zabreb, pp 81–86

Snijders CJ, Nordin M, Frankel VH (1995) Biomechanica van het spier-skeletstelsel. Lemma BV, Utrecht

Someren E (2004) Sleep propensity is modulated by circadian and behavior-induced changes in cutaneous temperature. J Therm Biol 29:437–444

Stępowski M (1973) Siedziska. Zasady prawidłowego doboru i użytkowania. Instytut Wydawniczy CRZZ, Warsaw

Stinson MD (2003) Seat-interface pressure: a study of the relationship to Gender, body mass index, and seating position. Arch Phys Med Rehabil 84(3):405–409

Stockert A (1989) Kostensenkung, eine bereichsubergreifende Aufgabe. VDI-Berichte 767

Svennberg K, Wadsö L (2005) Measurements of microclimates in beds in relation to the climatic requirements of house dust mites. www.byv.kth.se. Accessed 08 Jan 2008

Troussier B, Tesniere C, Fauconnier J, Grison J, Juvin R, Phelip X (1999) Comparative study of two different kinds of school furniture among children. Ergonomics 42:516–526

Tytyk E (2001) Projektowanie ergonomiczne. PWN, Warsaw, Poznań

Winkler T (2005) Komputerowo wspomagane projektowanie systemów antropotechnicznych. WNT, Warsaw

# Chapter 4
# Introduction to Engineering Design of Furniture

## 4.1 General Information

Each year, at furniture fairs that take place periodically, furniture factories present their latest models based on modern technology, construction and an attractive aesthetic and functional form. In many manufacturing companies from a few to over a dozen, prototypes or new designs appear during the year. In order to shorten the time as much as possible from the idea, to the preliminary draft to the model, and then testing technology and production, to placing the product on the market (Fig. 4.1), it is necessary to integrate computer-aided systems: of designing, manufacturing and management into one comprehensive system dedicated to the dynamic modelling of manufacturing processes (Smardzewski 2007). By reducing the time of preparing a new product, we extend the sales period and maximise the profit. In the process of developing a new product, in order to significantly reduce the costs of the project, IT support is needed to make furniture design and construction easier and simpler.

In technical literature, as well as in practice of furniture manufacturers, designing and construction are clearly distinguished, although in many circles these concepts are treated as synonymous. There are also many definitions of designing and construction provided by various authors. Tytyk (2006) after Bąbiński (1972), Gasparski (1978), Kotarbiński (1990), Krick (1971), Nadler (1988) and Tarnowski (1997) provides various meanings of the term designing, which he defines as follows:

- a manufacturing process, in which the ownership of the item, with the appropriate use of available methods and tools in certain conditions aim to include in the product of designing a model of the designed object of characteristics corresponding to the requirements specified in the design task (Bąbiński 1972),
- the adjustment of the way in which any system operates (Dietrych 1974),
- conceptual preparation of the relevant change (Gasparski 1978),
- the most important work of practical sciences (Kotarbiński 1990),

© Springer International Publishing Switzerland 2015
J. Smardzewski, *Furniture Design*,
DOI 10.1007/978-3-319-19533-9_4

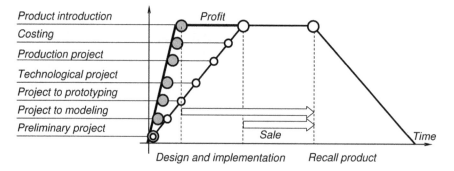

**Fig. 4.1** The profit flowing from shortening the development cycle of new furniture

- subsequent actions, starting with the identification of the problem, and ending at developing a satisfactory solution in terms of function, economy, etc., (Krick 1971),
- the creation of a system adequate to meet a certain need or fulfil a certain function (Nadler 1988) and
- a system of actions meeting needs, which aim to clearly define the product's construction and use based on the identified project task.

Taking into consideration these definitions, Jabłoński (2006) writes that the process of designing a product in the technical aspect consists in clearly determining such construction and usage requirements, which will guarantee the fulfilment of specific needs of the object by shaping the relevant characteristics.

Meanwhile, construction is defined as follows:

- choosing construction features which are the logical basis for identifying the construction (Dietrych 1974);
- part of the technical design process, the immediate goal of which is to detail the form of component parts of the designed object and the value of construction features (Tarnowski 1997).

Therefore, designing and constructing furniture consists in shaping the requirements of the anthropotechnical system, the integration of which during the shaping of the product's form and its component parts, will meet the projected or created needs of the user.


## 4.2  Methods of Furniture Design and Construction

The mutual interaction of all known and unknown design parameters requires taking into account the cooperation of competent partners during designing. The experience of many national and foreign concerns and companies manufacturing furniture shows that modern furniture constructions are created by multidisciplinary teams, which, in the course of work, solve any problems associated with placing a new product on the consumer market. High-quality furniture cannot be the result of conservative designing, based only on the work of an artistically talented person, but the result of creative marketing, design, cultural and technological discoveries emerging in interdisciplinary teams appointed for the time of preparing a new product.

An interdisciplinary design team should include the participation of a designer, a construction engineer, a technology engineer, a trader and a marketing specialist. Working in such a group has an interactive nature and is mostly assisted by CAx systems and thematic knowledge databases (Fig. 4.2).

Acronyms used in CAx technologies:

| CAD | Computer-aided design, |
| CAE | Computer-aided engineering, |
| CAM | Computer-aided manufacturing, |
| CAQ | Computer-aided quality control, |
| CE | Concurrent engineering, |
| CIM | Computer-integrated manufacturing, |
| CSG | Constructive solid geometry, |
| CRM | Customer relationship management, |
| DFM | Design for manufacturing, |
| EDM | Engineering data management, |
| ERP | Enterprise resource planning, |
| FEM | Finite elements method, |

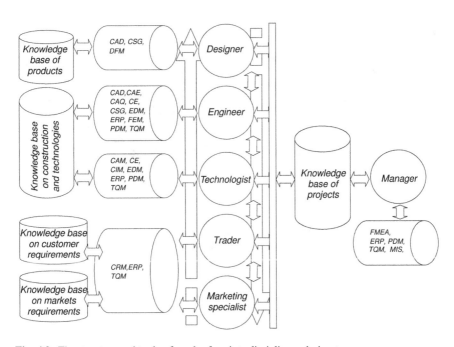

**Fig. 4.2** The structure and tools of work of an interdisciplinary design team

FMEA   Failure modes and effects analysis,
MIS    Management information system,
PDM    Product data management and
TQM    Total quality management.

The design space in such a team expands greatly and is conducive to shifting the project's centre of gravity, gradually, from aesthetic priorities, through construction and functional priorities, to the marketing visions, aiming to confront them in order to verify again the previous parameters and obtain the best market effect (Fig. 4.3).

The structural creation of innovative furniture systems based on anthropometric design process (ADP) should have well-defined phases and methods (Fig. 4.4).

A. Purpose and place of use of the furniture: hotel, cinema, restaurant, school, office, post office, bank, apartment, living room, bedroom, dining room, kitchen, bathroom, sitting, sleeping, relaxing, eating meals, learning, work, customer service and storage;
B. Object of design: wardrobe, bookcase, dresser, chest of drawers, glass case, bar, cabinet, dining table, table, desk, bedside table, attachment, chair, armchair, bed, couch and sofa;
C. Group of users: Europeans, Dutch, Poles, Americans, Chinese, children, teenagers and adults;
D. Characteristics of the population: adults between 30 to 50 years old and school children from 14 to 18 years old;

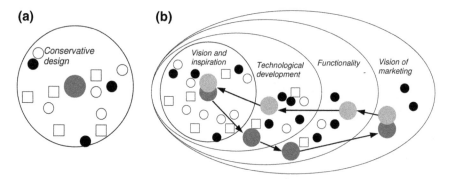

●  Problems identified and resolved,
○  Problems identified and unresolved,
□  Unidentified and unsolved problems,
●  The original location of the center of gravity,
●  Secondary location of the center of gravity.

**Fig. 4.3** The development of the design space and shifting the project's centre of gravity: **a** conservative designing without the ability to quickly identify and eliminate design problems, **b** multidisciplinary designing for rapid analysis of the occurrence of problems and their solution, *filled circle* Problems identified and resolved, *circle* Problems identified and unresolved, *square* Unidentified and unsolved problems, *dark grey circle* The original location of the centre of gravity, *light grey circle* Secondary location of the centre of gravity

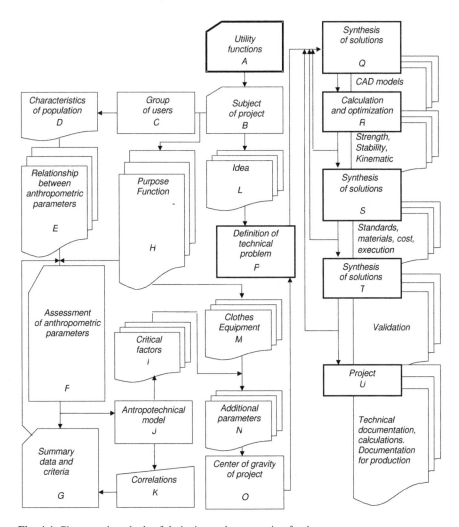

**Fig. 4.4** Phases and methods of designing and constructing furniture

   E. Relationship between anthropometric parameters: reference of the dimensions
      of the designed furniture to the appropriate anthropometric dimensions;

   F. Evaluation of anthropometric criteria: representativeness—characteristics of the
      population should correspond to anthropometric data, accuracy—statistical
      solutions must apply to precise calculations of the representative sample, and
      project error—evaluation of what percentage of the population should be
      excluded. It is usually accepted that from 1.5 to 10 % of the population for
      which a new construction of furniture is being designed should be excluded in
      order to avoid excessive increases in the cost of production of the future
      product;

G. A set of many types of data and criteria in order to adapt the characteristics of furniture to assumed use requirements, the precise realisation of the design, as well as creating a few sizes of the same product. This is especially important when designing school furniture, where there needs to be a compromise between the cost and anthropometric requirements;

H. Purpose, function, functionality and silhouette: resulting from user preferences and the type of clothes worn;

I. Critical factors: taking into account critical factors which are likely to have a significant impact on the health and safety of future users is knowledge flowing from the correlation between anthropometric dimensions, the type of population and function of furniture;

J. Anthropometric model of furniture;

K. Correlations;

L. Product idea: the source of inspiration and vision of the product;

M. Clothing and equipment;

N. Additional parameters: taking into account additional parameters which are likely to significantly improve the function and equipment of furniture;

O. The project's centre of gravity, leading direction: deciding on the definitive anthropometric requirements and going onto technical designing;

P. Defining the technical problems, including the disposal of the product and recycling;

Q. The synthesis of construction solution: a summary of the concept of the development of the construction, joints and dimensioning used;

R. Calculations and optimisation: carrying out calculation of strength, stiffness, stability of furniture and its elements, as well as optimisation of the results of these calculations;

S. Synthesis of technological solutions: a summary of the concept of technology development in terms of selecting the material, machine park, tools, etc.;

T. Synthesis of function and functionality: a summary of the concept of function and functionality development of furniture in terms of meeting the stated requirements; and

U. Product design.

Unlike the furniture produced using craft methods, furniture manufactured industrially should be done according to a specific, repetitive technology, in the conditions of mechanized unit production, small or large series. It imposes a number of important structural, technical, organisational and commercial requirements, unknown to crafts factories and furniture workshops. These requirements have become an essential stimulus to the overall mechanization and technical development of furniture production. They have also contributed to increased awareness of producers of the need to prepare partial or detailed project documentation. Engineering design methods, without which it is hard to imagine building, aviation and machine construction, have never been systematically or on a large scale introduced into furniture production. The implementation of these methods requires the gathering of detailed information concerning:

- the functions of furniture and maximum operational loads resulting from them, or also in extreme cases, unusual loads, especially when it concerns children's and teenager's furniture;
- elastic properties of materials used, taking into account the orthotropic characteristics of wood and wood-based materials, together with indicating maximum stresses; and
- elastic and strength properties of normalised glued and separable furniture joints and connectors, hinges and accessories carrying operational loads.

The development of the proper structural version of furniture, ensuring durable and safe use, is one of the components of the design process (Fig. 4.5). Certainly, however, designing the function of furniture has a unique practical importance as it provides the most effective possible fulfilment of the assumed functions of use. The purpose of aesthetic activities is shaping the proportions and spatial forms of furniture and the choice of surface colour, texture and drawing to the satisfaction of the most sensitive tastes of the potential user. This part of the project in many cases, however, clearly dominates the whole design works, shifting functional and durability features further away. In this situation, the functional requirements should have strong preference, sometimes at the expense of aesthetic or durability values.

In a properly planned process of manufacturing furniture, the stiffness of certain products is evaluated already at the designing and construction stage (Fig. 4.5). This helps eliminate any construction errors by setting the correct parameters for individual components, subassemblages and assemblages in accordance with the prescribed criteria of stiffness and durability. Formulating the engineering process of designing furniture in such a way enables to limit destructive tests of finished products and shorten the cycle of implementing furniture to production, limit the number of complaints in continuous production and save a significant amount of time and material.

The presented methods of designing clearly require teams of partners of an interdisciplinary education. In particular, this fosters ousting sequential methods by concurrent designing methods. Sequential designing (Fig. 4.6) limits the effectiveness of the work of constructors, with the necessity to deal with emerging and mounting technical conflicts in the design. With this approach to designing, only the first engineer enjoys the comfort of creatively using his working time to solving specific tasks. Other engineers, before they start to carry out their tasks, must become familiar with the project, eliminate the errors of predecessors and agree on the future modified solutions. The time spent on these activities is sometimes unproductive, lost, due to the incorrect selection of methods and tools assisting designing. Replacing this system with a system of concurrent designing reduces the designing time and increases efficiency in achieving the objectives of the project through access to a common knowledge database (Smardzewski 2007) (Fig. 4.7).

In concurrent designing, interdisciplinary designers participate, who can more quickly and efficiently satisfy customer requirements than is the case in traditional sequential designing. After all, often a product expected by the customer does not

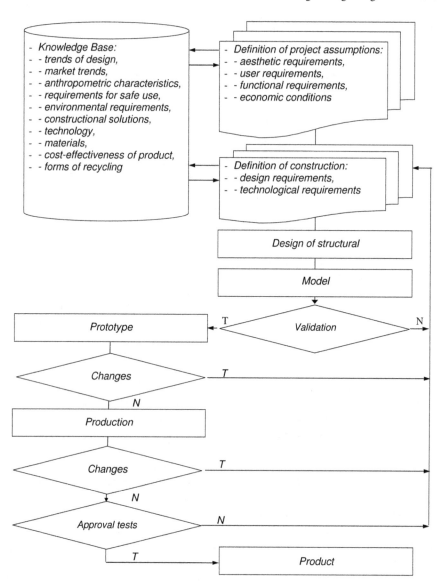

**Fig. 4.5** Constructing furniture

necessarily mean the best construction or technological values and does not always need to satisfy all the needs of the consumer, but should always optimally solve the current market expectations. Concurrent designing guarantees a high quality of designing activities because it significantly shortens the time for the emergence of a new model of furniture (mock-up, model, prototype), optimises technological processes and reduces to a minimum the time necessary to manufacture the product,

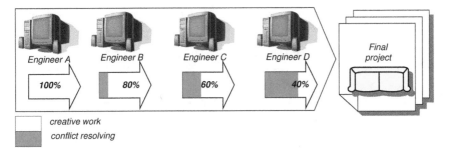

**Fig. 4.6**  Example of sequence designing

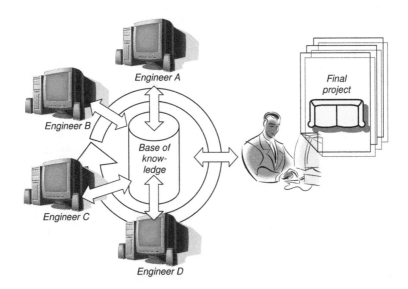

**Fig. 4.7**  Example of concurrent designing

increases customer satisfaction, informs and warns about the costs of the designed furniture at the stage of its virtual development, and minimises the costs both of the designing and construction process itself, as well as the cost of making furniture.

## 4.3   Designing and Cost Analysis

When designing furniture, costs must be taken into account and they should not surprise at the end of designing or during the validation of the model. The designer and constructor should be informed on a regular basis about the current costs of the designed product, so that they can, at appropriate times, make decisions about changing the design and/or technological concept.

In the course of designing, the constructor can be informed of the costs on the basis of:

- historical calculations of similar and already manufactured products;
- a simplified calculation based mainly on the current costs of materials and indicatory costs of labour, machinery and general costs; and
- information constituting a component of the knowledge base in the integrated IT management system.

Basing on historical calculations makes sense only if it concerns a product that is very similar to the currently designed product in terms of construction and technology. Using coefficients of dimension proportions enables to estimate the cost of a new product. However, it should be remembered that it is a calculation that is very approximate and sales price lists should not be drawn up on this basis.

A simplified calculation serves as an approximate estimation of the product's costs. Although it is inaccurate, it is usually enough for small- and medium-sized manufacturers to prepare a sales offer. The reference basis for this calculation is the cost of the material contained in the designed furniture. On the basis of an overall volume, area, length, weight and quantity of components used in the model of furniture, material costs are quite clearly specified. Labour costs, the cost of machinery, as well as departmental and overhead costs are specified by coefficients with respect to material costs. However, this element of the calculation is extremely unreliable, as material costs are not always repeatedly correlated with other costs. Usually such a cost estimation, fanned by the desire to maximise own profit and the uncertainty of the quality of calculations, leads to inflating the postulated cost of manufacturing a new product. In the case of rivalry of manufacturers in tender proceedings, the bidder is not certain as to the real level of profitability and balances on high-risk threshold of the venture.

The calculation of postulated costs, based on the product's bill of materials (BOM), is the closest to the real costs. Such a structure can be obtained automatically by designing a product in any CAD system for block modelling. Using the materials register of the integrated IT management system (ERP), for building indexes of a product's design construction in CAD environment, then the technological processes in ERP, one can obtain a reliable calculation of the designed product. Information about the prices of materials is included in the purchase price lists of raw materials, and the prices of semi-finished products are usually stored in Technical production cost (TPC) price lists. The cost of operation and costs of machinery are located in the technological processes, which constitute a component of the knowledge database in the manufacturing module of the integrated IT management system. The accuracy of calculating postulated costs depends, of course, on the phase of the development of the project (Fig. 4.8).

In initial projects, when the precise structure and technology of manufacturing the product is not known, the accuracy of the cost estimation does not exceed 55 % of the real costs. For modelling projects, in principle, one can precisely specify the material costs and estimates of remaining costs, which is why matching real costs reaches the level of 75 %. Only after manufacturing the product in production

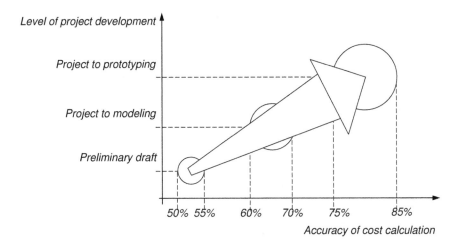

**Fig. 4.8** The accuracy of calculations depending on the degree of development of the project

version and checking the technological features in industrial conditions brings accuracy of estimation closer to the level of 85 %, and even 88 % of real costs.

One of the strengths of concurrent designing is the strong focus on costs. The linking of designers, engineers with one integrated CIM system provides concurrency of actions, the consistency of choices (in terms of both technology and costs), minimising the amount of fixes and a reduction of real costs.

Keeping a full calculation of costs while designing or manufacturing aims to create a database of information necessary for:

- determining the price of the final product and the prices of semi-finished products,
- developing the company's pricing policy,
- quoting products configured at the request of the customer,
- determining and controlling the costs of manufactured products and
- performing a comprehensive analysis of financial results.

Based on data from the production module ERP, and stored in material specifications, technological processes and production sites, both planned costs and real costs of realising production orders should be determined. As a result of breaking up all of the levels of the BOM structure, material components necessary for production are calculated, as well as the waste resulting from this production. If all the levels of the technological processes for each operation are developed, the following costs are calculated: production site, actual units of the product made, including use and non-use deficiencies, as well as actually realised times. Finally, for the production sites, based on user-defined formulas, the costs of their use are calculated.

In the furniture designing and manufacturing process, the following are most commonly calculated:

- The initial cost, carried out at any stage of the product's development, in order to determine the initial, estimated assessment of costs of the final product and/or semi-finished products, even before the product is approved by the customer and sent to production. The manufacturing cost saved as a result of this calculation should be compared with other such calculations in order to find the most beneficial solution of the design and method of manufacturing.
- The basic (standard) cost realised and recorded as valid at the stage of manufacturing any product. The basic cost established as a result of this calculation will provide the basis for comparisons with real costs.
- The real cost is performed on the basis of the data recorded in completed production orders. The recorded real cost of the order compared to the basic cost enables to determine the difference of costs, and in the future modifications of the material specification, the technological processes or costs of production sites.

In the system of cost record, the following basic classification sections are taken into account (Gabrusewicz et al. 2002):

- type-based system, which includes the costs of material consumption, remuneration, machinery depreciation, etc.—enables the evaluation of the internal structure of costs;
- function-based system, according to place of origin, including the costs of basic activity, departmental costs, sales costs, etc.—enables to control the development of costs at the place of their origin; and
- calculation-based system, which includes direct and indirect costs, according to cost carriers, which leads to the arithmetic determination of the own unit cost of objects or provided services.

Type-based system reflects costs concerning only basic operational activity. Basically, these are all costs incurred during a given period, regardless of whether they relate to periods prior to the reporting period, reporting periods or future periods.

Function- and calculation-based systems reflect complex costs, i.e. consisting of certain positions of simple costs per carrier and centre of liability and concerns basically just a given reporting period.

Calculation-based system of costs contains extended information about costs, and it indicates the relationship of costs incurred with the objective (product, service). Here, we distinguish (Table 4.1):

- direct costs—referred directly to a given product. These are as follows:
  - costs of direct materials,
  - costs of fuels and energy,
  - costs of direct remunerations together with surcharges and
  - other costs connected with preparing the necessary tools, the necessity of outsourcing, etc.,

**Table 4.1** Structure of own costs

| Own cost of the manufacturer | | | |
|---|---|---|---|
| Production cost | | | Non-production cost |
| Direct cost | Indirect cost | | |
| Cost of direct materials, direct labour cost and other direct costs | Overhand cost | Departmental cost | |
| | | Cost of operating machines and equipment | General cost of department |

- indirect costs—settled in a contractual manner. These include:
  - departmental costs—costs of the organisation and production support in particular departments,
  - overhead costs—costs of the general organisation and support of the company and maintaining its board and
  - sales costs—costs of packaging.

Calculation-based system of costs also lists losses in deficiencies, warranty costs, financial operations, etc. The cost analysis in calculation-based system of costs means determining the dynamics of specific positions on a bill and evaluating the structure of the system and its changes over time. The primary task of the calculation-based system of costs is to determine the amount and unit structure of own costs (Gabrusewicz et al. 2002; Karpiński 2004; Nowak et al. 2005).

Own cost constitutes the sum of production costs and non-production costs:

$$K_o = K_{wy} + K_{np} \tag{4.1}$$

where
$K_o$ own cost of the manufacturer,
$K_{wy}$ production cost and
$K_{np}$ non-production cost.

The cost of production consists of direct costs and indirect costs:

$$K_{wy} = K_b + K_p, \tag{4.2}$$

where
$K_b$ direct cost and
$K_p$ indirect cost.

Direct cost consists of sums of the costs of direct materials, direct labour and other direct costs:

$$K_b = K_{mb} + K_{rb} + K_{ib}, \qquad (4.3)$$

where

$K_{mb}$    cost of direct materials,
$K_{rb}$    direct labour cost and
$K_{ib}$    other direct costs.

Direct materials are materials collected from the warehouse or purchased in order to process and/or directly integrate with the element of furniture and/or furniture, e.g. wood, leather, fabric, chipboards, glue and lacquer. Direct labour is the documented work of workers directly involved in the production of a given product, e.g. cutting, drilling, lacquering, sanding and assembling, along with taking into account the insurance of employees. Other direct costs are those which consist of remunerations and materials directly related to the manufacture of the product, but that are not part of it, e.g. cleaning the job post, maintenance of runners, sandpaper, polishing paste and cleaning agents.

The costs of direct materials are calculated:

$$K_{mb} = \sum_{j=1}^{p} (mc + c_z)_j, \qquad (4.4)$$

where

$m$    quantity of raw material necessary to construct one item of the product,
$c$    price of material according to units of use,
$c_z$    cost of purchasing material and
$p$    number of types of materials.

Costs of direct labour are calculated:

$$K_{rb} = \sum_{i=1}^{n} \left( t_j s_g + t_{pz} s_o \right)_i q_{soc}, \qquad (4.5)$$

where

$t_j$    unit time of performing operation [h],
$t_{pz}$    preparation, completion time [h],
$s_g$    price rate for the operation at a given work position [\$/h],
$s_o$    price rate for preparation, completion operations [\$/h],
$q_{soc}$    coefficient of social costs and
$n$    number of operations included in the technological processes.

Other direct costs are calculated:

$$K_{ib} = \sum_{i=1}^{n} \left(t_j s_g q_{soc}\right)_i + \sum_{j=1}^{p} \left(m_{ib}c + c_z\right)_j,$$ (4.6)

where
$m_{ib}$ quantity of raw material not being part of the product unit.

Indirect costs are created by costs not related directly with the manufactured product, but with the sole fact of its production. These costs include departmental costs and overhead costs:

$$K_p = K_w + K_{oz},$$ (4.7)

where
$K_w$ departmental cost and
$K_{oz}$ overhead cost.

Departmental costs are generated by the department or division in relation to the realised production, and mainly result from the operational costs of machines and devices, as well as general costs of the division:

$$K_w = K_{rm} + K_{ow},$$ (4.8)

where
$K_{rm}$ cost of operating machines and equipment and
$K_{ow}$ general cost of department.

The operating costs of machines and equipment consist of the costs of fuel and technological fluids, electricity used for production and lighting, heating, maintenance and repair of machines and equipment, sharpening and repair of tools, operation of machines and equipment, as well as their depreciation. The overhead costs of the department include remunerations and insurance of the administration staff of a given department, the maintenance and repairs of buildings, cleaning supplies, departmental transport, cost of downtime, maintenance of laboratories and workshops and cost of using perishable items.

The cost of operation of the machines and equipment are calculated:

$$K_{rm} = K_{pm} + K_{pn} + K_{pst} + K_{rem} + K_{am},$$ (4.9)

where
$K_{pm}$ cost of operation of the machine,
$K_{pn}$ cost of operation of the tool,
$K_{pst}$ cost of operation of the technological means,

$K_{rem}$   cost of the machine's renovation and
$K_{am}$   cost of depreciation,

whereby

$$K_{pm} = \frac{Pq_p}{t_r}\left(t_{pms}s_s + \left(t_r - t_{pms}\right)s_j\right)t_m, \tag{4.10}$$

where
$P$      total nominal power of engines in the machine [kW],
$q_p$     coefficient of using nominal power of engines,
$t_r$     annual fund of machine's operation time [h],
$t_{pms}$  annual operation time of machine in peak hours [h],
$t_m$     unit time for machine performing operation [h],
$s_s$     price rate for 1 kWh of energy used in peak hours [$/h] and
$s_j$     price rate for 1 kWh of energy used outside peak hours [$/h].

$$K_{pn} = \frac{K_{pn} + n_o K_o}{t_{tr}(1 + n_o)}t_m, \tag{4.11}$$

where
$K_{pn}$   initial book value of tool [$],
$K_o$     cost of one sharpening of tool [$],
$t_{tr}$   initial duration of work of tool [h] and
$n_o$     number of tool sharpenings.

$$K_{pst} = \frac{K_{st}}{t_{st}}t_m, \tag{4.12}$$

where
$K_{st}$   initial book value of technological means [$] and
$t_{st}$   expected duration of work of technological means [h].

$$K_{rem} = K_{rpl} + K_{nb} + K_{ko}, \tag{4.13}$$

where
$K_{rpl}$  cost of planned renovations [$],
$K_{nb}$   cost of current repairs and inspections [$] and
$K_{ko}$   conservation cost [$].

$$K_{am} = \frac{K_{na}}{n_{am}}, \tag{4.14}$$

where
$K_{na}$   cost of purchasing fixed asset, machine [$] and
$a_{am}$   number of years.

Overhead costs are all costs related to the functioning of the factory as a whole. This includes costs of the management board, costs of overhead transport and the maintenance costs of sites occupied by the factory. Usually, they are calculated as follows:

$$K_{oz} = q_{oz} \sum_{i=1}^{n} \left( K_{rb} + K_{wy} \right)_{i}, \qquad (4.15)$$

where

$q_{oz}$    coefficient of surcharge of overhead costs,
$n$      number of divisions,
$K_{rb}$    direct labour cost and
$K_{wy}$    production cost.

In industrial practice, the responsibility of constructors and/or technologists is keeping and recording calculations of the technical production cost (TPC) for products both during the process of creating and modifying the material specifications (BOMs). Based on the postulated values of TPC, recorded for specific products and semi-finished products, planners of production should make comparisons with real technical costs of production calculated by them.

## 4.4 Errors in Furniture Design

The practice of the Laboratory of Testing and Validation of Furniture at the Department of Furniture Design of the Poznań University of Life Sciences shows that most furniture defects that are revealed during testing come from design errors. Yet there are few instances of damages to the product as a result of the quality of treatment or the quality of the machine park. In general, design errors can be divided into ones that result from the following assumptions:

- usage,
- functional,
- anthropometric,
- anthropotechnical,
- construction and
- technological.

The most common errors of usage assumptions include:

- adopting assumptions of climatic conditions of using furniture which do not comply with the real conditions of places where the product is exploited in the future,

- failure to take into account seasonality of the climate of the place of exploitation,
- the assumption of too liberal schemes of using furniture by future customers and
- the omission of requirements related to the infrastructure of the place of exploitation.

Among errors of functional nature, the following dominate:

- non-compliance of dimensions of places and surfaces for storing with the dimensions of the stored items,
- limited free access for fixed and moving pieces of the furniture, including a high degree of difficulty for single-person use of the item and
- the lack of functions resulting in an obvious way from the purpose of the furniture.

Errors relating to the anthropometric characteristics constitute a serious percentage of all errors made during the designing stage. The cause of this phenomenon lies in the routine of designers and basing on obsolete measurement data of the population for which the product is being created. During the certification tests, the following main errors were pointed out:

- the lack of variations in the dimensions of furniture not regulated according to the characteristics of the centile groups of relevant populations,
- improper dimensions of seating and worktops,
- dimensions of furniture mismatched to the ranges of the upper and lower extremities and
- collisions of dimensions of furniture forming sets.

Among the errors resulting from the person–technical object relation (anthropo-technics), the following usually appear:

- selecting the hardness of seats and mattresses routinely and not in accordance with the physiology of the human being,
- curvatures of the furniture parts directly in contact with the body of the user mismatched to the natural geometry of the body,
- methods of use of the moving pieces of the furniture not compatible with natural motor features of the body and
- wrong placement of important manipulating and usable elements.

Structural errors often come down to:

- the inaccurate selection of joints and connectors, as well as the inaccurate location of structural nodes,
- the underestimation of the global stiffness and local stiffnesses of individual parts of the furniture and
- the lack of stability of the furniture or its components.

The most commonly encountered technological errors include:

- incorrect selection of materials not taking into account: purpose, form, function and construction of the furniture,
- mismatching the finishing technology to the conditions of use and
- defects of performance resulting from simple technology, causing the reduction in use value and/or exposing the user to injury or damaging clothing.

Most often, these errors can be eliminated by preparing, reviewing and consulting with a wide group of professionals of all design assumptions.

## 4.5   Materials Used in Furniture Design

### 4.5.1   Wood

#### 4.5.1.1   General Characteristics of Wood

Wood is a material of natural origin, obtained by felling trees. Further processing of wood, cutting logs for lumber and sawing it into chocks and friezes, leads to obtaining semi-finished products necessary for the production of furniture. Due to the natural heterogeneity of wood, the presence of defects, sensitivity to changes in temperature and humidity of the environment, a tendency to warping and shrinking, the varied chemical composition and the diverse vulnerability for gluing and finishing with paint and lacquer, the selection of different species of wood when designing furniture should be particularly well thought out and based on thorough knowledge of the material. Because it is not possible to discuss in this work all the important properties of wood or criteria, which should guide the designer when selecting a specific species of wood, the author recommends that the reader become acquainted with other publications on the subject of wood, its structure and properties (Galewski and Korzeniowski 1958; Kollmann and Cote 1968; Kokociński 2002; Krzysik 1978; Požgaj et al. 1995), as well as works on gluing wood and finishing its surface (Pecina and Paprzycki 1997; Proszyk 1995; Tyszka 1987; Zenkteler 1996). However, other important aspects related to safety of using wood and its elastic properties needed when analysing the strength of furniture constructions have been presented below.

As a construction material, wood has always been and is the primary, and most popular material for furniture manufacturing. The conviction has been widely developed that it is an eco-friendly material, durable, safe and friendly for the health of the user. Unfortunately, not all species of wood have the appropriate references and not all of them should be applied in the same way. Therefore, when designing furniture, it is necessary to very carefully select particular species of wood, paying particular attention to species deriving from trees from tropical zones.

In the world, there are close to 30,000 species of woody plants, of which 5000 species can potentially be used in construction and ornaments (Kokociński 2002; Krzysik 1978). On the global market of wood, however, there are only 250 species registered, which are considered useful for industrial applications, including about 150 traditionally used in the manufacture of furniture.

The wood of various tree species recommended for use in furniture has significantly varying characteristics of mechanical, physical and chemical properties (Galewski and Korzeniowski 1958; Kokociński 2002; Krzysik 1978).

Side components of wood, so-called unstructured, saturating in varying intensity of the wood tissue or filling the porous spaces, are called physiologically active, having a significant impact on the health of workers processing the wood and users of the finished products made of wood. Physiologically active compounds, found in many species of wood, can be a cause of metabolic and functional disorders, allergic and asthmatic diseases, poisoning and irritation in the digestive tract, as well as carcinogenic effects (Kokociński and Romankow 2004). Therefore, choosing the right species of wood at the design stage of furniture should take into account not only the colour, drawing or relevant mechanical properties of the wood, but also its smell and potential harmful effect on the health of the user. The right choice of wood species constitutes a key element of the project, in which finishes using painting and lacquer coats are planned.

The main threat of wood on the skin can be caused by the juices or milky secretion of certain species of shrubs and trees, e.g. from the apocynaceae (*Apocynaceae Juss.*), euphorbiaceae (*Euphorbiaceae Juss.*), moraceae (*Moraceae Link*) and anacardiaceae (*Anacardiaceae* R.Br.), whereas the wood dust of araroba (*Andira araroba*) can cause rashes (Hausen 1981). Skin irritations in the form of burns due to contact with the same wood also occur. Wood species causing similar reactions are obeche (*Triplochiton scleroxylon*), larch (*Larix decidua*), swiss pine (*Pinus cembra* L.) and occasionally teak (*Tectona* L.f.) (Hausen 1981).

Table 4.2 provides information on selected major wood species indicated in the literature as toxic, irritant or sensitising (Hausen 1981).

### 4.5.1.2   Elasticity of Wood

As a natural raw material, wood is characterised as both anisotropic, as well as volatile in its properties in the function of space; therefore, it is referred to as a non-homogeneous material. Below are the physical theories of elasticity for an anisotropic body and understood as an equation binding the component of the stress tensor with the components of a strain tensor. The most general form of such equations in the linear theory of elasticity is the equation (Nowacki 1970):

$$\sigma_{ij} = A_{ijkl} \cdot \varepsilon_{kl}, \tag{4.16}$$

where

$\sigma_{ij}$   stress tensor,
$A_{ijkl}$   tensor of elasticity and
$\varepsilon_{kl}$   strain tensor.

**Table 4.2** Major species of wood showing toxic, irritant or sensitising properties (Hausen 1981)

| Botanical name | Trade name | Symptoms of illness | Common use | Inadvisable use |
|---|---|---|---|---|
| *Pericopsis elata* v. Meeuwen | Afrormosia, kokrodua, asamela | Skin irritation | To use inside and outside of buildings, furniture and veneers | |
| *Afzelia africana* Sm. | Afzelia, doussi, lengue, apa | Skin irritation, irritation of mucous membranes | To use inside and outside of buildings | Kitchens |
| *Gossweilerodendron balsamiferum* Harms | Agba, tola, tola branca | Irritation of mucous membranes | To use inside and outside of buildings, substitutes oak | |
| *Antiaris africana* Engl. | Antiaris, bonkonko, oro, kirundu, andoum, upas | Irritation of mucous membranes of the nose, throat and skin | Furniture and veneers | |
| *Turraeanthus africanus* Pellegr. | Avodire | Irritation of mucous membranes, nosebleeds | Luxurious finishing of building interiors, children's furniture and veneers | |
| *Distemonanthus benthamianus* Baillon | Ayan, movingui | Mild allergies | Doors, windows, office furniture | Kitchens laundry rooms |
| *Castanospermum australe* A. Cunn. | Black bean, moreton bay chestnut | Contains isoflavones | Finishing of building interiors, furniture and veneers | |
| *Fagus sylvatica* L. | Beech | Possibility of eczema caused by particulates and sawdust | Furniture, food containers, tools, parts of musical instruments | |
| *Betula papyracea* Ait | Birch | Possibility of recurring skin irritation during sanding of wood | Furniture and veneers | |
| *Acacia melanoxylon* R. Br. | Blackwood | Nosebleeds asthma, skin irritation | High-quality furniture, finishing of building interiors, musical instruments | |
| *Guibourtia tessmannii* J. Leonard | Bubinga, kevazingo | Skin irritation | High-quality furniture, finishing of building interiors, floors | |

(continued)

**Table 4.2** (continued)

| Botanical name | Trade name | Symptoms of illness | Common use | Inadvisable use |
|---|---|---|---|---|
| *Machaerium scleroxylon* Tul. | Caviuna vermelha, pao ferro, moradillo, jacaranda pardo, santos palisander | Possible cases of skin irritation | High-quality furniture, finishing of building interiors | |
| *Cedrela odorata* L. | Cedar, cedro | Skin irritation | Executive furniture, interior design | |
| *Castanea sativa* Mill. | Chestnut, Spanish chestnut | Possible cases of skin irritation | Furniture, kitchen furniture, veneers | |
| *Brya ebenus* DC. | Cocus, Jamaican ebony, | Skin irritation | Furniture, veneers | |
| *Cordia millenii* Baker | Cordia, canalete, freijo | Skin irritation | Furniture, finishing of building interiors | |
| *Hymenaea courbaril* L. | Courbaril, locust | Skin irritation | Furniture | |
| *Pseudotsuga menziesii* (Mirb.) Franco | Douglas fir, oregon pine, douglasie | Skin irritation and dermatitis, eczema | Furniture, veneers | |
| *Entandrophragma angolense* DC. | Gedu nohor, tiama, edinam, kalungi | Skin irritation | Furniture | |
| *Guarea thompsonii* Sprague & Hutch. | Guarea, bossé, obobo, | Irritation of mucous membranes of the nose, throat and skin | Furniture, finishing of building interiors | |
| *Liquidambar styraciflua* L. | Gum, American sweetgum, red gum, bilsted, amberbaum | Skin irritation | Furniture, interior design | |
| *Terminalia ivorensis* A. Chev. | Idigbo, framire, emeri, black afara | Skin irritation | Furniture, finishing of building interiors | Kitchen furniture |

(continued)

**Table 4.2** (continued)

| Botanical name | Trade name | Symptoms of illness | Common use | Inadvisable use |
|---|---|---|---|---|
| *Chlorophora excelsa* Bentham and Hooker | Iroko, kambala, mwule, odum | Skin irritation | Substitutes teak | |
| *Larix decidua* Miller | Larch | Skin irritation | Furniture | |
| Shorea | Lauan, red | Skin irritation | Furniture, interior design | |
| *Terminalia superba* Engler and Diels | Limba, afara, korina | Skin irritation, nosebleeds | Chairs, interior design | |
| *Diospyros celebica* Bakh. | Makassar Ebony, coromandel | Skin irritation and dermatitis, eczema | Luxurious furniture, elements of musical instruments | |
| *Khaya grandifoliola* DC. | African mahogany khaya, krala | Skin irritation | Furniture, executive offices, surfaces of worktops of office furniture | |
| *Swietenia macrophylla* King | American mahogany, tabasco, caoba | Skin irritation | Furniture, executive offices, interior design | |
| *Tieghemella heckelii* Pierre ex Chev. | Makore, baku | Irritation of mucous membranes of the nose and upper respiratory tract | Furniture, veneers, high-quality interior design, doors | |
| *Mansonia altissima* A. Chev. | Mansonia, bété | Skin irritation, cough, nosebleeds, headaches | Telecommunications engineering, furnishings of building interiors | |
| *Prosopis juliflora* DC. | Mesquite | Skin irritation | To use inside and outside of buildings, furniture and interior design | |
| *Pterocarpus angolensis* DC. | Muninga, kejaat | May cause allergies | Furniture, veneers, high-quality interior design | |
| *Triplochiton scleroxylon* K. Schum. | Obeche, samba, wawa, abachi | Asthma, skin irritation and rash | Veneers, interior design | |

(continued)

**Table 4.2** (continued)

| Botanical name | Trade name | Symptoms of illness | Common use | Inadvisable use |
|---|---|---|---|---|
| *Nerium oleander* L. | Oleander, laurier rose | Poisonings have toxic properties | Haberdashery | Kitchen, contact with food |
| *Olea europaea* L. | Olive wood | Severe skin irritation and paralysis | Interior design, jewelery, for turning | Direct contact with body |
| *Aspidosperma peroba* Fr. All | Peroba rosa | Skin irritation. Particulates and sawdust cause irritation of mucous membranes of the nose, irritation of larynx and eyes, weakness, drowsiness, sweating, fainting | Outdoor furniture, hand tools | |
| *Pinus radiata* D. Don | Radiata pine, Monterey pine | Skin irritation caused by the presence of resin and turpentine | Furniture | |
| *Pinus silvestris* L. | Pine | Skin irritation | Furniture, stairs and many more | |
| *Gonystylus bancanus* Baillon | Ramin, malawis | Skin irritation | Furniture and interior design | |
| *Dalbergia nigra* All | Brazilian rosewood, Jacarandá, Rio palisandre | Eczema of hands and face, skin irritation | Furniture, interior design, veneers, musical instruments | |
| *Dalbergia latifolia* Roxb. | East Indian rosewood, Indian palisandre | Eczema, skin irritation | High-quality furniture, interior design, elements of musical instruments | |
| *Entandrophragma cylindricum* Sprague | Sapelli, sapele, sapeli | Skin irritation | Furniture, interior design, veneers | |

(continued)

**Table 4.2** (continued)

| Botanical name | Trade name | Symptoms of illness | Common use | Inadvisable use |
|---|---|---|---|---|
| *Chloroxylon swietenia* DC. | Ceylon satinwood, East Indian satinwood | Skin irritation | Ship furniture, veneers, interior design | |
| *Fagara flava* Krug and Urban | West Indian satinwood, San Domingan satinwood | Skin irritation, toxic properties | High-quality furniture, interior design | |
| *Bowdichia nitida* Bentham | Sucupira | Skin irritation of various severity | Veneers for high-quality furniture | |
| *Tectona grandis* L. | Teak | Skin irritation, toxic properties | Furniture, products used outside of buildings | |
| *Entandrophragma utile* Sprague | Utile, sipo | Skin irritation | Furniture, interior design, veneers | |
| *Juglans regia* L. | Walnut | Skin irritation | High-quality furniture and veneers | |
| *Millettia laurentii* Wild. | Wenge, panga-panga | Skin irritation, toxic properties | Veneers for furniture in public buildings, hotels, high-quality furniture | |
| *Taxus baccata* L. | Yew | Skin irritation, toxic properties | Occasionally for veneers | |
| *Cedrela odorata* L. | Spanish cedar | Irritation of skin and mucous membranes | Joinery, interior design | Kitchen, food containers |

By entering the engineering markings of components of the stress and strain tensor: $\sigma_{11} = \sigma_x$, $\sigma_{22} = \sigma_y$, $\sigma_{33} = \sigma_z$, $\tau_{12} = \tau_{xy}$, $\tau_{23} = \tau_{yz}$, $\tau_{31} = \tau_{zx}$, $\varepsilon_{11} = \varepsilon_x$, $\varepsilon_{22} = \varepsilon_y$, $\varepsilon_{33} = \varepsilon_z$, $\gamma_{12} = \gamma_{xy}$, $\gamma_{23} = \gamma_{yz}$ and $\gamma_{31} = \gamma_{zx}$, and entering the markings: $A_{11} = A_{1111}$, ..., $A_{16} = A_{1131}$, ..., $A_{21} = A_{2211}$, ..., $A_{26} = A_{2231}$, ..., we obtain the following form of generalised Hooke's law (Litewka 1997):

$$
\begin{bmatrix} \sigma_x \\ \sigma_y \\ \sigma_z \\ \tau_{xy} \\ \tau_{yz} \\ \tau_{zx} \end{bmatrix} = \begin{bmatrix} A_{11} & A_{12} & A_{13} & A_{14} & A_{15} & A_{16} \\ A_{21} & A_{22} & A_{23} & A_{24} & A_{25} & A_{26} \\ A_{31} & A_{32} & A_{33} & A_{34} & A_{35} & A_{36} \\ A_{41} & A_{42} & A_{43} & A_{44} & A_{45} & A_{46} \\ A_{51} & A_{52} & A_{53} & A_{54} & A_{55} & A_{56} \\ A_{61} & A_{62} & A_{63} & A_{64} & A_{65} & A_{66} \end{bmatrix} \cdot \begin{bmatrix} \varepsilon_x \\ \varepsilon_y \\ \varepsilon_z \\ \gamma_{xy} \\ \gamma_{yz} \\ \gamma_{zx} \end{bmatrix} . \tag{4.17}
$$

If the symmetry of stress and strain tensors is considered, the number of components is 36. But when we take into account the differentiating alternation of free energy function in relation to the tensor's components:

$$
\frac{\partial^2 \overline{V}}{\partial \varepsilon_{ij} \partial \varepsilon_{kl}} = \frac{\partial^2 \overline{V}}{\partial \varepsilon_{kl} \partial \varepsilon_{ij}}, \tag{4.18}
$$

where
$\overline{V}$  elastic energy,

then the number of the tensor A components will decrease to 21. In specific anisotropic cases, like for example an orthotropic body, tensor A has 9 components, and for a transversally isotropic body—5 components. By expressing the strains of an anisotropic body in the general form, the following equation can be written as follows:

$$
\varepsilon_{ij} = a_{ijkl} \cdot \sigma_{kl}, \tag{4.19}
$$

where
$a$    compliance tensor and
$a_{ijkl}$  components of the compliance tensor determined by the measurement of the strains of planes of a three dimensional body, taking into account normal and shear strains (Fig. 4.9).

At the same time, the following dependencies apply: for the direction of X-axis

$$
v_{xy} = \frac{\varepsilon_y}{\varepsilon_x}; \quad v_{xz} = \frac{\varepsilon_z}{\varepsilon_x}; \quad \mu_{zx,z} = \frac{\varepsilon_x}{\gamma_{zx}}; \quad \mu_{zy,x} = \frac{\varepsilon_x}{\gamma_{zy}}, \tag{4.20}
$$

**Fig. 4.9** The state of strains in a 3-D element

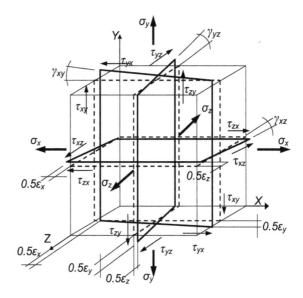

for the direction of $Y$-axis

$$v_{yx} = \frac{\varepsilon_x}{\varepsilon_y}; \quad v_{yz} = \frac{\varepsilon_x}{\varepsilon_y}; \quad \mu_{zy,y} = \frac{\varepsilon_y}{\gamma_{zy}}; \quad \mu_{xy,y} = \frac{\varepsilon_y}{\gamma_{xy}}; \quad \mu_{zx,y} = \frac{\varepsilon_y}{\gamma_{zx}}, \quad (4.21)$$

for the direction of $Z$-axis

$$v_{zx} = \frac{\varepsilon_x}{\varepsilon_z}; \quad v_{zy} = \frac{\varepsilon_y}{\varepsilon_z}; \quad \mu_{zx,z} = \frac{\varepsilon_z}{\gamma_{zx}}; \quad \mu_{zy,z} = \frac{\varepsilon_z}{\gamma_{zy}}; \quad \mu_{xy,z} = \frac{\varepsilon_z}{\gamma_{xy}}, \quad (4.22)$$

in the XY plane

$$\mu_{y,xy} = \frac{\gamma_{xy}}{\varepsilon_y}; \quad \mu_{xz} = \frac{\gamma_{xy}}{\varepsilon_x}; \quad \mu_{z,xy} = \frac{\gamma_{xy}}{\varepsilon_z}; \quad \mu_{zx,xy} = \frac{\gamma_{xy}}{\gamma_{zx}}; \quad \mu_{zy,xy} = \frac{\gamma_{xy}}{\gamma_{zy}} \quad (4.23)$$

in the YZ plane

$$\mu_{y,zy} = \frac{\gamma_{zy}}{\varepsilon_y}; \quad \mu_{x,zy} = \frac{\gamma_{zy}}{\varepsilon_x}; \quad \mu_{z,zy} = \frac{\gamma_{zy}}{\varepsilon_z}; \quad \mu_{xy,zy} = \frac{\gamma_{zy}}{\gamma_{xy}}; \quad \mu_{zx,zy} = \frac{\gamma_{zy}}{\gamma_{zx}}, \quad (4.24)$$

in the XZ plane

$$\mu_{y,zx} = \frac{\gamma_{zx}}{\varepsilon_y}; \quad \mu_{x,zx} = \frac{\gamma_{zx}}{\varepsilon_x}; \quad \mu_{z,zx} = \frac{\gamma_{zx}}{\varepsilon_x}; \quad \mu_{xy,zx} = \frac{\gamma_{zx}}{\gamma_{xy}}; \quad \mu_{zy,zx} = \frac{\gamma_{zx}}{\gamma_{zy}}. \quad (4.25)$$

Thus, it is easy to write, e.g. an equation of normal strains in the direction of the $X$-axis:

$$\varepsilon_x = \frac{\sigma_x}{E_x} - \nu_{yx}\varepsilon_y - \nu_{zx}\varepsilon_z + \mu_{zy,x} \cdot \gamma_{zy} + \mu_{zx,x} \cdot \gamma_{zx} + \mu_{xy,x} \cdot \gamma_{xy}. \tag{4.26}$$

By substituting the now well-known dependencies $\varepsilon = \sigma/E$ and $\gamma = \tau/G$, we obtain, respectively:

$$\varepsilon_x = \frac{\sigma_x}{E_x} - \nu_{yx}\frac{\sigma_y}{E_y} - \nu_{zx}\frac{\sigma_z}{E_z} + \mu_{zy,x}\frac{\tau_{yz}}{G_{yz}} + \mu_{zx,x}\frac{\tau_{xz}}{G_{xz}} + \mu_{xy,x}\frac{\tau_{xy}}{G_{xy}}, \tag{4.27}$$

or

$$\varepsilon_x = \frac{1}{E_x}\left(\sigma_x - \nu_{xy}\sigma_y - \nu_{xz}\sigma_z + \mu_{x,yz}\tau_{yz} + \mu_{x,xz}\tau_{xz} + \mu_{x,xy}\tau_{xy}\right). \tag{4.28}$$

For shear strains, e.g. $\gamma_{xy}$, the equation of the sum of partial strains will have the form:

$$\gamma_{xy} = \frac{\tau_{xy}}{G_{xy}} + \mu_{y,xy} \cdot \varepsilon_y + \mu_{x,xy} \cdot \varepsilon_x + \mu_{z,xy} \cdot \varepsilon_z + \mu_{zx,xy} \cdot \gamma_{xz} + \mu_{zy,xy} \cdot \gamma_{yz}, \tag{4.29}$$

hence finally

$$\gamma_{xy} = \frac{1}{G_{xy}}\left(\mu_{xy,x}\sigma_x + \mu_{xy,y}\sigma_y + \mu_{xy,z}\sigma_z + \mu_{xy,zx}\tau_{xz} + \mu_{xy,zy}\tau_{yz} + \tau_{xy}\right). \tag{4.30}$$

In the above equations, $E_x$, $E_y$ and $E_z$ are the linear elasticity modules at stretching, $G_{xy}$, $G_{xz}$ and $G_{yz}$ are shear elasticity modules in planes that are parallel to the lines of direction coordinates $x$, $y$, $z$, $\nu_{xy}$, $\nu_{yx}$, and $\nu_{zx}$, $\nu_{xz}$, $\nu_{yz}$, $\nu_{zy}$ are Poisson's ratios characterising elongation in the direction of the first axis and shortening in the direction of the second axis of the plane. Coefficients $\mu_{xz,yz}\ldots\mu_{xz,xy}$, called Chentsov coefficients (Ashkenazi 1958; Lekhnickij 1977), characterise shear strains in planes that are parallel to the coordinate system, caused by tangential stresses acting in the second planes parallel to the coordinate system. Coefficients $\mu_{yz,x}\ldots\mu_{xy,z}$, according to Rabinowicz (1946) called the coefficients of mutual impact of first degree, they express elongation in the direction of the axis of the coordinate system, caused by tangential stresses acting in planes parallel to the coordinate system. Coefficients $\mu_{x,yz}\ldots\mu_{z,xy}$, express shear strains in planes parallel to the global coordinate system, caused by normal stresses acting in the direction of the axis of the system. They can be called coefficients of mutual impact of the second degree (Ashkenazi 1958).

The equations above correspond only to the given Cartesian system of coordinates. Changing this system will automatically change the values of the coefficients, although their number remains constant. However, if at any point of the anisotropic elastic body, three mutually perpendicular planes of its internal structure can be led, such material can be called orthotropic. Wood, as an orthotropic body, in a spatial state of stresses is subject to normal strains (Fig. 4.10) and changes in shape (Fig. 4.11).

**Fig. 4.10** Normal strains of wood in the spatial state of stresses

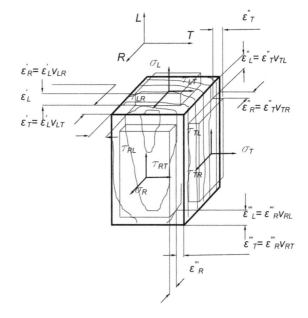

**Fig. 4.11** Shear strains of wood in the spatial state of stresses

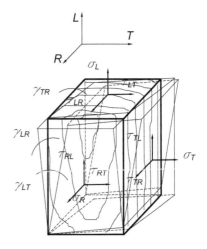

By summing up the value of partial strains in particular anatomical directions, we obtain expressions for total strains in the form:

$$\varepsilon_L = \frac{1}{E_L}\sigma_L - \nu_{TL}\frac{1}{E_T}\sigma_T - \nu_{RL}\frac{1}{E_R}\sigma_R,$$

$$\varepsilon_T = \frac{1}{E_T}\sigma_T - \nu_{LT}\frac{1}{E_L}\sigma_L - \nu_{RT}\frac{1}{E_R}\sigma_R, \qquad (4.31)$$

$$\varepsilon_R = \frac{1}{E_R}\sigma_R - \nu_{TR}\frac{1}{E_T}\sigma_T - \nu_{LR}\frac{1}{E_L}\sigma_L,$$

where

$\sigma_L,\ \sigma_R,\ \sigma_T$      vector of normal stresses, respectively, in the direction: longitudinal, radial and tangential;

$E_L,\ E_R,\ E_T$      linear elasticity modules of wood, respectively, in the direction: longitudinal, radial and tangential; and

$\nu_{LR},\ \nu_{LT},\ \nu_{RT},\ \nu_{TR},$
$\nu_{RL},\ \nu_{TL}$      Poisson's ratios, respectively, in anatomical directions: longitudinal–radial, longitudinal–tangential, radial–tangential, tangential–radial, radial–longitudinal and tangential–longitudinal.

And expressions for shear strains are given as follows:

$$\gamma_{LT} = \frac{1}{G_{LT}}\tau_{LT},\ \gamma_{LR} = \frac{1}{G_{LR}}\tau_{LR},\ \gamma_{TR} = \frac{1}{G_{TR}}\tau_{TR}, \qquad (4.32)$$

where

$\gamma_{LR},\ \gamma_{LT},\ \gamma_{TR}$   shear strains in anatomical planes: longitudinal–radial, longitudinal–tangential and tangential–radial;

$G_{LR},\quad G_{LT},$
$G_{TR}$   shear elasticity modules of wood in anatomical planes: longitudinal–radial, longitudinal–tangential and tangential–radial; and

$\tau_{LR},\ \tau_{LT},\ \tau_{TR}$   tangential stresses in anatomical planes: longitudinal–radial, longitudinal–tangential and tangential–radial.

Hence

$$
\begin{bmatrix} \varepsilon_L \\ \varepsilon_T \\ \varepsilon_R \\ \gamma_{LT} \\ \gamma_{LR} \\ \gamma_{TR} \end{bmatrix} =
\begin{bmatrix}
\dfrac{1}{E_L} & -\nu_{TL}\dfrac{1}{E_T} & -\nu_{RL}\dfrac{1}{E_R} & 0 & 0 & 0 \\[2mm]
-\nu_{LT}\dfrac{1}{E_L} & \dfrac{1}{E_T} & -\nu_{RT}\dfrac{1}{E_R} & 0 & 0 & 0 \\[2mm]
-\nu_{LR}\dfrac{1}{E_L} & -\nu_{TR}\dfrac{1}{E_T} & \dfrac{1}{E_R} & 0 & 0 & 0 \\[2mm]
0 & 0 & 0 & \dfrac{1}{G_{LT}} & 0 & 0 \\[2mm]
0 & 0 & 0 & 0 & \dfrac{1}{G_{LR}} & 0 \\[2mm]
0 & 0 & 0 & 0 & 0 & \dfrac{1}{G_{TR}}
\end{bmatrix} \cdot
\begin{bmatrix} \sigma_L \\ \sigma_T \\ \sigma_R \\ \tau_{LT} \\ \tau_{LR} \\ \tau_{TR} \end{bmatrix}. \qquad (4.33)
$$

The generalised Hooke's law in matrix convention has the form:

$$\sigma_i = A_{ij}\varepsilon_j, \tag{4.34}$$

while for the discussed case, we will obtain:

$$
\begin{bmatrix}
\sigma_L \\
\sigma_T \\
\sigma_R \\
\tau_{LT} \\
\tau_{LR} \\
\tau_{TR}
\end{bmatrix}
=
\begin{bmatrix}
\dfrac{C_L E_L}{C_T(v_{RT}+v_{LT}v_{RL})} & \dfrac{C_{TL}E_L}{C_T(v_{RT}+v_{LT}v_{RL})} & \dfrac{C_R E_L}{C_T} & 0 & 0 & 0 \\[2ex]
\dfrac{v_{LT}(1-v_{LR}v_{RL})E_T}{C_T} & \dfrac{(1-v_{LR}v_{RL})E_T}{C_T} & \dfrac{(v_{RT}+v_{LT}v_{RL})E_T}{C_T} & 0 & 0 & 0 \\[2ex]
\dfrac{C_{RL}E_R}{C_T(v_{RT}+v_{LT}v_{RL})} & \dfrac{(1-v_{LR}v_{RL})(1-v_{LT}v_{TL})E_R}{C_T(v_{RT}+v_{LT}v_{RL})} & \dfrac{(1-v_{LT}v_{TL})E_R}{C_T} & 0 & 0 & 0 \\[2ex]
0 & 0 & 0 & G_{LT} & 0 & 0 \\[1ex]
0 & 0 & 0 & 0 & G_{LR} & 0 \\[1ex]
0 & 0 & 0 & 0 & 0 & G_{TR}
\end{bmatrix}
$$

$$
\times
\begin{bmatrix}
\varepsilon_L \\
\varepsilon_T \\
\varepsilon_R \\
\gamma_{LT} \\
\gamma_{LR} \\
\gamma_{TR}
\end{bmatrix},
$$

$$\tag{4.35}$$

where:

$C_T$  $(1 - v_{LT}v_{TL})(1 - v_{LR}v_{RL}) - (v_{TR} + v_{LR}v_{LT})(v_{RT} + v_{LT}v_{RL})$,

$C_L$  $C_T(v_{RT} + v_{LT}v_{RL}) + v_{LT}v_{TL}(1 - v_{LT\setminus R}v_{RL})(v_{RT} + v_{LT}v_{RL}) + v_{RL}(v_{LT}(1 - v_{LT}v_{RL}))$
$\quad (1 - v_{LT}v_{TL}) - v_{LT}C_T)$,

$C_R$  $v_{TL}(v_{RT} + v_{LT}v_{RL}) + v_{RL}(1 - v_{LT}v_{TL})$,

$C_{TL}$  $v_{TL}(1 - v_{LR}v_{RL})(v_{RT} + v_{LT}v_{RL}) + v_{RL}((1 - v_{LR}v_{RL})(1 - v_{LT}v_{TL}) - C_T)$,

$C_{RL}$  $v_{LT}(1 - v_{LT}v_{RL})(1 - v_{LT}v_{TL}) - v_{LT}C_T$

Table 4.3 shows a list of elastic properties of selected wood species, commonly used in the furniture industry.

## 4.5.2   *Wood-Based Materials*

Wood, as the basic construction material in furniture, has found widespread use in designed elements: decorative, turned, curved, beam and board. However, surface elements, such as worktops of tables, shelves, partitions and side walls of bodies, require the use of higher quality materials and dimensional stability. For these reasons, in the furniture industry, it is common to use wood-based materials that provide freedom in forming surface shapes, greater stability of thickness than wood and higher material efficiency. The most frequently used wood-based materials in the furniture industry include:

**Table 4.3** Resilient properties of wood (Hearmon 1948, Bodig and Goodman 1973)

| Wood property | Wood species | | | | |
|---|---|---|---|---|---|
| | Oak | Beech | Ash | Pine | Alder |
| Density of wood (g/cm³) | 0.60 | 0.75 | 0.67 | 0.55 | 0.38 |
| Linear elasticity module (GPa) | | | | | |
| $E_L$ | 16.21 | 13.96 | 15.78 | 16.60 | 10.42 |
| $E_R$ | 12.02 | 22.84 | 15.09 | 11.17 | 0.809 |
| $E_T$ | 0.626 | 1.160 | 0.799 | 0.583 | 0.355 |
| Shear elasticity module (GPa) | | | | | |
| $G_{LT}$ | 0.698 | 1.082 | 0.889 | 0.693 | 0.313 |
| $G_{LR}$ | 0.842 | 1.645 | 1.337 | 1.181 | 0.632 |
| $G_{RT}$ | 0.311 | 0.471 | 0.471 | 0.070 | 0.144 |
| Poisson's ratio | | | | | |
| $v_{LR}$ | 0.360 | 0.450 | 0.460 | 0.420 | 0.440 |
| $v_{LT}$ | 0.330 | 0.510 | 0.510 | 0.510 | 0.560 |
| $v_{RT}$ | 0.780 | 0.750 | 0.710 | 0.680 | 0.570 |
| $v_{TR}$ | 0.370 | 0.360 | 0.360 | 0.310 | 0.290 |
| $v_{RL}$ | 0.060 | 0.075 | 0.051 | 0.038 | 0.031 |
| $v_{TL}$ | 0.030 | 0.044 | 0.030 | 0.015 | 0.013 |

- boards bonded from strips and panels,
- carpentry and honeycomb boards,
- chipboards,
- fibreboards and
- plywood.

### 4.5.2.1  Structure of Wood-Based Materials

The quality of the final product, which is a glued wooden board, to a large extent depends on the quality of the technological process. Making board furniture elements from solid wood requires gluing slatted wooden panels or laths together. The choice of the dimensions of these elements directly affects the form of distortion or twisting of finished boards. The mechanism of the formation of these phenomena can be briefly presented as follows. As a result of the hygroscopy of the material of individual elements, the boards absorb or expel steam in order to strive to the state of hygroscopical balance. Moisture content changes are accompanied by the phenomenon of shrinking and swelling of particular component parts of the board, causing the formation of stresses inside the material and glue-lines connecting the wood. The differences in the construction of individual elements of the board constitute the direct cause of differentiating strains and stresses, which consequently leads to distortions, twisting, cracks inside and on the surface of the wood and cracks in glue-line (Fig. 4.12).

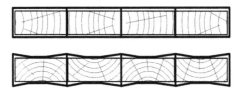

**Fig. 4.12** Change of the dimensions and shape of the board element

The moisture content of the wood, from which strips or laths are made, is shaped at the stage of their drying. Manufacturers of glued wooden boards are aware of the dependencies that occur between the quality of drying and later changes in the shape of finished boards. There is, however, a lack of general understanding of the impact of the parameters of air in objects, where semi-finished products and boards are stored or finished products are used, on changes in the moisture content of the wood, and thus, the change in the stresses and strains occurring in them. Assuming that the change of the dimension of the wood in the function of moisture content, between 0 and 30 %, is a linear dependency, changes of the dimensions of linear cross sections of board elements can be easily tracked. In order to do this, the following equation has to be used:

$$N_k = 0.01 N_p \left( \beta_{w(o)} DW_{w(o)} + 100 \right), \tag{4.36}$$

where

$N_k$          final measurement,
$N_p$          initial measurement,
$DW_{w(o)}$   moisture content increase and
$\beta_{w(o)}$   linear expansion of the wood calculated from the equation.

$$\beta_{w(o)} = \beta_R \cos^2 \alpha + \beta_T \sin^2 \alpha, \tag{4.37}$$

where
$\beta_T$   linear expansion of the wood in the tangential direction,
$\beta_R$   linear expansion of the wood in the radial direction and
$\alpha$     the angle between the tangent to the growth ring and the direction of extensibility (Fig. 4.13).

Figure 4.14 shows the effect of changes in the moisture content of the wood and the angle of inclination of growth rings on the change of thickness of individual elements forming the board. It should be noted that a change in the moisture content of the wood by 2 %, at the simultaneous change of the angle of inclination of growth rings from 20° to 90°, causes changes in the element by 0.1 mm. The variability of dimensions in the area of one element translates into the change of the dimensions and shape of the whole board element. In effect, for the worktop of a

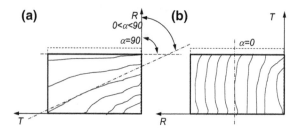

**Fig. 4.13** The measurement of the angle between the tangent to the growth ring and the direction of extensibility: **a** α < 90°, **b** α = 0°

**Fig. 4.14** The change of the initial dimension of the element 20 mm thick while drying and moisturising (anatomical direction from the radial to the tangential)

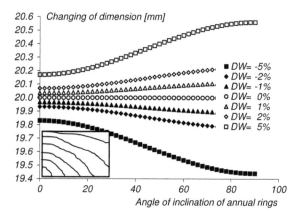

table this results in the impression of an uneven surface, on which rays of light are reflected in an disorderly manner.

Carpentry boards are one of the oldest semi-finished products used for board elements of furniture. They are built from the middle layer with single or double layers of veneers, plywood or sheets of hard fibreboards glued from both sides. The middle layer of the board may contain elements that are not connected or glued together. These elements can be laths, strips, veneer, cardboard or paper reinforced with synthetic resins (Oniśko 1994) (Figs. 4.15 and 4.16). Depending on the construction of the middle layer, carpentry boards can be divided into:

• solid (Fig. 4.15) and
• honeycomb (Fig. 4.16).

The frame of the honeycomb frame board can be made of wood, chipboard, fibreboard and possibly of another wood-based material. However, it cannot show defects, which would have a negative effect on the strength of the material or cause strains. The elements of the frame are connected using glue or clasps. A frameless honeycomb consists of only the middle layer and lining (Fig. 4.17). In the process of the production of this board, the glue is to fix the middle layer with outer layers,

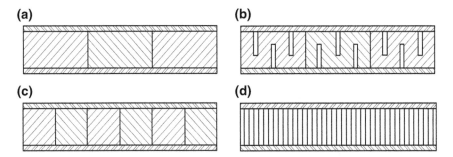

**Fig. 4.15** Types of carpentry boards of full middle layers: **a** lath middle, **b** cut lath middle, **c** strip middle and **d** middle from strips of veneer (Oniśko 1994)

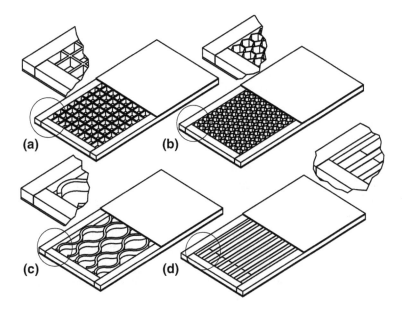

**Fig. 4.16** The construction of carpentry honeycomb (frame) boards: **a** with the middle made of hard fibreboard, **b** with the middle made of cardboard, **c** with the middle made of veneer and **d** with the middle made of strips (Szczuka and Żurowski 1995)

which is why the glue-line created should exhibit great resistance to stretching and bending (Zawierta 2007).

An important benefit resulting from the use of full carpentry boards and frameless honeycomb, in contrast to frame honeycomb, is the possibility of freely shaping the dimensions of board elements of furniture, including curved elements.

Chipboard is produced in the process of pressing piles of glued wood shavings in conditions of high temperature and pressure. Usually, they are produced as multi-layer boards. The inner layer of the board is made of chips of a thicker

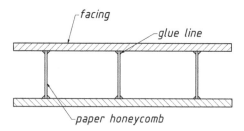

**Fig. 4.17** Cross section of honeycomb with *glue-line* (Zawierta 2007)

**Fig. 4.18** Structure of a chipboard

fraction, while the external layers are made of very fine and thin chips, so-called microchips (Fig. 4.18). Thanks to this, the surface of the board is characterised by low roughness and high adhesion strength. The chipboard is a direct raw material for the production of furniture units. In subsequent treatment processes, their wide and narrow planes are subjected to veneering with natural or synthetic veneers. This is a typical technological process used in the manufacture of house furniture. For public institutions, and in particular offices, banks, post offices, employee furniture is made of laminated boards. Laminated boards are a product made on the basis of a three-layer chipboard or MDF board covered unilaterally or bilaterally with papers saturated in thermosetting resins. In the process of lamination, the decorative film is pressed onto the board, while giving its surface the desired structure, which does not require further treatment. Chipboards in every form enable the design and production of rail, board and panel elements of furniture, giving them any shape in one, and sometimes even two planes.

Fibreboards are wood-based products made as the effect of pressing wood fibres with an organic addition of adhesive and hardening compounds in conditions of high pressure and temperature. They are a material of similar density and composition in the entire cross section, thanks to which they have an excellent workability in the process of cutting. Fibreboards produced nowadays can be divided into the following:

- Medium density fibreboard (MDF)—this is a medium density board of wood fibres. It constitutes the primary raw material for the production of furniture elements, including ornamental strips, worktops, grilles, panels and door frames and milled drawer fronts. It is the perfect substitution for wood in covering or dark finishes, and it goes well with finishes of transparent dark natural veneers. Due to the qualities of the surface, they are suitable for upgrading by covering with thin melamine films, painting and veneering with natural and synthetic veneer.

- Low density fibreboard (LDF)—a board, which is characterised by a smaller, in relation to MDF, density and is not usually used in furniture. However, this board is a base material for the production of wall panels used in dry areas.
- High density fibreboard (HDF)—a board of great hardness and increased density. HDFs are intended mainly for the production of floor panels. They are also a base material for coating with HPL and CPL laminates, resin papers and natural veneers. Lacquered HDFs are ideal for rear walls and bottoms of drawers of case furniture.

Like chipboards, fibreboards enable the design and production of rail, board and panel elements in furniture of shapes contained in one or two planes. Unlike chipboards, fibreboards are suitable for milling wide and narrow surfaces. Thanks to this, in certain applications they are the perfect substitute for wood both in aesthetics and construction terms.

Plywood is a wood-based material built of layers as a result of bonding, usually at an angle of 90°, most commonly with an odd number of veneers (Fig. 4.19). Glues based on synthetic resins are used for bonding, that is, urea, melamine, phenolic and resorcinol. Veneers for plywood are mainly made of the wood of birch, alder, beech, pine and spruce. Usually, a sheet of plywood is made of different species of wood, e.g. pine/alder and birth/pine. The type of wood used to build the outer layers determines the qualification of the plywood as a coniferous or deciduous type. The inner layers can be made of wood of the same or other type than the outer layer. Depending on the adhesive mass used, distinguished are moisture resistant plywood, waterproof plywood and so-called waterproof plywood with a clear glue-line (Table 4.4).

In furniture, plywood is used for:

- door panels, rear walls, bottoms of drawers and slides of case furniture;
- skeletons and casings of frames, containers and strips of upholstery frames of upholstered furniture; and
- seats, backrest boards, rails and worktops of skeletal furniture.

**Fig. 4.19** Construction of three-layer plywood

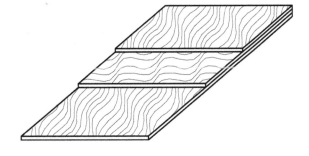

**Table 4.4** Examples of plywood markings according to adhesive bond type

| Glue-line type | PN-EN 636:2013-03 | DIN 68705-2:2003-10 | BS1 203:2001 | PN-83/D 97005.11 BN-73/7113-06 |
|---|---|---|---|---|
| Moisture resistant: urea–formaldehyde resin-based adhesive | For use in dry conditions | IF 20, BFU 20 | INT MR | Moisture resistant |
| Waterproof: with clear glue-line: urea–melamine–formaldehyde resin-based adhesive | For use in humid conditions | A 100 | BR | Semi-waterproof |
| Waterproof: phenol–formaldehyde resin-based adhesive | For use in exterior conditions | BFU100 AW 100 | WBP | Waterproof |

### 4.5.2.2  Elasticity of Wood-Based Materials

All of the materials presented above have three mutually perpendicular principal axes, defining their orthogonal anisotropy. Usually, it is assumed, however, that the thickness of boards is small in relation to other dimensions, and the elastic properties of the wood-based materials in question adopt extreme values in the mutually orthogonal directions, while perpendicular to the thickness of the board. In other words, in two mutually perpendicular directions, the elastic properties of wood-based materials assume extreme values that is the largest in one direction and the smallest in the other direction. These directions correspond to the axes of symmetry of elastic properties, and it has been adopted to call them the main directions of anisotropy. Materials with such properties are called orthotropic. Therefore, Hooke's law for the orthotropic body in a system of rectangular Cartesian coordinates, overlapping with the main directions of anisotropy (Fig. 4.20), can be written in the form:

**Fig. 4.20** A model of a multi-layered orthotropic wood-based material

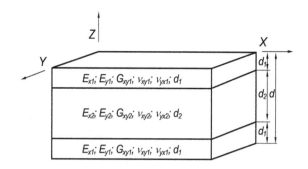

$$\begin{cases} \sigma_x = \frac{E_x}{1-v_xv_y}\left(\varepsilon_x + v_y\varepsilon_y\right) \\ \sigma_y = \frac{E_y}{1-v_xv_y}\left(\varepsilon_y + v_x\varepsilon_x\right) \\ \tau_{xy} = G_{xy}\gamma_{xy} \end{cases} \tag{4.38}$$

where

$E_x, E_y$  linear elasticity modules in the main directions of anisotropy,
$v_x, v_y$  Poisson's ratios in the main directions of anisotropy,
$G_{xy}$  shear elasticity module in the anisotropic plane,
$\varepsilon_x, \varepsilon_y$  normal strains in the main directions of anisotropy and
$\gamma_{xy}$  shear strains in the anisotropic plane,

whereby the following dependency applies here

$$\frac{v_{xy}}{E_x} = \frac{v_{yx}}{E_y}, \tag{4.39}$$

Tables 4.5, 4.6 and 4.7 show examples of values of mechanical properties of different types of wood-based materials.

Table 4.5  Properties of chipboards (Wilczyński and Kociszewski 2000)

| Property | Unit | Value |
|---|---|---|
| Density | kg/m³ | 550–680 |
| Young's modulus $E_y$ | MPa | 3080 |
| Young's modulus $E_x$ | MPa | 2530 |
| Poisson's ratio $v_{yx}$ | | 0.207 |
| Poisson's ratio $v_{xy}$ | | 0.282 |
| Shear modulus $G_{yx}$ | MPa | 794 |
| Bending strength $k_{gy}$ | MPa | 13.3 |
| Bending strength $k_{gx}$ | MPa | 11.9 |
| Splitting strength $k_r$ | MPa | 0.15–0.20 |

Table 4.6  Properties of MDF boards (Wilczyński et al. 2001; Schulte and Frühwald 1996)

| Property | Unit | Value |
|---|---|---|
| Density | kg/m³ | 510–710 |
| Young's modulus $E_y$ | MPa | 4000 |
| Young's modulus $E_x$ | MPa | 3850 |
| Poisson's ratio $v_{yx}$ | | 0.30 |
| Poisson's ratio $v_{xy}$ | | 0.30 |
| Shear modulus $G_{yx}$ | MPa | $G = E/2(1 + v)$ |
| Bending strength $k_{gy}$ | MPa | 34.5 |
| Bending strength $k_{gx}$ | MPa | 32.2 |
| Splitting strength $k_r$ | MPa | 0.31 |

**Table 4.7** Properties of plywood

| Property | Unit | Value |
|---|---|---|
| Density | kg/m$^3$ | 550–800 |
| Young's modulus $E_y$ | MPa | 3500–10,000 |
| Young's modulus $E_x$ | MPa | 500–700 |
| Poisson's ratio $v_{yx}$ | | 0.439 |
| Poisson's ratio $v_{xy}$ | | 0.031 |
| Shear modulus $G_{yx}$ | MPa | 822, $G = E/2(1 + v)$ |
| Bending strength $k_{gy}$ | MPa | 30–100 |
| Stretching strength $k_{rx}$ | MPa | 30–60 |
| Shearing strength $k_{gy}$ | MPa | 25–50 |

### 4.5.3   Leathers and Fabrics

The use of leathers is strictly related to the history of the origins of civilisation. From the cave paintings that date back to the Palaeolithic Era (approx. 20,000 years BC), we learn about the common, at the time, use of leathers as a raw material for the manufacture of outer clothing or bed covers. Scholars also point out that for the first time the term leather was used in the third part of the Book of Genesis. The ancient Arabs, commonly using leathers, developed a written procedure of tanning them. The instructions stated that: *first skin, cleaned from fat and impurities, should be stored in flour and salt for three days. Then, the root of the plant Chulga should be grated using large stones and dissolved in water, then cover the inner surface of the skin with preparation obtained and leave for one day, after which the skin will lose its hair. The skin should be left in this state for another two days, after which the treatment process is completed.* Thanks to this procedure, ancient Arabs achieved an excellent material for the manufacture of famous saddles.

Leather as an upholstery material is a special product due to its uniqueness and unique features of use. When choosing leather for the upholstered layer of the furniture piece, we usually accept traces of scars, scratches and stabs on its surface which came about due to natural causes, as well as differences in the texture and colour shades showing the uniqueness of nature. Undoubtedly, these features strengthen the belief in the authenticity of the natural origin of this product. Therefore, a set of lounge furniture made of hides of leathers of identical or similar colours, but with a unique surface texture, can be attractive.

In dermatological terms, the skin is an organ which covers and guards the organism. It consists of three layers: epidermis, dermis and subcutaneous tissue. The epidermis consists mainly of ageing epithelial cells, called keratinocytes, and creates several layers: basal, spinous, granular and cornified. In addition to keratinocytes in the epidermis, there are also colour cells—melanocytes, the cells responsible for immunological reactions—Langerhans cells and the cells of the nervous system—Merkel cells. In dermis, made of connective tissue, there are collagen fibres, elastin and cellular elements: fibroblasts, mastocytes, blood cells, vessels and nerves. The hypodermic tissue is created by the adipose and connective

tissue. The skin has skin appendages: sweat glands (eccrine and apocrine), seba-
ceous glands, nails and hair. The skin protects against bacteria, fungi, virus
infections, against mechanical, thermal, chemical and light radiation factors, as well
as ensures constant conditions for the internal environment of the organism (ho-
moeostasis). In addition to this, skin fulfils the following functions: perceptive
(heat, pain, touch), expressive (in articulating emotional states), resorption and
participates in storage and metabolism (Placek 1996).

As an organic material, such as wood, skin reacts to changes of external climatic
factors, including light intensity, temperature and humidity of the environment.
During treatment in tanneries, skin is subjected to complex processes, using
moisturising substances, giving it softness and elasticity, however, during use these
substances vaporise. Drying of the skin increases its fragility and susceptibility to
cracking and flaking. In order to ensure its original elasticity and extend its lifespan,
appropriate care and conservation agents should be used. Leather needs to be
conserved and cleaned every 3–6 months, depending on its kind and degree of
exploitation of the product.

From the commercial point of view, the following kinds of leathers are
distinguished:

- madras is the basic kind of single-colour leathers with an adjusted grain.
  Adjustment of the leather consists in polishing the grain and embossing
  appropriately regular cracks on its surface;
- modial, grain leather, two colour, with identical properties to madras;
- antic, grain leather, adjusted and embossed, usually two colour and shiny with a
  grain size that is smaller than madras;
- lissone, split leather made in the production process through cutting (splitting)
  grain leathers. A smaller thickness of split leathers causes them to be more prone
  to stretching and mechanical damage. Usually, it is used as a coordinate
  material, a supplement of grain leather on a furniture piece; and
- reno, the best kind of leather with a partially adjusted grain, preserving the
  natural drawing and texture of the surface.

The fabric is a textile product woven on a loom (Chyrosz and Zembowicz-
Sułkowska 1995). As a work of human hands, fabrics are much younger than
leathers. They were made in the Neolithic about 8000 years BC. Initially, they were
loosely laced grass fibres, thin creepers and leather straps. A true fabric of use and
design value, similar to modern fabrics, was woven in ancient Egypt during the
period of the Old Kingdom (27th–22nd century BC). It was a linen cloth commonly
used by the Egyptians in the manufacture of clothing and upholstering furniture.

The fabrics are made of two systems of threads, warps and wefts. Combining
these two systems according to a specific order (weave) creates the structure of the
fabric. The quality of the material depends on the main structural parameters of
fabric, which consist of:

- the weave, determining the appearance and purpose of the fabric. Freedom in linking weaves enables to obtain any number of fabric designs according to their purpose;
- the quantity (density) of the warp and weft, which affects the elasticity and permeability of fabrics;
- the weaving of the warp and weft is a percentage ratio of the difference in length of the thread between its length after straightening $l$ and the length in the fabric $l_o$ to the length of the thread in the fabric $W = 100\ \%(l - l_o)/l_o$. The size of the weaving has a significant impact on the elasticity and permeability of the fabric; and
- the thickness is the number of the yarn and its type. Thickness has a significant impact on the purpose of the fabric.

Taking into consideration the structure of fabrics, they can be divided into:

- single layer, characterised by the same appearance on both sides and
- multilayer, in which the appearance on one side may significantly differ from the appearance on the other side.

Due to the texture of the surface, we differentiate between fabrics that are as follows:

- smooth, distinguished by an even surface on both sides and a clearly visible structure (plait);
- with hair (fleece) cover, characterised by a fluffy cover, hiding the fabric structure on one of the sides;
- with looped (terry) cover, characterised by total or partial looped surface on one or both sides of the fabric; and
- with mixed cover, distinguished by a partial filling with looped cover and fleece.

Depending on the raw material, among others, the following fabrics are distinguished: cotton, linen, wool, silk and synthetic fibres.

## 4.6  Furniture Joints

The construction of a furniture piece is done by creating appropriate bonds between its particular elements, subassemblages and assemblages. Choosing the right kind of joints for the designed furniture piece depends mainly on the type and form of the construction, but it should always lead to ensure its high stiffness and strength, and ease of realisation technologically.

A fragment of the structure in which parts are joined using connectors, interfaces and/or glue is called a joint. An element of a joint for connecting two parts is called a connector, while an interface refers to properly formed fragments of connected parts (Fig. 4.21). The quality of furniture joints is usually determined by assigning

**Fig. 4.21** The joint in the
design of a furniture piece

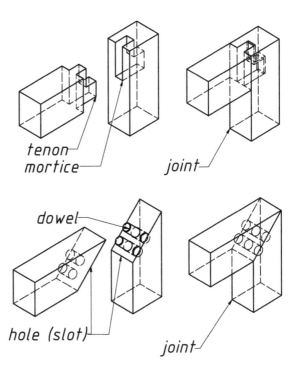

reliability, strength and stiffness characteristics. The reliability of joints is characterised by a number of measurable indicators. One of the most important indicators
of reliability is the probability of failure-free work, i.e. work without damages
within a given time period, or the probability of realising a given measurable work,
e.g. the number of hours of usage of the furniture piece, the number of cycles of
dynamic loads, the number of damages and abrasions or scratches. (Smardzewski
2005, 2009). If the problem were to be restricted to the analysis of the strength of a
given element of the furniture piece, then the designer should answer the question:
What is the probability of not exceeding the permissible level by the workload, or in
other words—the probability of not exceeding the appropriate level of permissible
stress in a given time by stress.

Practice shows that furniture manufacturers do not assume any probability of
overwork by the produced structure in a given time. Therefore, they do not know
the criteria of reliability of the produced furniture and cannot properly assess the
time of their correct exploitation until damages occur. Usually, the time period of
warranty validity on a product is determined on the basis of the designer's intuition
rather than a pragmatic statistical analysis.

Let us consider the results of studies of a large number of angular wall joints in
time $t$ (Smardzewski 2005, 2009). At the end of the test, let $n(t)$ of undamaged and
$m(t)$ of damaged joints remain. In this case,

$$R(t) = \frac{n(t)}{n}, \qquad (4.40)$$

is the probability of non-damage, that is failure-free work, while the probability of damage will be equal to

$$P(t) = \frac{m(t)}{n}. \qquad (4.41)$$

Because the probability of damage and non-damage are events that are mutually exclusive, then the sum of them will amount to

$$\frac{n(t)}{n} + \frac{m(t)}{m} = R(t) + P(t) = 1. \qquad (4.42)$$

The density of the probability of damages $f(t)$ (frequency of damages) of joints in the unit of time is a derivative of the function $P(t)$ in relation to time or other units:

$$f(t) = \frac{dP(t)}{dt} = \frac{1}{n}\frac{dm(t)}{dt} = -\frac{dR(t)}{dt}, \qquad (4.43)$$

thereby

$$P(t) = \int_0^t f(t)dt, \qquad (4.44)$$

therefore

$$R(t) = 1 - \int_0^t f(t)dt = \int_{\sim t}^{\infty} f(t)dt. \qquad (4.45)$$

The integral of the probability density in the probability theory, in the general sense, is called the distribution function of a given random variable. The average failure-free operating time of an element is determined on the basis of a known distribution of probability density $f(t)$ or on the basis of the results of statistical studies. In the first case, the expected operating time $E(t) = T$ amounts to

$$E(t) = T = \int_0^{\infty} tf(t)dt. \qquad (4.46)$$

By using this relation, it can be written as follows:

$$T = -\int_0^\infty tR'(t)\mathrm{d}t, \tag{4.47}$$

whereby after integration by parts:

$$T = -tR(t)\Big|_0^\infty + \int_0^\infty R(t)\mathrm{d}t. \tag{4.48}$$

It is obvious that at $t = 0$ and $t = \infty$, the first part of that equation will be equal to zero, thus

$$T = -\int_0^\infty tR'(t)\mathrm{d}t. \tag{4.49}$$

In the case of the examined statistical sets of furniture joints (Table 4.8):

$$t = \frac{\sum_{i}^{n} t_i}{n}, \tag{4.50}$$

where

$t_i$ operating time of the $i$-th connection until damage.

The behaviour of individual parts of the furniture piece influenced by operational loads depends not only on the fundamental laws of Newtonian mechanics, but also on the physical characteristics of materials used to make the construction. Joints of

| Table 4.8 The probability of failure-free operation of selected furniture joints | Joint type | Probability of failure-free work | |
|---|---|---|---|
| | | Unclenching | Clenching |
| | Dowel $d = 6$ mm | 0.81 | 0.85 |
| | Dowel $d = 8$ mm | 0.94 | 0.59 |
| | Confirmat screw $d = 5$ mm | 0.99 | 0.99 |
| | Eccentric without sleeve | 0.84 | 0.81 |
| | Eccentric with sleeve | 0.23 | 0.69 |
| | Trapezoidal | 0.62 | 0.21 |
| | VB35 without sleeve | 0.70 | 0.49 |
| | VB35 with sleeve | 0.92 | 0.69 |

furniture, like other components, are characterised by a limited resistance to loads causing both stresses and strains. For the designer, an important premise for choosing a specific connector or interface is the carrying capacity of the joint. The carrying capacity is the ability of taking up external loads by a material, joint or construction. The maximum load that can be transferred by the designed system is called the strength limit (Table 4.9).

Along with the appearance of external loads, constructions of furniture face strains, the size of which depends on the stiffness of joints used (Fig. 4.22).

The stiffness of the joint is determined by the coefficient $k$. It marks the strains caused by the external load. The best way of expressing joint stiffness is by the ratio of the value of the bending moment $M$ to the value of the rotation angle of the node $\varphi$ (Fig. 4.23):

$$k = \mathrm{tg}\alpha = \frac{M}{\varphi}\ [\mathrm{Nm/rad}]. \qquad (4.51)$$

In the literature, however, many other ways of defining the stiffness coefficient are encountered, for example, by measuring the displacement of $\delta_p$ of point $p$ on the direction of the force $P$ (Fig. 4.24). By conducting the experiment in such a way, the authors define the joint stiffness as follows:

$$k = \frac{P}{\delta_p}\ [\mathrm{N/m}]. \qquad (4.52)$$

Both expressions determine joint stiffness, and a comparison of the obtained results and the assessment of the quality of structural nodes are possible only if identical test methods or mathematical transformations are applied, which enable to express stiffness in the form of a quotient of the bending moment and rotation angle of the joint. The next page demonstrates the transformations of equations expressing linear displacements on expressions describing shear strains.

**Table 4.9** Stiffness and strength of selected furniture joints

| Joint type | Stiffness and strength of joints | | | |
|---|---|---|---|---|
| | Destructible moment (Nm) | | Stiffness coefficient (Nm/rad) | |
| | Unclenching | Clenching | Unclenching | Clenching |
| Dowel $d = 6$ mm | 17.1 | 39.5 | 121.2 | 167.5 |
| Dowel $d = 8$ mm | 23.9 | 49.9 | 167.8 | 277.5 |
| Confirmat screw $d = 5$ mm | 42.7 | 70.0 | 285.8 | 182.5 |
| Eccentric without sleeve | 17.4 | 37.1 | 81.2 | 184.7 |
| Eccentric with sleeve | 17.8 | 27.6 | 78.0 | 81.2 |
| Trapezoidal | 7.5 | 13.3 | 31.7 | 53.5 |
| VB35 without sleeve | 13.2 | 25.8 | 66.3 | 86.3 |
| VB35 with sleeve | 25.1 | 27.1 | 111.2 | 103.8 |

**Fig. 4.22** The strains of the
body of the furniture piece
under the influence of external
loads

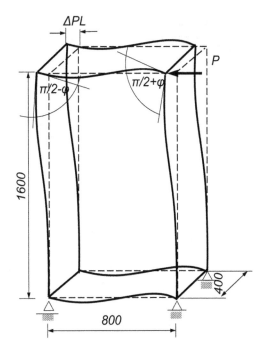

**Fig. 4.23** Method of
determining joint stiffness

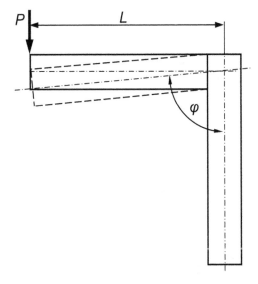

**Fig. 4.24** Alternative
methods of determining joint
stiffness

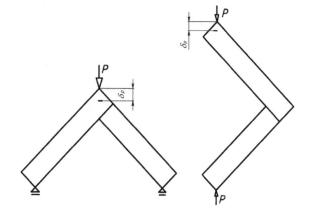

For the load scheme causing unclenching of the joint (Fig. 4.25), we obtain

$$k_R = \text{tg}\alpha_R = \frac{M_R}{\varphi_R}, \tag{4.53}$$

where for markings like in Fig. 4.25:

$$M_R = \frac{1}{2}p(L^2 + h_p^2)^{0,5} \sin\left(\arccos \frac{h - \delta_p}{\left(L^2 + h_p^2\right)^{0,5}}\right), \tag{4.54}$$

$$\varphi_R = \eta - \varepsilon, \tag{4.56}$$

whereby

$$\eta = \eta_1 + \eta_2, \tag{4.57}$$

$$\eta_1 = \arccos \frac{L - h_p}{\left((L - h_p)^2 + h_p^2\right)^{0,5}}, \tag{4.58}$$

$$\eta_2 = \arccos \frac{h_2}{\left((L - h_p)^2 + h_p^2\right)^{0,5}}, \tag{4.59}$$

$$h_2 = \frac{\sqrt{2}}{2}(L + h_p) - h_p\cos(\varepsilon) - \delta_p, \tag{4.60}$$

$$\varepsilon = 90° - \gamma, \tag{4.61}$$

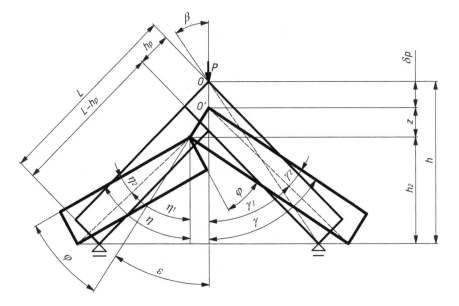

**Fig. 4.25** Load scheme causing unclenching of the joint

$$\gamma = \gamma_1 + \gamma_2, \tag{4.62}$$

$$\gamma_1 = \arccos \frac{\left(L^2 + h_p^2\right)^{0,5}\left(\frac{\sqrt{2}}{2}\left(L + h_p\right) - \delta_p\right)}{L^2 + h_p^2}, \tag{4.63}$$

$$\gamma_2 = \arccos \frac{L\left(L^2 + h_p^2\right)^{0,5}}{L^2 + h_p^2} \tag{4.64}$$

For the load causing clenching of the joint (Fig. 4.26), the stiffness coefficient can be calculated from the equation:

$$k_z = \mathrm{tg}\alpha_z = \frac{M_z}{\varphi_z}, \tag{4.65}$$

where for markings like in Fig. 4.26:

$$M_z = P\cos(\Delta\varepsilon_2)\left(\left(L - h_p\right)^2 + h_p^2\right)^{0,5}, \tag{4.66}$$

$$\varphi_z = \varphi_1 + \varphi_2, \tag{4.67}$$

**Fig. 4.26** Load scheme
causing clenching of the joint

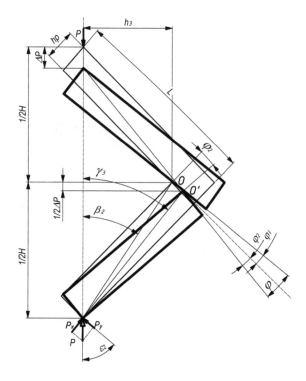

whereby

$$\Delta\varepsilon_2 = 45° + \arcsin\frac{h_p}{\left(\left(L - h_p\right)^2 + h_p^2\right)^{0,5}} - \varphi_2, \qquad (4.68)$$

$$\varphi_2 = \gamma_3 - \beta_2, \qquad (4.69)$$

$$\gamma_3 = \arccos\frac{\frac{1}{2}\left(a - \delta_p\right)\left(\left(L - h_p\right)^2 + h_p^2\right)^{0,5}}{\left(L - h_p\right)^2 + h_p^2}, \qquad (4.70)$$

$$\beta_2 = \arccos\frac{\frac{1}{2}a\left(\left(L - h_p\right)^2 + h_p^2\right)^{0,5}}{\left(L - h_p\right)^2 + h_p^2}, \qquad (4.71)$$

$$a = \sqrt{2}L, \qquad (4.72)$$

because

$$\varphi_1 = \varphi_2, \tag{4.73}$$

therefore

$$\varphi_z = 2\varphi_2. \tag{4.74}$$

Table 4.9 provides example stiffnesses and strengths of selected furniture joints.

Tables 4.8 and 4.9 show that furniture joints are characterised not only by different reliability, but also diverse stiffness and strength. Due to the type of joint and type of joined materials, the stiffness of structural nodes can vary from very small to matching the stiffness of joined elements or exceeding it many times. Small stiffness of joints $k_3 = M_3/\varphi_3$ (Fig. 4.27) causes that in the idealisation of the actual object, they should be treated as articulated joints. Stiffness determined by the quotient $k_1 = M_1/\varphi_1$ exceeds the stiffness of joined elements, which is why in analytical models, joints of such characteristics are considered to be perfectly stiff. Between the curves $k_1$ and $k_3$, there is a huge set of furniture joints showing characteristics of susceptible connections (semi-stiff). Calculating susceptible joints requires a detailed specification of the distribution of all the forces in the structural node and determining places of mutual effect of contact surfaces.

The joints can be divided into two main groups: with a mechanical connector, and shaped and shape-adhesive (Fig. 4.28).

**Fig. 4.27** Characteristics of the stiffness of furniture joints

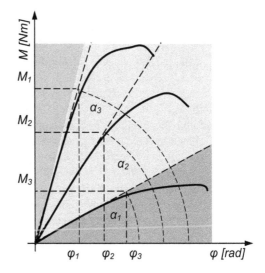

**Fig. 4.28** Division of
furniture joints

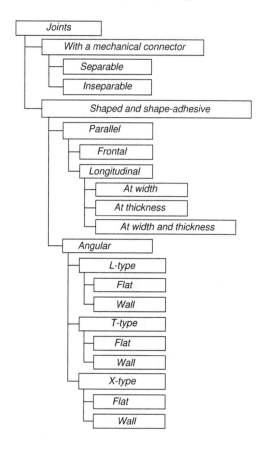

Joints with mechanical connectors form a large group of metal and plastic separable and inseparable structural nodes. Currently, the most representative can be considered joints with connectors such as staples, nails, bolts, screws, hooks and eccentric joints.

Shaped and shape-adhesive joints contain shaped interfaces in specific parts of furniture elements, which ensure their independent connection without or with the use of glue as a connector. Of course, shape-adhesive joints prevail in this group. Formed and perfected for generations, they provide the inseparability of the construction, therefore, a satisfactory stiffness and strength. Due to the mutual system of joined elements, these joints are applied in the design of skeletal furniture, case furniture and bearing structures of upholstered furniture.

### 4.6.1   Joints with Mechanical Connectors

Due to the ease of realisation and in most cases the possibility of disconnecting the elements, mechanical connectors are widely used in furniture practice (Fig. 4.29). Staples, like nails, allow easy assembly, which does not require any additional preparation of the elements before joining. The use of staples in the designs of case furniture in comparison with upholstered and skeletal furniture is quite limited and boils down to fastening the rear walls. In upholstered furniture, staples are used to connect most of the elements of frames and to fasten covers to upholstery frames.

Bolts with nuts and screws for metal are used mainly for fixing metal and plastic fittings or accessories. By binding them with wooden elements, hard species of wood should be used, or wood-based materials of a high density. Bending loads and transverse forces produce high pressures of the core of bolts on the peripherals of holes in joined elements, which consequently increases clearance in the node and reduces the strength of the entire structure. By screwing wooden or metal parts to wooden elements of the furniture piece, screws for wood should be used. Before embedding this connector, in order to avoid cracking of the material, it is recommended to make holes with a diameter that is smaller than the diameter of the screw core. In construction practice, it is proposed to use screws with a length of not more than 2/3 of the thickness of the joined elements. By mounting fittings or other elements to chipboards, due to their loose structure in the middle layer, special screws for chipboards are required. These screws are characterised by a greater diameter of the screw core, a larger skip and height of the threat. The holes for the screws are only slightly larger in diameter than the diameter of the screw core. Thanks to this, the screw can independently transform the hole into a nut when

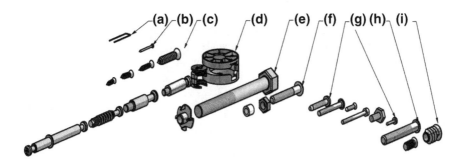

**Fig. 4.29** Mechanical connectors: **a** staple, **b** nail, **c** screws for wood, **d** eccentric joints—box with eccentric joint, core, **e** screw with blind nut, **f** screw with nut, **g** screws for metal, **h** confirmat screw and **i** screws for chipboards

being screwed in. Eccentric joints and confirmat-type screws are commonly used in the assembly of case furniture. An advantage of these connections is the possibility of repeated assembly and disassembly of furniture without significant deterioration to their stiffness and strength. A box with an eccentric joint is usually set in horizontal elements, while the core, through the sleeve or without it, is screwed into a vertical element of the furniture body. Binding elements takes place after turning the eccentric joints and causing assembly stress between the wedge and the core head. Drilling for this type of connectors requires special tools or specialised machine tools. If confirmat-type connectors are used (e.g. $\phi$ 5 × 50 mm), it is enough to drill a through hole with a diameter of 7 mm in the side wall and a blind hole with a diameter of 5 mm in the horizontal element, in order to embed the connector.

## 4.6.2  Shaped and Shape-Adhesive Joints

### 4.6.2.1  Frontal Parallel Joints

Frontal parallel joints are used to increase the length of beam elements. They are commonly used in the production of upholstery frames, strips of furniture boards, door rails and in joinery. This group of joints include bevelled lap joints, straight lap joints, slant lap joints, wedge joints, straight bridle joints, slant bridle joints, multi-wedge (finger) joints (Fig. 4.30).

### 4.6.2.2  Longitudinal Parallel Joints

Longitudinal parallel joints are used to increase the thickness, width or thickness and width of beam elements (Fig. 4.31). Usually, they are used in the production of upholstery frames, strips of furniture boards, door rails, joinery, as well as large-size building elements (beams, girts, trusses, girders, etc.). They are often used in conjunction with frontal parallel joints. This group of joints includes spline rectangular, spline trapezoidal, spline triangular, spline semicircular, spline rectangular with a ledge and spline framed (Fig. 4.32).

### 4.6.2.3  Flat L-Type Joints

Flat L-type joints are used to connect beam elements at an angle. In regard to the value of the slant angle of adjoining elements, we distinguish perpendicular flat L-type joints (Fig. 4.33) and slant flat L-type joints (Fig. 4.34). In the first case, the

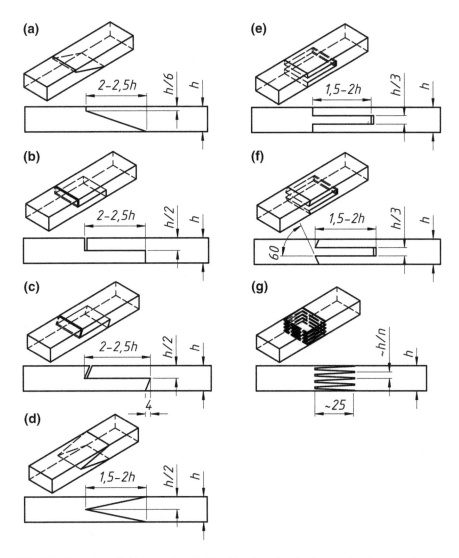

**Fig. 4.30** Frontal parallel joints: **a** bevelled lap joint, **b** straight lap joint, **c** slant lap joint, **d** wedge joint, **e** straight bridle joints, **f** slant bridle joints and **g** multi-wedge (finger) joint

slant angle amounts to 90°, and in the second −45°. However, there is a possibility of joining elements at any angle, but maintaining their sufficient stiffness and strength. The group of perpendicular flat L-type joint includes straight lap joints, single through mortise and tenon joints, single covered mortise and tenon joints,

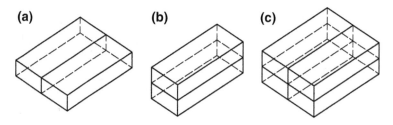

**Fig. 4.31** Longitudinal parallel joints: **a** at width, **b** at thickness and **c** at width and thickness

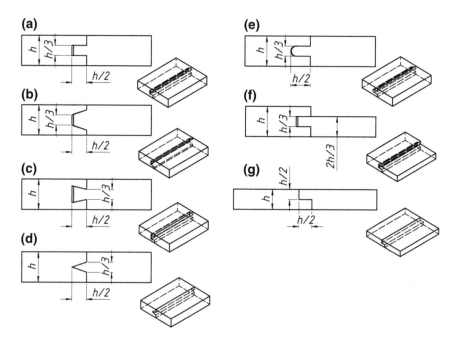

**Fig. 4.32** Longitudinal parallel joints at width: **a** spline rectangular, (**b**, **c**) spline trapezoidal, **d** spline triangular, **e** spline semicircular, **f** spline rectangular with a ledge and **g** spline framed

single semi-covered mortise and tenon joints, single separated mortise and tenon joints, double through mortise and tenon joints, double semi-covered mortise and tenon joints, double dowel joints. The group of slant flat L-type joints includes bevel lap joints, single bevel mortise and tenon joints, double bevel mortise and tenon joints, double bevel dowel joints, covered bevel mortise and tenon joints. Figure 4.35 shows the principles of measuring perpendicular flat L-type joints.

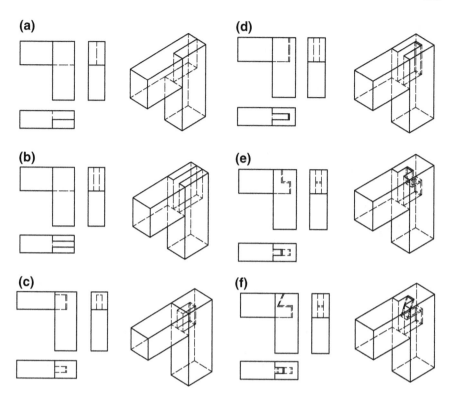

**Fig. 4.33** Perpendicular flat L-type joints: **a** straight lap joint, **b** single through mortise and tenon joint, **c** single covered mortise and tenon joint, **d** single semi-covered mortise and tenon joint, (**e, f**) single separated mortise and tenon joint perpendicular flat L-type joints, **g** single separated mortise and tenon joint, **h** double through mortise and tenon joint, **i** double semi-covered mortise and tenon joint and **j** double dowel joint

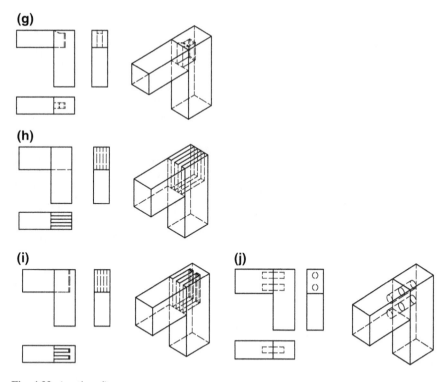

**Fig. 4.33** (continued)

### 4.6.2.4   Flat T-Type Joints

Flat T-type joints in furniture are used for connecting bearing structure elements of case and skeletal furniture. This group of joints includes single covered mortise and tenon joints, double dowel joints, double through mortise and tenon joints, single covered mortise and tenon joints, single through mortise and tenon joints with a cap, fin lap joints and straight lap joints (Fig. 4.36).

### 4.6.2.5   Flat X-Type Joints

Flat X-type joints dominate mainly in joinery products. In furniture, they are used occasionally to connect decorative elements of door panels and upholstery frames (Fig. 4.37).

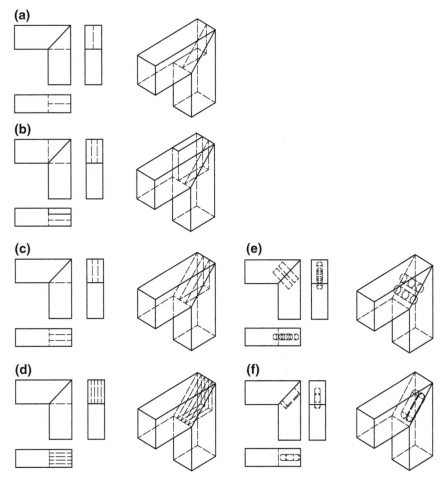

**Fig. 4.34** Slant flat L-type joints: **a** bevel lap joint, (**b, c**) single bevel mortise and tenon joint, **d** double bevel mortise and tenon joint, **e** double bevel dowel joint and **f** covered bevel mortise and tenon joint

### 4.6.2.6 Plate Frontal Parallel Joints

Plate frontal parallel joints are used to increase the length of board elements. Joints with dowel connectors or biscuit joints are most commonly used here (Fig. 4.38).

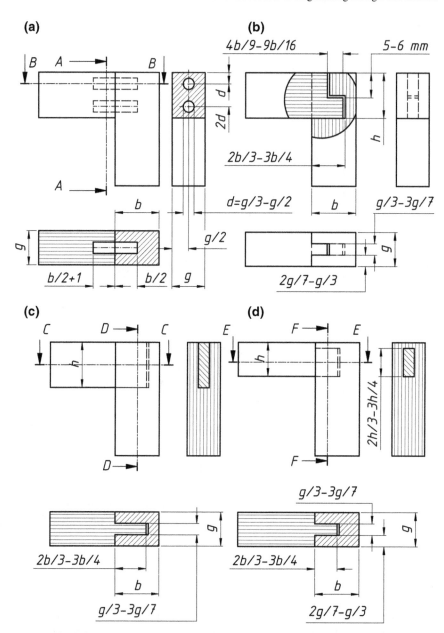

**Fig. 4.35** Principles of measuring perpendicular flat L-type joints: **a** double dowel joint, **b** single separated mortise and tenon joint, **c** single semi-covered mortise and tenon joint and **d** single covered mortise and tenon joint

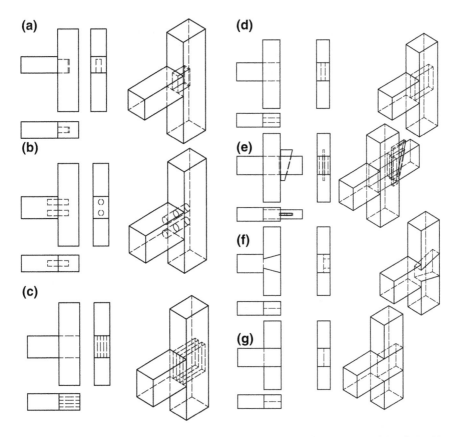

**Fig. 4.36** Perpendicular flat T-type joints: **a** single covered mortise and tenon joint, **b** double dowel joint, **c** double through mortise and tenon joint, **d** single covered mortise and tenon joint, **e** single through mortise and tenon joint with a cap, **f** fin lap joint and **g** straight lap joint

### 4.6.2.7  Plate Longitudinal Parallel Joints

Plate longitudinal parallel joints are used to increase the width of board elements. They are usually used in the production of table worktops, bottoms, top surfaces and side walls of case furniture. Joints with dowel connectors or biscuit joints are most commonly used (Fig. 4.39).

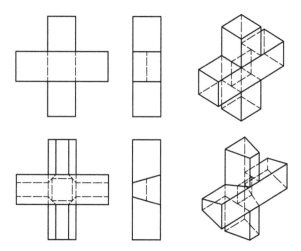

**Fig. 4.37** Flat X-type joints

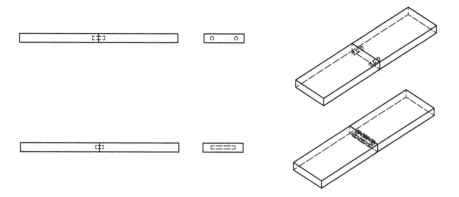

**Fig. 4.38** Plate frontal parallel joints with a dowel connector and biscuit joint

### 4.6.2.8  Plate L-Type Joints

Plate L-type joints are used to connect board elements at an angle. They are
commonly encountered in places of joining side walls with the bottom or top and
rear walls. This group includes perpendicular dowel joints, bevel dowel joints,
straight framed joints, slant framed joints, straight single spline joints, slant double

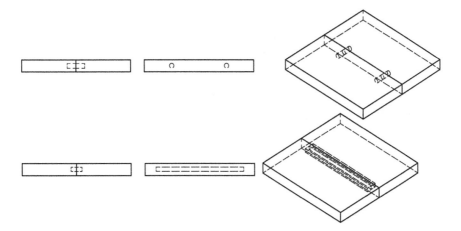

**Fig. 4.39** Plate longitudinal parallel joints with a dowel connector and biscuit joint

spline joints, bevel spline joints, straight spline joints, straight dovetail joints, slant dovetail joints, semi-covered straight dovetail joints and covered straight dovetail joints (Fig. 4.40).

#### 4.6.2.9   Plate T-Type Joints

Plate T-type joints are used to connect board elements of the bodies of case furniture, in particular side walls, bottom boards and top surfaces with horizontal and vertical elements, as well as connecting elements between one another. This group of joints includes straight dowel joints, regular mortise and tenon joints, full spline mortise and tenon joints, sharp spline joints, full spline joints, fin unilateral joints, fin bilateral joints, spline joints and spline framed joints (Fig. 4.41).

#### 4.6.2.10   Plate X-Type Joints

Plate X-type joints occur only in places of joining of elements and interior walls of the bodies of case furniture, as well as places where honeycomb mesh fills of woodwork boards meet (Fig. 4.42).

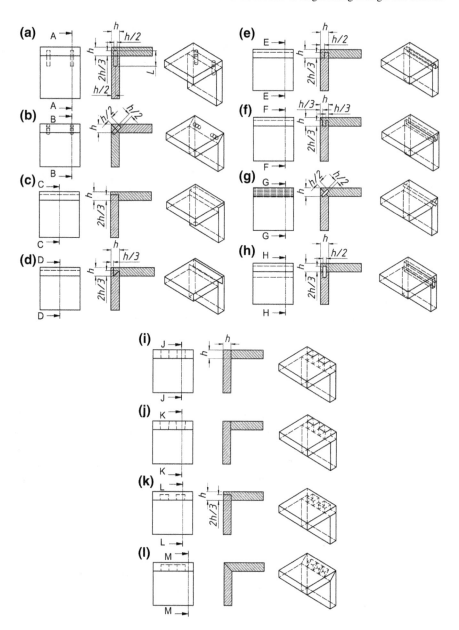

**Fig. 4.40** Plate L-type joints: **a** perpendicular dowel joint, **b** bevel dowel joint, **c** straight framed joint, **d** slant framed joint, **e** straight single spline joint, **f** slant double spline joint, **g** bevel spline joint, **h** straight spline joint, **i** straight dovetail joint, **j** slant dovetail joint, **k** semi-covered straight dovetail joint and **l** covered straight dovetail joint

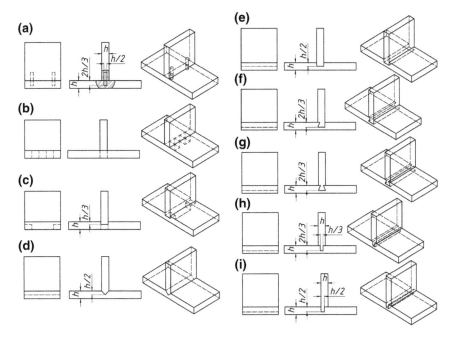

**Fig. 4.41** Plate T-type joints: **a** straight dowel joint, **b** regular mortise and tenon joint, **c** full spline mortise and tenon joint, **d** sharp spline joint, **e** full spline joint, **f** fin unilateral joint, **g** fin bilateral joint, **h** spline joint and **i** spline framed joint

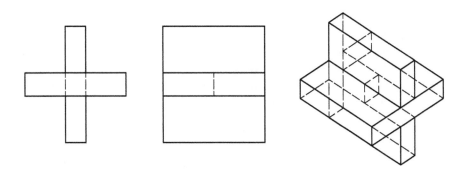

**Fig. 4.42** Plate X-type joints

# 4.7   Assemblages, Subassemblages and Elements of Furniture

The construction of a furniture piece is a type of system and binding of its elements in a way that is correct in terms of the principles of physics and economy. It comprises:

- elements,
- subassemblages and
- assemblages.

A single assemblage part in the structure of a furniture piece is called an element. It can be made of one or several different materials permanently connected with each other.

A subassemblage is created by an assembled set of elements, constituting a separate whole at the stage of assembling the furniture piece. In upholstered parts, connected upholstery layers form a subassemblage.

An assemblage is a set of subassemblages or subassemblages and elements, constituting a separate structural unit, which most commonly fulfils a specific function in the furniture piece.

## 4.7.1   Elements of Furniture

### 4.7.1.1   Board Elements

Board elements of furniture are characterised by a much larger width and length in relation to thickness. They are usually divided into flat and curved elements. They can occur in all kinds of furniture as the top, the bottom, the sides, partitions, shelves, wide fronts and sides of drawers, worktops, etc. (Fig. 4.43). Due to the type of raw material they are made of, board elements require finishing both wide and narrow surfaces. Wide surfaces are subjected to veneering with decorative paper, laminate, foil, PVC and natural veneer. Narrow surfaces can be finished off by gluing wooden or synthetic laminated boards or also by folding and gluing milled edge parts of the board sheet (Fig. 4.44). This type of treatment does not increase the thickness of the board; however, it eliminates the use of veneers and fringes on narrow planes. If the thickness of the bottom, top, partition, side wall or worktop of a table is greater than the thickness of the chipboard, then a profiled laminated wooden board is usually used.

A similar effect can be achieved by milling, folding and gluing cut elements of boards (Fig. 4.45). Thanks to this treatment, the desired visual effect is obtained, without significantly increasing the weight of the furniture piece. Curved board

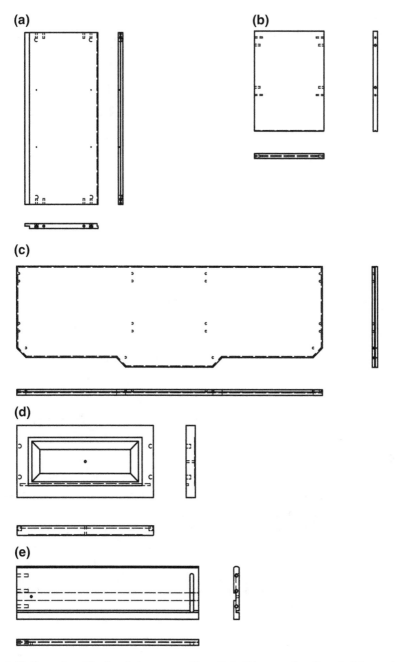

**Fig. 4.43** Examples of flat board elements: **a** side wall, **b** slide, **c** top (*top flange*), **d** drawer front and **e** side wall of drawer

**Fig. 4.44** Methods of
working narrow planes of
board elements

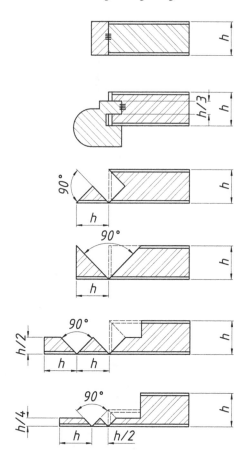

elements can be obtained by gluing many thin layers of board sheets: chipboards,
HDF, MDF, laminates and plywood (Fig. 4.46a). When using a woodwork board,
the corner is usually shaped from four-sided sanded strips, and then, it is all laminated with veneer (Fig. 4.46b). Chipboards and fibreboards are the perfect material
for shaping repeatedly bent surfaces. The designed arches of folds of boards are
obtained by incising grooves with a depth of up to 2/3 of the board's thickness,
bending the board on the same or the opposite side of the incision, and at the same
time laminating both surfaces with laminate, plywood or veneer (Fig. 4.46c, d).

### 4.7.1.2  Beam and Rod Elements

Elements of furniture with dimensions similar to a polygonal transverse cross
section, much smaller than the length dimension, are called beam elements.

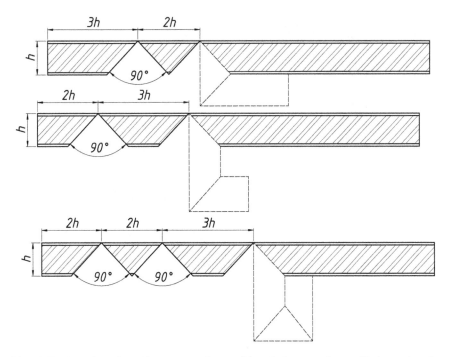

**Fig. 4.45** Examples of working narrow planes of board elements along with increasing the thickness of the element

Rod elements are characterised by circular, elliptic, oval or similar to circular cross sections. They mostly occur as straight and curved. Figure 4.47 shows some examples of finished beam and rod elements: laminated board of the top board, laminated board of the slide, decorative strip of the side wall, bottom horizontal rail and support of a dresser (rod element). Some of them are a decoration and complement the design of the furniture piece, and others such as legs, columns and frames constitute the strength and stiffness of the designed system. Curved elements can be made as a result of bending wood, rattan or bamboo shoots after hydro-thermal treatment, and these are usually bars, rims, semi-rims, support legs, seat frames, connectors, etc. (Fig. 4.48). They can also be cut from a board or lumber according to a curved template (Fig. 4.49). Straight beam elements can also be obtained by gluing boards cut out from chipboards or also by incising the surfaces of boards, and then folding and gluing into uniform closed profiles (Fig. 4.50).

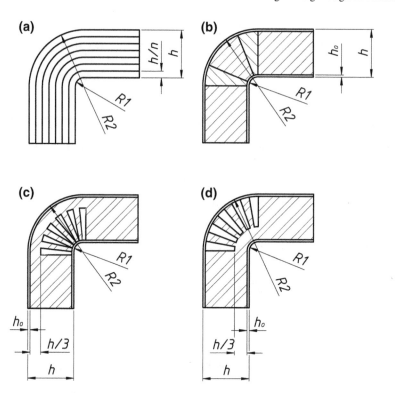

**Fig. 4.46** Examples of curved board elements: **a** element glued from layers, **b** woodwork board with profiled strips and (**c**, **d**) chipboard with incised grooves

### 4.7.1.3  Elastic Elements

In furniture, often elements are used which essential feature is high pliability, much greater than the pliability of other elements. Most structural elements, due to the impact of load, change their dimensions; however, this is a very undesired side effect, since their main purpose is to preserve the designed stiffness and strength. The susceptible elements are by nature prone to strains and are used in those construction assemblages, which are to provide the feeling of comfort and convenience to the user. The most frequently used susceptible elements in furniture include springs, elastomer foams and wooden of plywood fittings (Fig. 4.51).

The main task of susceptible elements is given as follows:

- taking over external forces acting on the furniture piece or its elements, especially forces of shock nature,
- accumulation of energy in order to use it later and
- the even support of the human body in contact with a seat or bed.

**Fig. 4.47** Straight beam and rod elements: **a** laminated board of the top board, **b** laminated board of the slide, **c** decorative strip of the side wall, **d** bottom horizontal rail and **e** support of a dresser (rod element—turned)

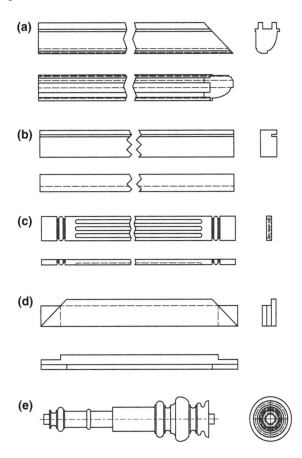

## 4.7.2   Subassemblages of Furniture

Subassemblages of furniture can be built from rod elements, beam elements, boards and panels. Usually, they are connected in a permanent and inseparable way, although in the group of furniture for disassembly there are many subassemblages connected using connectors which ensure freedom in assembling and disassembling them. They occur as 2-D and 3-D structures. The most common subassemblages can include drawers, including cases, doors of a frame panel construction, socle and upholstery frames.

Drawers, although they have a similar purpose, can be made both from wooden and wood-based materials, as well as from metal and plastic. Several types of drawers have been shown in Fig. 4.52.

**Fig. 4.48** Curved bent rod
elements: **a** connection and
**b** semi-rim

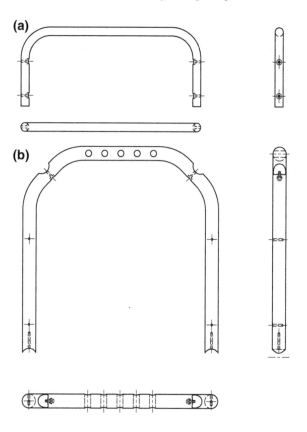

A frame panel door is a typical subassemblage of furniture produced with the great participation of wood. The frame construction is usually made up of beam elements connected with dowels, simple single plugs or splines. The panel can constitute (Fig. 4.53) a milled MDF board, glass, milled wooden board, stained glass and steel sheet with holes.

Socles perform the function of frames of case or upholstered furniture supporting the entire construction. Their composition can consist of wooden or metal rod elements or also of boards and beam elements (Fig. 4.54).

The function supporting the bed or seat in upholstered furniture is performed by upholstery frames. Typical constructions are upholstery frames shown in Fig. 4.55. Wooden bearing structures are made mainly of coniferous or deciduous lumber. As it can be seen, the bearing structures are frames, to which upholstery layers are mounted or applied and/or handles, backrests and seats are attached.

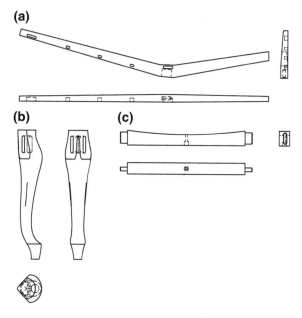

**Fig. 4.49** Curved bent beam elements: **a** support leg, **b** front leg and **c** bar

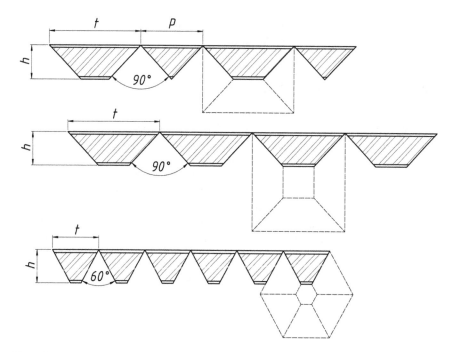

**Fig. 4.50** Straight beam elements as uniform closed profiles

**Fig. 4.51** Elastic elements
used in furniture construction:
**a** spring (one or two cone,
barrel, cylindrical),
**b** elastomer foam and
**c** plywood fitting

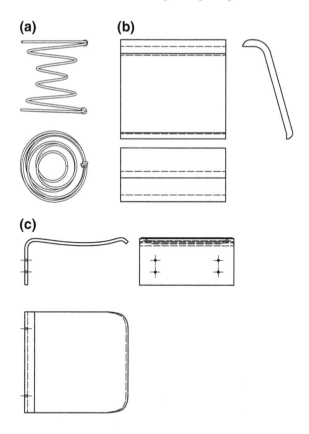

## 4.7.3  Assemblages of Furniture

An assemblage of a furniture piece can be both the finished product, as well as the
assembly of elements along with subassemblages. Figure 4.56 shows the simplest
assemblages of skeletal furniture, i.e. the wooden skeleton of a chair prepared for
staining and painting, as well as a metal skeleton of an office armchair. Each of
these structures is not a finished product and requires completion with a subassemblage or element, in order to assemble the product. Similar characteristics are
shown by the top part of a dresser (Fig. 4.57). Although it is a finished product
(stained and lacquered), however, this cannot function on its own without the
bottom part—cabinet. A dresser cabinet (Fig. 4.58) is a finished product and can be

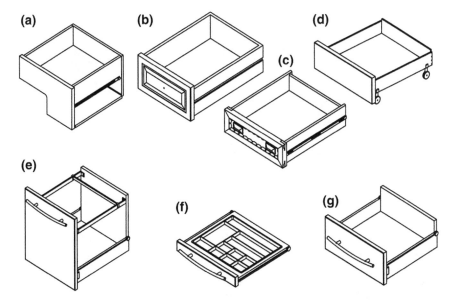

**Fig. 4.52** Examples of drawers: **a** of steel construction with a chipboard front, **b** folding type of a frame panel construction, **c** from boards and with a frame panel front, **d** container for bedding in the armchair, **e** files drawer, **f** case and **g** drawer with front and back from chipboard and steel sheet sides

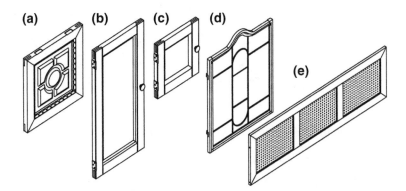

**Fig. 4.53** Frame panel door with filling: **a** milled MDF board, **b** glass, **c** milled wooden board, **d** stained glass and **e** steel sheet with holes

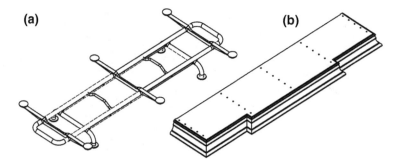

**Fig. 4.54** Socles: **a** metal socle of an office desk and **b** wood and chipboard socle of a house furniture

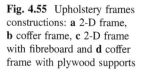

**Fig. 4.55** Upholstery frames constructions: **a** 2-D frame, **b** coffer frame, **c** 2-D frame with fibreboard and **d** coffer frame with plywood supports

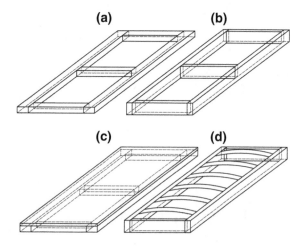

used without additional equipment; however, along with the extension, it constitutes only an assemblage of the dresser. Figure 4.59 shows some finished products, which form one solid usable composition which is a finished product for the user. With this approach, to understand the final product, chairs and a table are the only sets of furniture for eating meals.

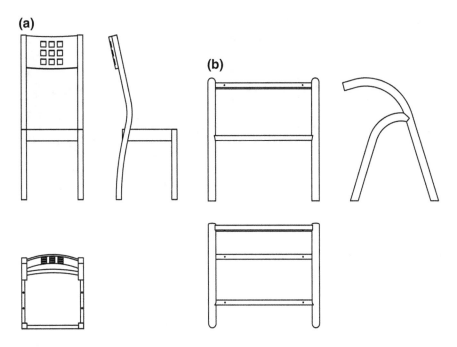

**Fig. 4.56**   Assemblages of furniture: **a** skeleton of a wooden chair and **b** skeleton of a metal chair

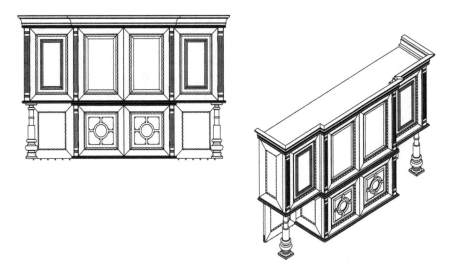

**Fig. 4.57**   Assemblage of a furniture piece—top part of a dresser

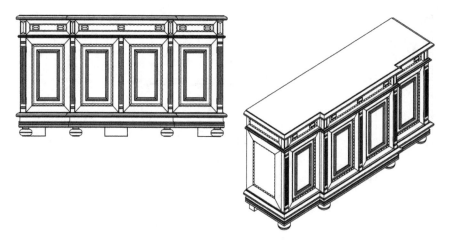

**Fig. 4.58**  Assemblage of a furniture piece, finished product—cupboard of a dresser

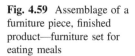

**Fig. 4.59**  Assemblage of a furniture piece, finished product—furniture set for eating meals

## 4.8   Construction of Case Furniture

Construction of case furniture depends mainly on the purpose of the furniture piece. The designer always has a different approach when designing furniture for children, schools, offices, hotels, bedrooms, dining rooms or garages. Joints, types of elements and construction materials are chosen in terms of future conditions of use and variability of loads. A few examples have been provided in this chapter of construction solutions used in the design of case furniture.

## 4.8.1   *Joints of Elements of Furniture Body*

To connect board elements of case furniture, usually joints from the flat L-type and wall T-type group are used. Depending on the needs, these are separable and inseparable joints. Figure 4.60 shows the design of a glass case, in which the side walls have been connected with flanges using coupling eccentric connectors only. A characteristic feature of the chosen type of connector is minimal visibility of the eccentric box on the surface of the flange and the possibility of repeated assembly and disassembly of the construction. These types of joints are used in the production of kitchen furniture, office furniture (especially containers and boxes), as well as bedroom furniture.

Another, probably the most widely used type of joint is a wooden dowels. It is an inseparable joint, preferred in designing constructions supplied to the user in an assembled state (Fig. 4.61). Usually, this type of structural node can be encountered in furniture made only from wood or also in furniture, where it is the dominant material. It should be noted here that this kind of connecting elements of a furniture piece follows the best traditions of the carpentry craft.

**Fig. 4.60**  Glass case connected using eccentric joints

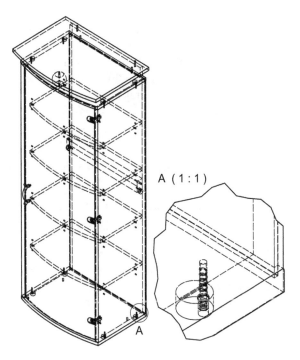

A ( 1 : 1 )

A

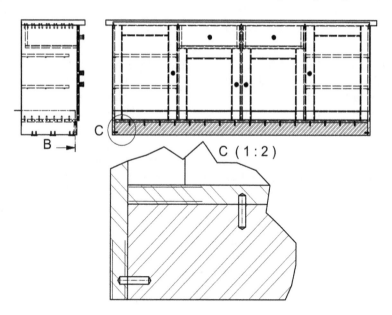

**Fig. 4.61** Cabinet connected using dowel connectors

Modern furniture intended for DIY assembly not only have eccentric joints, but also dowel connectors. However, the task of wooden dowels is not connecting elements, but enabling easy determination of the mutual position of individual elements during DIY assembly by an inexperienced user.

Figure 4.62 shows the dimensional proportions of connectors most commonly used in connections of elements of the bodies of case furniture.

## 4.8.2  Joints of Opening Doors with Elements of the Body

The doors in case furniture enable to close a space limited by the case of the cabinet and also ensure access to its interior. Currently, in industrial practice, a huge number of various types of hinges and fittings are used, which ensure the rotating motion of a door. Figure 4.63 illustrates the way in which a box hinge in a door and a guide in a side wall is set. Depending on the construction of both of these elements, the door can open by an angle from 110° to 360°.

Depending on the way of positioning of the door in a front build, we distinguish constructions with lift-off doors (Fig. 4.64a, b) and constructions with mortise doors (Fig. 4.64c). For each of these constructions, a different guide and hinge with a different shape of the arm are required.

**Fig. 4.62** Examples of
L-type bodies of case
furniture: **a** dowel in a rack
structure, **b** dowel in a flange
structure, **c** spline in a rack
structure, **d** spline in a flange
structure, **e** eccentric in a rack
structure, **f** eccentric in a
flange structure, **g** spline in a
rack structure and **h** spline in
a flange structure

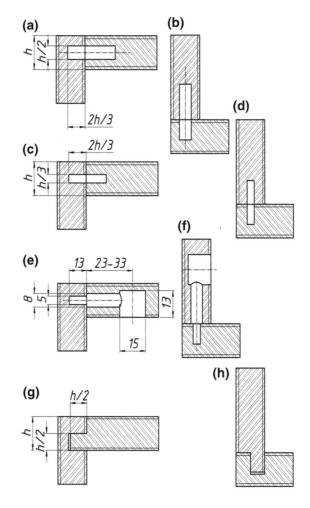

When designing the mounting system of a door in the furniture body, the
designer should foresee the method of solving the doorstop. The easiest scheme of a
collision-free work of a door is leaving a small gap between the edges (Fig. 4.65a).
The door is then rested on narrow planes of the bottom and the top boards (in the
construction of lift-off door) or limiting dowels (in the construction of mortise
door). To eliminate the gap between the wings of the door, often wooden or plastic
doorstop strip is used (Fig. 4.65b), screwed to one of the wings. Solving the way in
which the door is supported is identical as in the previous construction. The use of a
durable (fitted to the body) doorstop strip (Fig. 4.65c) solves both the issue of the
gap between the wings of the door, as well as their support system. If the door has

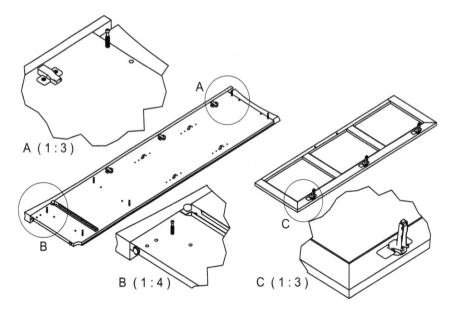

**Fig. 4.63**  Way of setting: *A* guide, *B* runner and *C* box hinge in a door

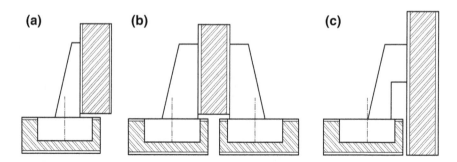

**Fig. 4.64**  Construction of with furniture with a door: **a** lift-off set on an independent side wall, **b** lift-off set on a shared vertical element and **c** mortise set on an independent side wall

been made of wood or is finished with wooden laminated boards, then the doorstop can be created by the appropriate milling of these boards (Fig. 4.65d).

The number of hinges mounted to the door wing depends on the weight of the given element and the anticipated operational load. From the mechanical point of view, the use of more than two hinges on one door wing brings the construction to an over stiff scheme with zero passive forces forming overvalues of the static system. In practice, however, the rule of selecting the number of hinges depending on the height of the door is adopted (Fig. 4.66).

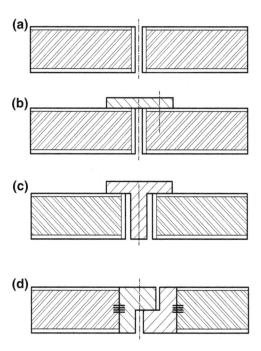

**Fig. 4.65** Examples of doorstop constructions: **a** simple, **b** with doorstop strip on the wing, **c** with doorstop strip in the body and **d** framed

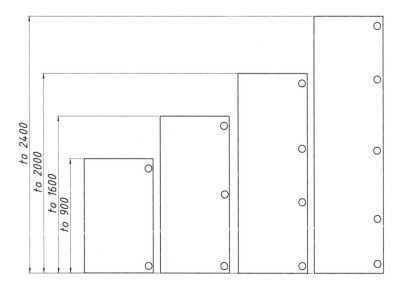

**Fig. 4.66** Method of selecting the number of hinges (dimensions in mm)

### 4.8.3  Joints of Sliding Doors with Elements of the Body

Sliding doors have the same function as opening doors; however, they do not require an equally large space to use them. They are commonly found in built in furniture, office, school and hotel, while less common in home furniture. The most common door construction solutions have been shown in Fig. 4.67.

### 4.8.4  Joints of Louvered Doors with Elements of the Body

The functionality of louvered doors is the highest among all other door constructions, because not only do they not restrict freedom of movement in the space surrounding the furniture piece, but they also allow free access to all the resources gathered in the cabinet. A louvered door is a flexible surface usually formed by gluing wooden strips on a fabric or other thin connector. In metal- and plastic-louvered doors, the proper connectors on adjacent lamellas form hinged joints. In order to function properly, the louvered doors require fitting or forming a guide in the side wall (Fig. 4.68) and installing a false rear wall preventing damage to the sliding mechanism of the louvered door by the stored items.

### 4.8.5  Constructions of Drawers

Depending on the dimensions, the drawer can be used to store small items of a low mass, as well as large and heavy things. The drawers of office furniture (Fig. 4.69)

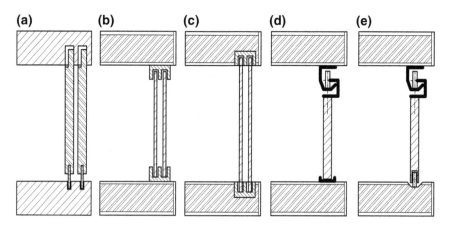

**Fig. 4.67** Constructions of sliding doors: **a** in grooves in the top, **b** in runners screwed to the top, **c** in mortise runners in the board, **d** on the upper reel and lower runner and **e** on reels

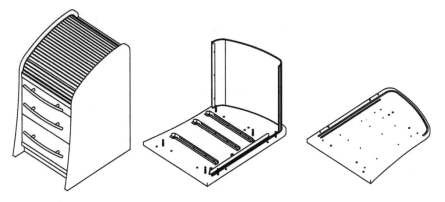

**Fig. 4.68**  An example of fitting louvered doors in the body of a furniture piece

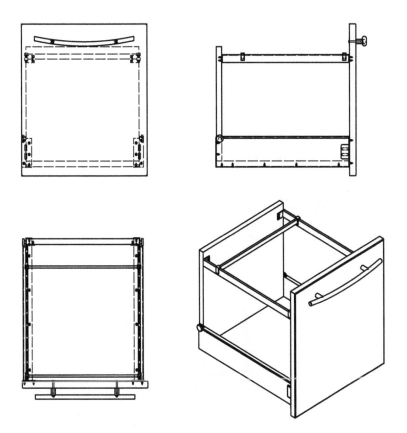

**Fig. 4.69**  Construction of a metal drawer

should be characterised by high durability to static loads and guarantee the full extension of the container beyond the outline of the furniture piece. The basic design of a drawer is made of metal, and only the front and back is made individually by the manufacturer, from wood or chipboard. Home furniture and, in particular, furniture for DIY assembly are fitted with folding-type drawers, made of a front, bottom and strips of sides and a rear wall (Fig. 4.70).

Furniture made of wood requires the use of solid solutions and the best materials. In the subassemblages of case furniture, the fronts of drawers constitute a frame panel subassemblage, and the side walls and rear wall are made from wood or plywood, connected inseparably with dowel, spline or single covered mortise and tenon joints (Fig. 4.70).

Elements connecting the drawer (case) with the body of the furniture piece and at the same time ensuring its extension are runners made from wood, metal or plastic. They can connect the container from the bottom, from the side wall, and from the top at the edge of the side wall (Fig. 4.71).

### 4.8.6  Joints of Rear Wall

The stiffness of the furniture body depends only on the stiffness of immovable component elements of the construction, and the thickness of boards affects it to the greatest degree. Rear wall, usually made from thin fibreboards, chipboards or plywood, improves the stiffness of the furniture piece, taking over the majority of shield loads during operational loads. The strength of the furniture is then

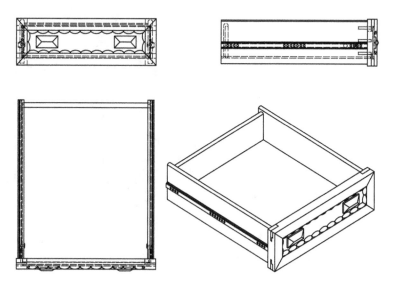

**Fig. 4.70** Construction of a drawer made of wooden elements

**Fig. 4.71** Example of
connecting the runner with a
drawer

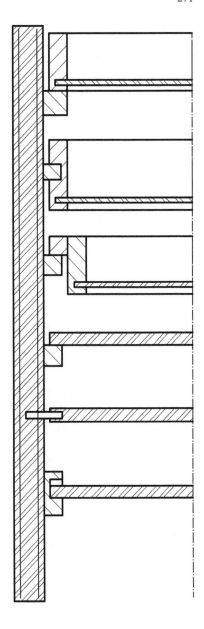

determined by how the board is attached to the body (Fig. 4.72). The most
advantageous contructional solution is gluing the board to the groove (Fig. 4.72c),
and then, the rear wall acts as an elastic shield that was set on the entire circumference. It definitely needs to be avoided to attach the board to the front
(Fig. 4.72a), and at such a scheme, the board reflects the load state of the cut shield
supported discreetly around the entire circumference.

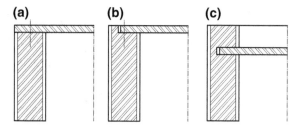

**Fig. 4.72** Example of mounting the rear wall to the furniture body: a) to the front, b) in the frame and c) in the groove

## 4.9   Constructions of Skeletal Furniture

The basic constructional parts creating skeletal elements are beam elements, rod elements and sporadically board elements. The connections of beam elements belong to the angular corner, semi-cross and flat cross group, while the connections of boards are mostly parallel joints. Among these constructions, there is also a certain regularity. The tables are designed as constructions for assembly and disassembly, while chairs in the vast majority constitute closed systems.

### 4.9.1   Constructions of Tables

The main components of tables are the worktop and frame, which consists of rails and table legs. A typical connection of the legs of a table with the rails is a single covered L-type mortise and tenon joint (Fig. 4.73).

**Fig. 4.73** A table of rail construction connected permanently using mortise and tenon joints

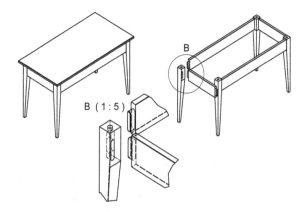

It ensures high durability of the furniture piece; however, it restricts the freedom of transport, and in the event of large dimensional furniture, it forces the designers to seek alternative solutions. This kind of connection can be used in constructions of small sizes and a complex build, requiring special equipment or tools during assembly. Rails and legs of a table can also be connected using screws and special connectors. With such a solution, a furniture piece is suitable for packing into a cardboard box, increasing the use of transport means and reducing the risk of damages during transportation. Examples of inseparable and separable connection of rails with the legs using the connections are shown in Fig. 4.74.

Worktops of tables are made of carpentry boards, chipboards, MDF boards, as well as glued wooden boards. Wide external surfaces are finished with veneers, and narrow surfaces with veneers, edges or laminated boards made of wood and plastic. Worktop can be mounted to the frame using wooden dowels set in holes of the rails and board (Fig. 4.75). However, this is the least popular way and most sensitive to the accuracy of the machining. Other connection methods involve the use of additional strips fixed to the top board of the table and bolted to the rails with screws or bolts (Fig. 4.75b–d).

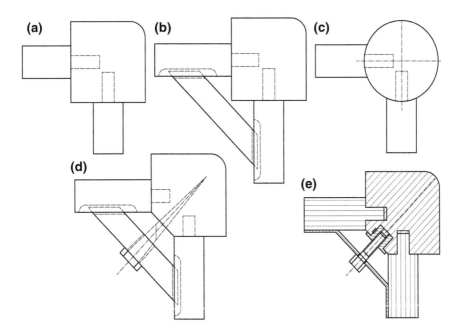

**Fig. 4.74** Examples of connecting rails with legs using the following joints: **a** straight single mortise and tenon joints, **b** straight single mortise and tenon joints with a connector, **c** straight single mortise and tenon joints used for a rod element, **d** straight single mortise and tenon joints with a screw connector and **e** straight single mortise and tenon joints with a bolt connector

**Fig. 4.75** Examples of
mounting working tops of
tables using: **a** dowel joint,
**b** screw joint and strip **c** screw
and spline joint, and **d** bolt
and spline joint

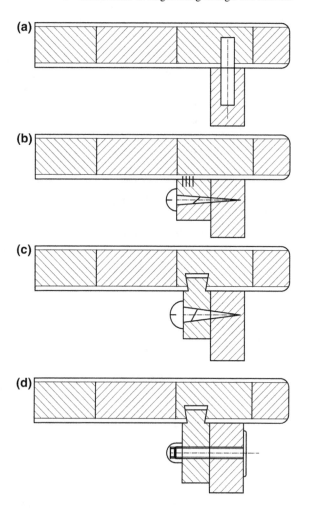

## 4.9.2   Constructions of Chairs

Below, we will consider two ways of constructing chairs, both inseparable despite
the application of typical separable and inseparable connections. Figure 4.76 shows
the construction of a rattan chair, in which all the elements are connected with metal
screws. However, due to the fact that the screws are covered by filling in the holes,
and then reinforced with a weave of leather strips (or rattan), the construction is
completely inseparable.

Similarly, inseparable is the construction of a carpentry chair as shown in
Fig. 4.77. This construction uses only single covered L-type mortise and tenon
joints. However, the designed model can be strengthened by connectors glued by
means of finger joints in the side surfaces of the rails (Fig. 4.78).

**Fig. 4.76** Construction of a
rattan chair with screw joints

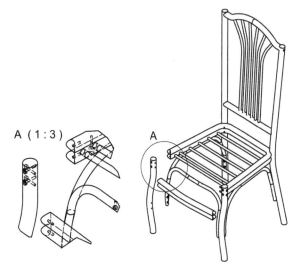

**Fig. 4.77** Construction of a
carpentry chair with mortise
and tenon joints

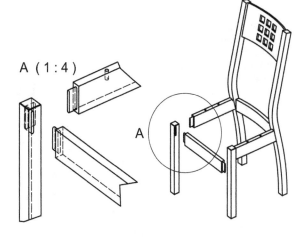

**(a)**    **(b)**

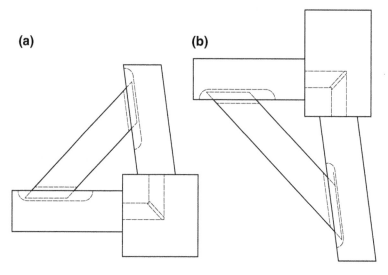

**Fig. 4.78** Examples of construction nodes: **a** of the rail and front leg, and **b** of the rail and support leg

**Fig. 4.79** Construction of upholstery furniture: *1* footrest slide mechanism, (*2, 7*) backrest, *3* seat, (*4, 8*) side, *5* front of the footrest, *6* front of the drawer, (*9, 11*) bolt, and *10* transverse rod

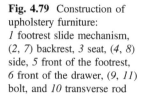
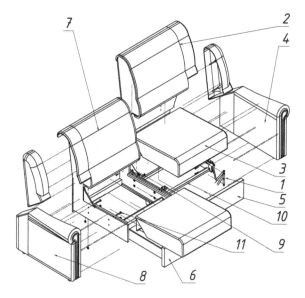

## 4.10   Constructions of Upholstered Furniture

A typical construction of upholstered furniture intended for sleeping and/or relaxing consists of (Fig. 4.79):

- a set of foams glued to the frame,
- a frame,
- a mattress assemblage or seat made of layers:

  - support and base,
  - spring (elastic) and
  - lining,

- a cover being the tapestry layer of the mattress or seat.

**Fig. 4.80** Construction of seat backrest assemblage of a single-person armchair: (*1*, *2*) side, *3* transverse joining skirt, *4* dowel connectors, *5* supporting block, *6* upholstery belts, *7* longitudinal joining skirt and *8* upholstery cardboard

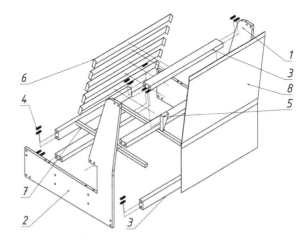

**Fig. 4.81** Construction of side of an armchair: *1* side unit, *2* blind nut, *3* baseboard, *4* overlay board, *5* rear sideboard, *6* armrest board, *7* bevel board, *8* supporting sideboard, *9* vertical skirt, *10* slant armrest board, *11* supporting block, *12* curve cardboard unit, *13* side cardboard unit and *14* armrest cardboard unit

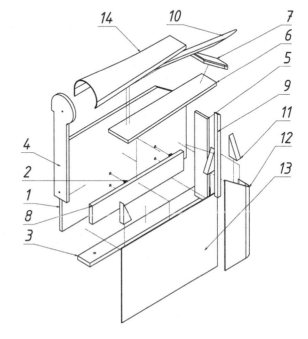

The frame is an essential construction assemblage of upholstered furniture, and it is responsible for its durability, strength and stability. Mostly, furniture frames are made of chipboards, OSB boards, wooden strips (mainly deciduous—beech), hard fibreboard and upholstery cardboard. Figures 4.80 and 4.81 show the frame structure of a seat backrest assemblage and side of a single-person armchair. As it can be seen, some of the construction elements have been connected using dowel joints. In industrial practice, however, inserted staple-type connections are mostly used.

The support and base layer is a part of the upholstery assemblage, on which the remaining layers of the mattress or seat pillow are rested. It has the form of a frame, which gives the desired stiffness and constitutes a bearing part of the product. The base in such a construction can be:

hard      made for example of wooden strips, chipboards, hard fibreboards (Fig. 4.82e);

flexible  made of rubber belts, textile rubber tapes, sackcloth belts (Fig. 4.82a); and

spring    made of flattened spiral sinusoidal springs and springing grids (Fig. 4.82b–d).

The strength of the upholstery frames is connected with the cross-sectional dimensions of rails and type of the material used. In these constructions, it is recommended to use rails with the cross section $38 \times 43$ mm. Research confirms that not only traditional frames made from solid wood of different species, but also frames made of elements glued from layers, even bases made of wood-based materials meet the strength expectations, often increasing the stiffness of an upholstery base and equalling the strength level of the whole base structure (Kapica et al. 1983, 1985).

Spring layers in upholstered furniture are an essential part of the product both in terms of volume, as well as usage. In furniture for sleeping and/or relaxing, they may constitute the fillings:

spring           made using spring units: bonnell, schlarafia, pocket springs (Fig. 4.83a);

without          made from flexible elastomers, latex foam, coconut mat, seagrass,
springs          etc. (Fig. 4.83b)

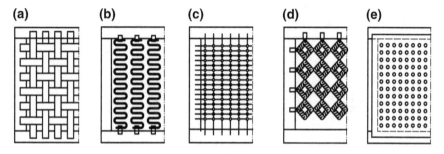

**(a)**        **(b)**        **(c)**        **(d)**        **(e)**

**Fig. 4.82** Types of support layer bases: **a** sackcloth belts, **b** sinusoidal springs, **c** springing grid, **d** flattened springs and **e** hard fibreboard

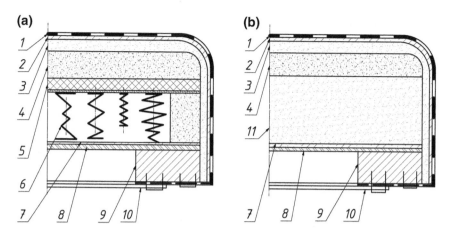

**Fig. 4.83** Construction of upholstery system: **a** spring, **b** without springs: *1* upholstery fabrics, *2* wadding, *3* highly flexible foam, *4* soft polyurethane foam, *5* coconut mat (stubble and latex mat), *6* springs (biconical, cylindrical, barrel, etc.), *7* covering fabrics or felt, *8* base layer, *9* upholstery frame, *10* underlay fabrics and *11* hard polyurethane foam

Spring layer gives the upholstery system appropriate elastic properties, which in effect leads to a dynamic adaptation of the base shape to the silhouette of the user, and at the same affects in a significant way the comfort and ergonomics of use of the product. Construction of the spring layer should be dependent on the individual needs of the user, and above all on his weight, age and preferences (most commonly adopted position during sleeping and relaxing). The stiffness of spring systems (and thus the feeling of comfort of use) is affected by the use of springs with different internal diameters. In addition, the softness of upholstery units (depending on the purpose of furniture) can be easily modelled by the appropriate selection of inside diameter of springs, their number and spacing (Dzięgielewski and Smardzewski 1995a). The softness of biconical springs, and at the same time of bonnell type spring units, depends on the diameter of the smallest coil (Kapica and Smardzewski 1994), and these units, despite their widespread use, do little to comply with ergonomic requirements for lounge furniture (Smardzewski 1993). According to Kapica (1988), upholstery systems without springs made as a layer system show generally greater hardness than upholstery systems with bonnell spring unit. The most widely used spring materials include:

- bonnell spring units, made with biconical springs with five coils and wire diameter 2.0–2.2 mm, connected at the top and bottom part with spiral springs of wire diameter 1.3–1.4 mm. Mostly, springs have dimensions: external diameter 80–85 mm, height of the springs 120–130 mm and distance between the springs in the unit 20–40 mm;
- pocket spring units, made of cylindrical springs of wire diameter 1.2–2.5 mm, placed additionally in bags made of canvas or other upholstery cloth and woven

upholstery belts, joined with upholstery cord (Dzięgielewski and Smardzewski 1995b);

- polyurethane foams, belonging to the group of elastomers, are characterised by high durability, chemical and physical resistance, and most of all resistance to abrasion;
- latex foams are made in the process of vulcanization of foamed natural rubber milk. Due to low natural ventilation of the material, in latex boards special ventilation channels are made;
- coconut mats are made in two technologies as needle coconut mats and vulcanised mats; and
- seagrass—combed and dried leaves of palm trees, additionally can be subject to needling on a base fabric.

Properties of polyurethane foams primarily depend on the structure that is the size of the cells and their shape (density), the construction of the cells (open or closed), as well as material constant values (Saha et al. 2005). Brandel and Lakes (2001) have showed that there is a possibility of thermomechanical modification of the polyurethane foams structure in which there is a change of Poisson ratio—from positive to negative. This positively affects the physical properties of these materials and causes almost equalling the value of linear pliability module with the value of figural pliability module. In mattresses, foams with a density from 14.5 to 65.0 kg/m$^3$, stiffness from 1.0 kPa to approximately 7.0 kPa, permanent pliability from 4 to 20 % and resiliency of 37–80 % are used. Figure 4.84 shows the order of glueing foams on the seat frame. Foams most widely used in furniture industry include:

- classical (standard), e.g. type T;
- classical flame resistant, e.g. type C;

**Fig. 4.84** Order of gluing foams on the seat frame: *1* frame, *2* highly flexible foam, *3* medium foam and *4* soft foam

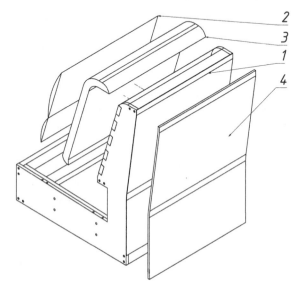

- highly elastic (highly resilient), e.g. type HR;
- highly elastic—flame resistant (highly resilient flame resistant), e.g. type CMHR; and
- low elasticity (thermoelastic).

Thermoelastic foams have so-called memory, which is the ability to slowly return to the original shape after strains. In addition, they are characterised by the ability to absorb vibrations and shocks, which has been confirmed by research: 20 % effectiveness of rebounding a ball and at 50–60 % effectiveness for regular polyurethane foams. The foam has the ability to adapt to the contours of the body, which causes that its surface gently affects the human body, especially in such sensitive areas as heels, hips, elbows and head. Foam properties largely depend on ambient temperature: in low temperature—e.g. about −1 °C—the foam becomes very stiff, while optimal stiffness is observed at 13–29 °C. Susceptibility to strains increases along with increased relative humidity of air.

The lining layer in furniture functions as insulation (heat shield) and affects the softness of the system, increasing the comfort of use. Usually, fabrics weighing 50–400 g/m$^2$, as well as canvas, packing fabrics, wadding and sheep's wool are used.

Tapestry layer is the last, outer layer (cover), fulfilling the function of a tensioner of the upholstery system and raising the aesthetic appearance of the product. This layer is typically made from fabrics such as terry, stretch, jacquard, chenille, velour, nubuck, cotton, as well as knitted fabrics, natural and ecological leathers.

# References

Ashkenazi EK (1958) Anizotropija miechaniczieskich swoistw driewiesiny i faniery. Gostiechnizdat, Moscow, Leningrad

Cz Bąbiński (1972) Elementy nauki o projektowaniu. WNT, Warsaw

BN-73/7113-06 Export grade plywood for general purpose

Bodig J, Goodman JG (1973) Prediction of elastic parameters for wood. Wood Sci Tech 4:249–264

Brandel B, Lakes RS (2001) Negative Poisson's ratio polyethylene foams. J Mater Sci 36:5885–5893

BS 1203:2001 Hot-setting phenolic and aminoplastic wood adhesives. Classification and test method

Chyrosz M, Zembowicz-Sułkowska E (1995) Materiałoznawstwo odzieżowe. Wydawnictwo Szkolne i Pedagogiczne, Warsaw

Dietrych J (1974) Projektowanie i konstruowanie. WNT, Warsaw

DIN 68705-2:2003-10 Plywood—Part 2: Blockboard and laminboard for general use

Dzięgielewski S, Smardzewski J (1995a) Meblarstwo. Projekt i konstrukcja. Państwowe Wydawnictwo Rolnicze i Leśne, Poznań

Dzięgielewski S, Smardzewski J (1995b) Badania formatek sprężynowych mebli tapicerowanych. In: Zbornik Prednasok z vedecko-odborneho seminara Calunnicke Dni, Katedra Nabytku a Drevarskich Vyrobkov, Drevarska Fakulta Technickej Univerzity vo Zvolene, p 7–25

Gabrusewicz W, Kamela-Sowińska A, Poetschke H (2002) Rachunkowość zarządcza, PWE Warszawa

Galewski W, Korzeniowski A (1958) Atlas najważniejszych gatunków drewna. Państwowe Wydawnictwo Rolnicze i Leśne, Warsaw

Gasparski W (1978) Projektowanie—koncepcyjne przygotowanie działań. PWN, Warsaw

Hausen B (1981) Woods injurious to human health. Walter de Gruyter, Berlin, New York
Hearmon RFS (1948) Elasticity of wood and plywood. Forest Products Research, Department of Scientific and Industrial Research, London, Special Raport No. 7
Jabłoński J (2006) Ergonomia produktu. Ergonomiczne zasady projektowania produktów, Wydawnictwo Politechniki Poznańskiej
Kapica L (1988) Badania porównawcze konstrukcji zespołów tapicerowanych mebli do leżenia. Przemysł Drzewny 1:2–5
Kapica L, Dzięgielewski S, Proszyk S, Bilski A (1983) Drewniane ramy tapicerskie z elementów warstwowo klejonych. Przemysł Drzewny 10:9–14
Kapica L, Smardzewski J (1994) Wpływ obróbki cieplnej sprężyn dwustożkowych do formatek bonnell na ich właściwości deformacyjne. Przemysł Drzewny 7:3–5
Kapica L, Korgól B, Malinowski J (1985) Podłoża tapicerskie z tworzyw drzewnych dla mebli do leżenia. Przemysł Drzewny 9:6–8
Karpiński T (2004) Inżynieria Produkcji. WNT, Warsaw
Kokociński W (2002) Anatomia drewna. Wydawnictwo Pawdruk, Poznań
Kokociński W, Romankow J (2004) Substancje szkodliwe dla zdrowia pracowników drzewnictwa znajdujące się w pyłach i wiórach drzewnych. Przemysł Drzewny 10:20–23
Kollmann F, Cote W Jr (1968) Principles of wood science and technology, I-solid wood. Springer, New York
Kotarbiński T (1990) Elementy teorii poznania logiki formalnej i metodologii nauk. Ossolineum, Wrocław
Krick EV (1971) Wprowadzenie do techniki i projektowania technicznego. WNT, Warsaw
Krzysik F (1978) Nauka o drewnie. PWN, Warsaw
Lekhnickij SG (1977) Teorija upugosti anizotropnovo tela. Izdatielstwo Nauka, Glavnaja Redakcja Fizyko-Matiematicheskojj Literatury, Moscow
Litewka A (1997) Wytrzymałość materiałów. Wydawnictwo Politechniki Poznańskiej, Poznań
Nadler G (1988) Metodologia i projektowanie systemowe. Projektowanie i systemy. PAN, Komitet Naukoznawstwa, Ossolineum, Warsaw
Nowacki W (1970) Teoria sprężystości. PWN, Warsaw
Nowak E, Gabrusewicz W, Czubakowska K (2005) Podstawy rachunkowości zarządczej. Państwowe Wydawnictwo Ekonomiczne, Warsaw
Oniśko W (1994) Technologia Tworzyw Drzewnych 1. WSiP, Warsaw
Pecina H, Paprzycki O (1997) Pokrycia lakierowe na drewnie. PWRiL, Poznań
Placek W (1996) Encyklopedia Badań Medycznych. Wydawnictwo Medyczne MAKmed, Gdańsk
PN-83/D 97005.11 Plywood for general use. Requirements
PN-EN 636:2013-03 Plywood. Specifications
Pożgaj A, Chovanec D, Kurjatka S, Babiak M (1995) Štruktúra a vlasnosti dreva. Príroda a.s, Bratislava
St Proszyk (1995) Technologia tworzyw drzewnych. Part II. Wykończanie powierzchni, WSiP, Warsaw
Rabinowicz AN (1946) Ob uprugikh postojannych i procznosti aviacionnykh materialov. Trudy CAGI 582:1–56
Saha M, Mahfuz H, Chakravarty U, Uddin M, Kabir M, Jeelani S (2005) Effect of density, microstructure, and strain rate on compression behavior of polymeric foams. Mater Sci Eng A 406:328–336
Schulte M, Frühwald A (1996) Shear modulus, internal bond and density of medium density fibre board (MDF). Holz als Roh u. Werkstoff 54:9–55
Smardzewski J (1993) Model ergonomiczny formatki sprężynowej. Przemysł Drzewny 11:6–8
Smardzewski J (2005) Niezawodność konstrukcji mebli skrzyniowych. Przemysł Drzewny 6:24–27
Smardzewski J (2007) Komputerowo zintegrowane wytwarzanie mebli. Państwowe Wydawnictwo Rolnicze i Leśne, Poznań
Smardzewski J (2009) The reliability of joints and cabinet furniture. Wood Research 54(1):67–76
Szczuka J, Żurowski J (1995) Materiałoznawstwo Przemysłu Drzewnego. WSiP, Warsaw

Tarnowski W (1997) Podstawy projektowania technicznego. Wspomaganie Komputerowe CAD, CAM. WNT, Warsaw

Tyszka J (1987) Powierzchniowe uszlachetnianie wyrobów z drewna. WNT, Warsaw

Tytyk E (2006) Podstawy metodologii projektowania ergonomicznego. In: Jabłoński J (ed) Ergonomia produktu. Ergonomiczne zasady projektowania produktów. Wydawnictwo Politechniki Poznańskiej, Poznań

Wilczyński A, Kociszewski M (2000) Anizotropia właściwości sprężystych płyty wiórowej. In: Materiały XIII Sesji Naukowej p.t. Badania dla meblarstwa, Poznań, p 101–105

Wilczyński M, Tydryszewski K, Kociszewski M (2001) Anisotropy of mechanical properties of particleboard and medium density fibreboard in bending. In: Materiały XIV Sesji Naukowej p. t. Badania dla meblarstwa, Poznań, p 135

Zawierta Z (2007) Technologia produkcji i wykorzystanie płyt komórkowych bezramkowych w meblarstwie. Typescript Department of Furniture Design, Agricultural Academy of Poznań

Zenkteler M (1996) Kleje i klejenie drewna. Wydawnictwo Akademii Rolniczej w Poznaniu, Poznań

# Chapter 5
# Technical Documentation

## 5.1 Types and Contents of Technical Documentation

The immediate product of design and construction activities is technical documentation. It is a collection of drawings and other documents that are created during designing. Bearing in mind the order of documentation arising, the variability of the scope and form of furniture, as well as separate treatment in the evaluation and approval system of projects, the following types of technical documentation of furniture are distinguished:

- Preliminary draft—consists of drawings and other documents presenting the concept of the furniture concerning its purpose, function, form, structure, materials and finishes used. The final result of this draft may be a mock-up made in a reduced scale or to the scale of 1:1. Mock-up—a type of model of the furniture piece or its part—made from substitute materials, usually to the scale of 1:1, in order to check the visual form, proportions, and even its functionality;
- Technical draft for modelling—contains drawings and other documents concerning the concept of the furniture piece in terms of its function, construction, materials and finishes used, enabling the realisation of its model. The final result of this project should be a model. The model is the first copy of the furniture piece made from appropriate materials to the scale of 1:1; and
- Technical draft for prototypes—comprises drawings and other documents of executive nature, constituting the final stage of designing the furniture piece and developed in such a way that enables the realisation of the prototype or test series. The final result of this project is the prototype. A prototype is a model of the furniture piece made in production conditions, taking into account the specific technical, technological, material and organisational characteristics of the company.

© Springer International Publishing Switzerland 2015
J. Smardzewski, *Furniture Design*,
DOI 10.1007/978-3-319-19533-9_5

The design documentation typically includes:

- technical description,
- technical requirements (conditions),
- view drawings of the furniture piece,
- assembly drawing,
- executive drawings,
- drawings of the more important details,
- drawings of fittings and non-standard accessories,
- drawings of packaging,
- drawings (instructions) of packing the furniture,
- drawings (instructions) of assembling furniture,
- strength calculations and
- photographs, mock-ups and models.

The technical description should include:

- the name and symbol of the furniture piece, suite or set;
- the name of the designer or the names of the members of the design team and the name of the institution where the design was created;
- components of the furniture suite or set overall dimensions and more important functional dimensions;
- purpose and description of functionality; and
- realisation.

The description of the purpose and functionality of the furniture piece is in every case a necessary component of the different types of technical documentation and should contain information on:

- the general purpose of the furniture piece in the furniture—housing interior relation, the surrounding objects, as well as how to meet social demand;
- the detailed purpose of the furniture piece in the relation furniture—individual user.

The description of the purpose of multifunctional furniture should include information about how to use the furniture piece. And the description of realisation include the details that were not provided in the drawings, and relating to materials, construction solutions, veneering, finishing, upholstering methods, type of packaging, etc. For upholstered furniture, the elasticity class should be provided. Illustrative materials can also be included, but only in a situation when they are as follows:

- complicated multifunctional furniture,
- complex interior design and
- furniture of innovative structural-material solutions.

These materials include sketches, illustrative drawings, charts and similar documents explaining the functional program and colour solutions. For a set of segment furniture, it is recommend to make vertical projections of the rooms using programmed arrangement systems. The calculations should justify the construction and material solutions adopted in the project, e.g. the strength of joints and cross sections of construction elements.

A mock-up is done if the solutions to the shape and design of the furniture piece are particularly complicated, difficult to present in a drawing. It is also a supplement to the preliminary technical documentation.

A photograph of the model mainly shows the aesthetic and structural features of the furniture piece. In the case of multifunctional furniture piece or set, it is recommended to take a number of images that is necessary for a clear presentation of individual functions or functional programs.

Technical requirements (conditions) complement the technical drawing. In technical conditions concerning the furniture piece, the requirements should be specified which were not shown or cannot be shown in a drawing, and which are relevant to the development of quality of the furniture piece. Usually they should include:

- a reference to the norms related to a particular type of furniture;
- a description of the place and climatic conditions of using the furniture piece (office, kitchen, school, etc.);
- a description of the quality requirements for materials, including the definitions of permissible defects or repetitive systems: colour, drawing, polishing or the tone of wood or other materials (the proper material standards and quality certificates should be referenced);
- the requirements concerning accuracy of realisation, including the quality of joints, colour and system of drawings of wood on contiguous elements, the thickness of adhesive bond and its colour, clearances of moving parts, smoothness of surfaces, size of bending, phasing and rounding the edges, etc. These requirements should be clearly differentiated for frontal, exterior, interior and invisible surfaces;
- requirements in the scope of assembly, working and finishing fittings surfaces, and requirements concerning packaging, storage and transport of the furniture piece.

The technical requirements in many cases form the basis for settling disputes in the event of a complaint.

The number and type of drawings included in the technical documentation for particular stages of designing depend on the complexity of the shape and structure of the furniture piece. Depending on the type of documentation, its component parts are shown in Table 5.1.

Particular parts of the documentation have individual requirements. And so:

- The view drawing of the furniture piece is done on a sheet of A3 or A4 format in reduced scales. Usually this drawing shows the external appearance of the furniture piece according to the principles of rectangular and perspective viewing. At the same time, it is recommended that the perspective drawing presenting the visual form of the product was done on a separate sheet. For readability of the documentation, however, it is permitted to place the

**Table 5.1** Component parts of individual types of design documentation

| Component parts of documentation | Type of documentation | | |
|---|---|---|---|
| | Preliminary draft | Technical draft for modelling | Technical draft for prototypes |
| View drawing | + | + | + |
| Assembly drawing, | − | + | + |
| Drawings of the more important details | + | − | − |
| Drawings of fittings and non-standard accessories | − | + | + |
| Technical description | + | + | + |
| Illustrative materials | + | − | − |
| Calculations | − | + | + |
| Mock-up | + | − | − |
| Photographs | − | − | + |

perspective drawing over the drawing board of the sheet, on which the views of the furniture piece in rectangular viewings have been presented. In the drawing of the furniture piece's views, done according to the principles of rectangular viewing, it is recommended to place the most important dimensions and functional parameters in addition to the overall dimensions.

- The assembly drawing should be done in accordance with the principles of rectangular viewing using the necessary cross sections, on sheets constituting multiple A4 units. Assembly drawings can be done in two varieties, in a reduced scale with drawn out structural details or to a scale of 1:1 applied to box furniture of uncomplicated construction. If a scale greater than 1:5 is applied, the assembly drawing should include the views of the furniture piece, realised in accordance with the principles of rectangular projection, to a scale of 1:10. The scale of 1:1 is applied with the use of dimensional shortcuts in relation to case furniture and skeletal furniture of a complicated construction.
- Drawings of the more important structural details are drawn up to a scale of 1:1 and on A4 and A3 sheets, in order to better understand the structure of the furniture piece.
- The drawings of fittings and non-standard furniture accessories are realised as executive drawings according to the principles of the technical machine drawing to a scale of 1:1 and 1:2.

In the developed technical documentation, on the outer page of its cover, at least the following inscriptions and symbols should be included:

- the name of the furniture piece of furniture set, documentation marking;
- symbol or name of institution where the documentation was created;
- the name of the designer, author of the design or name of members of the design team; and
- date of completion of developing the documentation.

## 5.2    Formats of Sheets

The format of the drawing sheet is called the format of this drawing's copy after being cut in accordance with PN-76/N-01601. The basic format of the sheet is A4 with the measurements after being cut 210 × 297 mm. The formats A3, A2, A1 and A0 are derived from the multiplication of the A4 format, while the formats A5 and A6 are derived by dividing it into two and four equal parts (Fig. 5.1). All formats marked by the letter A are basic formats. In addition to them, formats can also be used which are created by multiplying the shorter sides of the basic formats in accordance with Table 5.2.

Marking the derivative format consists of marking the basic format, the digit indicating the multiplexing of the shorter side of the basic format, e.g. A3 × 7. Border deviations of sides are accepted for both basic and derivative formats, which do not exceed the values shown in Table 5.3.

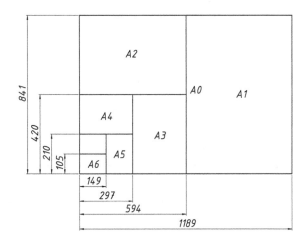

**Fig. 5.1** Basic formats of drawing sheets

**Table 5.2** Formats of derivative sheets

| Multiplexing | Formats [mm] | | | | |
|---|---|---|---|---|---|
| | 10 | 11 | 12 | 13 | 14 |
| 2 | 1189 × 1682 | | | | |
| 3 | 1189 × 2523 | 841 × 1783 | 594 × 1261 | 420 × 891 | 297 × 630 |
| 4 | | | 594 × 1682 | 420 × 1189 | 297 × 841 |
| 5 | | | 594 × 2102 | 420 × 1486 | 297 × 1051 |
| 6 | | | | 420 × 1783 | 297 × 1261 |
| 7 | | | | 420 × 2080 | 297 × 1471 |
| 8 | | | | | 297 × 1682 |
| 9 | | | | | 297 × 1892 |

**Table 5.3** Border deviations of dimensions of sides of formats

| Dimensions of sides [mm] | | Deviations [mm] |
|---|---|---|
| from | to | ±1.5 |
| – | 150 | |
| 150 | 600 | ±2.0 |
| 600 | – | ±3.0 |

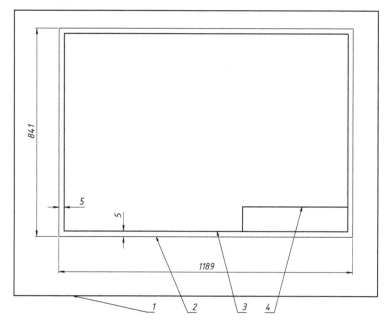

**Fig. 5.2** The basic format with cutting lines and border: *1* sheet before cutting, *2* cutting lines to the dimensions of the basic sheet, *3* border and *4* block

Sheets before cutting, that is in a state in which they are attached to the tablet or placed in the plotter, before realisation of the drawing, should have cutting lines marked, which limit the surface of the sheet after cutting (Fig. 5.2). At a distance of 5 mm from the cutting line, the sheet should have a border limiting the surface of the sheet for the drawing and title block.

## 5.3  Title Blocks

Blocks used in technical furniture drawing can be divided into two types (BN-90/7140-03/02):

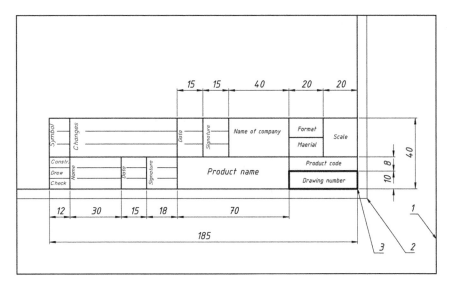

**Fig. 5.3** Basic block: *1* edge of the sheet of paper, *2* cutting line to the dimensions of the basic sheet and *3* border

- basic blocks and
- assembly blocks.

The basic block (Fig. 5.3) should include in the drawings of the assemblages, subassemblages and elements, providing information that identifies the subject of the drawing and the constructor, and including at least:

- the name of the product, assemblage, subassemblage or element;
- the number of the drawing indicating the hierarchy of the object in the structure of the product;
- scale;
- the name of the company;
- personal data of the persons responsible for the documentation, dates and signatures; and
- annotations about changes.

The reduced basic block is filled in, in accordance with the provisions (BN-90/7140-03/02). In the box "scale", one main scale should be provided which is valid for a given sheet, especially where no scale has been indicated separately. The company name and name of the object should not be entered.

The assembly block is placed in assembly drawings. It consists of the basic block and list of parts (Fig. 5.4), which should include the following information:

- the number of the position of the marked part in the drawing;
- the name of the assemblage, subassemblage, element, fittings or connector;

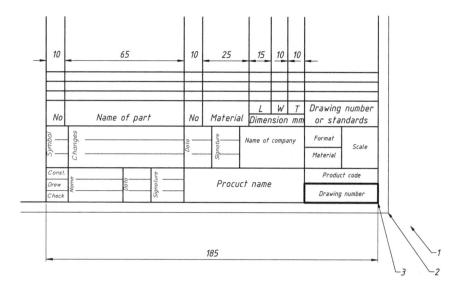

**Fig. 5.4** Assembly block: *1* edge of the sheet of paper, *2* cutting line to the dimensions of the basic sheetand *3* border

- the number of pieces of a given part in the product and subassemblage;
- type of material;
- number of drawing or norm.

The list of parts of the assembly drawing of the product should specify all assemblages, subassemblages and elements which do not constitute parts of the separate subassemblages. Subassemblages that make up the assemblage should be included in the list of parts of the assembly block of a drawing of a given subassemblage. It is recommended that on the list, the elements constituting the subassemblage or subassemblages constituting the assemblage are included in accordance with the expected technological process for this product. This is in order to locally, logically and technologically relate simple elements, fittings and accessories with elements that are more complex. All parts of the furniture piece must be entered in singular form regardless of the number of pieces in the product, assemblage or subassemblage. In the event of the repeated occurrence of two or more identical positions in a given column, characters of repeatability cannot be used.

In the production plant, each completed element, which the drawing documentation concerns, should have a specific drawing number, which should be written in the last column of the list of parts. For the remaining elements, the norm number or catalogue number should be provided. If the element has no counterpart in the norm number or catalogue symbol, then in the last box of the list there should be a dash. For elements of a circular cross section in the column "width", before the diameter measurement, the symbol ø should be written, while in the column "thickness", there should be a short dash. In the case of connectors, it is permissible to enter measurement markings in the column "name of part", and for metal fittings, the column "material" should not be completed, but a short dash inserted.

The drawing block is placed with its right corner in the lower right corner of the drawing sheet's border, and the appropriate number is inserted.

## 5.4   Numbering of Drawings

Methods of numbering drawings depend on many factors, and the numbering system adopted in one construction office or production plant may be completely useless in another. Therefore, there are a dozen or so, more or less common numbering systems of drawings, and the few attempts of unifying them have not given any results so far. In order to avoid a compilation of advantages and disadvantages of the most common related groups of numbering systems, it should be noted that a properly built numbering system should:

- have short numbers of drawings and
- be transparent, easily understandable to the user of the drawing.

The numbering of a drawing can consist of only digits or digits and letters (Fig. 5.5). In the most commonly used digit numbering systems, the numbers of drawings are composed of several parts: the first of which specifies the type of product and its size or other variable characteristic feature; the second—the number of the assemblage or subassemblage; and the third—the number of the element or part in this regard.

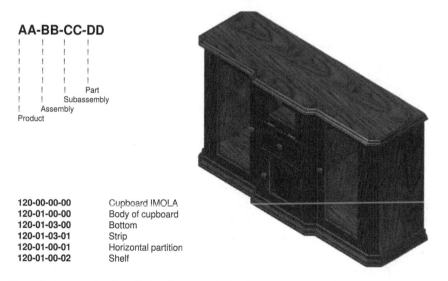

| AA-BB-CC-DD | |
| --- | --- |
|        Part | |
|      Subassembly | |
|   Assembly | |
| Product | |

| | |
| --- | --- |
| 120-00-00-00 | Cupboard IMOLA |
| 120-01-00-00 | Body of cupboard |
| 120-01-03-00 | Bottom |
| 120-01-03-01 | Strip |
| 120-01-00-01 | Horizontal partition |
| 120-01-00-02 | Shelf |

**Fig. 5.5**  An example of numbering drawings of a furniture piece, an assemblage, a subassemblage and element

For example, in the number 1748-01-02, the part 1748 means the type of product, the second 01 describes the assemblage, e.g. top surface, and the third part 02, the element in the assemblage 01—e.g. side-laminated board of the table's top surface. If it is necessary to split some assemblages into subassemblages, and those into elements, then between the parts specifying the number of the assemblage and the element, an additional number of the subassemblage is inserted, e.g. 1803-01-02-03. This number describes an internal stile 03, an upholstery frame 02, a seat 01 and an armchair 1803 made using spring units 18. If the assemblage consists of individual parts aside from the subassemblages, then in the numbering of their drawings, the part of the subassemblage has the digits 00, e.g. 1803-01-00-03.

In digital letter numbering, the part specifying the product, and sometimes the assemblage, is replaced by letters, e.g. SZG25-W6-01, which could mean the board 01 of the top flange W6 of a wardrobe SZG of the subsequent number 25.

In the era of computer use as archival devices, it is more convenient to use digital numbering instead of mixed numbering. This simplifies sorting and searching archival collections, and provides clarity of the description of product groups.

## 5.5   Storing of Drawings

Only blueprints are stored, because originals on tracing paper get destroyed by repeatedly folding and unfolding them. That is why they are stored in the form of rolls. Paper drawings are transferred or sent in files, albums, envelopes, etc.

Drawings intended to be attached in notebooks or folders are folded into A4 format. Methods of folding sheets from A0 × 2 to A3 have been shown in Fig. 5.6. The drawing is first folded along perpendicular lines to the base of the drawing, and then along the lines parallel to it. The order of folding is marked on drawings by numbers, the point lines are convex folds, while dashed lines are concave folds.

Drawings that are not intended to be attached, but stored in folders or envelopes, should be folded into A4 format according to the guidelines provided in Figs. 5.7 and 5.8.

For drawings done on a computer, the same storage rule applies of technical documents plotted on tracing paper. Computer-aided design (CAD) systems, however, ensure storage of construction documentation both in paper and digital form (Fig. 5.9).

Drawings in digital form are stored during a session with the computer at a given workstation, and until they are completed, they can be exchanged directly between different workstations. Ready drawings go directly to the server for archiving. In order to ensure the safety of copies of drawings and construction and technological documentation stored on the server, a copy of the server disc should be done onto a streamer before the end of the working day. The documentation gathered on the server can be used by all active users of a computer network in the production plant, from construction and technological and production departments to individual

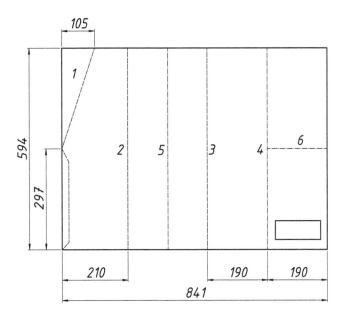

**Fig. 5.6** Examples of folding drawings to be attached to a folder (1–6 folding sequence)

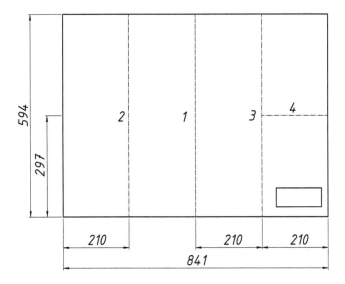

**Fig. 5.7** Examples of folding drawings in a vertical system (1–4 folding sequence)

departments of the plant connected to the network or numeric processing centres NC. Most CAD systems also enable archiving drawings on magnetic or optical carriers. This makes it easy to transfer and exchange drawings with those stations

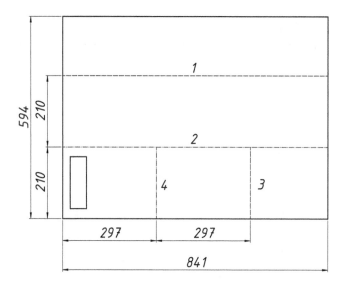

**Fig. 5.8** Examples of folding drawings in a horizontal system (1–4 folding sequence)

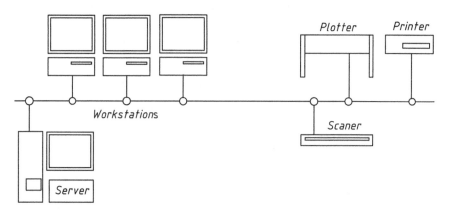

**Fig. 5.9** Scheme of CAD stations working in a network system

that are not connected to the company computer network. By using the description as for paper documentation, optical discs can also be labelled by a soft marker on a paper label or assign the disc a name, by giving, e.g. the number 1703-01-02.

Construction documentation stored in digital form takes up less space, and it is not destroyed as a result of the passage of time and can be modified or adapted repeatedly for design purposes. Saving and storing construction documentation in this way gains and will continue to gain an increasingly higher position among elements that determine the standard of services provided by design offices of furniture factories.

A well-functioning construction department should have an efficient recording system of drawing documentation. Files of drawings and prints (on blueprints) and a registry of archive numbers are created to serve this purpose.

In particular, it is important to keep a file of prints, because in the event of changes, all holders of the documentation should receive new, updated drawings, and at the same time return old prints to the archive. Therefore, prints of each drawing should be numbered sequentially, and their numbers, together with information of when and to whom they were issued, shall be entered on the card of prints of a given drawing.

The originals of the drawing are entered into the archive only in order to apply changes. Every loan and return of a drawing must be noted down on the drawing card. If a paper archive and a registry of archive numbers are kept, then each new drawing entering the archive must obtain an archive number consisting of a description of the sheet format and the subsequent number from the registry book of the drawing. For example, no. 3-3754 marks a drawing in A3 format saved under the catalogue number 3754. Drawings, depending on the format, are stored in appropriately numbered drawers. Such a system of storing construction documentation requires extensive facilities and a skilled administration staff. Unlike paper documentation, realising drawings in the CAD system simplifies and speeds up the archiving process. Thanks to computer technology and combining work station by a network, it is possible to scan old paper documentation and store it on a disc of the server in the form of raster files (Fig. 5.10), as well as currently saving documentation in the digital archive. In this archive, catalogues are created with names that correspond to the conditions in which this documentation was transferred to the archive. Each of the drawings obtains a unique name during saving, while only authorised persons have access to it after entering the correct password. Changing the drawing's content, resulting from editorial or substantive needs, is done by

**Fig. 5.10** Archiving scheme of drawing documentation

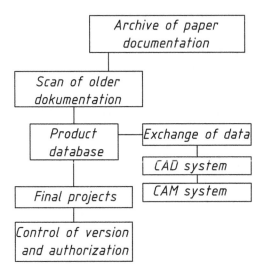

editing the drawing on one of the CAD working stations, and saving it again in the archive database gives all users access to the revised and current version.

From the digital archive, the documentation may be printed on printers or plotters, or, if possible, sent to numerical machine-tool stations and there, through the appropriate transfer of information, translated into machine language NC and CNC. This form of drawing documentation management is obtained by reducing the time of its development, replacing sequential design with concurrent design.

Efficient management of data related to the product, its modifications and versions are provided by modern PDM and TDM systems. Product data management systems (PDM) are responsible for the management of data resources developed on the basis of the structure of the product. Team data management systems (TDM) constitute a coherent data environment in the area of group work, which is lately also often referred to as project management (Chlebus 2000). The function of PDM/TDM systems is as follows:

- Data vault and document management (DVDM),
- Workflow and process management (WPM),
- Product structure management (PSM),
- Classification and retrieval of information (CR),
- Program management (PM),
- Communication and notification (CN),
- Data transport and translation (DTT),
- Image services (IS) and
- System administration (AD).

Most documents drawn up by the departments of technical–technological preparation of production arise usually as a result of the introduction of changes or modifications. If the production of furniture is of a repeatable, serial nature, the production is planned on the basis of model structures, and there are few changes in the documentation, and made rarely. Taking into account the individual requirements of the customer, a strong discretization of production, variant products or the configuration of the product, force constant modification of structures and updating documentation. Data models, tailored to the respective phases of the product's development, processed in the PDM system, and generated using CAx systems, are stored in the PDM base, from which they are transferred to the database in the ERP system (Chlebus 2000). The basis of logical and physical integration of data is the structure of the product seen both CAx and CIM systems (Fig. 5.11).

The network characteristic of the PDM systems facilitates smooth work for many users in real mode and free use of the potential of the application and database (Fig. 5.12). It is also possible to configure project assemblages through access to one's domain of data and CAx tools, which should be configured according to the needs of the project manager and the users themselves.

The management of data about the product by the system requires the user to meet specific requirements. The most important of these include: working in an environment of relational databases, support for raster and vector graphics, group- and task-oriented work, preparing technological processes, systematics and

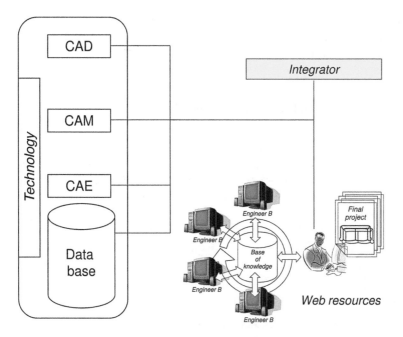

**Fig. 5.11**   Scheme of logical and physical integration of data

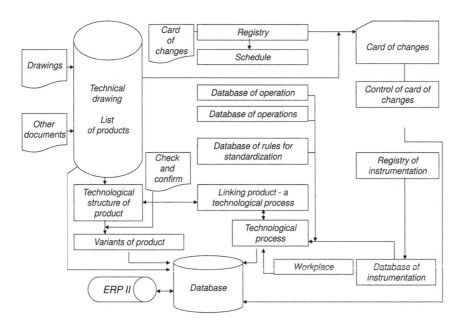

**Fig. 5.12**   The nature of the PDM/TDM systems

unification of data, connections between CAx and CIM systems, creating variants and management of changes in the structure of the product and processes, integrating product data and the elimination of data redundancy.

## 5.6  Normalised Elements of Drawing

Among the numerous types of drawing lines, which constitute the subject of the standard PN-82/N-01616 in industry standards regarding the furniture technical drawing, according to BN-90/7140-03/05, the types of lines presented in Fig. 5.13 are used. In addition to this, there are several line thicknesses selected, depending on the drawing sheet format.

Different types of lines have the following uses:

- A thick continuous line is used to draw:

  - the outline of the box "drawing number" in the title block.

- A medium continuous line is used to draw:

  - visible edges, outlines and blending lines,
  - visible places of contact of parts,
  - displaced quads,
  - plane trace of the cross section,
  - drawing simplifications of joints,

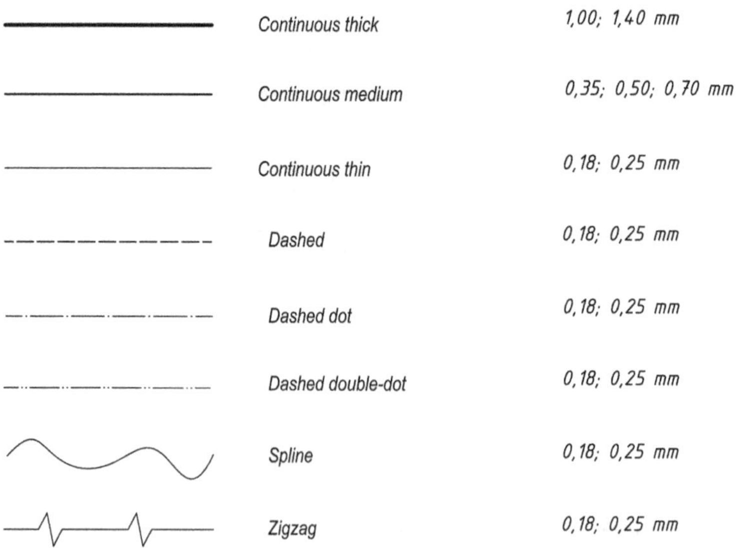

| | | |
|---|---|---|
| ———————— | Continuous thick | 1,00; 1,40 mm |
| ———————— | Continuous medium | 0,35; 0,50; 0,70 mm |
| ———————— | Continuous thin | 0,18; 0,25 mm |
| – – – – – – | Dashed | 0,18; 0,25 mm |
| –·—·—·—·— | Dashed dot | 0,18; 0,25 mm |
| –··—··—··— | Dashed double-dot | 0,18; 0,25 mm |
| ∼∿∼ | Spline | 0,18; 0,25 mm |
| —/\—/\— | Zigzag | 0,18; 0,25 mm |

**Fig. 5.13**  Types of drawing lines

- drawing borders and
- outline of the title block and its main boxes.

• A thin continuous line is used to draw:

- dimensional lines and auxiliary dimensional lines,
- hatching cross sections,
- reference lines along with the "shelves" and marking details,
- local quads,
- axis of circles and other geometric figures with a diameter less than 12 mm,
- columns and rows in boxes of the drawing block,
- markings of the direction of veneer fibres on cross sections and views,
- markings of adhesive bonds,
- lines limiting an enlarged detail,
- drawing simplifications of fittings (on views) and the thread line and
- veneer line on cross sections.

• A thin spline (curved) line is used to draw:

- invisible edges, outlines and blending lines and
- invisible places of contact of parts.

• A thin dot line is used to draw:

- imaginable axis of objects (e.g. axis of symmetry),
- the axis of circles and other figures with a diameter or dimensions above 12 mm and
- arbitrary markings of repeated joints.

• A thin double dot line is used to draw:

- extreme or significantly different positions of moving parts and
- the outline of parts presented when expanded.

• A zigzag continuous line is used to draw:

- interruptions or tears of projections,
- limitations of the projection from the view (tears) and
- partial limitation of the view or cross section.

• A thin continuous polyline for drawing interruptions or tears of projections.

In a furniture technical drawing, for most joints, fittings and accessories, drawing simplifications are applied (BN-90/7140-03/03, Table 5.4) reducing the time of documentation being drawn up. Joints should be drawn in simplified or arbitrary markings depending on the scale of the drawing and the number of repetitions. The arbitrary presentation is recommended to be used particularly in the case of a series of repeated same joints and in drawings made in reduced scales. The dimensions of these joints should correspond to actual basic dimensions, such as the diameter and length of the core, and the dimensions of screw heads. Examples of applying drawing simplifications have been provided in Fig. 5.14.

**Table 5.4** Simplified and arbitrary markings of selected joints in a furniture technical drawing (BN-90/7140-03/03)

| Name of joint | Type of joint | Simplified presentation | Arbitrary presentation |
|---|---|---|---|
| Bolt | With hexagonal head | | |
| | With cylindrical head with hexagonal socket | | |
| | With mushroom head with neck | | |
| Metal screw | With cylindrical head | | |
| | With round head | | |
| | Slotted with conical head | | |
| | Cross recessed with conical head | | |
| | Slotted with conical raised head | | |
| | Cross recessed with conical raised head | | |
| Hexagonal | Nut | | |
| Screw for wood | With hexagonal head | | |
| | With round head | | |
| | Slotted with conical head | | |
| | Cross recessed with conical head | | |
| | Slotted with conical raised head | | |
| | Cross recessed with conical raised head | | |
| Screw for chipboard | Confirmat screw slotted | | |
| | Confirmat screw cross recessed | | |
| Nail | Straight | | |
| | Curved | | |

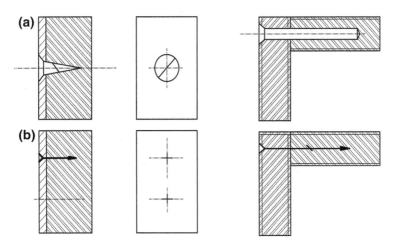

**Fig. 5.14** Examples of markings: a) simplified and b) arbitrary

When describing the joints, presented in simplification using reference lines and inscriptions, information details should be provided about the name, dimensions, number of norm or number corresponding to the position number in the title block. Joints of a perpendicular axis to the plane of the drawing are arbitrarily presented always in the form of two short sections of a thin continuous line crossed over at a right angle. Furniture fittings should be drawn in a simplified manner, without marking construction details, slants, phasings, etc., on them, presenting only their outlines resulting from overall dimensions. Detailed information about the fittings, such as the number or catalogue symbol, drawing number, and for a special fittings also characteristic dimensions, is entered in the title block, by providing the number of the box on the shelf of the reference line. On the reference line, it is also permissible to provide more information about the fittings. This applies to cases in which a list of parts in the title block is not provided. Symmetrical fittings, such as handles, support pegs under shelves, grooves for keys and escutcheon plates, can be presented in the form of an intersection point of two perpendicular axes made in a point line or thin continuous line depending on the characteristic dimensions of the presented fittings.

The position of the fittings drawn in a simplified manner is provided by dimensioning the distance of the fittings from the edge of the furniture element, on which they occur. For fittings presented using a point, the distance of the inter-section point of the axis of symmetry from the edge of the element is provided. Other drawing simplifications concerning, among others, joint parts provided in PN-81/N-01613, riveted and soldered joints in PN-76/N-01635, welded in PN-79/M-01134 and PN-64/M-01138.

Standardised drawing elements also include graphic markings of wood-based materials, upholstery materials and others used commonly in furniture production (BN-90/7140-03/04). The more important of these markings have been provided in Table 5.5.

**Table 5.5**  Graphic marking of materials (BN-90/7140-03/04)

| Name of material | Graphic marking of cross-section |
|---|---|
| Wood cross-section | |
| Wood longitudinal cross-section | |
| Carpentry board cross-section | |
| Carpentry board longitudinal cross-section | |
| Chipboard | |
| Honeycomb board | |
| Plywood | |
| HDF | |
| MDF | |
| Metals | |
| Plastics, rubbers | |
| Glass and other transparent materials | |
| Polyurethane foam | |
| Coconut mat, stubble and latex | |
| Lining material, covering | |
| Fabrics | |
| Sinusoidal spring | |
| Spring unit top view | |
| Spring unit vertical cross-section | |

In natural and artificial veneers, the outer layer on cross sections made to the scale of 1:2, 1:1 or more should be presented using thin continuous lines, about 20 mm in length, running inside the cross section at a distance of 1 mm from the contour lines (Fig. 5.15).

**Fig. 5.15** Examples of using
graphic markings

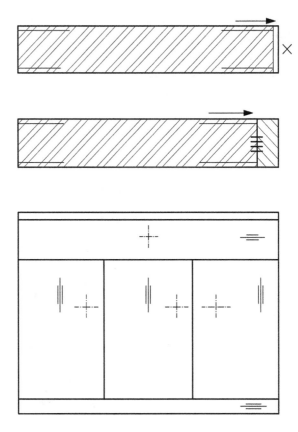

Veneers should not be presented on cross sections made in other reduced scales. Subveneers are also not presented on the cross sections.

The direction of the fibres of natural veneers or a clearly visible drawing of wood of synthetic veneers should be marked on the outer side in the following manner (Fig. 5.15):

- longitudinal direction in relation to the plane of the cross section using an arrow with a length of about 10 mm;
- transverse direction in relation to the plane of the cross section using a cross with a length of the arms of about 4 mm.

The direction of the fibres of a natural veneer or drawing of wood of a synthetic veneer on views should be presented using a system of three lines (Fig. 5.15). The middle line should be approximately 2 times longer than the outer lines. The direction of the fibres in solid wood or the outer veneers of plywood is marked the same as the direction of fibres in the veneers.

By drawing cross sections of materials, they should be hatched by thin continuous lines. Diagonal hatching is carried out at an angle of 45° to the line of the border of the drawing with a slope to the right or left. If two transverse cross

**Table 5.6** The symbols of markings used in clothing technical drawings (Czurkowa and Ulawska-Bryszewska 1998)

| Drawing symbol | Description | Application |
|---|---|---|
| | Piercing with needle | In sewing marking |
| | Sewing direction | In markings of symbol of stitches and seams |
| | Zigzag stitch | In marking of sewing direction |
| | Surge stitch with one thread | In markings of edges finishing, decorating etc. |
| | Surge stitch with two threads | In markings of edges finishing, decorating etc. |
| | Blind hem stitch | In marking of sewing direction |
| | Overlockstitch | In markings of hem finishing |
| | Protrusion | In marking of protrusions |
| | Outside surface of material | In markings of the outside surface |
| | Inside of material | In markings of the inside surface |
| | Material thickness | In markings of material thickness |

sections lie next to each other, then the second cross section should be hatched with a slope rotated by 90°, and in the case of three cross sections, the smallest of them is hatched more densely.

In drawings realised at a large reduction, or when there is no room for hatching in the drawing, blackening the cross sections is permitted. If the blackened areas on the cross section are in contact with each other, then they should be drawn with a small space in between.

In furniture technical drawing, next to previously provided standardised elements, also other rules of machine drawing apply, in particular:

- rectangular views and locating views according to PN-78/N-01608,
- axometric views according to PN-82/N-01619,
- views and cross sections according to PN-79/M-01124,
- writing according to PN-80/N-1606,
- dimensioning according to PN-70/M-01143, PN-70/M-01141 and
- tolerating dimensions and shape according to BN-81/7140-11.

When preparing technical documentation of upholstered furniture, drawings of templates and systems of the fabric and leather cuts should also be done. These are so-called tutorial drawings of technological processing indicating how to make a given piece of cover for an upholstered furniture piece. Drawing symbols of stitches and seams are included in the Polish Standards. Table 5.6 shows the selected symbols of markings used in drawing up tutorial drawings.

**Table 5.7** Symbols of seams (Czurkowa and Ulawska-Bryszewska 1998)

| Drawing symbol | Name of the seam | Characteristics of the seam |
|---|---|---|
| | Regular joining seam | Two layers of material sewn |
| | Lapped joining seam | Two layers of material sewn |
| | Gorge joining seam | Materials sewn in one plane |
| | Overlapping joining seam | Overlapping materials sewn |
| | French joining seam | Two layers of material sewn inverted, backstitched seam |
| | French joining seam | Two layers of material sewn, backstitched |
| | Recessed joining seam | Three layers of material sewn, external layers inverted |
| | Pin tuck joining seam | Decorative sewing of the material |
| | Trimming edge seam | Trimmed edge |
| | Fringing edge seam | Backstitched, rolled edge of the material |
| | Overcasting edge seam | Edges of two layers of material overcast |
| | Glued seam | Two layers of material glued |

Table 5.7 presents graphic symbols of representative seams and their names. They are cross sections showing the structure of the seam. The thick lines represent loose layers of fabrics (leathers), bendings of these lines represent material overlaps, and the thin lines represent needle sewings. Figure 5.16 illustrates a few examples of machine stitches and seams.

## 5.7 Technical Documentation of Case Furniture

The subject of the drawing documentation is a cupboard for RO documents (Figs. 5.17, 5.18, 5.19 and 5.20):

- The project was drawn up at the Department of Furniture of the University of Life Sciences in Poznań;

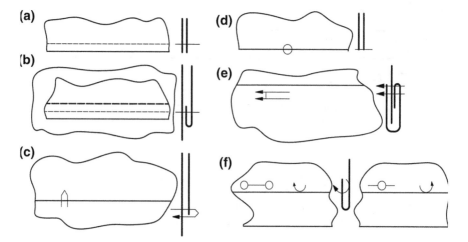

**Fig. 5.16** Examples of machine stitches and seams: **a, b** regular stitch, **c** zigzag stitch, **d** surge stitch, **e** surge stitch and **f** blind hem stitch

- The set includes a bookcase, a desk, a container and side table;
- The cupboard for documents has a height of 1800 mm, a width of 800 mm and a depth of 420 mm. The distance between movable shelves provides free storage of office binders, documents in formats A4, B5, etc.;
- the cupboard is designed for use in offices, libraries, post offices, banks and other similar public institutions. The furniture piece is fitted with two drawers with hooks for files. Full extension of drawers enables free access to stored collections; and
- the body of the furniture is made of chipboard covered with laminate, veneered at the edge with PVC, the doors with a tempered glass filling—made of beech wood, frontal elements finished with side-laminated board covered with a transparent water-dilutable lacquer with a glossiness of 15°.

## 5.8  Technical Documentation of Skeletal Furniture

The subject of the drawing documentation is the chair Adam (Figs. 5.21, 5.22 and 5.23):

- The project was drawn up at the Department of Furniture of the University of Life Sciences in Poznań;

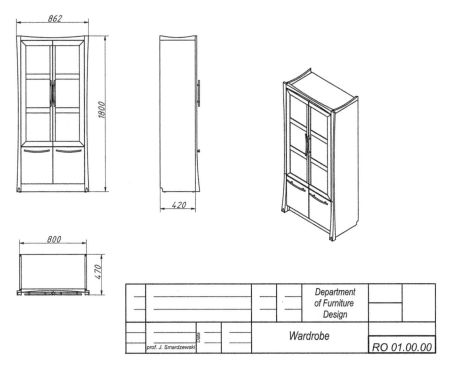

**Fig. 5.17**  Offer drawing of the wardrobe

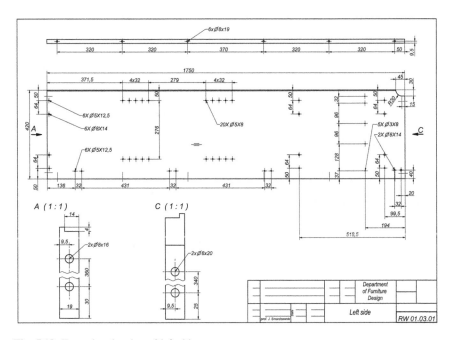

**Fig. 5.18**  Executive drawing of left side

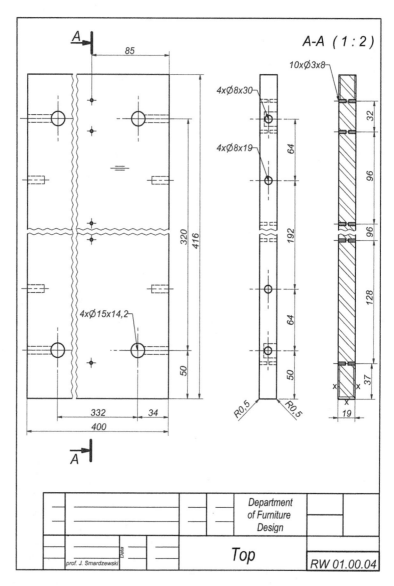

**Fig. 5.19** Executive drawing of top

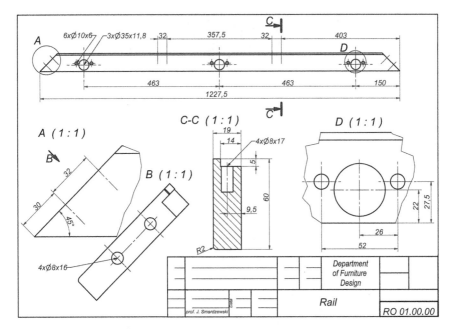

**Fig. 5.20** Executive drawing of vertical rail

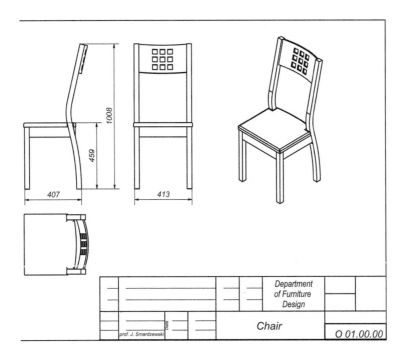

**Fig. 5.21** Offer drawing of the chair

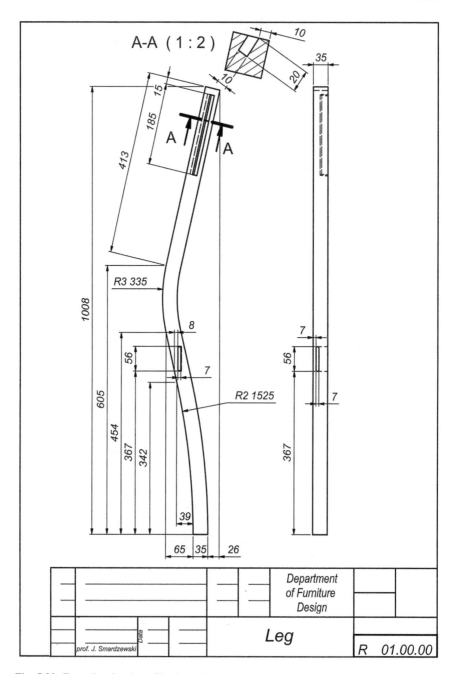

**Fig. 5.22** Executive drawing of backseat leg

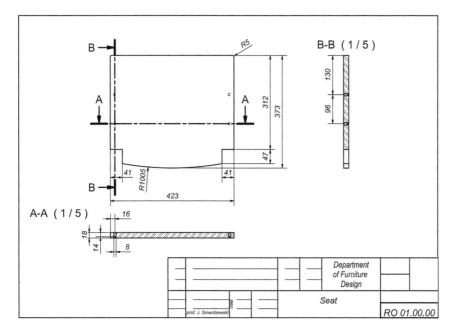

**Fig. 5.23** Executive drawing of seat

- The chair is a product designed for a set together with Adam tables;
- The chair has a backrest height of 1008 mm, a seat height of 459 mm, a seat width of 413 mm and a seat depth of 407 mm;
- The product is intended for use by persons representing the 50th centile, in housing rooms, restaurants, bars, cafes, schools, as well as offices and offices of public institutions; and
- The skeleton of the chair is made of oak wood covered with a transparent water-dilutable lacquer with a glossiness of 5°. The seat is made of layers of polyurethane foams covered with jacquard fabric.

## 5.9 Technical Documentation of Upholstered Furniture

The subject of the drawing documentation is the 2-seater sofa Europe (Figs. 5.24, 5.25, 5.26, 5.27 and 5.28):

- The project was drawn up at the Department of Furniture of the University of Life Sciences in Poznań;
- The set includes a pouffe, single armchair and 3-seater sofa;

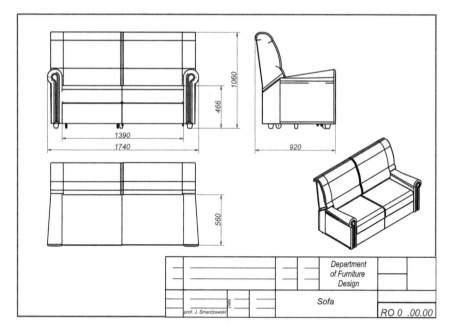

**Fig. 5.24** Offer drawing of the sofa

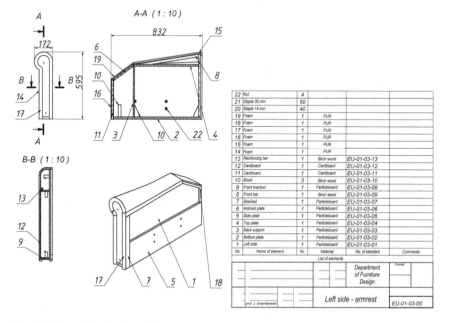

**Fig. 5.25** Assembly drawing of armrest

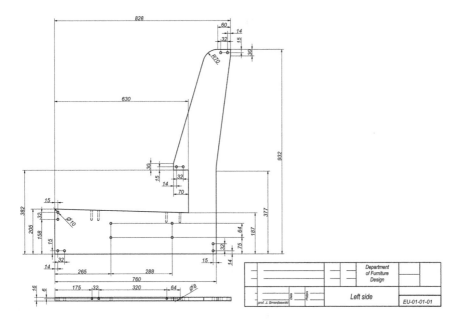

**Fig. 5.26** Executive drawing of left side

- The width of the sofa is 1740 mm, the seat width 1390 mm, the seat depth 560 mm, the sofa depth 920 mm and the backrest height 1060 mm. After unfolding the bed has the dimensions 2050 × 1390 mm. The bedding container has the dimensions 600 × 500 × 300 mm;
- The furniture piece is intended for use by persons representing the 50th centile in housing and hotel rooms. The furniture piece is designed mainly for sitting and relaxing, and it also has the additional feature of lying down and storing bedding. However, the sofa should not be used as the primary piece of furniture for sleeping; and
- The furniture piece is made from grain leather and coordinates made from split. The seat consists of a Bonnell spring unit and highly elastic system of polyurethane foams. The frame of the construction is made of chipboard and beech skirts.

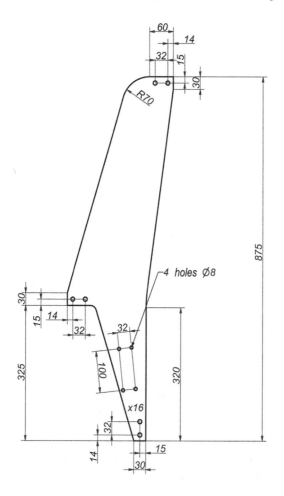

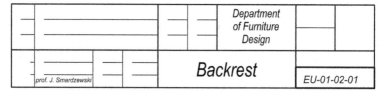

**Fig. 5.27** Executive drawing of backrest

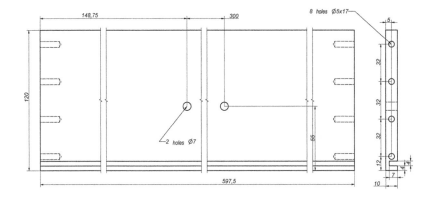

**Fig. 5.28** Executive drawing of front of the drawer front

# References

BN-81/7140-11 System of tolerances and fits for furniture
BN-90/7140-03/02 Furniture technical drawing. Title blocks
BN-90/7140-03/03 Furniture technical drawing. Drawing simplifications
BN-90/7140-03/04 Furniture technical drawing. Graphic signs
BN-90/7140-03/05 Furniture technical drawing. Drawing lines
Chlebus E (2000) Techniki komputerowe CAx w inżynierii produkcji. WNT, Warsaw
Czurkowa MH, Ulawska-Bryszewska I (1998) Rysunek zawodowy dla szkół odzieżowych. Wydawnictwa Szkolne i Pedagogiczne, Warsaw
PN-64/M-01138 Machine technical drawing. Welded joints
PN-70/M-01143 Machine technical drawing. Dimensioning. Order rules
PN-70/M-01141 Machine technical drawing. Dimensioning. General rules
PN-76/N-01601 Technical drawing graphic form of sheets
PN-76/N-01635 Machine technical drawing. Drawing simplifications. Riveted, soldered, glued and sewn joints
PN-78/N-01608 Technical drawing. Rectangular viewing
PN-79/M-01134 Machine technical drawing. Drawing simplifications. Principles of marking joints
PN-79/M-01124 Machine technical drawing
PN-80/N-1606 Technical drawing. Writing
PN-81/N-01613 Technical drawing. Drawing simplifications. Joined parts
PN-82/N-01616 Technical drawing. Drawing lines
PN-82/N-01619 Technical drawing. Axonometric

# Chapter 6
# Stiffness and Strength Analysis of Skeletal Furniture

## 6.1 Properties of Skeletal Furniture

Constructions of skeletal furniture belong to the group of multiple statically indeterminate spatial systems. The analytical solution to the distribution of internal
forces or displacements of nodes from the point of view of accounting is a very
laborious task. A simple stool with a bar can be a system that is 30-fold internally
statically indeterminate, while an armchair with armrests usually constitutes a
system 62-fold indeterminate. This means that in order to calculate it, a system of
62 equations with 62 unknowns must be built. Therefore, in solving spatial systems
numerical methods are applied.

However, due to the symmetry of the construction of skeletal furniture (Fig. 6.1)
and the symmetry of load, analytical solutions can be reduced to stiffness–strength
calculations of side frames of this furniture. The side frame of a furniture piece can
be created from beam elements connected together in a stiff or articulated way. If
we connect four items together articulately, then instead of a construction we shall
obtain a mechanism (Fig. 6.2b) that does not bring external loads. In this situation,
the applied joints provide the free rotation of each of them in respect of one another,
thereby causing significant changes of the angles contained between the components of the system. Using perfectly stiff connections in the location of the articulate
joints, we transform the mechanism into a construction, which sustains slight
deformations caused only by the deformation of its components, while the nodes
still remain undeformed (Fig. 6.2a).

Of course, not all systems containing articulate joints are mechanisms. Among
the many furniture constructions, there are also those which are built from
three-element systems. If, in fact, three of the structural elements of the furniture are
connected articulately (Fig. 6.3a–c), then we shall obtain a skeleton shifting the
external load, while the only type of force occurring in these elements shall be axial
forces of compressing or stretching nature. By using two three articulated frames, a
garden chair like in Fig. 6.3b can be built. In the discussed constructions, the

© Springer International Publishing Switzerland 2015
J. Smardzewski, *Furniture Design*,
DOI 10.1007/978-3-319-19533-9_6

**Fig. 6.1** Symmetry of construction of skeletal furniture

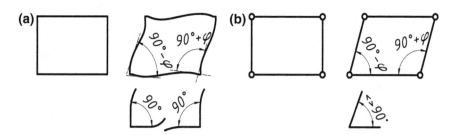

**Fig. 6.2** Deformations of subassemblages of furniture with nodes that are **a** stiff, **b** articulate

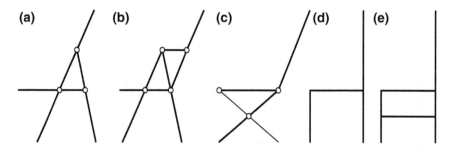

**Fig. 6.3** Side frames of furniture: **a–c** trusses, **d, e** frames

**Fig. 6.4** The impact of how
chairs are supported on
deformations of side frames:
**a** unmovable support under
the *front leg*, **b** movable
support under the *front leg*

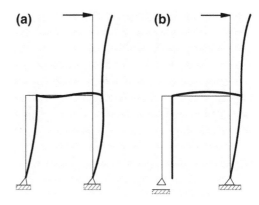

stiffness of frames depends on the arrangement of the elements in the shape of a
triangle and not on the stiffness of joints. Constructing such joints is not a complicated task. Therefore, below we will deal with frame constructions, in which
particular elements are connected together in a stiff way (Fig. 6.3d, e).

The simplest side frame construction of a chair is a beam of a curved axis
(Fig. 6.4). The stiffness and strength of such a frame depends not only on the
stiffness and strength of elements and structural nodes, but also on the way the
tested construction is supported. The overstiffness of the system by an unmovable
support of the front and rear leg (Fig. 6.4a) results in the bending of both these
elements and a double overbending of the rail. This obviously improves the overall
stiffness of the structure.

In a system that is externally statically determinate (Fig. 6.4b), only the support
leg and rail of the seat are subject to bending. As a consequence, this system is far
more susceptible and less durable from the previous one.

It should also be noted that a greater number of double-bent structural elements
have a positive effect on the increase of stiffness of the system. And so, the side
frame of the chair from Fig. 6.5a is characterised by a four times smaller stiffness in
relation to the stiffness of the frame illustrated in Fig. 6.5b. Similarly, the construction of the table presented in Fig. 6.5c has a more than four times smaller
stiffness than a table frame reinforced by an additional bar (Fig. 6.5d). The cause of
this regularity is the fact that in systems internally statically determinate (Fig. 6.5a,
c), on the length of each beam, the signs of normal stresses, deriving from bending,
are not changed. However, in internally statically indeterminate systems (Fig. 6.5b,
c), the elements that make up the frame, subject to double bending, change signs of
normal stresses on the same side of the beam. In this way, the construction increases
its stiffness.

From the engineering point of view, designing skeletal furniture should consist
in, among others, assigning reaction of forces of the base, internal forces acting on
particular elements and structural nodes, as well as determining displacements of
selected construction points (Fig. 6.6).

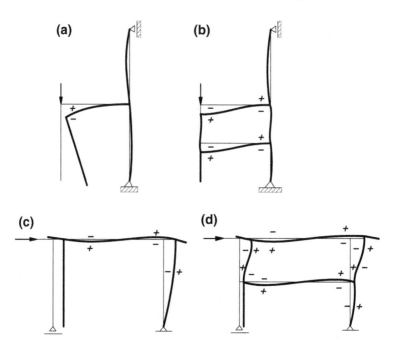

**Fig. 6.5** The impact of the number of elements of the side frame on the stiffness of **a, b** a chair, **c, d** a table

**Fig. 6.6** The side frame of a chair as a system: **a** externally and internally statically determinate, **b** externally statically determinate and internally statically indeterminate

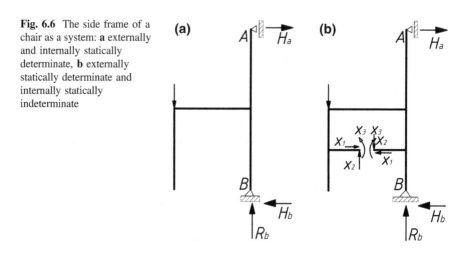

By analysing the construction of side frames of skeletal furniture, as flat systems, usually externally statically determinate (Fig. 6.6), we are dealing with three reactions (Fig. 6.7a): force $F_x$ parallel to the $X$-axis, force $F_y$ parallel to the $Y$-axis, and the moment $M_{xy}$ in the $XY$ plane. Each of these reactions appears in the case of support having the character of fixing (Fig. 6.7b). Unmovable support (Fig. 6.7c)

**Fig. 6.7** Any flat system of forces: **a** $F_x$, $F_y$ vectors of forces, respectively, parallel to the system of coordinates, $\delta_x$, $\delta_y$, linear displacements in the direction of the forces, $M_{xy}$ the vector of the moment of forces perpendicular to the plane $XY$, $\varphi_{xy}$ rotation in the plane $XY$, **b** fixing, **c** unmovable support, **d** movable support in the direction of the $X$-axis

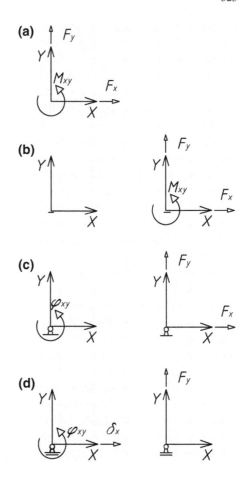

causes only reactions in the form of axial forces and enables free rotation in relation to the point of support, while a movable support (Fig. 6.7d) generates a vector of reaction force directed vertically, at the same time ensuring freedom of rotation and shifting in the horizontal direction.

In spatial systems, we are dealing with six reactions (Fig. 6.8a): the forces $F_x$, $F_y$, $F_z$ parallel, respectively, to the axes $X$, $Y$, $Z$, moments $M_{xy}$, $M_{yz}$, $M_{xz}$ in the plane $XY$, $YZ$, $XZ$. Each of these reactions appears in the case of support having the character of fixing (Fig. 6.8b). Unmovable support (Fig. 6.8c) causes only reactions in the form of axial forces, at the same time enabling free rotation in relation to the point of support. Shifting support (Fig. 6.8d) generates one vector of force reaction directed vertically, while enabling the freedom of rotation in relation to each of the axes and shifts in the plane $ZX$.

By shifting the above systems on particular constructional solutions of skeletal furniture frames, it should be noted that in the cross sections of the rods of products in which articulate connections were used, only axial forces occur (Fig. 6.9). This

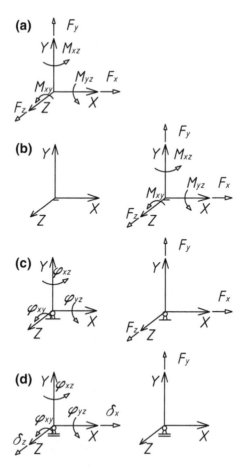

**Fig. 6.8** Any spatial system of forces: **a** $F_x$, $F_y$, $F_z$ vectors of forces, respectively, parallel to the system of coordinates, $\delta_x$, $\delta_z$, linear displacements in the direction of the forces, $M_{xy}$, $M_{xz}$, $M_{yz}$, vectors of moments of forces perpendicular to the planes $XY$, $XZ$, $YZ$, $\varphi_{xy}$, $\varphi_{xz}$, $\varphi_{yz}$, rotations, respectively, in the plane $XY$, $XZ$, $YZ$, **b** fixing, **c** unmovable support, **d** movable support in the direction of the $X$ and $Z$ axes

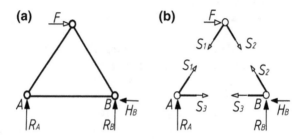

**Fig. 6.9** Truss: **a** static scheme, **b** internal forces in rods

**Fig. 6.10** Flat frame: **a** static scheme, **b** internal forces in rods

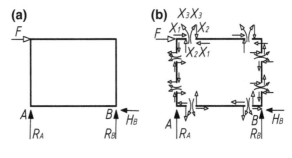

**Fig. 6.11** Spatial frame: **a** static scheme, **b** internal forces in rods

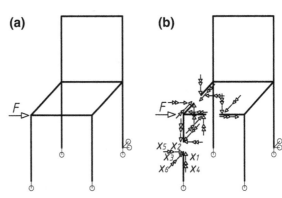

means that components in such constructions should not be too slender in order not to lose their own stability.

In each of the cross sections of flat frames internally statically determinate, we release three overvalues (Fig. 6.10). Their presence causes that the constructor, in conducting stiffness–strength calculations, should take into account both the work of elements and structural nodes under load causing shearing and bending.

In spatial constructions, the shearing of any component in thought causes the release of six overvalues (Fig. 6.11). Each cross section may therefore be compressed, sheared, bent and twisted. Taking this into account, in the process of dimensioning cross sections and connectors of particular joints, the lowest strength of wood and glue joints determining strength in the entire construction is taken into account.

## 6.2   Operational Loads on Chairs and Stools

The relevant safety requirements for using chairs and stools have been provided in PN-EN 1335-2:2009. The tests described in this norm assume that a chair is used for 8 h a day by people weighing up to 110 kg. The locations of applying forces in durability and load stability tests are determined by using a special template

**Fig. 6.12** Template for determining points of loading the seat or backrest according to PN-EN 1728:2012: *A* load on a seat for a chair, *B* load on the backrest for a chair, *C* load on a seat for a stool (mm)

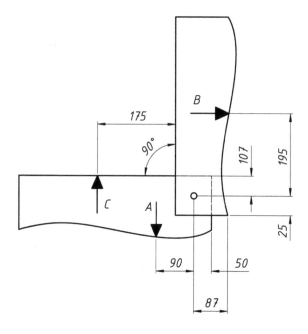

(Fig. 6.12). This template consists of two profiled segments connected articulately. Profiled surfaces are designed in such a way as to sink into the upholstery. To this end, a segment loading the seat should have a total weight of 20 kg, applied at the point of loading the seat.

The most commonly encountered schemes of load on chairs and stools are systems enabling the determination of:

- loss of balance by the front edge,
- loss of balance to the front,
- loss of balance to the sides of chairs without armrests,
- loss of balance to the sides of chairs with armrests,
- maximum distance of backrest displacement,
- loss of balance to the back of chairs with an untilting backrest,
- loss of balance to the back of chairs with a tilting backrest,
- rolling resistance,
- the durability of the seat and the backrest,
- the durability of backrests that rotate around a horizontal axis,
- the durability of the armrests.

An example of a static load on a seat and backrest is shown in Fig. 6.13. Forces should be located in points of load specified by the template (Fig. 6.12). By using an element or device for loading, the tests should be carried out as shown in Fig. 6.13a. The tested seat is loaded with a force of 750 N, while the static force acting on the backrest should not be less than 410 N.

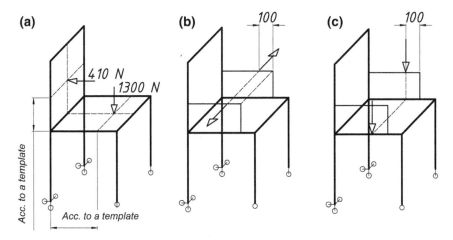

**Fig. 6.13** Determining the strength and durability of home and school furniture for sitting according to PN-EN 1728:2012, PN-EN 1729-2:2006: **a** load on the seat and the backrest, **b**, **c** load on the armrests and side headrests with **b** horizontal forces, **c** vertical forces (mm)

By testing the durability of armrests, a force directed outwards in the point in which it is easiest to cause damage should be applied, no less than 100 mm from both ends of the armrest structure (Fig. 6.13b). In the case of vertical loads, the force is applied to the armrests in the location where it is most likely to cause damage (Fig. 6.13c), but no closer than 100 mm from one of the ends of the armrests.

Impact tests usually aim to determine the durability of backrests, armrests and seats. By testing the durability of a backrest, the product should be set with the front legs at the safety blocks preventing from moving forwards and then struck from the outside in the middle of the construction of the top part of the backrest with an impact hammer weighing 6.5 kg (Fig. 6.14a). In carrying out the impact test of the backrest (Fig. 6.14b), the blow should be carried out from the outside on the surface of one armrest, in the place where causing damage is most likely. The durability of the seat is determined by a free fall of the device to hit the seat (Fig. 6.14c) from a specific height on the point of loading the seat marked using a template.

By testing the loss of stability of office chairs to the front, a vertical force of 600 N must be applied, from a distance of 60 mm from the edge of the front load-carrying construction, in places where the loss of stability is most likely. Then, a horizontal force of 20 N should be applied, acting on the outside from the point in which the base of the element shifting load is in contact with the top surface of the seat (Fig. 6.15a).

In attempting to assess stability, the sides of chairs without armrests should be applied with a vertical force of 600 N on the seat, at a distance of 60 mm from the edge of the load-carrying construction of the side closer to the blocked points of supports, in those places where the loss of stability is most likely, and then apply a

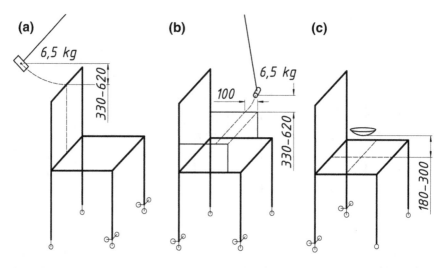

**Fig. 6.14** Impact tests of home and school furniture for sitting according to PN-EN 1728:2012, PN-EN 1729-2:2006: **a** backrest, **b** armrests and side headrests, **c** seat (mm)

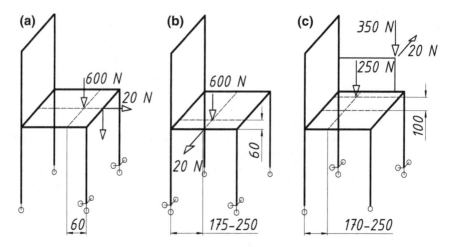

**Fig. 6.15** Determining stability of office chairs PN-EN 1335-3:2009: **a** loss of stability to the front, **b** loss of stability to the side, **c** loss of stability to the side of a chair with armrests (mm)

horizontal force of 20 N directed on the outside, from the point in which the base of the element shifting load is in contact with the top surface of the seat (Fig. 6.15b).

In testing a chair with armrests, a vertical force of 250 N should be applied at a distance of 100 mm from the middle plane, on the side of the blocked points of support (Fig. 6.15c), and at a distance from 175 to 250 mm to the front of the rear edge of the seat, as close as possible to the side edge. Then, a vertical force of 350 N should be applied acting on the armrest in points that are up to 40 mm from

**Fig. 6.16** Determining the
maximum distance of
displacement of the backrest
of office chairs according to
PN-EN 1335-3:2009 (mm)

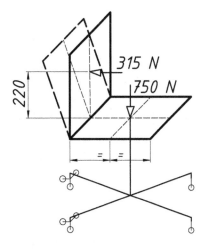

the outer edge, directed inwards, but not going beyond the middle of the armrest in
the most unfavourable place of its length. Subsequently, a load with a value of 20 N
should be applied, directed outwards from the point where the base of the loading
element comes into contact with the upper surface of the armrest.

When testing the maximum distance of displacement of the backrest of office
chairs, the base of the chair should be secured against lifting, by placing a weight of
75 kg on the seat. Then, to the backrest, at a point lying at a height of 220 mm, a
force with a value of 315 N should be applied (Fig. 6.16). The distance of displacement of the backrest is the distance measured horizontally between the point
of back support on a loaded backrest and the axis of rotation of the chair.

By specifying the conditions of stability loss to the front of home and school
furniture for sitting, a vertical force should be applied with a value from 200 to
600 N, at a distance of 60 mm from the edge of the front load-carrying construction,
in places, where the loss of stability is most likely to occur. Then, a horizontal force
of 20 N should be applied, acting outwards from the point in which the base of the
element shifting load is in contact with the top surface of the seat (Fig. 6.17a). In
attempting to assess stability to the sides of chairs without armrests, a vertical force
of 200–600 N should also be applied on the seat, at a distance of 60 mm from the
edge of the load-carrying construction of the side closer to the blocked points of
supports, in those places where the loss of stability is most likely to occur. Then, a
vertical force of 20 N should be applied directed outwards from the point in which
the base of the element shifting load is in contact with the top surface of the seat
(Fig. 6.17b). In testing the stability of a chair backwards, a vertical force of 200–
600 N should be applied, at a distance of 180–300 mm from the front edge of the
seat, and a horizontal force of 50–180 N, at a height of 120–180 mm, counting from
the top surface of the seat (Fig. 6.17c).

The strength and durability of home and school furniture for sitting is determined
by applying a vertical force of 1300–2000 N, a horizontal force of 410–700 N
(Fig. 6.18a) and by loading the front bar with a force equal to 1000 N (Fig. 6.18b).

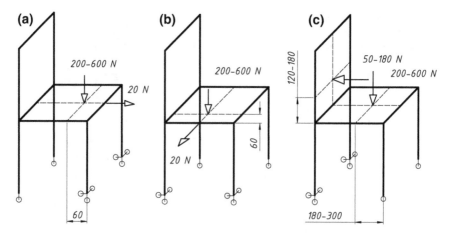

**Fig. 6.17**  Determining the stability of home and school furniture for sitting according to PN-EN 1728:2012, PN-EN 1729-2:2006: **a** loss of stability to the front, **b** loss of stability to the side, **c** static load on the seat and the backrest (mm)

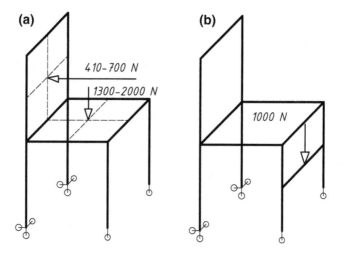

**Fig. 6.18**  Determining the strength and durability of home and school furniture for sitting according to PN-EN 1728:2012, PN-EN 1729-2:2006, load on **a** the seat and the backrest, **b** bars

An important operational load of furniture for sitting is also loading the legs, enabling us to determine the durability of the rack construction. To this end, a vertical force of 1300–1600 N should be applied to the seat and a horizontal force at the back of the seat in its axis, directed forwards (Fig. 6.19a, b) through an element shifting local loads. In furniture for sitting without legs (e.g. chairs with castors or

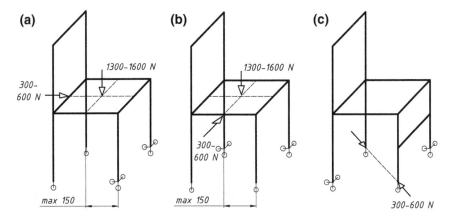

**Fig. 6.19** Loads of legs of home and school furniture for sitting according to PN-EN 1728:2012, PN-EN 1729-2:2006: **a** front, **b** side, **c** opposite (mm)

sliders, which are attached directly to the construction), two forces of 300–600 N going in opposite directions should be applied simultaneously to the pair of corners of the product which lie opposite to one another (Fig. 6.19c).

## 6.2.1   Internal Forces in Chair Frames

### 6.2.1.1   Statically Determinate 2-D Structures

By adopting one of the load schemes discussed above, the distribution of internal forces in the system internally statically determinate (Fig. 6.20) can be indicated as follows.

The sums of moments in relation to point $B$ show that:

$$R_A l + F_1 l_1 - F_2(h_1 + h_2) = 0, \tag{6.1}$$

therefore,

$$R_A = \frac{F_2(h_1 + h_2) - F_1 l_1}{l}. \tag{6.2}$$

Sum of moments in relation to point $A$,

$$R_B l - F_1(l - l_1) - F_2(h_1 + h_2), \tag{6.3}$$

leads to the determination of reactions:

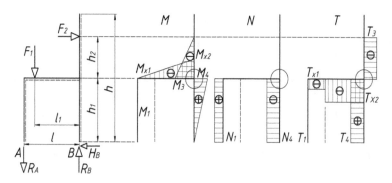

**Fig. 6.20** Graph of internal forces: bending moments, cutting forces and normal forces

$$R_B = \frac{F_1(l - l_1) + F_2(h_1 + h_2)}{l}. \tag{6.4}$$

While the sums of projections on the $X$-axis show that:

$$H_B = F_2. \tag{6.5}$$

On this basis, for all elements of the construction, a distribution of internal forces can be determined. And so, the bending moments of particular cross sections are illustrated in Fig. 6.20, and their extreme values are, respectively,

$$M_{x1} = R_A(l - l_1), \tag{6.6}$$

$$M_{x2} = R_A l + F_1 l_1, \tag{6.7}$$

$$M_3 = F_2 h_2, \tag{6.8}$$

$$M_4 = H_B h_1. \tag{6.9}$$

Normal forces can be expressed as follows:

$$N_1 = R_A, \quad N_4 = R_B, \tag{6.10}$$

and cutting forces as follows:

$$T_1 = 0, \quad T_{x1} = R_A, \quad T_{x2} = R_A + F_1, \quad T_3 = F_2, \quad T_4 = H_B. \tag{6.11}$$

On the basis of the characteristics of $M$, $T$, $N$ in a selected structural node of a furniture piece (Fig. 6.21), the state of internal forces can be determined, which is necessary for strength calculations aiming to dimension the connectors of furniture joints.

**Fig. 6.21** The internal forces acting on the node excluded in thought

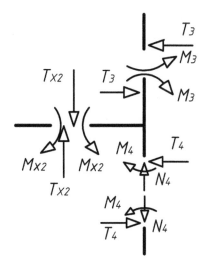

### 6.2.1.2 Statically Indeterminate 2-D Structures

The majority of side frames of skeletal furniture belong to systems internally statically indeterminate (Fig. 6.22). The degree of static indeterminate of a frame is specified by the equation:

$$s = (r + h) - 3t, \qquad (6.12)$$

**Fig. 6.22** The side frame of a chair: **a** statically indeterminate, **b** released from bonds by imaginary cutting the bar

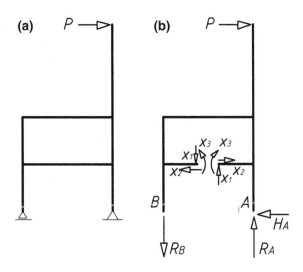

where

$s$   number of unknowns exceeding the number of equilibrium equations (for 2-D systems, we have three equilibrium equations),

$r$   number of passive forces (support rods),

$t$   number of shields (rods) of the system,

$h$   number of glue-lines joining the shields together.

For example, from Fig. 6.22, we have, respectively,

$$r = 3,$$
$$t = 7,$$
$$h = 21,$$

hence,

$$s = (3 + 21) - 3 \times 7 = 3.$$

The system is therefore threefold internally statically indeterminate.

In order to solve the system of the established degree of static indetermination, it is transformed into a statically determinate system, hereafter called the basic system. A basic system is formed from the statically indeterminate system by removing $n$ nodes (adding the system $s$ degrees of freedom) and placing in their place attributable generalised forces $X_1$, $X_2$, ..., $X_3$ called overvalues. In order to obtain the basic system, we can reject both external nodes (reactions) and internal nodes (by cutting the rod). Figure 6.22 shows the side frame of the chair before and after rejecting the internal nodes. Rejecting $s$ nodes causes that in the primary system in place of the deleted nodes, $s$ displacements can be formed, which could not be formed in an indeterminate system. The primary system becomes identical with a given statically indeterminate system if the relative displacements caused by simultaneous action of all supernumerary sizes and active loads are equal to zero. This condition can be written in the form of a so-called canonical system of equations by the method of forces, which takes the form:

$$
\begin{aligned}
X_1\delta_{11} + X_2\delta_{12} + \cdots + X_S\delta_{1S} + \delta_{10} &= 0 \\
X_1\delta_{21} + X_2\delta_{22} + \cdots + X_S\delta_{2S} + \delta_{20} &= 0 \\
\cdots\cdots\cdots\cdots\cdots\cdots\cdots\cdots\cdots\cdots\cdots\cdots\cdots\cdots \\
X_1\delta_{31} + X_2\delta_{32} + \cdots + X_S\delta_{SS} + \delta_{S0} &= 0.
\end{aligned}
\tag{6.13}
$$

Here, coefficients $\delta_{ik}$ are displacements caused by overvalues $X_k = 1$ for directions of adopted overvalues $X_i$, while coefficients $\delta_{i0}$ are identically located and directed displacements caused by external causes (load, assembly errors, etc.). In determining the displacements, $\delta_{ik}$ the following dependence should be used:

$$\delta_{ik} = \int\limits_{x} \left( \frac{M_i M_k}{EJ} + \frac{N_i N_k}{EA} + \kappa \frac{T_i T_k}{GA} \right) dx, \qquad (6.14)$$

where

$\kappa = \frac{A}{J^2} \int_A \frac{S^2}{b^2} dA$   coefficient dependent on the shape of the cross section of the rod ($k = 1, 2$—rectangle),

$S$   static moment of the field of cross section found above the line parallel to the neutral axis,

$J$   moment of inertia of the cross section,

$B$   width of the cross section,

$M_i, N_i, T_i$   internal forces caused by internal load, $X_i = 1$, located in the point of cutting the construction in thought,

$M_k, N_k, T_k$   internal forces caused by the load, $X_k$.

Values of integrals are as follows:

$$\int\limits_{x} M_i M_k dx, \quad \int\limits_{x} N_i N_k dx, \quad \int\limits_{x} T_i T_k dx, \qquad (6.15)$$

we calculate by graphic integration.

If the function $\overline{\Phi}$ is linear, then the integral $\int_a^b \Phi \overline{\Phi} dx$ is equal to the product of the area $\Omega$ of the function $\Phi$ and the ordinate $\eta$ of the function $\overline{\Phi}$ in cross section, in which lies the centre of gravity of the field of function $\Phi$ (Fig. 6.23). To simplify, in Table 6.1, fields of surfaces of the more important 2-D figures and locations of centres of gravity have been put together, and Table 6.2 shows ready equations for calculating certain integrals of this type.

Based on the theory about the reciprocity of displacements, coefficients located symmetrically in relation to the global diagonal of the system of canonical equations are equal to one another, that is,

$$\delta_{ik} = \delta_{ki}. \qquad (6.16)$$

To calculate the coefficients $\delta_{i0}$, the following equation is used:

$$\delta_{i0} = \int\limits_{x} \left( \frac{M_i M_0}{EJ} + \frac{N_i N_0}{EA} + \kappa \frac{T_i T_0}{GA} \right) dx, \qquad (6.17)$$

where

$M_0, N_0, T_0$   internal forces caused in the primary system statically determinate by the action of external causes.

**Fig. 6.23** Scheme of graphic integration

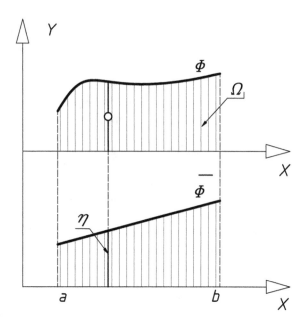

**Table 6.1** The areas and location of centres of gravity of certain 2-D figures

| Location of the centre of gravity | Area |
|---|---|
| $a$    0,5L    0,5L | $\Omega = aL$ |
| $a$    1/3L    2/3L | $\Omega = \dfrac{1}{2}aL$ |
| $a$    0,5L    0,5L | $\Omega = \dfrac{2}{3}aL$ |
| $a$    parabola    3/8L    5/8L | $\Omega = \dfrac{2}{3}aL$ |
| $a$    parabola    1/4L    3/4L | $\Omega = \dfrac{1}{3}aL$ |
| $a$    function of the third degree    1/5L    4/5L | $\Omega = \dfrac{1}{4}aL$ |

**Table 6.2** Equations for calculating some integrals $\int_x M_i M_k \mathrm{d}x$

| $M_k$ \ $M_i$ | $a$ ▭ $b$ | $f$ ◣ | $f$ ◿ | $f$ ◣ | $a$ ◢ $a$ |
|---|---|---|---|---|---|
| $a_1$ ▭ $b_1$ | $\dfrac{L}{6}\left[\begin{array}{c}a_1(2a+b)\\+b_1(2b+a)\end{array}\right]$ | $\dfrac{Lf}{3}(a_1+b_1)$ | $\dfrac{Lf}{12}(3a_1+5b_1)$ | $\dfrac{Lf}{12}(3a_1+3b_1)$ | $\dfrac{aL}{2}(a_1+b_1)$ |
| $f_1$ ◣ | | | $\dfrac{8}{15}f_1fL$ | $\dfrac{7}{15}f_1fL$ | $\dfrac{1}{5}f_1fL$ | $0$ |
| $f_1$ ◣ | | | | $\dfrac{8}{15}f_1fL$ | $\dfrac{3}{10}f_1fL$ | $-\dfrac{1}{6}f_1aL$ |
| $f_1$ ◿ | | | | | $\dfrac{1}{5}f_1fL$ | $-\dfrac{1}{6}f_1aL$ |
| $a_1$ ◢ $a_1$ | | | | | | $\dfrac{1}{3}a_1aL$ |

In many cases, for frame systems, the impact of cutting and normal forces is ignored, by applying the following equation for calculations:

$$\delta_{ik} = \int_x \frac{M_i M_k}{EJ}\,\mathrm{d}x. \qquad (6.18)$$

After determining all coefficients $\delta_{ik}$ $\delta_{i0}$, the system of equations should be solved by determining the sought overvalues $X_1$, $X_2$, ..., $X_S$. Definitive values of reactions $R$, bending moments $M$, cutting forces $T$ and longitudinal forces $N$, anywhere in the statically indeterminate system, are determined from the general superposition equations:

$$\begin{aligned}
R &= R_0 + X_1 R_1 + X_2 R_2 + \cdots + X_S R_S,\\
M &= M_0 + X_1 M_1 + X_2 M_2 + \cdots + X_S M_S,\\
T &= T_0 + X_1 T_1 + X_2 T_2 + \cdots + X_S T_S,\\
N &= N_0 + X_1 N_1 + X_2 N_2 + \cdots + X_S N_S,
\end{aligned} \qquad (6.19)$$

in which values corresponding to static sizes $R$, $M$, $T$, $N$ have been marked by the indexes 1, 2, ..., $S$, determined for particular states $X_i = 1$, while static sizes caused by an external factor in the statically determinate system are marked by index 0. If for particular supernumeraries $X_i$ only the course of moments $M$ have been determined, and the impact of cutting and normal forces were omitted, then the value of cutting forces, moments and normal forces are calculated by solving the primary statically determinate system loaded by external forces and all the already specified overvalues $X_i$ (Fig. 6.24).

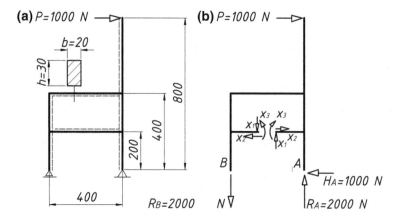

**Fig. 6.24** Calculation scheme of a side frame: **a** geometry of construction, **b** statically determinate system

*Example*

Let us establish the distribution of internal forces $M$, $T$, $N$ in the side frame of a chair, as in Fig. 6.24a, made from pinewood ($E = 12,000$ MPa), whose rails have a constant cross section in the shape of a rectangle with the dimensions $b \times h = 10 \times 30$ mm. The degree of statistical indetermination of the system is as follows:

$$s = (3 + 21) - 3 \times 7 = 3.$$

The statically indeterminate system is replaced by a statically determinate one by incising the muntin (Fig. 6.24b). Then, we introduce the virtual load $X_1 = 1$, $X_2 = 1$ and $X_3 = 1$, and we calculate the reactions and course of moments and cutting and normal forces (Fig. 6.25). Therefore,

$$\Sigma X = 0 \Rightarrow H_A = P = 1000 \text{ N},$$
$$\Sigma M_A = 0 \Rightarrow R_B = 2000 \text{ N}, \tag{6.20}$$
$$\Sigma Y = 0 \Rightarrow R_A = R_B.$$

Then, we calculate the coefficients $\delta_{ik}$ of canonical equation given below:

$$\begin{cases} X_1\delta_{11} + X_2\delta_{12} + X_3\delta_{13} + \delta_{10} = 0 \\ X_1\delta_{21} + X_2\delta_{22} + X_3\delta_{23} + \delta_{20} = 0 \\ X_1\delta_{31} + X_2\delta_{32} + X_3\delta_{33} + \delta_{30} = 0. \end{cases} \tag{6.21}$$

Omitting the negligible share in shearing and normal forces, we obtain:

$$\delta_{11} = \frac{1}{EJ}\left[4\left(\frac{1}{2}0.2 \cdot 0.2 \cdot \frac{2}{3} \cdot 0.2\right) + 2(0.2 \cdot 0.3 \cdot 0.2)\right] = \frac{1}{EJ}0.034666, \quad (6.22)$$

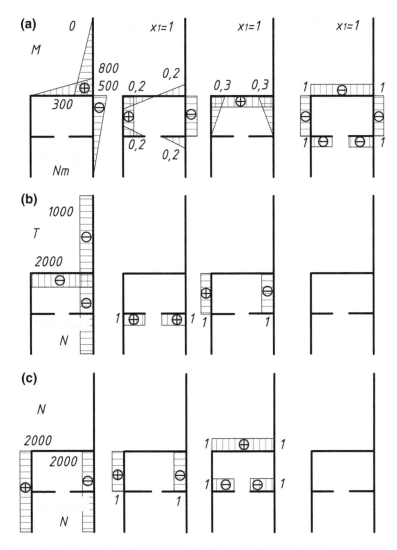

**Fig. 6.25** Charts of internal forces in a statically determinate system: **a** bending moments, **b** cutting forces, **c** normal forces

$\delta_{12} = \delta_{21} = 0$ due to the asymmetry of the charts,
$\delta_{13} = \delta_{31} = 0$ due to the asymmetry of the charts.

$$d_{22} = \frac{1}{EJ}\left[2\left(\frac{1}{2}0.3 \cdot 0.3 \cdot \frac{2}{3} \cdot 0.3\right) + 0.4 \cdot 0.3 \cdot 0.3\right] = \frac{1}{EJ}0.054, \qquad (6.23)$$

$$\delta_{32} = \delta_{23} = \frac{1}{EJ}\left[2\left(\frac{1}{2}0.3 \cdot 0.3 \cdot (-1) + 0.4 \cdot 0.3 \cdot (-1)\right)\right] = \frac{1}{EJ}(-0.21), \quad (6.24)$$

$$\delta_{33} = \frac{1}{EJ} \cdot 2(0.3(-1)(-1) + 0.4(-1) \cdot (-1)) = \frac{1}{EJ}1.4, \qquad (6.25)$$

$$\delta_{10} = \frac{1}{EJ}\left[\begin{array}{l}\frac{0.4}{6}((-800)(2 \cdot (-0.2) + 0.2) + 0(2 \cdot 0.2 + (-0.2))) \\ +\frac{0.3}{6}((-500)(2 \cdot (-0.2)) + (-0.2) + (-200)(2 \cdot (-0.2)) + (0.2))\end{array}\right]$$
$$= \frac{1}{EJ}31.666,$$

$$(6.26)$$

$$\delta_{20} = \frac{1}{EJ}\left[\frac{1}{2}(-800) \cdot 0.4 \cdot 0.3 + \frac{0.3}{6}(-500)(2 \cdot 0.3 + 0) + (-200)(2 \cdot 0 + 0.3)\right]$$
$$= \frac{1}{EJ}(-66),$$

$$(6.27)$$

$$\delta_{30} = \frac{1}{EJ}\left[\frac{1}{2}(-800) \cdot 0.4 \cdot (-1) + \left[(-200) \cdot 0.3 + \frac{1}{2}(-300) \cdot 0.3\right] \cdot (-1)\right]$$
$$= \frac{1}{EJ}265.$$

$$(6.28)$$

Canonical equation has the form:

$$\begin{cases} x_1 \cdot \dfrac{0.0347}{EJ} + \dfrac{31.7}{EJ} = 0 \\[2mm] x_2 \cdot \dfrac{0.054}{EJ} - x_3\dfrac{0.21}{EJ} - \dfrac{66}{EJ} = 0 \\[2mm] -x_2 \cdot \dfrac{0.21}{EJ} + x_3\dfrac{1.4}{EJ} + \dfrac{265}{EJ} = 0. \end{cases} \qquad (6.29)$$

Because the stiffness of all the elements is the same $EJ$ = const, we obtain:

$$\begin{cases} 0.0347x_1 + 31.7 = 0 \\ 0.054x_2 - 0.21x_3 - 66 = 0 \\ -0.21x_2 + 1.4x_3 + 265 = 0. \end{cases} \qquad (6.30)$$

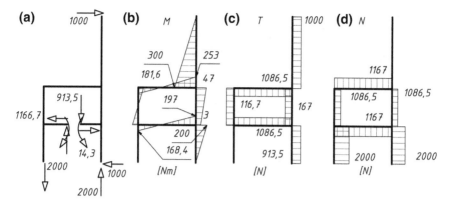

**Fig. 6.26** Distribution of internal forces in the side frame of a chair: **a** basic state, **b** bending moments, **c** cutting forces, **d** normal forces

By solving the system of equations, we obtain the supernumerary values:

$$\begin{cases} x_1 = -913.5 \\ x_2 = 1166.7 \\ x_3 = -14.3. \end{cases} \tag{6.31}$$

Graphs of moments in the statically indeterminate system, therefore, have the course as shown in Fig. 6.26.

## 6.2.2 Stresses in Cross Sections of Elements of Chair and Stool Frames

### 6.2.2.1 Stresses in Rectilinear and Slightly Curved Elements

After specifying the internal forces, the maximum stresses are calculated by using known equation of the strength of materials:

$$\sigma_{max} = \frac{N}{A} \pm \frac{M}{W}, \tag{6.32}$$

$$\tau_{max} = k\frac{T}{A}, \tag{6.33}$$

where
$N$ force normal to the cross section,
$N$ force tangent to the cross section,
$M$ bending moment,

**Table 6.3** Geometry characteristics of cross sections

| Type of cross-section | Area | Axial moment of inertia | Indicator of strength of the cross-section | Coefficient $k$ |
|---|---|---|---|---|
| *(rectangular section, width $b$, height $h$)* | $bh$ | $\dfrac{bh^3}{12}$ | $\dfrac{bh^2}{6}$ | $\dfrac{3}{2}$ |
| *(circular section, diameter $2r$, height $h$)* | $\pi r^2$ | $\dfrac{\pi r^4}{4}$ | $\dfrac{\pi r^3}{4}$ | $\dfrac{4}{3}$ |

$A$    area of the cross section,
$W$   indicator of the strength of the cross section,
$k$    coefficient determining the shape of the cross section.

During the designing of tables or chairs, the characteristics of rectangular and round cross sections are most commonly used. Characteristics of these cross sections are provided in Table 6.3.

#### 6.2.2.2  Stresses in Highly Curved Prismatic Isotropic Elements

This problem boils down to solving the issue of bending a simple curved rod. Curved rods mean rods which already in an unformed state have a curved linear axis (Zielnica 1996). We shall limit the following considerations to rods of symmetrical cross sections relative to the plane of the rod's axis. Such a rod is shown in Fig. 6.27.

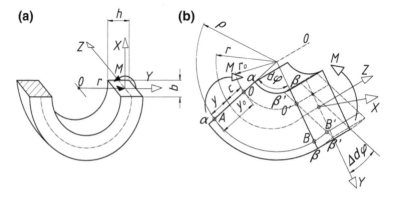

**Fig. 6.27** A rod of a symmetrical cross section relative to the plane of the rod's axis: **a** load state, **b** state of stress in the elementary section of a rod

The state of stress prevailing in a rod can be analysed on an infinitely small element (Fig. 6.27), which is cut out from the rod, leading two planes perpendicular to its axis, passing through the centre of curvature and inclined towards each other at an angle $d\varphi$. As a result of bending the rod by moment $M$, the transverse cross section $\beta$-$\beta$ rotates in relation to the cross section $\alpha$-$\alpha$ by the angle $\Delta d\varphi$, taking the position of $\beta'$-$\beta'$. Some of the fibres are subject to lengthening. Only the length of the fibre 0–0' passing through the intersection point of the cross sections is not changed. This fibre belongs to a layer that is neutral to the element. It should be noted here that the length of particular fibres of an unformed element is not created equal. Therefore, the adopted Bernoulli's principle and Hooke's law are preserved only if the neutral layer 0–0' is situated at a certain distance $c$ from the axis of the rod. Any fibre, distant by $y_o$ from the neutral layer, changes its length by the section $BB' = y_o \Delta d\varphi$. The relative lengthening of the fibre is determined by the equation:

$$\varepsilon_x = \frac{BB'}{AB} = \frac{y_o \Delta d\varphi}{\rho d\varphi},\tag{6.34}$$

where $\rho$ indicates the distance from the centre of the curvature of the rod axis to the considered fibre. The coordinates of normal stresses in the considered fibre of the element—in accordance with Hooke's law—shall be as follows:

$$\sigma_x = \varepsilon_x E = \frac{y_o \Delta d\varphi}{\rho d\varphi} E.\tag{6.35}$$

Let us next consider the balance of forces acting on the cut element (Fig. 6.28). Balance will be maintained if the following conditions are met: $\Sigma X = 0$, $\Sigma M = 0$. Condition $\Sigma X$ can be written as follows:

$$\int_A \sigma_x dA = 0.\tag{6.36}$$

**Fig. 6.28** The balance of forces acting on the sheared element of the rod

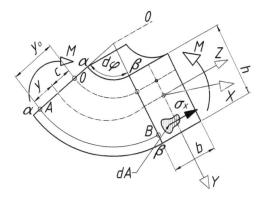

By substituting $\sigma_x$ with the equation:

$$\sigma_x = \varepsilon_x E = \frac{y_o \Delta d\varphi}{\rho d\varphi} E, \qquad (6.37)$$

we obtain:

$$\int_A \frac{y_o \Delta d\varphi}{\rho d\varphi} E dA = 0, \qquad (6.38)$$

$$\frac{\Delta d\varphi}{d\varphi} E \int_A \frac{y_o}{\rho} dA = 0. \qquad (6.39)$$

Because the equation:

$$\frac{\Delta d\varphi}{d\varphi} E \neq 0, \qquad (6.40)$$

hence,

$$\int_A \frac{y_o}{\rho} dA = 0. \qquad (6.41)$$

By substituting the following in place of $y_o$ in this equation,

$$y_o = \rho - r_o, \qquad (6.42)$$

and after appropriate transformations, we obtain:

$$r_o = \frac{A}{\int_A \frac{dA}{\rho}}, \qquad (6.43)$$

where $r_o$ is the distance from the centre of the curvature of the rod to its neutral axis.

Knowing $r_o$, the distance $c$ from the neutral axis of the transverse cross section of the rod to its centre of gravity can be determined from the equation:

$$c = r - r_o. \qquad (6.44)$$

The second of these conditions of balance, i.e. $\Sigma M$ relative to the neutral axis, is written as follows:

$$-M + \int_A \sigma_x y_o \mathrm{d}A = 0. \tag{6.45}$$

After converting the equation and using in it the predefined relations, we obtain:

$$M = \int_A \frac{y_o}{\rho} \frac{\Delta \mathrm{d}\varphi}{\mathrm{d}\varphi} E(\rho - r_o)\mathrm{d}A, \tag{6.46}$$

$$M = \frac{\Delta \mathrm{d}\varphi}{\mathrm{d}\varphi} E \left( \int_A y_o \mathrm{d}A - r_o \int_A \frac{y_o}{\rho} \mathrm{d}A \right). \tag{6.47}$$

In the above equation, the expression

$$\int_A \frac{y_o}{\rho} \mathrm{d}A, \tag{6.48}$$

has a value equal to zero.

Therefore, we can write:

$$M = \frac{\Delta \mathrm{d}\varphi}{\mathrm{d}\varphi} E \int_A y_o \mathrm{d}A, \tag{6.49}$$

whereby

$$\int_A y_o \mathrm{d}A = S, \tag{6.50}$$

indicates the static moment of the field of transverse cross section counted with respect to the neutral axis. Because the distance from the centre of gravity of the cross section to the neutral axis was marked previously by $c$, the static moment $S$ can also be presented as follows:

$$S = A \cdot c. \tag{6.51}$$

By comparing the previous equation, we obtain:

$$\int_A y_o \mathrm{d}A = Ac. \tag{6.52}$$

The equation of the moment $M$ can be now written as follows:

$$M = \frac{\Delta d\varphi}{d\varphi} EAc, \tag{6.53}$$

which gives:

$$\frac{\Delta d\varphi}{d\varphi} = \frac{M}{EAc}. \tag{6.54}$$

Therefore, the stress function is presented as follows:

$$\sigma_x = \frac{y_o}{\rho} \frac{M}{Ac}. \tag{6.55}$$

By substituting the subsequent equations,

$$\rho = r + y, \tag{6.56}$$

$$c = r - r_o, \tag{6.57}$$

we obtain the equation written in the main axes $xy$ of the cross section:

$$\sigma_x = \frac{M}{Ac} \frac{c + y}{r + y}, \tag{6.58}$$

which can be used in calculations of the normal stresses' values in curved rods, bent in the curvature plane. The individual symbols in the equation mean:

$M$   bending moment acting in the discussed cross section of the rod,
$A$   area of the cross section of the rod,
$r$   radius of the curvature of the rod axis,
$c$   distance of the neutral axis of the cross section of the rod to its centre of gravity,
$y$   coordinate of the point, in which we calculate the value of stress.

### 6.2.2.3   Stresses in Highly Curved Elements of Round Cross Sections

If the construction of the furniture piece consists of elements of large curvatures and round cross sections (Fig. 6.29), which takes place in the case of curved or curved glued furniture, then the stresses can be determined from the equations given by Korolew (1973):

**Fig. 6.29** Indications for a
bent rod of circular cross
section

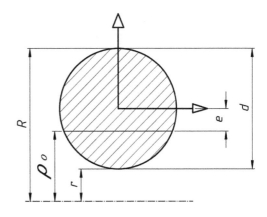

$$\sigma = \frac{M}{Ae(\rho_o + y)}, \qquad (6.59)$$

where

$$\rho_o = \frac{d}{4(2R - d - 2\sqrt{Rr})}, \qquad (6.60)$$

$$e = \frac{R + r}{2}\rho_o, \quad y = \frac{\alpha}{R}, \qquad (6.61)$$

$M$   bending moment,
$d$   diameter of the cross section of the rod,
$A$   area of the cross section,
$R$   radius of the curvature of the rod,
$r$   distance of extreme fibres from the centre of the rod curvature.

## 6.2.3 Strength of Joints

### 6.2.3.1 Strength of Adhesive Joints

Adhesion Strength of Loaded Joints

The model of adhesion strength of loaded glued joint was presented by Godzimirski (2002) (Fig. 6.30). In furniture, this type of joint does not, however, have any great significance, aside from few cases of front longitudinal joints. Normal stresses in the glue-line of such a connection can be calculated in a simplified manner from the equation:

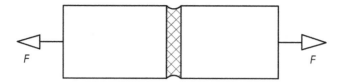

**Fig. 6.30** Adhesion load of glued joints

$$\sigma = \frac{F}{A}, \tag{6.62}$$

where
$F$  normal force to the surface of the glue-line,
$A$  area of the glue-line.

However, it should be noted that the narrowing occurring in the glue-line proves the presence of tangential stresses. Therefore, when making precise calculations of the strength of such joints, it is better to use numerical methods.


Peel Strength of Loaded Joints

When using narrow edges made of plastic and veneers for finishing narrow surfaces of furniture elements, these glued joints need to ensure a high resistance to peeling (Fig. 6.31).

Peeling of veneers from a massive board element is associated with the formation of large normal stresses in the glue-line of irregular distribution and bulges at the edge of the joint. In analysing the calculation model of a joint loaded for peeling, it is assumed that a thinner element is subjected to bending, and the deformations of the board, due to its great stiffness, can be omitted (Misztal 1956; Godzimirski 2002) (Fig. 6.32).


**Fig. 6.31** Peeling load of glued joints

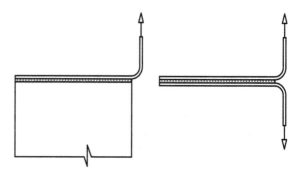

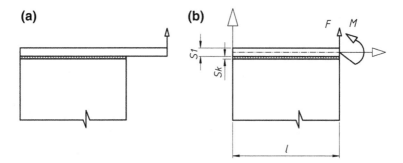

**Fig. 6.32** A calculation model of a joint formed from two adherents, with a significantly different thickness, loaded for peeling: **a** primary state, **b** loaded glue-line

**Fig. 6.33** Internal forces acting on the elementary section of a beam on an elastic base

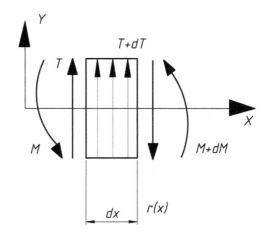

Usually, to describe deformations of such a thin element, calculation methods are used which are appropriate to describe deflections of beams working on an elastic base (Nowacki 1970; Godzimirski 2002). By extracting an elementary section of such a beam, the state of internal forces is illustrated in Fig. 6.33.

For the considered model of a glued joint, the maximum normal stresses appearing on its edges are as follows:

$$\sigma_{max} = \frac{M}{b}\sqrt{\frac{12E_k}{ES_1^3 S_k}} + \frac{\sqrt{2}F}{b}\sqrt[4]{\frac{12E_k}{ES_1^3 S_k}}, \qquad (6.63)$$

where
$E$   linear elasticity module of the veneer,
$E_k$   linear elasticity module of the glue,
$S_1$   thickness of the veneer,
$S_k$   thickness of the glue-line,

*b*   width of the glue-line,
*M*   bending moment,
*F*   normal force.

Shearing Strength of Loaded Joints

The literature shows that descriptions of tests of the state of stresses in the joints of connections of anisotropic bodies are nonlinear, and in its assumptions, they adopt many simplifications. Niskanen (1955, 1957) was the first to give the distribution of stresses in the joint of connected wooden elements. By using the model of a solid anisotropic two-sheared shield, the author ignored the impact of the elastic properties of the glue-line, believing that a thin glue-line occurring at gluing wood does not have a significant impact on deformations in connectors. These tests were also undertaken by Wilczyński (1988), when analysing the impact of dimensional proportions of connected elements on the size and form of the distribution of tangential stresses. Therefore, the considered connection was treated as a solid wooden element, simplifying the solution to the problem greatly. In order to accurately determine the distribution of stresses in the glue-lines of joints, numerical methods were used (Apalak and Davies 1993, 1994; Biblis and Carino 1993; Godzimirski 1985; Groth and Nordlund 1991; Ieandrau 1991; Janowiak 1993; Kline 1984; Lindemann and Zimmerman 1996; Nakai and Takemura 1995, 1996a, b; Pellicane 1994; Pellicane et al. 1994; Smardzewski 1994, 1995). Wilczyński's (1988) tests seem particularly interesting, as a result of which the state of tangential stresses in the joints of two-sheared wooden shields was computer-indicated using the finite element method, using a rectangular, four-node orthotropic element and by verifying these calculations with a laboratory measurement by using electrical strain gauges.

Assuming a constant distribution of stresses in the direction of the width of elements and adopting a glue-line as a single-layer system, the lap joint (Fig. 6.34) is reduced to the form of a flat task of the elasticity theory (Fig. 6.35).

The balance of forces in the connection elements for the bottom element can be written as follows:

$$\tau_x = \frac{d\sigma_{x2}}{dx} S_2, \tag{6.64}$$

hence, the stress in the lower lap is expressed as follows:

$$\sigma_{x2} = \frac{1}{S_2} \int_x \tau_x dx, \tag{6.65}$$

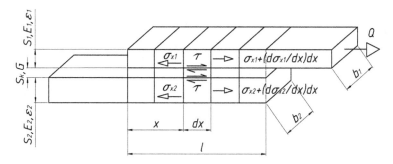

**Fig. 6.34** Geometry of the lap joint

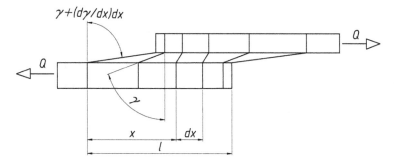

**Fig. 6.35** Figural deformations of an adhesive joint and linear of laps

and for the upper element,

$$\sigma_{x1} b_1 S_1 + \tau dx b_1 - \left( \sigma_{x1} + \frac{d\sigma_{x1}}{dx} dx \right) b_1 S_1 = 0, \qquad (6.66)$$

hence,

$$\tau_{x1} = \frac{d\sigma_{x1}}{dx} S_1, \qquad (6.67)$$

$$\sigma_{x1} = \frac{1}{S_1} \int_x \tau_x dx, \qquad (6.68)$$

where

$S_1, S_2$    thickness of the upper and lower laps,
$\sigma_{x1}, \sigma_{x2}$   normal stress in the upper and lower laps,
$\tau$          tangential stress in the glue-line.

**Fig. 6.36** Deformations of
the elementary, extreme
section of the glue-line

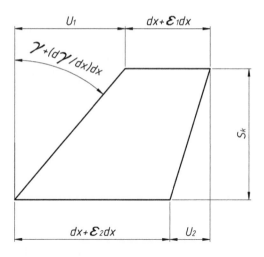

In analysing the deformation of a single extreme element of an adhesive joint
from Fig. 6.36, we obtain the system:

$$\begin{cases} U_1 + dx + \varepsilon_1 dx = U_2 + dx + \varepsilon_2 dx, \\ \quad U_1 = \left(\gamma + \frac{d\gamma}{dx} dx\right) S_k, \\ \quad U_2 = \lambda S_k, \end{cases} \tag{6.69}$$

from which we have:

$$\frac{d\gamma}{dx} = \frac{\varepsilon_2 - \varepsilon_1}{S_k}, \tag{6.70}$$

where

$\varepsilon_1, \varepsilon_2$   normal strains of the upper and lower laps,
$\gamma$     shear strain of the glue-line.

Bearing in mind the balance of forces acting between the single element of the
lap and glue-line, the deformations of the upper lap can be expressed as follows:

$$E_1 b_1 S_1 \frac{d\varepsilon_1}{dx} = \frac{d\tau_1}{dx}, \tag{6.71}$$

and shear strain in the form:

$$\gamma G b_1 = \frac{d\tau_1}{dx}. \tag{6.72}$$

Therefore, the normal strains $\varepsilon_1$, $\varepsilon_2$ of the upper and lower laps are expressed as follows:

$$\frac{d\varepsilon_1}{dx} = \frac{\lambda G}{E_1 S_1}, \tag{6.73}$$

$$\frac{d\varepsilon_2}{dx} = \frac{\lambda G}{E_2 S_2}, \tag{6.74}$$

where

$G$      shear modulus of the glue-line,
$E_1$, $E_2$    linear elasticity module of the upper and lower laps.

By differentiating on both sides the equation:

$$\frac{d\gamma}{dx} = \frac{\varepsilon_2 - \varepsilon_1}{S_k} \tag{6.75}$$

and substituting the above equations, we obtain a differential equation of the second order with a constant coefficient, expressing the change in the shear strain $\gamma$ on length $l$ of the glue-line:

$$\frac{d^2\gamma}{dx^2} - b^2\gamma = 0, \tag{6.76}$$

where

$$b^2 = \frac{G}{S_k}\left(\frac{1}{E_2 S_2} + \frac{1}{E_1 S_1}\right). \tag{6.77}$$

By using the typical substitution for this equation:

$$\gamma = e^{rx}, \tag{6.78}$$

we shall obtain the characteristic equation for the differential equation:

$$r^2 - b^2 = 0, \tag{6.79}$$

having two specific solutions:

$$\begin{aligned} r_1 &= b, \\ r_2 &= -b. \end{aligned} \tag{6.80}$$

The general integral of the differential equation of the second order takes the form:

$$\gamma = C_1 e^{rx} + C_2 e^{-rx}, \tag{6.81}$$

in which constant integrations $C_1$ and $C_2$ should be determined from border conditions for deformations of the glue-line:

$$
\begin{cases}
C_1 = \frac{\varepsilon_2}{S_k r} + C_2 \rightarrow dla \rightarrow x = 0, \quad \varepsilon_1 = 0 \\
C_1 e^{rl} - C_2 e^{-rl} = \frac{\varepsilon_1}{S_k r}
\end{cases}
\tag{6.82}
$$

By solving the above system relative to $C_1$, $C_2$ and introducing the shear strain of a glue-line $\gamma$, the value of tangential stresses in a glue-line can be written in the general form as follows:

$$
\tau_x = \frac{G}{S_k r \sinh(rl)} (\varepsilon_1 \cosh(rx) + \varepsilon_2 \cosh(r(l - x))),
\tag{6.83}
$$

where

$$
r = \sqrt{\frac{G}{S_k} \left( \frac{1}{E_2 S_2} + \frac{1}{E_1 S_1} \right)},
\tag{6.84}
$$

$$
\varepsilon_1 = \frac{Q}{b_1 S_1 E_1},
\tag{6.85}
$$

$$
\varepsilon_2 = \frac{Q}{b_2 S_2 E_2}.
\tag{6.86}
$$

If $E_1 = E_2 = E$ and $b_1 = b_2 = b$, then the value of tangential stresses in the glue-line is described by the equation:

$$
\tau_x = \frac{QG}{S_k r E b \sinh(rl)} \left( \frac{1}{S_1} \cosh(rx) + \frac{1}{S_2} \cosh(r(l - x)) \right),
\tag{6.87}
$$

where

$$
r = \sqrt{\frac{G}{S_k E} \left( \frac{1}{S_2} + \frac{1}{S_1} \right)}.
\tag{6.88}
$$

Depending on the susceptibility of $\varepsilon_1$ and $\varepsilon_2$ of elements of the connection, the maximum stresses can concentrate for $x = 0$ or $x = 1$ (Fig. 6.37).

The average value of these stresses on the whole length of the glue-line is well described by the equation:

$$
\tau_{sr} = \frac{Q}{bl}.
\tag{6.89}
$$

**Fig. 6.37** The distribution of stresses in a lap joint: **a** normal stresses in a lap, **b** tangential stresses in the glue-line

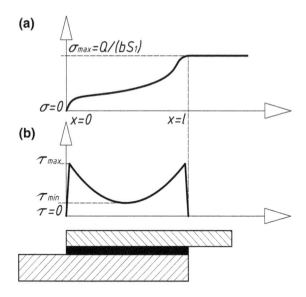

The correctness of the results calculated with the use of the above equations will depend on the correctness of the constant flexibilities of wood and glue-line, determined in the course of experimental studies. However, it is known that the elastic properties of wood depend, among others, on the anatomic direction, and hence, the method of cutting elements glued together is significant for the distribution of stresses in joints. The system of wooden fibres in relation to the axis of stretching projects on the value of extensions in laps and hence on the shear strain and distribution of tangential stresses in the glue-line. By analysing this state of stresses, the main methods of gluing woods in the following planes should be considered:

- tangential and radial LT-LR (Fig. 6.38a),
- tangential LT-LT (Fig. 6.38b),
- radial LR-LR (Fig. 6.38c).

In the discussed cases, in order to determine the distribution of tangential stresses in wooden lap joints, taking into account the natural system of wood fibres, it is necessary to determine the value of the linear elasticity module of any lap in the transformed system of coordinates, rotated by the angle $\varphi$ relative to the axis of stretching. At the same time, the values of constant elasticities of wood in non-transformable systems should be determined. The results of such tests enable to determine the impact of the type of material, plane of gluing, direction of fibres in relation to the axis of stretching on the distribution of tangential stresses in the glue-line. Moreover results determine possibilities to use typical solutions for isotropic bodies in case of wooden adhesive joints.

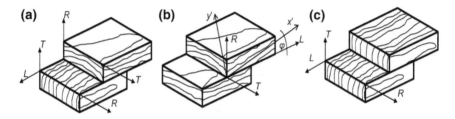

**Fig. 6.38** Methods of joining wood by gluing planes together: **a** LR and LT, **b** LT and LT, **c** LR and LR

Torsional Strength of Loaded Joints

Angular joints are the dominant structural solutions of modern skeletal wooden furniture, whose shape and proportions have changed little since the days of ancient Egypt, Greece and Rome (Dzięgielewski and Smardzewski 1995). For several decades, attempts have been made to provide their strength in the form of mathematical relationships. The first calculations, aiming to determine analytically the strength of rectangular glue-lines of angular wooden joints, were conducted by Rónai (1969), however mistakenly assuming that the state of tangential stresses in the glue-line corresponds to the state of stresses in any cross section of a prismatic rod under torsion. It was not until the work of Haberzak (1975) and later works (Matsui 1990a, b, 1991; Smardzewski 1994, 1995) that helped to establish that in a complex state of loads, the greatest tangential stress focuses in the corners of the glue-line.

Mortise and tenon joints and bridle joints are some of the most common in the constructions of furniture frames (Fig. 6.39). Therefore, let us separate corner from the frame (Fig. 6.39a), and let us indicate in its cross sections the forces $N$, $T$, $M$. Reducing these forces into the middle of the glue-line 0, we shall obtain that it is subjected to resultant force:

$$\overrightarrow{F} = \overrightarrow{N} + \overrightarrow{T}, \tag{6.90}$$

and the moment:

$$M = M_1 + F \cdot l. \tag{6.91}$$

The state of stresses in such a node was provided in Haberzak's (1975) work in a simplified manner, but useful for engineering practices.

Because force $F$ passes through the middle of the glue-line, it causes (arbitrarily) an average state of stresses on the entire surface of the glue-line with a value of:

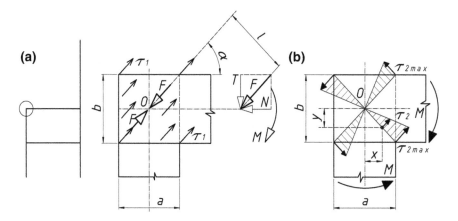

**Fig. 6.39** Tangential stresses in the glue-line caused by **a** normal and cutting forces, **b** torsion moment

$$\tau_1 = \frac{F}{nab} \le k_t^k, \tag{6.92}$$

where

$a, b$   dimensions of the glue-line,
$n$   number of the glue-lines,
$k_t^k$   shearing strength of the glue-line.

While moment $M$, attempting to bend the frames of the joint, and simultaneously twist the glue-line, is counteracted by tangential stresses $\tau_2$ emerging on the surfaces, different in different points of these surfaces. By assuming that their distribution is the same as in Fig. 6.39b, then we can write:

$$\tau_2 = \frac{\sqrt{x^2 + y^2}}{\sqrt{\left(\frac{a}{2}\right)^2 + \left(\frac{b}{2}\right)^2}} \tau_{2\max}. \tag{6.93}$$

By continuing to maintain the state of balance between the external moment and the moment originating from the sum of stresses $\tau_2$, we obtain:

$$M = n\tau_2 \int_A \int \sqrt{x^2 + y^2} \, dxdy. \tag{6.94}$$

Therefore, the value of the maximum stress occurring in the corners of the glue-line can be determined according to the formula:

$$\tau_{2max} = \frac{M}{J} \frac{\sqrt{\left(\frac{a}{2}\right)^2 + \left(\frac{b}{2}\right)^2}}{n}, \tag{6.95}$$

where
$M$   maximum bending moment,
$J$   moment of inertia of the cross section of the glue-line in relation to the middle
of rotation 0.

$$J = \int_{-\frac{1}{2}b}^{\frac{1}{2}b} \int_{-\frac{1}{2}a}^{\frac{1}{2}a} (x^2 + y^2) dx dy = ab \frac{(a^2 + b^2)}{12}. \tag{6.96}$$

Therefore, the greatest static stress is the sum of component stresses $\tau_1$ i $\tau_2$ in one of the corners of the glue-line. According to Fig. 6.40, it amounts to:

$$\tau_{max} = \sqrt{\tau_1^2 + \tau_{2max}^2 - 2\tau_1 \tau_{2max} \cos \delta} \leq k_t^k, \tag{6.97}$$

whereas for glue-line with the dimensions of $a < b$,

$$\delta = 90° + \beta + \gamma, \tag{6.98}$$

where

$$\beta = \text{arctg} \frac{a}{b},$$
$$\gamma = 90° - \alpha, \tag{6.99}$$
$$\alpha = \text{arcctg} \frac{T}{N},$$

and for glue-line with proportions $a > b$,

$$\delta = 90° + \alpha + \gamma, \tag{6.100}$$

where

$$\gamma = 90° - \beta. \tag{6.101}$$

When designing or checking bridle joints, the number n of glue-lines should be established for its provided dimensions a × b and for an established distribution of internal forces. The solutions proposed above are based on the elementary strength of materials and are completely correct from an engineering point of view. They take into account only a linear variability of tangential stresses caused by the bending moment and the lack of variability of the distribution of stresses originating from loads of axial forces, while these distributions should be nonlinear functions.

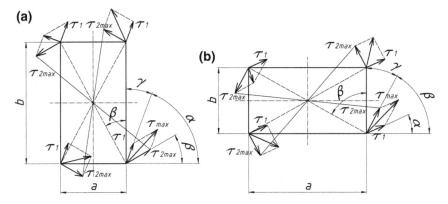

**Fig. 6.40** The distribution of maximum tangential stresses in the corners of the glue-line of proportions: **a** $a < b$, **b** $a > b$

On this assumption, Smardzewski (1994) suggested a description of the distribution of tangential stresses in a square glue-line of a joint subject to clean torsion, using the appropriate equations of the theory of elasticity.

However, it is known that in angular connections, the glue-line is in a complex state of loads. Establishing in them the distribution of tangential stresses requires constructing a suitable mathematical model. To this end, let us consider the single bridle joint, loaded by shearing forces and the bending moment (Fig. 6.41). In this joint, the distribution of tangential stresses in the glue-line is caused by the forces $T = F\sin\beta$ i $N = F\cos\beta$. Therefore, we can write down the first components of the state of stresses in the form of:

$$\tau_x = \frac{G_k}{S_k r_x} \cdot \frac{1}{\sinh(r_x b)} \cdot (\varepsilon_{x1} \cosh(r_x x) + \varepsilon_{x2} \cosh(r_x(b - x))), \qquad (6.102)$$

and

$$\tau_y = \frac{G_k}{S_k r_y} \cdot \frac{1}{\sinh(r_y h)} \cdot (\varepsilon_{y1} \cosh(r_y y) + \varepsilon_{y2} \cosh(r_y(h - y))), \qquad (6.103)$$

where

$$r_x = \left[ \frac{G_k}{S_k} \left( \frac{1}{S_2 E_{x2}} + \frac{1}{S_1 E_{x1}} \right) \right]^{0.5}, \qquad (6.104)$$

$$r_y = \left[ \frac{G_k}{S_k} \left( \frac{1}{S_2 E_{y2}} + \frac{1}{S_1 E_{y1}} \right) \right]^{0.5}. \qquad (6.105)$$

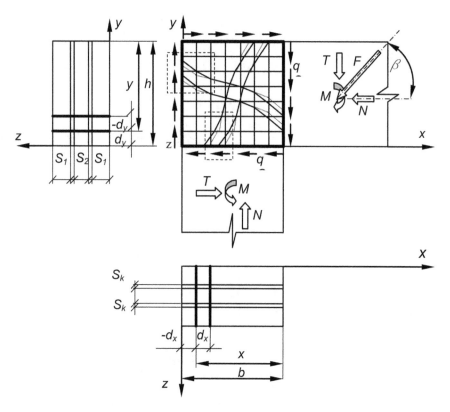

**Fig. 6.41** Deformations in a joint subject to torsion

Establishing the distribution of tangential stresses caused by the bending moment $M = 2qbh$ requires the analysis of forms of deformations of connected elements (Fig. 6.42).

The deformations of elementary sections of joint shown in Fig. 6.42, of the dimensions $dx$, $dy$, result from both the shear strains of elements and the glue-line. By undertaking to properly describe the state of tangential stresses in the joint, two adjacent elements were selected, with a width $dx$ (Fig. 6.42) and of elastic properties $G_i$, $E_i$ and thickness $S_i$ (1 in the bottom index concerns the upper element, 2—lower element, k—glue-line). These segments were then separated by planes perpendicular to the surface of joint at a distance of $y - dy$ and $y + dy$, by also entering the loads substituting the impact of the cut-out parts. An additional assumption was also adopted here that the loads between the elements and the layer of glue are shifted along the edges $dx$ and $dy$. By writing the equation of balance for the considered parts in the form:

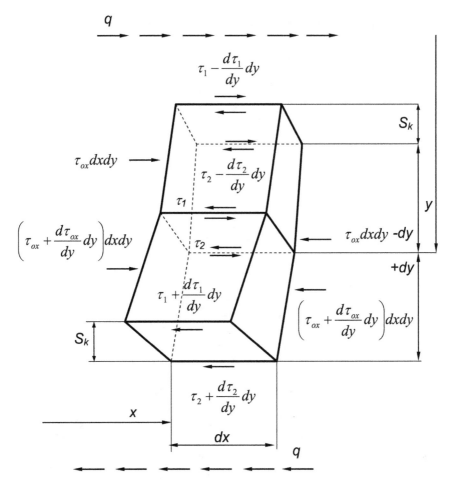

**Fig. 6.42** The distribution of internal forces in elementary sections of joint

$$\tau_{ox}\mathrm{d}x\mathrm{d}y = \left(\tau_{1x} - \frac{\mathrm{d}\tau_{1x}}{\mathrm{d}y}\mathrm{d}y\right) \cdot S_1\mathrm{d}x - \tau_{1x}S_1\mathrm{d}x = \left(\tau_{2x} - \frac{\mathrm{d}\tau_{2x}}{\mathrm{d}y}\mathrm{d}y\right)S_2\mathrm{d}x - \tau_{2x}S_2\mathrm{d}x,$$

$$(6.106)$$

and neglecting the small sizes of the higher orders, we shall obtain the dependence between tangential stresses $\tau_{ix}$ in the element:

$$\tau_{ox} = S_1\frac{\mathrm{d}\tau_{1x}}{\mathrm{d}y} \quad \text{and} \quad \tau_{ox} = -S_2\frac{\mathrm{d}\tau_{2x}}{\mathrm{d}y}. \qquad (6.107)$$

By further assuming the condition of the continuity of displacements, which requires that the adjacent walls of separated elements were also adjacent after deformation (Fig. 6.43), we can write the equation:

$$x'_o = x_2 + x_0 - x_1, \qquad (6.108)$$

where the displacement of a segment of glue in the upper element:

$$x'_o = \frac{\tau_{ox}}{G_k} S_k, \qquad (6.109)$$

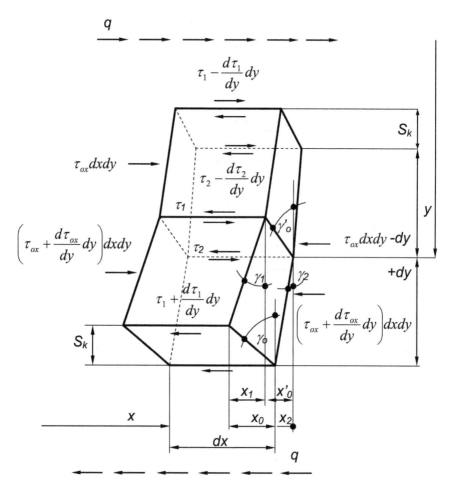

**Fig. 6.43** Deformations of elements' section and the layer of glue-line

the displacement of a segment of glue in the lower element:

$$x_o = \left(\tau_{ox} + \frac{d\tau_{ox}}{dy}dy\right)\frac{S_k}{G_k},$$ (6.110)

displacement of the upper element section:

$$x_1 = \left(\tau_{1x} + \frac{d\tau_{1x}}{dy}dy\right)\frac{dy}{G_{xy1}},$$ (6.111)

displacement of the lower element section:

$$x_2 = \left(\tau_{2x} + \frac{d\tau_{2x}}{dy}dy\right)\frac{dy}{G_{xy2}}.$$ (6.112)

By entering the above equations to the previous equation, and neglecting the small sizes of the higher orders, we shall obtain the equation:

$$\frac{d\tau_{ox}}{dy} = \frac{G_k}{S_k}\left(\frac{\tau_{1x}}{G_{xy1}} - \frac{\tau_{2x}}{G_{xy2}}\right).$$ (6.113)

from which by differentiating on both sides in relation to $y$, we shall obtain the differential equation of the second degree in the form:

$$\frac{d^2\tau_{ox}}{dy^2} = \frac{G_k}{S_k}\left(\frac{1}{G_{xy1}S_1} + \frac{1}{G_{xy2}S_2}\right)\tau_{ox}$$ (6.114)

By using further substitutions,

$$k^2 = \frac{G_k}{S_k}\left(\frac{1}{G_{xy1}S_1} + \frac{1}{G_{xy2}S_2}\right)\tau_{ox} \quad \text{and} \quad \tau_{ox} = ce^{ry},$$ (6.115)

we obtain the characteristic equation:

$$e^{ry}\left(r^2 - k^2\right) = 0,$$ (6.116)

that has two specific solutions:

$$r_1 = k \quad \text{and} \quad r_2 = -k.$$ (6.117)

The general equation of the differential equation of the second order, expressing tangential stresses $\tau_{ox}$ in the glue-line, can therefore be written as follows:

$$\tau_{ox} = c_1 e^{ky} + c_2 e^{-ky}. \tag{6.118}$$

Constants of integration $c_1$ and $c_2$ are determined by assuming the appropriate border conditions. Therefore,

$$\begin{cases} \dfrac{\partial \tau_{ox}}{\partial y} = \dfrac{G_s}{S_s} q \left( \dfrac{1}{G_{xy1}S_1} - \dfrac{1}{G_{xy2}S_2} \right) = c_1 k e^{k \cdot 0} - c_2 k e^{-k \cdot 0} \quad \text{for} \quad y = 0 \\[2ex] \dfrac{\partial \tau_{ox}}{\partial y} = \dfrac{G_s}{S_s} q \left( \dfrac{1}{G_{xy1}S_1} - \dfrac{1}{G_{xy2}S_2} \right) = c_1 k e^{kh} - c_2 k e^{-kh} \quad \text{for} \quad y = h \end{cases}, \tag{6.119}$$

By solving the above system of equations and entering $c_1$ and $c_2$, we shall obtain the general form of the equation describing the distribution of tangential stresses along the edge $y$ of the glue-line in the following form:

$$\tau_{ox} = \frac{M \cdot k}{2bh} \cdot \frac{1}{\sinh(kh)} \cdot [\cosh(ky) + \cosh(k(h - y))]. \tag{6.120}$$

For direction x, the distribution of stresses $\tau_{oy}$ can be written in the analogous equation:

$$\tau_{oy} = \frac{M \cdot k}{2bh} \cdot \frac{1}{\sinh(kb)} \cdot [\cosh(kx) + \cosh(k(b - x))], \tag{6.121}$$

where
$M$    bending moment,
$b, h$   dimensions of the glue-line.

Finally, the tangential stresses in a rectangular glue-line, caused by a complex state of load, can be written in the vector form:

$$\vec{\tau}_{xy} = \vec{\tau}_x + \vec{\tau}_y + \vec{\tau}_{ox} + \vec{\tau}_{oy}, \tag{6.122}$$

or also, respectively, for specific parts of the glue-line (Fig. 6.44) in the form:

$$\begin{aligned} \tau_{xy} &= \sqrt{(\tau_y - \tau_{oy})^2 + (\tau_x + \tau_{ox})^2}, \quad 0 \le x \le \tfrac{1}{2}b \wedge 0 \le y \le \tfrac{1}{2}h, \\[1ex] \tau_{xy} &= \sqrt{(\tau_y - \tau_{oy})^2 + (\tau_x - \tau_{ox})^2}, \quad 0 \le x \le \tfrac{1}{2}b \wedge \tfrac{1}{2}h \le y \le h, \\[1ex] \tau_{xy} &= \sqrt{(\tau_y + \tau_{oy})^2 + (\tau_x + \tau_{ox})^2}, \quad \tfrac{1}{2}b \le x \le b \wedge 0 \le y \le \tfrac{1}{2}h, \\[1ex] \tau_{xy} &= \sqrt{(\tau_y + \tau_{oy})^2 + (\tau_x - \tau_{ox})^2}, \quad \tfrac{1}{2}b \le x \le b \wedge \tfrac{1}{2}h \le y \le h. \end{aligned} \tag{6.123}$$

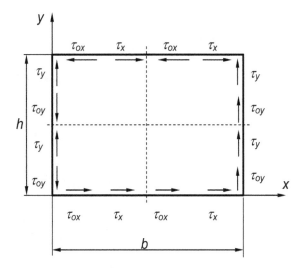

**Fig. 6.44**  The components of vectors of tangential stresses in a glue-line

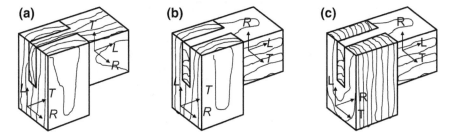

**Fig. 6.45**  Orientation of wooden elements in an angular connection with respect to the adhesion of connectors by planes: **a** tangential with tangential LT-LT, **b** tangential with radial LT-LR, **c** radial with radial LR-LR

The correctness of the results obtained with the use of those equations will depend on the correctness of the constant flexibilities of wood and glue-line, determined in the course of experimental studies. It should also be noted that in addition to constant elasticity related to linear deformations in the directions $x$ and $y$ of the local system of coordinates, the values of coefficients of elasticity associated with figural deformations should be determined. In practice, there are a few basic ways of connecting wooden elements in angular and cross-connections (Fig. 6.45).

Deformations of Wooden Joints in the Complex Load State

Assuming that we are dealing with a 2-D element placed in a Cartesian coordinate system XOY, in which the Z-axis is perpendicular to the other two, strains $\varepsilon$ and $\gamma$ can be written in a general manner by equations (Ashkenazi et al. 1958;

**Fig. 6.46** Strains of a 2-D
element

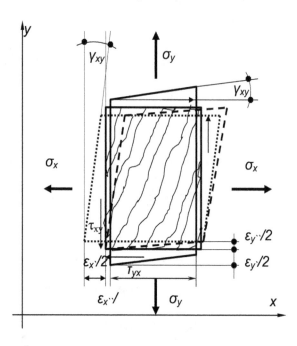

Ambarcumian 1967; Lekhnickij 1977; Leontiev 1952; Mitinskij 1948; Norris 1942;
Nowacki 1970; Rabinowicz 1946; Timoshenko and Goodier 1951):

$$\left.\begin{array}{l} \varepsilon_x = a_{11}\sigma_x + a_{12}\sigma_y + a_{13}\tau_{xy} \\ \varepsilon_y = a_{21}\sigma_x + a_{22}\sigma_y + a_{23}\tau_{xy} \\ \gamma_{xy} = a_{31}\sigma_x + a_{32}\sigma_y + a_{33}\tau_{xy} \end{array}\right\}. \qquad (6.124)$$

In this case, the equations consist of nine coefficients of susceptibility $a_{ij}$. The
value of these coefficients can be determined experimentally by analysing the
character of deformation of a 2-D element in a Cartesian coordinate plane
(Fig. 6.46).

To present the relations between stresses and strains in an element stretched
biaxially and sheared in the $XY$ plane, the principle of superposition can be applied,
by summing up the deformations of a rectangular element shown in Fig. 6.46.
Under the influence of stresses $\sigma_x$, the element shall sustain extension $\varepsilon_{x'} = \sigma_x/E_x$ in
the direction of the $X$-axis, while narrowing $\varepsilon_{y''} = v_{xy}\varepsilon_x$ towards the $Y$-axis and shear
strain $\gamma_{xy'''} = \mu_{x,xy} \cdot \varepsilon_x$ in the $XY$ plane. Due to stresses $\sigma_y$, the appropriate elongation
in the direction of the $Y$-axis will amount to $\varepsilon_y = \sigma_y/E_y$, while narrowing towards
the $X$ $\varepsilon_{x''} = v_{yx} \cdot \varepsilon_y$ and the shear strain $\gamma_{xy''} = \mu_{y,xy} \cdot \varepsilon_y$. The effect of tangential
stresses $\tau_{xy}$ will cause shear strain $\gamma_{xy'} = \tau_{xy}G_{xy}$, elongation in the direction of $X$
equal to $\varepsilon_{x'''} = \mu_{xy,x}\gamma_{xy}$ and elongation in the direction of $Y$, respectively,

$\varepsilon_y''' = \mu_{xy,y} \cdot \gamma_{xy}$. By summing up the appropriate strains of the element, caused by the joint effect of stresses $\sigma_x$, $\sigma_y$, $\tau_{xy}$, we shall obtain:

$$\varepsilon_x = \frac{\sigma_x}{E_x} - \nu_{yx}\frac{\sigma_y}{E_y} + \mu_{xy,x}\frac{\tau_{xy}}{G_{xy}},$$

$$\varepsilon_y = -\nu_{xy}\frac{\sigma_x}{E_x} + \frac{\sigma_y}{E_y} + \mu_{xy,y}\frac{\tau_{xy}}{G_{xy}}, \qquad (6.125)$$

$$\gamma_{xy} = \mu_{x,xy}\cdot\frac{\sigma_x}{E_x} + \mu_{y,xy}\frac{\sigma_y}{E_y} + \frac{\tau_{xy}}{G_{xy}}.$$

According to Rabinowicz (1946), the coefficients of susceptibility $a_{ij}$ should be called technical coefficients, while the above equation can be written in the form:

$$\begin{bmatrix} \varepsilon_x \\ \varepsilon_y \\ \gamma_{xy} \end{bmatrix} = \begin{bmatrix} \frac{1}{E_x} & -\frac{\nu_{yx}}{E_y} & \frac{\mu_{xy,x}}{G_{xy}} \\ -\frac{\nu_{xy}}{E_x} & \frac{1}{E_y} & \frac{\mu_{xy,y}}{G_{xy}} \\ \frac{\mu_{x,xy}}{E_x} & \frac{\mu_{y,xy}}{E_y} & \frac{1}{G_{xy}} \end{bmatrix} \cdot \begin{bmatrix} \sigma_x \\ \sigma_y \\ \tau_{xy} \end{bmatrix}. \qquad (6.126)$$

Taking into account the symmetry of matrix of deformations $a_{ij} = a_{ji}$, we shall obtain obvious dependencies:

$$\frac{\nu_{xy}}{E_x} = \frac{\nu_{yx}}{E_y}; \quad \frac{\mu_{xy,x}}{G_{xy}} = \frac{\mu_{x,xy}}{E_x}; \quad \frac{\mu_{xy,y}}{G_{xy}} = \frac{\mu_{y,xy}}{E_y}, \qquad (6.127)$$

which enable us to determine normal strains $\varepsilon_x$, $\varepsilon_y$ in the main directions of the plane's axis and shear strain $\gamma_{xy}$ in the plane XOY in the following form:

$$\varepsilon_x = \frac{1}{E_x}\left(\sigma_x - \nu_{xy}\sigma_y + \mu_{x,xy}\tau_{xy}\right),$$

$$\varepsilon_y = \frac{1}{E_y}\left(\sigma_y - \nu_{yx}\sigma_x + \mu_{y,xy}\tau_{xy}\right), \qquad (6.128)$$

$$\gamma_{xy} = \frac{1}{G_{xy}}\left(\tau_{xy} + \mu_{xy,x}\sigma_x + \mu_{xy,y}\sigma_y\right).$$

Functions of strains and stresses written in the form of constitutive equations express the relationships between the coordinates of tensors of strains and stresses. However, it is necessary to determine the coordinates of the tensor of susceptibility in any directions of the global plane of coordinates. This is connected with determining the state of stresses or strains in selected directions of anisotropy of the tested body. Therefore, let us consider the method of transforming elastic properties of wood after entering a new plane of coordinates.

The next equation allows us to calculate the values of normal and shear strains of wood in any direction tilted to the main axes of the plane of coordinates at an angle $\varphi$.

$$\begin{bmatrix} \varepsilon'_x \\ \varepsilon'_y \\ \gamma'_{xy} \end{bmatrix} = \begin{bmatrix} a'_{11} & a'_{12} & a'_{13} \\ a'_{21} & a'_{22} & a'_{23} \\ a'_{31} & a'_{33} & a'_{33} \end{bmatrix} \cdot \begin{bmatrix} \sigma'_x \\ \sigma'_y \\ \tau'_{xy} \end{bmatrix}. \tag{6.129}$$

Assuming the conditions for the uniaxial state of stresses:

$$\begin{aligned} \varepsilon'_x = a'_{11}\sigma'_x; \quad \varepsilon'_y = a'_{21}\sigma'_x; \quad \gamma'_{xy}=a'_{31}\sigma'_x \Leftrightarrow \sigma'_y = \tau'_{xy} = 0 \\ \varepsilon'_y = a'_{22}\sigma'_y; \quad \varepsilon'_x = a'_{12}\sigma'_y; \quad \gamma'_{xy} = a'_{32}\sigma'_y \Leftrightarrow \sigma'_x = \tau'_{xy} = 0 \\ \gamma'_{xy} = a'_{33}\tau'_{xy}; \quad \varepsilon'_x = a'_{13}\tau'_{xy}; \quad \varepsilon'_y = a'_{23}\tau'_{xy} \Leftrightarrow \sigma'_x = \sigma'_y = 0 \end{aligned} \tag{6.130}$$

we obtain:

$$a'_{11} = \frac{1}{E'_x} = a_{11}\cos^4\varphi + a_{22}\sin^4\varphi + (2a_{12} + a_{33})\sin^2\varphi\cos^2\varphi + 2a_{13}\sin\varphi\cos^2\varphi + 2a_{23}\sin^3\varphi\cos\varphi,$$

$$a'_{22} = \frac{1}{E'_y} = a_{11}\sin^4\varphi + a_{22}\cos^4\varphi + (2a_{12} + a_{33})\sin^2\varphi\cos^2\varphi - 2a_{13}\sin^3\varphi\cos\varphi - 2a_{23}\sin\varphi\cos^3\varphi,$$

$$a'_{33} = \frac{1}{G'_{xy}} = (a_{11} - 2a_{12} + a_{22})4\sin^2\varphi\cos^2\varphi + [(a_{23} - a_{13})4\sin\varphi\cos\varphi + a_{33}](\cos^2\varphi - \sin^2\varphi) \tag{6.131}$$

and

$$v'_{xy} = a'_{21}\cdot E'_x, \quad v'_{yx} = a'_{12}\cdot E'_y, \quad \mu'_{xy} = a'_{13}\cdot G'_{xy}, \quad \mu'_{xy,y} = a'_{23}\cdot G'_{xy}, \tag{6.132}$$

$$\mu'_{x,xy} = a'_{31}\cdot E'_y, \quad \mu'_{y,xy} = a'_{32}\cdot E'_y. \tag{6.133}$$

Hence, we ultimately obtain equations describing modules of wood elasticity in any direction $\varphi$, in the form of:

$$\begin{aligned} E'_x &= \frac{E_xE_yG_{xy}}{G_{xy}(E_y\cos^4\varphi + E_x\sin^4\varphi) + E_x(-2G_{xy}v_{yx} + E_y)\sin^2\varphi\cos^2\varphi + 2E_xE_y(\mu_{xy,x}\cos^2\varphi + \mu_{xy,y}\sin^2\varphi)\sin\varphi\cos\varphi} \\ E'_y &= \frac{E_xE_yG_{xy}}{G_{xy}(E_y\sin^4\varphi + E_x\cos^4\varphi) + E_x(-2G_{xy}v_{yx} + E_y)\sin^2\varphi\cos^2\varphi - 2E_xE_y(\mu_{xy,x}\sin^2\varphi + \mu_{xy,y}\cos^2\varphi)\sin\varphi\cos\varphi} \\ G'_{xy} &= \frac{E_xE_yG_{xy}}{4G_{xy}(E_y - E_x(2v_{yx} - 1))\sin^2\varphi\cos^2\varphi + E_xE_y((\mu_{xy,y} - 1)\sin\varphi\cos\phi + 1(\cos^2\varphi - \sin^2\varphi)} \end{aligned} \tag{6.134}$$

In order to determine the deformation of an anisotropic body, the following equation can be used:

$$\begin{bmatrix} \varepsilon'_x \\ \varepsilon'_y \\ \gamma'_{xy} \end{bmatrix} = \begin{bmatrix} a'_{11} & a'_{12} & a'_{13} \\ a'_{21} & a'_{22} & a'_{23} \\ a'_{31} & a'_{33} & a'_{33} \end{bmatrix} \cdot \begin{bmatrix} \sigma'_x \\ \sigma'_y \\ \tau'_{xy} \end{bmatrix} \tag{6.135}$$

or in the case of a uniaxial state of stresses, the equation:

$$\varepsilon'_x = \frac{1}{E'_x} \cdot \sigma'_x, \tag{6.136}$$

in which the values of factors $E_x$, $E_y$, $G_{xy}$, $v_{yx}$, $\mu_{xy,x}$, $\mu_{xy,y}$ should be determined in the primary XOY system.

As the analysis of the state of stresses showed in lap joints, wooden elements are subject to stretching. Therefore, in order to appoint dependencies of strains from a uniaxial stress $\sigma_x$ and angle of cutting samples $\varphi$ in the plane of orthotropy, the following equation should be used:

$$\varepsilon'_{xi} = a'_{11} \cdot \sigma'_{xi} \quad \text{or} \quad \varepsilon'_{xi} = \frac{1}{E'_{xi}} \cdot \sigma'_{xi} \tag{6.137}$$

in which the parameters $a'_{11}$ are subject to transformation. Because the wooden elements joined at lap after gluing may form different planes of anatomical build, it was decided to provide a description of normal strains of wood for the following planes:

- the radial plane LR,
- the tangential plane LT.

Normal strains $\varepsilon'_L$ in the LR plane amount to:

$$\varepsilon'_L = \left[ \frac{1}{E_L}\cos^4\phi + \frac{1}{E_R}\sin^4\phi - \left(\frac{2v_{RL}}{E_R} - \frac{1}{G_{LR}}\right)\sin^2\phi\cos^2\phi \right.$$
$$\left. + \frac{2\mu_{LR,L}}{G_{LR}}\sin\phi\cos^3\phi + \frac{2\mu_{LR,R}}{G_{LR}}\sin^3\phi\cos\phi \right]\frac{Q}{S_i b_i}, \tag{6.138}$$

while normal strains $\varepsilon'_L$ in the $LT$ plane are equal to:

$$\varepsilon'_L = \left[ \frac{1}{E_L}\cos^4\varphi + \frac{1}{E_T}\sin^4\varphi - \left(\frac{2v_{TL}}{E_T} - \frac{1}{G_{LT}}\right)\sin^2\varphi\cos^2\varphi \right.$$
$$\left. + \frac{2\mu_{LT,L}}{G_{LT}}\sin\varphi\cos^3\varphi + \frac{2\mu_{LT,T}}{G_{LT}}\sin^3\varphi\cos\varphi \right]\frac{Q}{S_i b_i}, \tag{6.139}$$

where
$Q$ value of the external load.

Based on the analysis of the state of tangential stresses in rectangular angular joints, it was shown that wooden elements, as a result of external forces, are in a complex state of stresses. This state causes that except to normal strains in the adherent (glued element) shear strains appear. Therefore, prior to establishing tangential stresses in the glue-line of a wooden joint, it would be necessary to determine the appropriate, due to the direction of stresses, global deformations in the selected plane of the adherent (Fig. 6.47).

**Fig. 6.47** Shear strains of the
adherent in angular joints

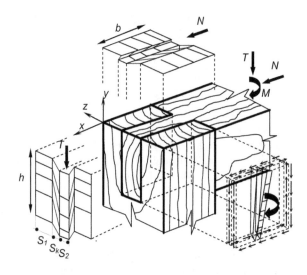

Normal strains $\varepsilon'_{xi}$ and $\varepsilon'_{yi}$ of any adherent in the direction of X or Y, caused by
stresses $\sigma'_x$ or $\sigma'_y$ for particular types of planes of wood glued together, are expressed
by equations described in the previous chapter. The causes of the emergence of
such strains are axial forces in structural nodes. In addition to these forces in
angular joints, also bending moments appear, causing states of tangential stresses in
glue-lines and shear strains of elements of joints. Load of an angular joint in the
plane of a glue-line resembles a complex state of stresses in an orthotropic body,
shown in Fig. 6.47. For such a case, shear strains are expressed by the equation:

$$\gamma'_{xy} = a'_{31}\sigma'_x + a'_{32}\sigma'_y + a'_{33}\tau'_{xy}. \tag{6.140}$$

Coefficients $a'_{31}, a'_{32}, a'_{33}$ allow us to determine the value of shear strains for the
examined planes of gluing wood:

description of the deformations $\gamma'_{LR}$

$$
\begin{aligned}
\gamma'_{LR} = &\left[\left(-\frac{v_{RL}}{E_R} - \frac{1}{E_L}\right)2\sin\varphi\cos^3\varphi + \left(\frac{1}{E_R} + \frac{v_{LR}}{E_L}\right)2\sin^3\varphi\cos\varphi + \left(\frac{\mu_{R,LR}}{E_R} - \frac{\mu_{L,LR}}{E_L}\right)2\sin^2\varphi\cos^2\varphi \right. \\
&\left. + \left(\frac{\mu_{LR,L}}{G_{LR}}\cos^2\varphi + \frac{\mu_{LR,R}}{G_{LR}}\sin^2\varphi + \frac{1}{G_{LR}}\sin\phi\cos\phi\right)(\cos^2\varphi - \sin^2\varphi)\right]\sigma'_L \\
&+ \left[\left(-\frac{v_{RL}}{E_R} - \frac{1}{E_L}\right)2\sin^3\varphi\cos\varphi + \left(\frac{1}{E_R} + \frac{v_{LR}}{E_L}\right)2\sin\varphi\cos^3\varphi + \left(\frac{\mu_{L,LR}}{E_L} - \frac{\mu_{R,LR}}{E_R}\right)2\sin^2\varphi\cos^2\varphi \right. \\
&\left. + \left(\frac{\mu_{LR,L}}{G_{LR}}\sin^2\varphi + \frac{\mu_{LR,R}}{G_{LR}}\cos^2\varphi - \frac{1}{G_{LR}}\sin\phi\cos\phi\right)(\cos^2\varphi - \sin^2\varphi)\right]\sigma'_R \\
&+ \left[\left(\frac{1}{E_L} + \frac{2v_{RL}}{E_R} + \frac{1}{E_R}\right)4\sin^2\varphi\cos^2\varphi + \left(\left(\frac{\mu_{LR,R}}{E_R} - \frac{\mu_{LR,L}}{E_L}\right)4\sin\varphi\cos\varphi + \frac{1}{G_{LR}}\right)(\cos^2\varphi - \sin^2\varphi)\right]\tau'_{LR},
\end{aligned}
$$

$$\tag{6.141}$$

description of the strains $\gamma'_{LT}$

$$\gamma'_{LT} = \left[ \left( \frac{-v_{TL}}{E_T} - \frac{1}{E_L} \right) 2 \sin\varphi \cos^3\varphi + \left( \frac{1}{E_T} + \frac{v_{LT}}{E_L} \right) 2 \sin^3\varphi \cos\varphi + \left( \frac{\mu_{T,LT}}{E_T} - \frac{\mu_{L,LT}}{E_L} \right) 2 \sin^2\varphi \cos^2\varphi \right.$$
$$+ \left. \left( \frac{\mu_{LT,L}}{G_{LT}} \cos^2\varphi + \frac{\mu_{LT,L}}{G_{LT}} \sin^2\varphi + \frac{1}{G_{LT}} \sin\phi\cos\phi \right) (\cos^2\varphi - \sin^2\varphi) \right] \sigma'_L$$
$$+ \left[ \left( -\frac{v_{TL}}{E_T} - \frac{1}{E_L} \right) 2 \sin^3\varphi \cos\varphi + \left( \frac{1}{E_T} + \frac{v_{LT}}{E_L} \right) 2 \sin\varphi \cos^3\varphi + \left( \frac{\mu_{L,LT}}{E_L} - \frac{\mu_{T,LT}}{E_T} \right) 2 \sin^2\varphi \cos^2\varphi \right.$$
$$+ \left. \left( \frac{\mu_{LT,L}}{G_{LT}} \sin^2\varphi + \frac{\mu_{LT,T}}{G_{LT}} \cos^2\varphi - \frac{1}{G_{LT}} \sin\varphi\cos\varphi \right) (\cos^2\varphi - \sin^{2\varphi}) \right] \sigma'_T$$
$$+ \left[ \left( \frac{1}{E_L} + \frac{2v_{TL}}{E_T} + \frac{1}{E_T} \right) 4 \sin^2\varphi \cos^2\varphi + \left( \left( \frac{\mu_{LT,L}}{E_T} - \frac{\mu_{LT,L}}{E_L} \right) 4 \sin\varphi \cos\varphi + \frac{1}{G_{LT}} \right) (\cos^2\varphi - \sin^2\varphi) \right] \tau'_{LT}.$$

$$(6.142)$$

Presented description of normal and shear strains of adherents must be used in equations describing tangential stresses in glue-line which connect these adherents. The mathematical models obtained in such a way allow us to determine the relation between the layout of wood fibres in elements of a joint and the form and values of tangential stresses in the glue-line.

## Influence of Technological Errors of Glue-Line on the Strength of Joint

Gluing wood is one of a few methods of inseparable and stiff connecting wooden elements. Usually, these connections are done in a method that allows the use of the greatest shearing strength of the created glue-line, and during designing, they are placed in such locations of the construction in which axial and transverse forces dominate, avoiding places subject to bending and torsional loads.

Constructions of wood joinery and skeletal furniture belong to the group of 2-D or 3-D systems. From the strength point of view structural nodes are located in irrational places, in corners. Of course this method are used for hundreds of years, ensuring technology of assembly and aesthetics of implementation. In spite of how furniture is used, glue-lines are subject to damage even before the state of stresses in wooden elements reaches acceptable values. Therefore, the correct design of the construction requires conducting destructive strength tests each time. The conclusions and guidelines resulting from them are the basis for modernising implemented or developing new constructions. These types of methods of design are not the fastest and certainly not the cheapest. That is why, the leading role in designing should be played by strength calculation of wooden glue-lines. Unfortunately, the methods of these calculations are not, as yet, recognised well and described. In many analyses of the strength of wooden joints, a continuity and homogeneity of the glue-line is assumed. Such an assumption is very convenient from the analytical point of view, however, for practical reasons, too risky. In the process of bonding structural elements, there are often deficiencies relative to the technological assumptions, which leads to:

- the formation of loosenesses, that is heterogeneities, in which the adhesive occurs in a state that is not completely hard, due to the inaccurate mixing of its ingredients,
- the occurrence of gas pockets in cohesive layers, often as a result of too vigorous mixing of finished adhesives or all of its ingredients,
- the formation of unglued areas, when the gas pockets applied with the adhesive begin to move in an unhardened glue-line towards the adherent, where they join air contained in micropores, open fibres or wood vessels,
- the formation of adhesive spills as a result of applying too much adhesive on the surfaces of connectors joined together.

In several works of analytical character, the issue of heterogeneity has already been undertaken, limiting it to the properties attributed to glue-line connecting metal adherents. Apalak and Davies (1994) designed angular joints of metal elements connected with epoxy adhesives, taking into account the sizes and form of hardened excesses of glue, hereinafter called spills. On the basis of the studies conducted, they claimed that spills significantly reduce tensions in glue-line and increase the strength of the joints. Kuczmaszewski (1995) came to similar conclusions on the basis of studies of metal lap joints, who in his studies took into account the discontinuity of a glue-line visible in the form of gas pockets and unglued areas. The heterogeneity of glue-line was also the subject of studies of Francis and Gutierrez-Lemini (1984), who, using the finite element method, modelled cracks and gas pockets in the glue-line, stating the concentration of stresses around the source of discontinuity of the glue-line.

This chapter will describe the issues related to the influence of heterogeneity of glue-line of wooden joints on the distribution of tangential stresses in the glue-line, and, in particular, with the importance of gas pockets and unglued areas.

In the solutions of Wnuk (1981) and Łączkowski (1988), concerning nonlinear models of the mechanics of cracking, it was assumed that around the gas pocket with a radius $R$ under the influence of load, a gap is formed in the shape of a circular ring with a radius $R$ and $(R + c)$. The front of the gap is surrounded by a plastic area with a radius $r_{pl}$ (Fig. 6.48).

**Fig. 6.48** The calculation scheme of energy balance—gas pocket

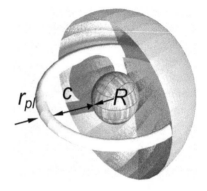

By treating the area of spheres with a radius $R + c + r_{pl}$ as isolated, they presented the balance of energy processes taking place during cracking in the following form:

$$U = U_o - U_{od} + U_p + U_{pl}, \tag{6.143}$$

where
$U_o$    elastic energy of the homogeneous centre,
$U_{od}$    offload energy released after the crack has been created,
$U_p$    surface energy, constituting residues of offload energy,
$U_{pl}$    work of structural deformation, constituting a part of offload energy.

Offload energy for gas pocket with the module $E = 0$ amounts to:

$$U_{od} = \frac{\sigma^2}{2E_s} \left[ \frac{4}{3} \pi (R + c + r_{pl})^3 - \frac{4}{3} \pi R^3 \right], \tag{6.144}$$

and surface energy:

$$U_p = 2\pi \left[ (R + c)^2 - R^2 \right] \Theta, \tag{6.145}$$

where
$\Theta$    proper surface energy.

The difficulty associated with identifying the radius of the plastic area $r_{pl}$ was solved by Kuczmaszewski (1995), by adopting, after Wnuk (1981), the equation:

$$r_{pl} = (R + c) \left[ \frac{R_{ek}}{\sqrt{R_{ek}^2 - \sigma^2}} - 1 \right], \tag{6.146}$$

where
$R_{ek}$    plasticity limit of the glue,
$\sigma$    normal stress.

Therefore, the energy of plastic deformation has been calculated from the equation:

$$U_{pl} = R_{ek} \cdot \varepsilon_e \cdot 2\pi^2 (R + c)^3 \left[ \frac{R_{ek}}{\sqrt{R_{ek}^2 - \sigma^2}} - 1 \right], \tag{6.147}$$

where
$\varepsilon_e$    relative structural elongation

By entering the obtained equations to the equation of energy balance, the final form is obtained:

$$
U = U_o - \frac{\sigma^2}{2E_k}\left\{\frac{4}{3}\pi\left[(R+c)\left(\frac{R_{ek}}{\sqrt{R_{ek}^2-\sigma^2}}\right)\right]^3 - \frac{4}{3}\pi R^3\right\} + 2\pi\left[(R+c)^2-R^2\right]\cdot\Theta
$$
$$
+ R_{ek}\cdot\varepsilon_e\cdot 2\pi^2(R+c)^3\left[\frac{R_{ek}}{\sqrt{R_{ek}^2-\sigma^2}}-1\right]^2.
$$

$$(6.148)$$

To determine the critical value of the radius of the gas pocket $R_{kr}$, at which the initiation of a crack gap takes place, the first derivative of the balance equation should be compared to zero and $c$ substituted by zero. As a result of these calculations, the following equation is obtained:

$$
R_{kr} = \frac{2\Theta}{\frac{\sigma^2}{E_k}\left[\frac{R_{ek}}{\sqrt{R_{ek}^2-\sigma^2}}\right]^3 - 3R_{ek}\varepsilon_e\pi\left[\frac{R_{ek}}{\sqrt{R_{ek}^2-\sigma^2}}-1\right]^2}.
$$

$$(6.149)$$

By assuming of surface energy, plastic elongation $\varepsilon_{pl}$ and other elastic features, on the basis of characteristics $\sigma = f(\varepsilon)$, the variability of critical radius, which initiates a process of cracking in the glue-lines of wooden joints can be specified.

Under the influence of load impact $\sigma$, the radius of unglued area $R$ increases to $R + c$ (Fig. 6.49). By adopting the assumption that deformations at the border of phases are the same, Kuczmaszewski (1995) assumed after Wnuk (1981) that loaded energy now takes the form: the calculation scheme of energy balance is as follows:

**Fig. 6.49** The calculation scheme of energy balance—unglued area

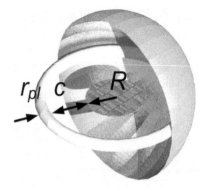

$$U_{od} = \frac{\sigma^2}{3} \pi (R + c + r_{pl})^3 \left( \frac{1}{E_x} + \frac{1}{E_k} \right). \tag{6.150}$$

where
$E_x$, $E_k$  elasticity modules of the adherent and glue-line.

Due to the clear difference of phases between the glue and the adherent on wooden joints, the ingredient of surface energy in the balance equation was replaced by work needed to separate these phases, that is, work of adhesion $W_s$ on the surface of the circular ring with a radius $R$ and $R + c$. Therefore,

$$U_p = W_s \cdot \pi \left[ (R + c)^2 - R^2 \right], \tag{6.151}$$

energy of plastic deformation:

$$U_{pl} = R_{ex} \cdot \varepsilon_e \cdot \pi^2 (R + c)^3 \cdot \left[ \frac{R_{ek}}{\sqrt{R_{ek}^2 - \sigma^2}} - 1 \right]^2 + R_{ek} \cdot \varepsilon_e$$

$$\cdot \pi^2 (R + c)^3 \cdot \left[ \frac{R_{ek}}{\sqrt{R_{ek}^2 - \sigma^2}} - 1 \right]^2, \tag{6.152}$$

and the energy equation takes the form:

$$U = U_o - \frac{\sigma^2 \pi}{3} \cdot (R + c)^3 \cdot \left[ \frac{R_{ek}}{\sqrt{R_{ek}^2 - \sigma^2}} \right]^3 \cdot \left[ \frac{1}{E_x} + \frac{1}{E_s} \right] + W_s \cdot \pi \cdot \left[ (R + c)^2 - R^2 \right]$$

$$+ \varepsilon_e \cdot \pi^2 (R + c)^3 \cdot \left[ \frac{R_{ek}}{\sqrt{R_{ek}^2 - \sigma^2}} - 1 \right]^2 \cdot (R_{ex} + R_{ek}). \tag{6.153}$$

where
$R_{ex}$  plasticity limit of the adherent.

Like before, the critical value of the radius of heterogeneity (unglued areas) is obtained after differentiating the energy equation $U$ with respect to $R + c$ and the corresponding transformations:

$$R_{kr} = \frac{2W_s}{\sigma^2 \left( \frac{1}{E_x} + \frac{1}{E_k} \right) \left( \frac{R_{ek}}{\sqrt{R_{ek}^2 - \sigma^2}} \right)^3 - 3\varepsilon_e \pi \left( \frac{R_{ek}}{\sqrt{R_{ek}^2 - \sigma^2}} - 1 \right)^2 (R_{ex} + R_{ek})}. \tag{6.154}$$

Assuming a work of adhesion $W_s$, for different materials (Proszyk et al. 1997), and a elastic characteristics of adherents, the variability of the critical radius which initiates the process of cracking in the glue-lines of wooden joins can be determined.

The process of destroying adhesive joints is generally violent by nature, regardless of the elastic characteristics of the glue. Hardened glue in the form of a glue-line stiffened additionally through elements of connectors shows greater stiffness than the same glue hardened in the form of a sample for studying the shear module of elasticity (Smardzewski and Dzięgielewski 1994; Dzięgielewski and Wilczyński 1990). Due to the fact that processes of decohesion begin within the ends of the glue-line, where the greatest tangential stresses occur, the existence of heterogeneity in this area can only facilitate the process.

Gas pockets present in glue-line contribute to an increase in stresses around the source of heterogeneity. This phenomenon initiates the process of decohesion, thus reducing the strength of the connection. Unglued areas, such as gas pockets, constitute a potential source for initiating the processes of friable cracking and decrease the strength of the connection. Increasing the value of stresses in glue-line to the limit of plasticity of the glue suddenly decreases the critical radius value of heterogeneity, which reaches values many times smaller than the thickness of the glue-line. In such conditions, even the smallest type of heterogeneity in the form of a gas pocket, unglued area or looseness constitutes the source of friable cracking. Due to the significant influence of heterogeneity of the adhesive on the strength of joints, the processes of preparing adhesive mass and elements meant for gluing should be done with extreme care.

### 6.2.3.2  Strength of Shape-Adhesive Joints

Strength of Dowel Joints

Joints with dowel connectors are particularly recognised in furniture due to the many significant benefits, including:

- reducing material consumption by the size of tenons,
- the simple technological process,
- the uncomplicated construction.

The literature on the subject recognises the model of a bent 2-D angular dowel joint (Fig. 6.50) (Dzięgielewski and Smardzewski 1995; Smardzewski 1998a). This model assumes that the total load is transferred only by the dowels, while there are no interactions between the connected elements. The bending moment in this case can be broken down into two equal, in terms of value, normal forces acting in the axes of the dowels; therefore, the neutral axis of bending is in the middle of the distance between the dowels. Then, the strength conditions of a structural node can be expressed as follows:

**Fig. 6.50** The scheme of a dowel joint with the axis of bending located in the axis of symmetry of the dowels

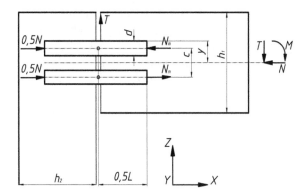

shearing strength of the dowels:

$$k_t^d > \frac{4T}{n\pi d^2},\qquad(6.155)$$

bending strength of the dowels:

$$\sigma_{max} = \frac{M}{J}y \le k_g^d,\qquad(6.156)$$

where

$$J = 2\left(J_o + Aa^2\right),\qquad(6.157)$$

$$A = \frac{\pi d^2}{4},\quad J_o = \frac{\pi d^4}{64},\qquad(6.158)$$

hence,

$$k_g^d \ge \frac{16M}{\pi d^4 + 4\pi d^2 c^2}(c+d),\qquad(6.159)$$

where
$M, T$    bending moment, cutting force,
$d$       dowel diameter,
$a$       distance of dowel axis from the neutral axis,
$c$       spacing between the dowel axes,
$\rho$       distance of extreme fibres from the neutral axis,
$n$       number of connectors,
$k_t^d$     shearing strength of the wood,
$k_g^d$     bending strength of the wood.

By choosing the greater diameters $d$ of the calculated values, we proceed to determine the length of a dowel. It depends mainly on the strength of the glue-line, which occurs on the side of the dowel, on the wall. The state of loading a dowel with normal forces is illustrated in Fig. 6.50. It shows that the maximum normal force concentrates only in one dowel and amounts to:

$$N_{max} = \frac{N}{2} + \frac{M}{c}, \tag{6.160}$$

where
$N$  normal force.

Therefore, we calculate the shearing stress for the smaller length of the dowel embedded in the two elements, by using the equation:

$$\tau_{max} = \frac{N_{max}}{n\pi d L_{min}} \leq k_t^k, \tag{6.161}$$

from which we determine the length of the dowel:

$$L = 2L_{min} \geq 2 \frac{\frac{N}{2} + \frac{M}{c}}{n\pi d k_t^k}, \tag{6.162}$$

where
$k_t^k$  shearing strength of the adhesive glue-line,
whereby

$$k_t^k > \frac{0.5N + \frac{M}{c}}{0.5\pi d L}, \tag{6.163}$$

where
$L$  length of the dowel.

The conditions adopted in the model are met when between the joined elements there is a gap preventing or limiting immediate contact of the elements, or when the load of the node is insignificant in relation to the stiffness of the glue-line. The results in accordance with the above analytical description were obtained in the works (Smardzewski and Dzięgielewski 1994; Smardzewski 1990). However, in these works, only those connections specified in the range of loads were studied, not providing, in both cases, what part of the interim strength of the joint was constituted by the upper limit of the applied force. Additionally, the authors (Smardzewski and Dzięgielewski 1994) reported that in the studied samples, due to technological reasons, there was a gap between the connected elements.

**Fig. 6.51** The scheme of a dowel joint with the axis of bending located below the axis of symmetry of the dowels

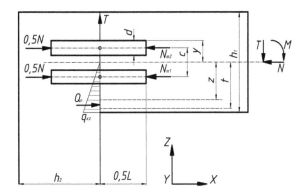

As can be seen from the author's studies, as the load value gets closer to the interim strength limit of the joint, there is a displacement of the neutral axis of bending in the direction of the lower part of the joint. Such a state of the structural node takes into account the mathematical model for which the scheme of forces is shown in Fig. 6.51. Then, contact stresses arise between the joined elements, the resultant of which is as follows:

$$Q_n = q_{xz}h_1\partial y. \tag{6.164}$$

where
$\partial y$ the width of the contact surface.

This value remains in close connection with the location of the neutral axis z and the distance of the resultant $Q_n$ from the centre of the distance between the dowels:

$$Q_n(t - z) = N_{m2}(z + 0.5c) + N_{m1}(z - 0.5c), \tag{6.165}$$

while the location of the neutral axis is expressed by the equation:

$$z = \frac{2Q_n t - c(N_{m2} - N_{m1})}{N_{m2} + N_{m1} + 2Q_n}, \tag{6.166}$$

where in turn

$$\frac{1}{3}q_{xz}(0.5h_1 - z)^2\partial y = M. \tag{6.167}$$

The strength conditions in such a model can be expressed as follows:

$$k_t^d > \frac{2T}{\pi d^2}, \tag{6.168}$$

**Fig. 6.52** The scheme of a
dowel joint with the point of
elastic contact of the elements

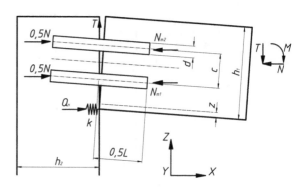

$$k_t^d > \frac{32 q_{xz} h_1 \partial y (z - 0.5c)}{\pi d^4 + 4\pi d^2 (4z^2 + c^2)} (z + 0.5(c + d)), \tag{6.169}$$

$$k_t^k > \frac{0.5N + \frac{M}{c}}{0.5\pi dL}. \tag{6.170}$$

A further build-up of the load leads to a case where the growing stress between elements is focused along the lower edge of the contact surface of the elements (Fig. 6.52).

When considering the system of forces in the plane XZ, such a stress can be expressed as a reaction $Q_n$ of a spring of the stiffness $k$. Then,

$$Q_n z = N_{m2}(0.5c + 0.5h_1 - z) + N_{m1}(0.5h_1 - 0.5c - z), \tag{6.171}$$

$$z = \frac{h_1(N_{m2} + N_{m1}) + c(N_{m2} - N_{m1})}{2(Q_n + N_{m1} + N_{m2})}, \tag{6.172}$$

$$Q_n z = M, \tag{6.173}$$

while the strength conditions are in accordance with the previous equations.

## Strength of Mortise and Tenon Joints

As can be seen from the previous considerations, covered mortise and tenon joints maintain elasticity as long as the glue-line joining the wooden elements is not damaged. Its strength is calculated in the same way as for bridle joints, by entering the relevant dimensions and the number of glue-lines. However, it should be noted that the damage of glue-line will cause displacement of external load on the stresses between the mortise and tenon. In these conditions, the following has to be established (Fig. 6.53):

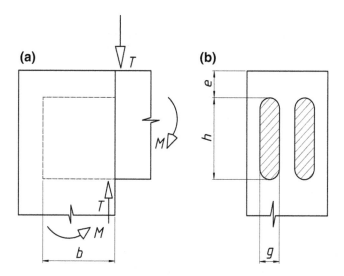

**Fig. 6.53** The geometry of mortise and tenon joints: **a** state of load, **b** transverse cross section of tenons

shearing strength of the tenon:

$$\tau_{max} = \frac{T}{ngh} \leq k_t^d, \tag{6.174}$$

bending strength of the tenon:

$$\sigma = \frac{M}{J_0}y \leq k_g^d, \tag{6.175}$$

shearing strength of part of the mortise:

$$\tau_{max} = \frac{T}{ne(2b+g)} \leq k_t^d, \tag{6.176}$$

where
$T, M$ cutting force and bending moment read from the graphs $N$, $T$, $M$,
$n$ number of tenons,
$J_x$ moment of inertia of the $n$-tenons' cross section
$\rho$ distance of extreme fibres from the neutral axis,
$g$ thickness of the tenon,
$b$ length of the tenon,
$h$ height of the tenon,
$e$ depth of the tenon,
$k_t^d$ shearing strength of the wood across the fibres,
$k_g^d$ bending strength of the wood.

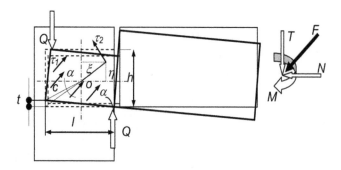

**Fig. 6.54** Distribution of stresses in the glue-line of a loosely fitted mortise and tenon

Mortise and tenon joints can be treated as an adhesive connection only when the elements of the mortise and tenon are in contact with each other only along the wide planes, in which a rectangular glue-line has formed (Fig. 6.54).

In this situation, the strength of a connection depends only on tangential stresses in the glue-line. For this case, the centre of bending $O$ of the joint is located in the geometric centre of the tenon and as long as the rotation angle $\alpha$ meets the conditions of inequality:

$$0 < \alpha < \frac{2t}{\sqrt{l^2 - 4t^2}}, \tag{6.177}$$

where

$t$  fit between the mortise and tenon,
$l$  length of the tenon,

the strength of the joint will be specified by the equation:

$$\tau_{max}^2 = \tau_1^2 + \tau_{2max}^2 - 2\tau_1 \tau_{2max} \cos \gamma, \tag{6.178}$$

where
tangential stresses caused by external moment $M$,

$$\tau_{2max} = \frac{M}{\frac{n}{c} \iint_A (\eta^2 + \xi^2) d\eta d\xi}, \tag{6.179}$$

tangential stresses caused by axial forces $T$, $N$,

$$\tau_1 = (T^2 + N^2)^{0.5} / (nhl), \tag{6.180}$$

where
c    distance from the centre of torsion,
h    height of the tenon,
n    number of the glue-lines.

When the acceptable values of the moment bending the connection $M$, exceed, causing that the angle $\alpha$ satisfies the inequality:

$$\alpha \geq \frac{2t}{\sqrt{l^2 - 4t^2}}, \tag{6.181}$$

some elements of the surface of the mortise and tenon begin to put pressure on each other, causing a resistance moment $M_Q$, which reduces the external moment value $M$ to $M_\tau$, causing tangential stresses in the glue-line $M_\tau = M - M_Q$.

Assuming that the length of the surface of compressions represents about 10 % of half of the length of the tenon 0.5 l (Fig. 6.55), we shall obtain the relationship:

$$M_Q = 0.05l\delta\sigma[(l-2t)-0.05l], \tag{6.182}$$

where
$\sigma$    shearing strength of the wood,
$\delta$    thickness of the tenon,

through which the stresses causing shearing of the glue-line can be easily determined:

$$\tau_{2\max} = 6(M-M_Q)(h^2 + l^2)^{-0.5}/(nhl). \tag{6.183}$$

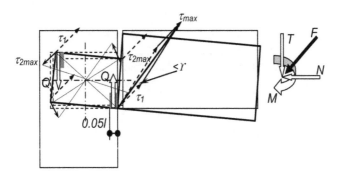

**Fig. 6.55** The distribution of stresses in a glue-line loosely fitted to the mortise putting pressure on the tenon

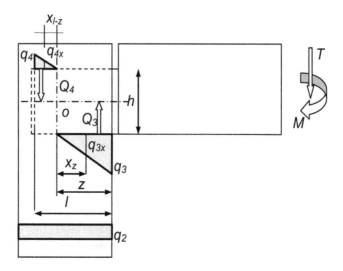

**Fig. 6.56** The distribution of pressures in a shape-adhesive joint

This equation shows that the value of tangential stresses forming in the glue-line, which can cause damage to the connection, depends mainly on the strength of wood for compression. Only a portion of the external load is transferred by the glue-line to the upper limit of the wood's strength. The rest of the part is moved by a pair of internal forces $Q$. The damage of wood in places of mutual pressure consequently causes an increase in the angle $\alpha$ and angles of shear strain of the glue-line; therefore, the increase of tangential stresses is caused by an uncompensated external moment $M$.

A perfectly adjusted height $h$ of the mortise and tenon causes their exact adjacency with narrow planes on the length $l$ of the joint's edge (Fig. 6.56). Loading the joint with the bending moment $M$ and force $T$ causes contact pressure $q_4$ and $q_3$ in adjacent surfaces, of resultants $Q_4$ and $Q_3$. Due to the lack of symmetry in the distribution of these pressures, the location of the bending centre will also change.

In drawing up balance equations for a 2-D system of forces and taking into account the similarity of triangles of pressures, the position of the centre of bending is determined by solving the system of equations:

$$\begin{cases} 2T = \delta[q_3 z - q_4(l-z)] \\ 3(M+Tz) = \delta[q_4(l-z^2) + q_3 z^2] \\ q_4 = \frac{l-z}{z} q_3, \end{cases} \qquad (6.184)$$

and equating the quadratic equation to zero:

$$-4Tz^2 + z(3Tl - 4T - 6M) - 4(-4T)(3Ml + 2Tl) = 0, \tag{6.185}$$

thus, we obtain a new location of the centre of bending at a distance $z$ equal to:

$$z_{1,2} = \frac{1}{-8T} \left\langle (3Tl - 4T - 6M)^2 \pm \left\{ T^2 \left[ 9l^2 + 8(l+2) \right] + 4M[3T(7l+4) + 9M] \right\}^{0.5} \right\rangle. \tag{6.186}$$

In the provided situation, tangential stresses in the glue-line $\tau_{2max}$ will appear once the elastic strains of wood $\varepsilon$ caused by mutual pressures of joint elements increase. The larger these strains are, the higher the value of tangential stresses in the glue-line. The value of moment $M_\tau$, causing tangential stresses in a glue-line, will therefore be the difference of the acting external moment $M$ and moment of the pair of forces $Q_3$ and $Q_4$.

$$M_t = M - M_Q, \tag{6.187}$$

where

$$M_Q = (\delta/3)[q_4(l-z)^2 + q_3 z^2] - Tz. \tag{6.188}$$

The similarity of the triangles of pressures (Fig. 6.56) shows further that:

$$q_{4x} = \frac{q_4}{l-z} x_{l-z},$$
$$q_{3x} = \frac{q_3}{z} x_z. \tag{6.189}$$

Therefore, in order to calculate the value of tangential stresses in the glue-line of shape-adhesive mortise and tenon joints, the size of deformations should be determined caused by compressing wood in places of mutual pressures of elements of the joint. The stresses associated with these deformations are generally expressed by Hooke's law $\sigma = E\varepsilon$.

For load distributed increasingly linearly (Fig. 6.57), we can provide the general equation:

$$\sigma = \frac{\int_a^b f(x)\mathrm{d}x}{b-a} E\varepsilon, \tag{6.190}$$

**Fig. 6.57** Elementary section
of pressure function

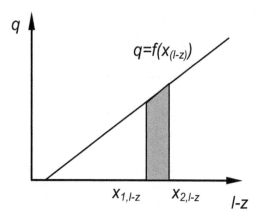

from which we will eventually obtain for $q_4$:

$$q_4 = E\varepsilon(l-z)\left(x_{2,l-z} - x_{1,l-z}\right)\frac{1}{\int_{x_{1,l-z}}^{x_{2,l-z}} x_{l-z}\mathrm{d}x}, \qquad (6.191)$$

and for $q_3$:

$$q_3 = E\varepsilon z\left(x_{2,z} - x_{1,z}\right)\frac{1}{\int_{x_{1,z}}^{x_{2,z}} x_z\mathrm{d}x}. \qquad (6.192)$$

Finally, the equation of the pressure moment shall obtain the form:

$$M_Q = \frac{1}{3}\delta E\varepsilon\left[(l-z)^3\left(x_{2,l-z} - x_{1,l-z}\right)\frac{1}{\int_{x_{1,l-z}}^{x_{2,l-z}} x_{l-z}\mathrm{d}x} + z^3\left(x_{2,z} - x_{1,z}\right)\frac{1}{\int_{x_{1,z}}^{x_{2,z}} x_z\mathrm{d}x}\right] - Tz, \qquad (6.193)$$

and hence, the maximum tangential stresses in the glue-line caused by the reduced moment $M_\tau$ for the range of elastic deformations of wood can be written as follows:

$$\tau_{2\max} = (M - M_Q)\frac{1}{\frac{n}{c}\iint_A\left(\eta^2 + \xi^2\right)\mathrm{d}\eta\mathrm{d}\xi}, \qquad (6.194)$$

where

$$c = \left[z^2 + (h/4)^2\right]^{0.5}. \qquad (6.195)$$

Taking into account the largest resultant vector of tangential stresses (Fig. 6.58), maximum stresses shall be written in the form of the equation:

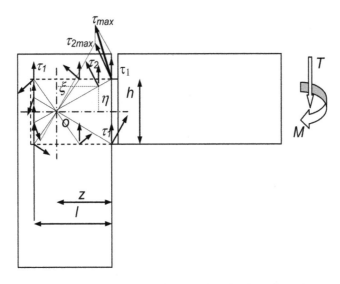

**Fig. 6.58** Distribution of stresses in the glue-line of tightly fitted mortise and tenon

$$\tau_{max}^2 = \tau_1^2 + \tau_{2max}^2 - 2\tau_1 \tau_{2max} \cos\left(90° - \text{arctg}\,\frac{h}{2z}\right), \qquad (6.196)$$

in which $\tau_1 = T/nhl$.

When the external load $M$ exceeds the value, at which moment $M_Q$ causes to exceed the range of elastic strains of wood, the load of the glue-line will suddenly increase. It will begin to shift the bending moment $M_\tau$ dependent only on the strength of wood to compression $\sigma$. In this case, the equation describing the maximum tangential stress caused by this moment shall take the form:

$$\tau_{2max} = \frac{c}{n}\left[M - \frac{1}{3}\sigma\delta\left((l-z)^2 + z^2\right) - Tz\right]\frac{1}{\int_{z-l}^{z}\int_{-h/2}^{h/2}\left(\eta^2 + \xi^2\right)d\eta d\xi}. \qquad (6.197)$$

In the place of the glue-line, in which the value of the vector of tangential stresses is the largest, processes of decohesion are initiated which reduce the strength of the joint and therefore the strength of the whole furniture construction.

Another commonly practised way of gluing is three-sided contact of elements of the tenon to elements of the mortise (Fig. 6.59). In this case, the pressures will also occur in the lower part of vertical planes of contact. By writing an equation of balance for such a system and entering geometric relationships between the graphs of pressures, the position of the centre of bending of the tenon can be determined by solving the system of equations:

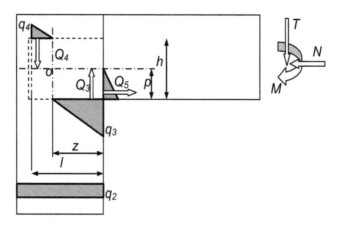

**Fig. 6.59** Distribution of pressures in a shape-adhesive joint

$$
\begin{cases}
q_3 = \dfrac{3z\left[M_Q + \frac{1}{3}(Tz + 4N)\right]}{\delta l(l-z)^2} \\[2mm]
q_3 = \dfrac{2Tl + 3\left[M_Q + \frac{1}{3}(Tz + 4N)\right]}{l\delta z}
\end{cases}
\tag{6.198}
$$

from which we obtain:

$$
z = \frac{l(3M_Q + 2Tl + 4N)}{2(3M_Q + 2Tl + 4N) - Tl}.
\tag{6.199}
$$

If in this case we assume that the state of tangential stresses in the glue-line will depend on the values of the external moment $M$ and the moment $M_Q$ caused by pressures on the surface of the wood, then the equation of maximum tangential stresses shall obtain the form:

$$
\tau_{2max} = \frac{c}{n}[M - M_Q] \frac{1}{\int_{-(l-z)}^{z} \int_{-h/2}^{h/2} (\eta^2 + \xi^2)\,d\eta\,d\xi},
\tag{6.200}
$$

in which

$$
M_Q = \frac{1}{3}\delta \left[
\begin{array}{l}
E\varepsilon(l-z)^3 (x_{2,l-z} - x_{1,l-z}) \frac{1}{\int_{x_{1,l-z}}^{x_{2,l-z}} x_{l-z}\,dx} + E\varepsilon z^3 (x_{2,z} - x_{1,z}) \frac{1}{\int_{x_{1,z}}^{x_{2,z}} x_z\,dx} \\[2mm]
+ E_2\varepsilon_2 p^3 (x_{2,p} - x_{1,p}) \frac{1}{\int_{x_{1,p}}^{x_{2,p}} x_p\,dx}
\end{array}
\right] - Tz - Nh.
$$

$$
\tag{6.201}
$$

In the scope of elastic deformations of wood, the glue-line will transfer smaller external moments than loads; thus, the strength of the structural node will be much higher than the simple strength of glue-line subject to shearing. Upon crossing the

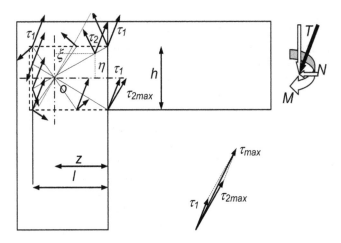

**Fig. 6.60** The distribution of tangential stresses in glue-line tightly fitted to the tenon putting pressure on the mortise

border of elastic deformations for wood and achieving strength of wood to compression, damage to the joint can occur unexpectedly due to the sudden increase of the moment of twisting the glue-line. In this situation, the greatest tangential stresses caused by this moment can be written as follows:

$$\tau_{2max} = \frac{c}{n}\left[M - \frac{1}{3}\left(\sigma\delta l(l-z) - Tz - Nh\right)\right]\frac{1}{\int_{-(l-z)}^{z}\int_{-h/2}^{h/2}\left(\eta^2 + \xi^2\right)d\eta d\xi}. \quad (6.202)$$

In both of the cases discussed above, the maximum tangential stresses generating the process of decohesion of the glue-line occur in one, strictly defined place, dependent on the method of applying external load and the geometry of the mortise and tenon (Fig. 6.60). The value of this stress is expressed by the equation:

$$\tau_{max}^2 = \tau_1^2 + \tau_{2max}^2 - 2\tau_1\tau_{2max}\cos\left(90° + \text{arctg}\frac{2z}{h} + \text{arctg}\frac{T}{N}\right), \quad (6.203)$$

where

$$\tau_1 = \left(T^2 + N^2\right)^{0.5}/nhl. \quad (6.204)$$

### Influence of Wood Species and Shape of Glue-Line on the Strength of Mortise and Tenon Joints

Despite the new technologies of wood-based materials available for multidimensional building constructions, glue-lines still play a dominant role (Pellicane 1994).

Innovative wooden structures, such as the Norwegian Viking Ship stadium in Hamar (Aasheim 1993) or the German pedestrian overpass over the River Altmühl in Essing (Brüninghoff 1993), show how great a load they bear and how safely they should be designed. Solutions to improve the behaviour of loaded elements and constructional joints made of wood are constantly sought, and methods of controlling stresses are being developed. One of the oldest and most commonly used joints of elements of wooden structures is shape-adhesive joints and among them mortise and tenon joints. Nakai and Takemura (1995) conducted studies on the torsion stiffness of mortise and tenon joints, proving that the mathematical formulas developed by them correctly describe the stiffness of tenons of rectangular and elliptical cross sections. In subsequent works Nakai and Takemura (1996a, b), by analysing the distribution of tangential stresses using the method of electrical strain gauges, numerically verified the value of tangential stresses and indicated the necessity to avoid loads, which could cause cracks in the tenon. These tests, however, did not take into account the presence of glue-lines and their impact on stiffness and strength of the joint. A few other works were devoted to the issue of the distribution of tangential stresses in wooden glue-lines of shape-adhesive joints (Haberzak 1975; Matsui 1990a, b, 1991; Pellicane et al. 1994; Smardzewski 1996, 1998a, b). The character of the distribution of stresses was analysed in tests of stretching, bending and twisting loads. Mathematical models describing these distributions were also developed. Hill and Eckelman (1973) dealt with the deformation and bending stresses in mortise and tenon joints. Eckelman (1970) and Smardzewski (1990) also developed computer programs for analysing the stiffness and strength of furniture with susceptible joints, including mortise and tenon joints. These studies, however, did not take up the subject matter of mutual stresses of the joint's elements, including the mortise and tenon, nor the impact of the species of wood and type of adhesive used on this phenomenon.

It is well known that mathematical modelling is a rational alternative to costly and time-consuming laboratory tests. This chapter presents the possibilities of using numerical methods for the analysis of contact stresses in mortise and tenon joints. In particular, the size of normal stresses was described in places of mutual pressures of the tenon and mortise and the impact of these stresses on the value of changes in the species of wood and type of adhesive used, or no adhesive in the joint.

Due to the common use of mortise and tenon joints in the construction of chairs, for examination, a chair with a muntin was chosen, establishing the connection of a horizontal strip and rear leg as the structural node (Fig. 6.61). From the practice of a university laboratory, tests and validation of furniture, it results that the most common cause of damage to a chair construction is its improper use and loading with forces that significantly exceed the mass of one user.

By taking into account the symmetry of a chair as a 3-D structure, one side frame was selected for strength analysis, loading the front edge of the seat with a concentrated force of 800 N, corresponding to the impact of a user weighing approximately 160 kg (Fig. 6.62). On the basis of the distribution of internal forces (Fig. 6.63), the appropriate bending moments, cutting forces and normal forces were transferred onto the chosen structural node (Fig. 6.64).

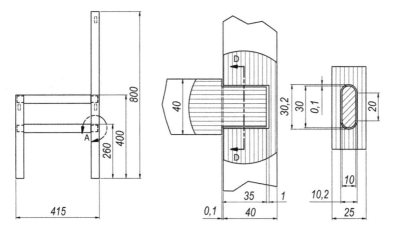

**Fig. 6.61**  Dimensions of **a** the frame, **b** mortise and tenon joints of a wooden construction (mm)

**Fig. 6.62**  Static scheme of
the construction

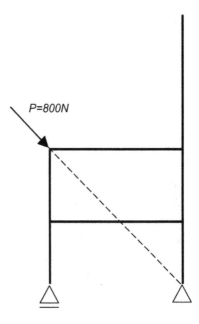

The numerical model of the mortise and tenon joints was developed in the
environment of the program Algor® (Fig. 6.65). To this end, for modelling the
wooden parts of the joint, 20-node, orthotropic solid elements were used, while for
modelling glue-lines, isotropic elements were used. The fit between tenon and
mortise is equal 0.1 mm (Fig. 6.61). Inside the gap between the side surface of
the mortise and tenon, a glue-line was formed. Additionally, in order to trace the
process of pressure of the tenon on the mortise, in all gaps, perpendicular to the
surface of the tenon and mortise, gap-type stress elements were distributed with a

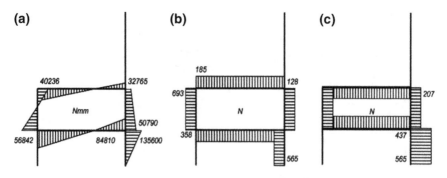

**Fig. 6.63** Distribution of internal forces in the frame: **a** bending moments, **b** cutting forces, **c** normal forces

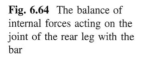

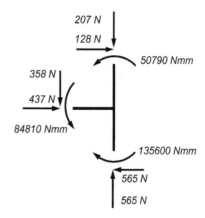

**Fig. 6.64** The balance of internal forces acting on the joint of the rear leg with the bar

stiffness of 16,000 N/mm. The developed solid model was supported and loaded with external forces as in Fig. 6.66, which correspond to the load of the node by internal forces from Fig. 6.64.

In the selection of wood species for comparative material, it was decided to perform numerical calculations for representatives of deciduous wood and coniferous wood. In order to do this, four of the most popular wood species in the production of furniture were selected: beech, ash, pine and alder (Table 6.4), assuming their elastic properties on the basis of the studies of Hearmon (1948) and Bodig and Goodman (1973).

In industrial practice, for the assembly of furniture, polyvinyl chloride adhesive (PVC) is commonly used, while urea–formaldehyde adhesive is less frequently used. In determining the module of linear elasticity of these adhesives, both own research and data from the literature were used. By indicating the stiffness of layer-glued beam elements in the bending test, Dzięgielewski and Wilczyński (1990) established the value of the linear elasticity module of PVC glue at the level of 2845–33,450 MPa. According to Wilczyński (1988), the linear elasticity module

**Fig. 6.65** Mesh of finite
elements of the bridle mortise
and tenon joints

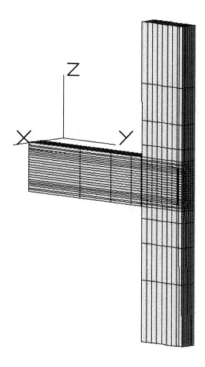

**Fig. 6.66** Scheme of loading
the joint with external forces

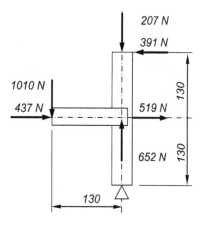

of PVC glue, established in the torsion test of prismatic glued samples, amounts to
358 MPa. A similar value to PVC glue, equal to 465.74 MPa, was also obtained by
Smardzewski (1998a), in conducting tests on samples in the shape of a paddle,
subjected to tension. Clada (1965) dealt with determining the stiffness of urea–
formaldehyde glue, according to whom the value of Young's modulus varies from
4940 to 5200 MPa. Based on this, the following values of the linear elasticity
module of glue were selected in MPa: 33,450, 4940, 465, 100, 50, 10 and 0.

**Table 6.4** Elastic properties of chosen wood species (Hearmon 1948; Bodig and Goodman 1973)

| Property of wood | Species of wood | | | |
|---|---|---|---|---|
| | Beech | Ash | Pine | Alder |
| Density of wood [g/cm$^3$] | 0.75 | 0.67 | 0.55 | 0.38 |
| Linear elasticity module [MPa] | | | | |
| $E_L = E_Z$ | 13,969 | 15,788 | 16,606 | 10,424 |
| $E_R = E_X$ | 2284 | 1509 | 1117 | 809 |
| $E_T = E_Y$ | 1160 | 799 | 583 | 355 |
| Shear modulus [MPa] | | | | |
| $G_{LT} = G_{ZY}$ | 1082 | 889 | 693 | 313 |
| $G_{LR} = G_{ZX}$ | 1645 | 1337 | 1181 | 632 |
| $G_{RT} = G_{XY}$ | 471 | 471 | 70 | 144 |
| Poisson's ratio | | | | |
| $v_{LR} = v_{ZX}$ | 0.450 | 0.460 | 0.420 | 0.440 |
| $v_{LT} = v_{ZY}$ | 0.510 | 0.510 | 0.510 | 0.560 |
| $v_{RT} = v_{XY}$ | 0.750 | 0.710 | 0.680 | 0.570 |
| $v_{TR} = v_{YX}$ | 0.360 | 0.360 | 0.310 | 0.290 |
| $v_{RL} = v_{XZ}$ | 0.075 | 0.051 | 0.038 | 0.031 |
| $v_{TL} = v_{YZ}$ | 0.044 | 0.030 | 0.015 | 0.013 |

The values 100, 50 and 10 MPa correspond to those types of adhesives, which have a technological application mistakes and characterised by lower mechanical properties. The use of the indication 0 MPa only shows the lack of glue-line in the joint. In the course of numerical calculations, the change in the thickness of the glue-line $\Delta g$ was determined, caused by pressure of the tenon and mortise and the change of normal stresses in points A and B, indicating the possibility of mutual pressure of elements of the tenon and mortise (Fig. 6.67).

The analysis of deformed meshes of the numerical model shows that the tenon sustained rotation and bending. Its upper part was subjected to sliding out and moving downwards, while the lower edge deeper with smaller movement downwards (Fig. 6.68). The effect of mutual stress of joint elements is, therefore, a result of both bending of the tenon and twisting of the glue-line. As it is shown in Fig. 6.69, the change in the species of wood, of which the joint was made, does not determine the change of the strength of the glue-line. These changes are mainly caused by the change of the value of linear elasticity modulus of the glue-line. Adhesives characterised by a very high linear elasticity module, above 4940 MPa, are not subject to significant geometrical changes and do not change the thickness of the glue-line. The thickness of the glue-line is reduced by 10 % for glue-line of module $E = 465$ MPa, by 50 % for glue-line of module $E = 100$ MPa and by 80 % for glue-line of module $E = 50$ MPa. For all of these glue-lines, however, it can be assumed that the stress of wooden elements in point A occurs only by compression of the glue-line. Reducing the elasticity module of glue to 10 MPa or no glue in the gap between elements results in a direct stress of the tenon and mortise.

**Fig. 6.67** Determining control points of stresses in a joint

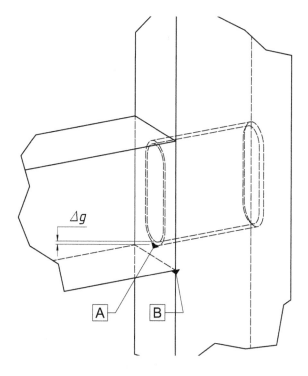

**Fig. 6.68** Illustration of deformations in the joint, scale of 10:1

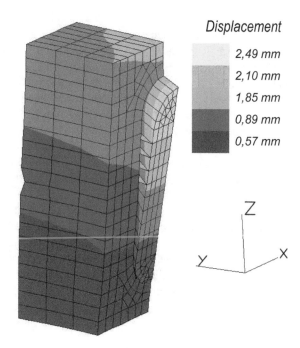

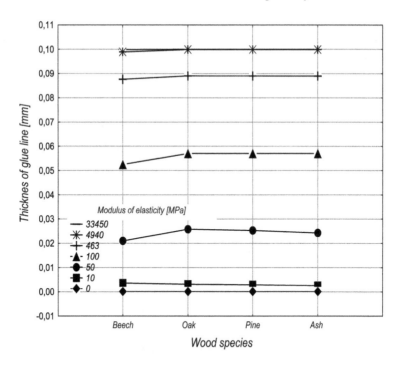

**Fig. 6.69** The impact of the linear elasticity module of glue and wood species on the thickness of the glue-line in the place of the stress of mortise and tenon

Figure 6.70 shows clearly that the tenon is in a state of bending stresses, and in point B, normal stresses $\sigma_{yy}$ are triggered caused by mutual pressure of the bar on the leg. When analysing this impact, it can be noticed that in the bar, of a longitudinal course of fibres, stresses occur which are six times greater than in the leg, where the fibres run perpendicular to the direction of the pressure. Furthermore, stresses in pinewood and alder wood are approximately 9–13 MPa higher than those in beech wood and ash wood (Fig. 6.71). The greatest stresses $\sigma_{yy}$ in the leg are formed, if alder wood is used. In relation to ash wood, these stresses are higher by 1–1.5 MPa and in relation to beech wood and pinewood by 0.5–1.0 MPa. The type of adhesive used has a significant impact on the value of stresses $\sigma_{yy}$ in the place of the bar's pressure on the leg.

Figure 6.71 shows that the proportional growth of stresses occurs together with the fall of the value of the linear elasticity module of the glue-line up to 50 MPa, while these stresses do not reach the acceptable limits for particular species of wood. Below this value, stresses both in the bar and in the leg rise suddenly, reaching a maximum in the absence of a glue-line in the connection. For an element of a bar made of pinewood, the maximum stress reaches a value of 93 MPa, which is higher than the strength of this wood to compression along the fibres

**Fig. 6.70** Illustration of contact stresses $\sigma_{yy}$ (in MPa) at the junction of the leg with bar: **a** tenon, **b** mortise

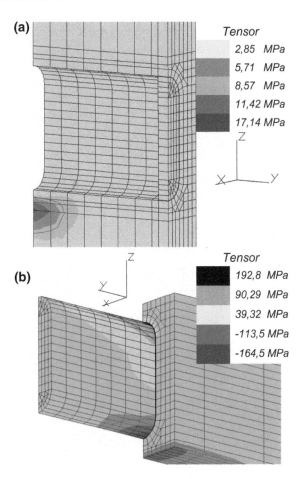

$R_{c\parallel}$ = 80 MPa. For the remaining wood species, maximum contact stresses were as follows: for alder 88 MPa > $R_{c\parallel}$ = 51 MPa, for ash 83 MPa $\gg R_{c\parallel}$ = 63 MPa and for beech 79 MPa < $R_{c\parallel}$ = 84 MPa. In the leg, the stresses were directed perpendicular to the fibres. Also here it was noticed that the value of stresses exceeded compression strength of the wood across the fibres, which is as follows: for the alder wood 15.1 MPa > $R_{c\perp}$ = 2 MPa, for beech 15.05 MPa > $R_{c\perp}$ = 7 MPa, for pine 14.67 MPa > $R_{c\perp}$ = 4.4 MPa and for ash 14.61 MPa > $R_{c\perp}$ = 7 MPa.

Much greater stress pressure $\sigma_{zz}$ appeared in the same connection, and the place particularly exposed to damage is proved to be point A, in which the edge of the tenon after inserting into the mortise is pressed on the edge of the mortise (Fig. 6.72). Based on Fig. 6.73, it is clear that in the mortise, higher stresses $\sigma_{zz}$ are generated than in the tenon. When analysing this impact, it can be noticed that in the mortise of a longitudinal course of fibres, stresses $\sigma_{zz}$ occur which are 65–130 % greater than in the tenon, where the fibres run perpendicular to the direction of the pressure. Furthermore, stresses in beech wood are approximately 15–45 MPa higher

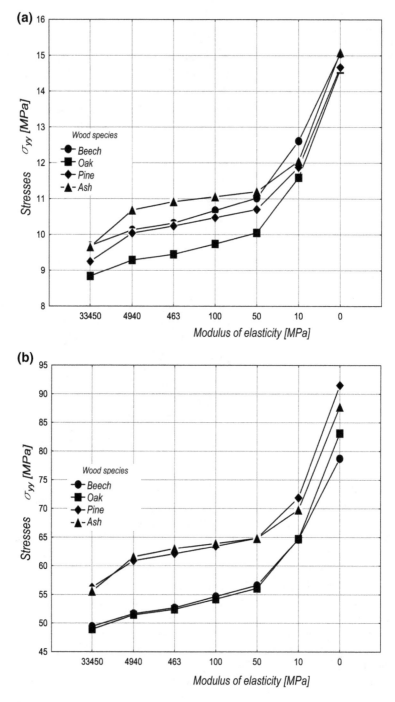

**Fig. 6.71** Impact of linear elasticity module of the glue and wood species on the value of stresses $\sigma_{yy}$: **a** leg at the point of stress on the bar, **b** bar at the point of stress on the leg

**Fig. 6.72** Illustration of contact stresses $\sigma_{zz}$ in MPa at the junction of the mortise and tenon: **a** tenon, **b** mortise

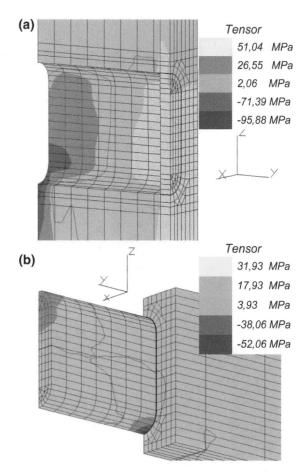

than in alder wood (Fig. 6.73). In relation to ash wood, these stresses are higher by 8 MPa and in relation to pinewood by 9–30 MPa. In the element of the tenon, the stresses in beech wood are approximately 17–34 MPa higher than in alder wood, 8–12 MPa than in ash wood and 12–23 MPa than in pinewood. The type of adhesive used has a significant impact on the value of stresses $\sigma_{zz}$ in the point $B$, place of the tenon pressure on the mortise. Figure 6.73 shows the nonlinear fall of values of stresses $\sigma_{zz}$, together with a decrease in the values of the linear elasticity module of the glue-line to the 50 MPa. It should be noted here that these stresses teeter on the verge of acceptable values for each particular species of wood. Below, the border value of the module of linear elasticity of adhesive equal to 50 MPa, stresses in the tenon and mortise grow suddenly, reaching a maximum in the absence of a glue-line. For an element with a mortise made of beech wood, the maximum stress reaches a value of 165 MPa, which is higher than the strength of this wood to compression along the fibres $R_{c\parallel} = 84$ MPa. For the remaining wood species, maximum contact stresses were as follows: for ash 157 MPa > $R_{c\parallel} = 63$ MPa, for pine

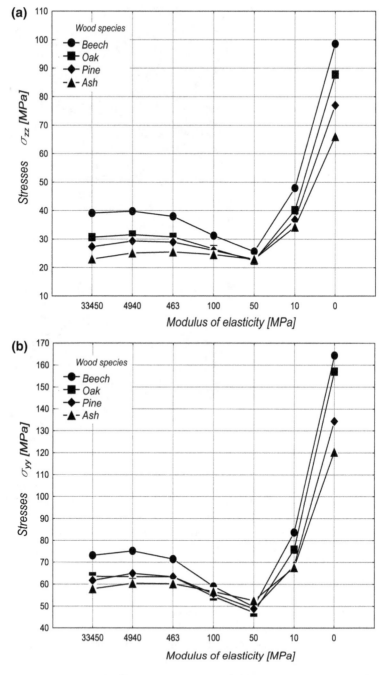

**Fig. 6.73** Impact of linear elasticity module of the glue and wood species on the value of stresses $\sigma_{zz}$: **a** tenon at the point of stress on the mortise, **b** mortise at the point of stress on the tenon

134 MPa > $R_c \parallel$ = 80 MPa and for alder 120 MPa > $R_c \parallel$ = 51 MPa. In the tenon, the stresses were directed perpendicular to the fibres. Also here it was noticed that the value of stresses exceeded compression strength of the wood across the fibres, which is as follows: for the beech wood 100 MPa > $R_c\perp$ = 7 MPa, for ash 88 MPa > $R_c\perp$ = 7 MPa, for pine 78 MPa > $R_c\perp$ = 4.4 MPa and for alder 65 MPa > $R_c\perp$ = 2 MPa.

In the indicated places of the construction, in which glue line were used of Young's modulus above 50 MPa, stresses are similar or slightly exceed the acceptable values for particular species of wood. However, this does not mean that the construction gets completely destroyed. These stresses are concentrated on the edges and corners and have a local character. Moreover, the concentration of stresses is eased by the compression of the glue-line. Without a doubt, the most dangerous, because damaging, are stresses resulting from the connections of weak glue-line or with a damaged glue-line. They constitute a direct cause of damage of the node in the entire construction.

A separate structural problem is to determine the impact of the shape of glue-line on its strength and form of distribution of tangential and normal stresses in the glue-line. To this end, the construction of a chair has been chosen with a bar, selecting a horizontal bar and rear leg connection as the structural node (Fig. 6.74). Due to the perpendicular position of the bar and leg in the global plane of coordinates and the resulting different longitudinal course of wood fibres, two separate local systems have been used. For a bar, it was assumed that wood fibres will run parallel to the $Y$-axis, and the radial–tangential plane will lie on the $XZ$ plane of the global system of coordinates. For the leg, the fibres will be oriented in the direction of the $Z$-axis, and the radial–tangential plane shall lie on the $XY$ plane. Between the tenon and the mortise around the entire perimeter, a gap with a thickness of 0.1 mm has been formed. An oval glue-line has been created in this gap.

By taking into account the symmetry of the construction, one side frame was selected for strength analysis, loading the front edge of the seat with a concentrated force of 800 N (Fig. 6.62). The value, direction and rotation of this force corresponded to extreme conditions of use of the furniture by a user weighing approximately 160 kg. By analysing the static side frame of a chair, the concentrated forces appropriate for it have been transferred onto the structural node, which correspond to internal bending moments, shearing forces and normal forces (Fig. 6.63). The numerical model of the mortise and tenon joints was developed in the environment of the program Algor® (Fig. 6.75). To this end, for modelling the wooden parts of the joint, 20-node, orthotropic solid elements were used, while for modelling glue-lines—isotropic elements with a thickness of 0.1 mm.

For beech wood (*Fagus silvatica* L.), from which the model of the chair frame was made from, numerical calculations were conducted. Elastic properties of wood are provided in Table 6.4, based on the studies of Hearmon (1948) and Bodig and Goodman (1973). Taking into account the different orientation of individual anatomic directions of wood in elements of joints, this table provides values that correspond to two local coordinate planes. Another material component, which is

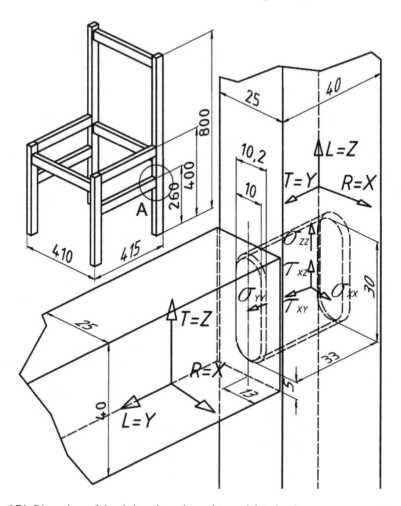

**Fig. 6.74** Dimensions of the chair and mortise and tenon joints (mm)

widely used for furniture assembly, is polyvinyl chloride glue (PVAC), for which the value of Young's modulus was selected equal to 465.74 MPa.

Numerical calculations included two models of joints, which differed in shape of glue-line. The first model constituted a joint, in which the glue-line in the form of two parallelepipeds of the dimensions $0.1 \times 20 \times 33$ mm, was set up on opposite surfaces of the tenon. In Fig. 6.76, by numbers from 1 to 11 vertically and from 1 to 8 horizontally, the nodes of the mesh of finite elements have been indicated. The second model represented a joint with a glue-line in an oval shape, with a thickness of 0.1 mm, formed on the entire side of the tenon. In this case, the vertical numbers of mesh nodes, from 1 to 23, also included hardened parts of the glue-line on both parts of the tenon (Fig. 6.76). During the course of numerical calculations, the change of the shape and thickness of glue-line was determined, caused by the

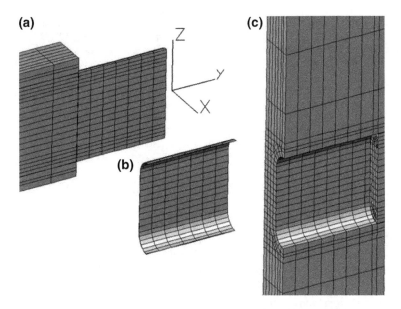

**Fig. 6.75** Mesh of finite elements of the mortise and tenon joints **a** tenon, **b** glue-line, **c** mortise

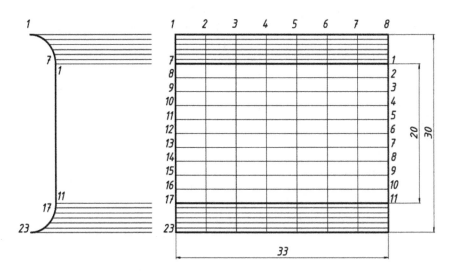

**Fig. 6.76** Dimensions of the glue-line and determination of the points of measuring stresses

pressure of the tenon on the mortise, as well as change of reduced stresses, tangential stresses and normal stresses in the nodes of the mesh on the surface of the glue-line.

**Fig. 6.77** Deformations of glue-line: **a** rectangular glue-line, **b** oval glue-line

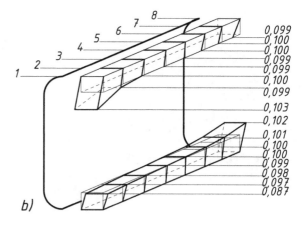

An analysis of deformed meshes shows that in connection, the tenon sustained rotation and bending. The consequence of this deformation is a visible disproportion of normal and shear strains of the glue-line along its edges. In a rectangular glue-line, shear strain dominates, of which the largest are concentrated in the corners and on the edge adjacent to the base of the tenon (Fig. 6.77a). These deformations were formed as a result of bending the tenon, in which the extreme fibres were subject to elongation or shortening. In other parts of the tenon, normal strains were not significant enough to affect the deformations of the glue-line. Therefore, in this part, the glue-line sustained proportional twisting between two stiff adherents. By testing the deformation of an oval glue-line, it was noted that both normal and shear strains occur in them. Figure 6.77b shows the change in the thickness of the glue-line, caused by pressure of the mortise and tenon. At the beginning and at the end of the tenon, as a result of compression, the glue-line reduces its thickness from 0.1 to 0.087 mm or 0.099 mm, while as a result of stretching, it increases its thickness to 0.102 or 0.103 mm. Pressure of the tenon on the mortise by the glue-line significantly reduces shear strain and thus decreases the value of stresses forming.

The disproportion of figural deformations in the glue-line also causes migration of its centre of rotation and uneven distribution of stresses. By assigning the complex state of stress, which is created in the glue-line, the uniaxial state has been characterised by reduced stress $\sigma_R$ according to the equation:

$$\sigma_{R_{max}} = \frac{1}{\sqrt{2}}\left((\sigma_{XX} - \sigma_{YY})^2 + (\sigma_{XX} - \sigma_{ZZ})^2 + (\sigma_{ZZ} - \sigma_{YY})^2 + 6\left(\tau_{XY}^2 + \tau_{ZY}^2 + \tau_{XZ}^2\right)\right)^{0.5},$$

$$(6.205)$$

where

$\sigma_{ij}$  normal stress for the chosen direction,
$\sigma_{ij}$  tangential stress for the chosen plane.

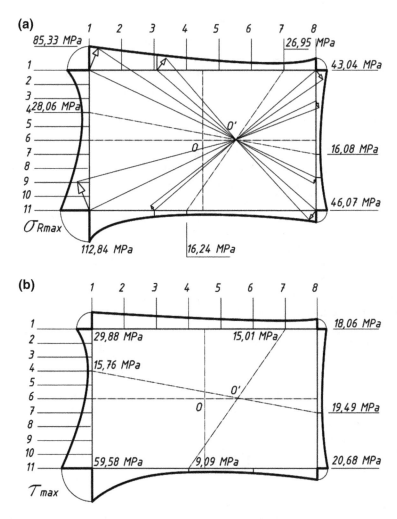

**Fig. 6.78** The distribution of stresses in a rectangular glue-line: **a** reduced stresses according to Mises, **b** tangential stresses $\tau_{xz}$, $\tau_{xy}$

It was determined that for rectangular glue-line, the biggest reduced stresses according to Mises occur in the lower left corner of the glue-line, reaching a value of 112.84 MPa (Fig. 6.78a). It should be noted here that the stresses in other corners are from 1.33 to 2.62 times smaller. The components of reduced stress are appropriate normal and tangential stresses. Figure 6.78b presents the distribution of tangential stresses $\tau_{XZ}$, $\tau_{XY}$. These stresses are responsible for shearing the glue-line and therefore for the strength of mortise and tenon joints. The biggest tangential stresses, like the biggest reduced stresses, occur in the left lower corner of the glue-line. Their value is 59.58 MPa. In the remaining corners, stresses constitute from 30 to 50 % of the value of maximum stresses. Taking into account the technical strength of glue-line of PVAC glue to shearing, ranging from 17 to 23 MPa, the damaging process is initiated in the corners of the bond. However, this does not mean that in these particular points of the glue-line, along with reaching the acceptable values by the tangential stresses, damage to the entire joint must occur. High concentration of reduced or tangential stresses in the corners of the glue-line results only in the decohesion in a few places on a small area. At a small distance from those corners, the level of stresses falls rapidly below the acceptable values. By calculating the average value of reduced and tangential stresses on the entire surface of the glue-line, the following values were obtained, respectively, $\sigma_R = 32.46$ MPa and $\tau = 16.28$ MPa. Therefore, the tangential stresses, at extremely unfavourable operational load of a furniture construction, teeter on the verge of shearing strength of the adhesive.

In addition, in these glue-lines, it was also observed that the centre of rotation of a glue-line does not lie in its geometric centre. By connecting opposite sides of the glue-line by sections in point of the lowest value of reduced and tangential stresses (Fig. 6.78b), the actual position of the centre of rotation $0'$ was obtained. This point shifted to the right on the horizontal axis of symmetry of the glue-line. The displacement of the centre of rotation is largely due to the bending of the tenon.

For oval glue-line, it was determined that the greatest reduced stresses according to Mises also occur in the lower left corner of a developed glue-line (that is under the axis of the tenon), reaching a value of 52.96 MPa (Fig. 6.79a). It should be noted here that the stresses in the corners of the opposite edge are from 7.8 to 18.3 times smaller. Figure 6.79b presents also distribution of tangential stresses $\tau_{XZ}$, $\tau_{XY}$. The largest of these, with a value of 18.92 MPa, occur in the point of transition of the glue-line's shape from the rectangular part into a round one (node of number 17). It is also interesting that the stresses at the horizontal edges of the glue-line are negligibly small.

By calculating the average value of reduced and tangential stresses on the entire surface of the glue-line, the following values were obtained: $\sigma_R = 14.31$ MPa and $\tau = 4.35$ MPa. Tangential stresses, at a chosen unfavourable operational load of a furniture construction, constitute only 25 % of the values of acceptable stresses, corresponding to the acceptable strength of the adhesive to shearing. Slightly higher average values of reduced stresses are caused by the presence of normal stresses $\sigma_{ZZ}$ and $\sigma_{YY}$. During bending of the joint and twisting of the glue-line, the tenon, sustaining bending and rotation, put pressure on the lower and upper parts of the glue-line, causing very high normal stresses $\sigma_{ZZ}$ in it. For this reason, small and safe tangential and reduced stresses were formed, contributing to a smaller stress of material, than in the rectangular glue-line.

In this way, it was shown that the shape of the glue-line clearly differentiates the mortise and tenon joints. Stress of the tenon on a mortise by the layer of glue-line changes the form and sizes of its deformations. In rectangular glue-line, only shear strains occur, which generate tangential stresses of values that exceed the ultimate strength. In oval glue-line, shear strains are clearly limited by the pressure of the tenon on the mortise, through which the level of dangerous shearing stresses significantly decreases.

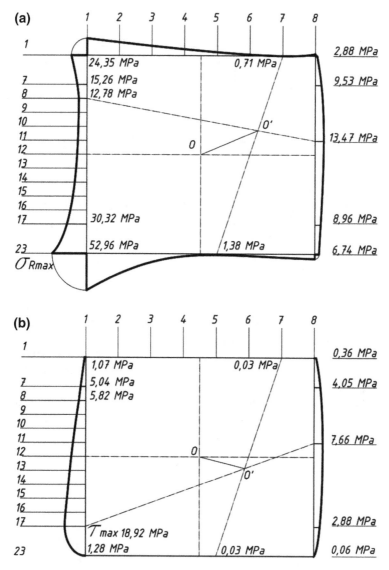

**Fig. 6.79** The distribution of stresses in an oval glue-line: **a** reduced stresses according to Mises, **b** tangential stresses $\tau_{xz}$, $\tau_{xy}$

Strength of Finger Joints

State of stresses in a finger joint subjected to bending was presented by Tomusiak (1988), specifying the relationship of tangential stresses in the glue-line from the value of bending moment caused by concentrated force. From the point of view of the constructor, it is also necessary to take into account the effect of tangential

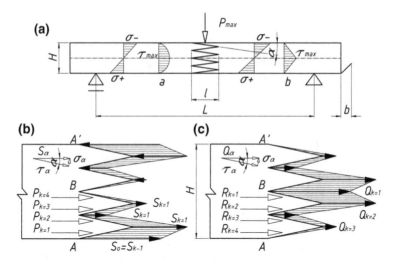

**Fig. 6.80** Finger joint: **a** beam—static scheme, **b** distribution of normal stresses, **c** distribution of tangential stresses

stresses, whose distribution, for the purposes of the calculations below, was brought to the form of a triangle (Fig. 6.80).

Vectors of stresses $S_\alpha$ and $Q_\alpha$ constitute the sum of vectors of tangential stresses $\tau_\alpha^S$, $\tau_\alpha^Q$ and vectors of normal stresses $\sigma_\alpha^S$, $\sigma_\alpha^Q$, whereas in points $A$ and $A'$ stresses $S_{k=1} = S_{max}$, and $Q_{k=4} = Q_{min} = 0$, and in point $B$ stresses $S_{k=4} = S_{min} = 0$, and $Q_{k=1} = Q_{max}$. Stresses on the surface of the glue-line $A = bh(1/\sin \alpha)$ can be reduced to the form of concentrated forces $P_k$ and $R_k$. Their values are expressed by the equations:

$$P_k = \frac{S_{k-1} + S_k}{2h} \frac{bH}{\sin \alpha},$$

(6.206)

$$R_k = \frac{Q_{k-1} + Q_k}{2h} \frac{bH}{\sin \alpha},$$

(6.207)

where

$$S_{k-1} = S_\alpha \frac{n - 2k - 2}{n},$$

(6.208)

$$S_k = S_\alpha \frac{n - 2k}{n},$$

(6.209)

$$Q_{k-1} = Q_\alpha \frac{n - 2k - 2}{n},$$

(6.210)

$$n = H/h, \tag{6.211}$$

$$k \leq n/2, \tag{6.212}$$

$b, H$   dimensions of the cross section of the beam,
$h$       scale of finger joints,
$\alpha$       angle of inclination of the side of the finger joint.

Resultant vector of all forces $P_k$ and $R_k$ is the sum of component vectors:

$$\vec{W}_P = \vec{P}_{k=1} + \vec{P}_{k=2} + \vec{P}_{k=3} + \vec{P}_{k=4} \Rightarrow W_P = 2S_\alpha \frac{bH}{n \sin \alpha}, \tag{6.213}$$

$$\vec{W}_R = 2(\vec{R}_{k=1} + \vec{R}_{k=2} + \vec{R}_{k=3} + \vec{R}_{k=4}) \Rightarrow W_R = 4Q_\alpha \frac{bH}{n \sin \alpha}. \tag{6.214}$$

In order to determine the values of the stress $\tau_\alpha^S$ caused by bending moment, the moment of external forces should be compared with the moment of the two forces $W_P$:

$$\frac{4}{3} S_\alpha \frac{bH^2}{n \sin \alpha} = \frac{PL}{4}, \tag{6.215}$$

where
$P$   external force,
$L$   spacing between supports of the bent beam.

Because tangential stresses in general form can be expressed by the equation,

$$\tau_\alpha^S = S_\alpha \cos \alpha, \tag{6.216}$$

expression determining the value of tangential stresses at any point $k$ of the glue-line, caused by bending moment, will therefore take the form:

$$\tau_k^S = \frac{3}{16} PL(n - 2k) \frac{\sin \alpha \cos \alpha}{bH^2}. \tag{6.217}$$

And tangential stresses $\tau_\alpha^Q$ caused by cutting forces are described by the equation:

$$\tau_k^Q = \frac{1}{4} P(n - 2k) \frac{\sin \alpha \cos \alpha}{bH}. \tag{6.218}$$

Knowing vectors of the component stresses, the vector of resultant stresses $\tau_k^C$ can be determined in the form:

$$\vec{\tau}_k^C = \vec{\tau}_k^S + \vec{\tau}_k^Q. \tag{6.219}$$

All the vectors values $\tau_k^C$, $\tau_k^S$, $\tau_k^Q$, as well as average values of the stresses between the points $k$ and $k+1$, can be written as follows:

$$\tau_{k-1}^C = \tau_{k-1}^S + \tau_{k=4}^Q,$$

$$\tau_{k=1}^C = \tau_{k=1}^S + \tau_{k=3}^Q,$$

$$\tau_{k=2}^C = \tau_{k=2}^S + \tau_{k=2}^Q,$$

$$\tau_{k=3}^C = \tau_{k=3}^S + \tau_{k=1}^Q,$$

$$\tau_{k=4}^C = \tau_{k=1}^Q \tag{6.220}$$

and

$$\tau_{k-1,k=1}^{CV} = 0.5\left(\tau_{k-1}^C + \tau_{k=1}^C\right),$$

$$\tau_{k=1,k=2}^{CV} = 0.5\left(\tau_{k=1}^C + \tau_{k=2}^C\right),$$

$$\tau_{k=2,k=3}^{CV} = 0.5\left(\tau_{k=2}^C + \tau_{k=3}^C\right),$$

$$\tau_{k=3,k=4}^{CV} = 0.5\left(\tau_{k=3}^C + \tau_{k=4}^C\right). \tag{6.221}$$

### 6.2.3.3 Strength of Connector Joints

Strength of Nail Joints

Nails are connectors least used in furniture making. They can be only used to connect invisible and inseparable elements. Load-carrying capacity of nails for pulling is calculated in cases of connecting strips and boards which are skeletons of upholstered furniture (Fig. 6.81).

According to the norm PN-B-03150:2000, the calculated load-carrying capacity of nails is calculated by applying the lowest value calculated using the equations:

$$R_d = f_{1d}dl, \tag{6.222}$$

$$R_d = f_{2d}d^2, \tag{6.223}$$

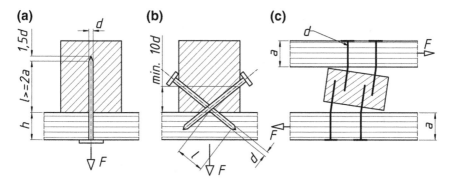

**Fig. 6.81** Load schemes, taking into account the load-carrying capacity of the nail for pulling: **a** axial load, **b** load at an angle to the axis of the nail, **c** load causing bending and pulling nails

where

$d$  nail diameter,

$l$  working length of the nail, and it has to be assumed that $l \leq 12d$ for nails with smooth cores and $l \geq 8d$ for other types of nails,

$$f_{1d} = \frac{k_{\mathrm{mod},1} f_{h,1,k}}{\gamma_M}, \tag{6.224}$$

$$f_{2d} = \frac{k_{\mathrm{mod},2} f_{h,2,k}}{\gamma_M}, \tag{6.2245}$$

$$f_{h,1,k} = \left(18 \times 10^{-6}\right) \rho_k^2, \tag{6.226}$$

$$f_{h,2,k} = \left(300 \times 10^{-6}\right) \rho_k^2, \tag{6.227}$$

$\rho_k$  density of wood,

$k_{\mathrm{mod}}$  partial modification coefficient (Table 6.5),

$\gamma_M$  partial safety coefficient (Table 6.6).

There are general rules for selecting the diameter of nails, which recommend using connectors with diameters from 1/6 to 1/11 of the thickness of the thinnest of the joined elements. To join elements made of hard fibreboard and plywood with a thickness of 8 mm, it is recommended to use nails with a diameter of 2–4 mm, and for chipboards with a thickness of up to 25 mm, it is recommended to use nails with a diameter of 2.5–5.0 mm.

When choosing the length of the nail, the necessary depth of setting of the connector should be taken into account, assuming additionally 1 mm for each connection of joined elements and 1.5 of the nail diameter that corresponds to the length of its sharp end (Fig. 6.82).

**Table 6.5** Values of the coefficient $k_{mod}$ (PN-B-03150:2000)

| Material/class of load duration | Class of use | | |
|---|---|---|---|
| | 1 | 2 | 3 |
| Hard wood, glued from layers, plywood | | | |
| Permanent | 0.60 | 0.60 | 0.50 |
| Long term | 0.70 | 0.70 | 0.60 |
| Medium term | 0.80 | 0.80 | 0.70 |
| Short term | 0.90 | 0.90 | 0.80 |
| Momentary | 1.10 | 1.10 | 0.90 |
| Chipboards (PN-EN 312-6:2000), Oriented Stress Boards class 3, 4 (PN-EN 300:2000) | | | |
| Permanent | 0.40 | 0.40 | |
| Long term | 0.50 | 0.50 | |
| Medium term | 0.70 | 0.70 | |
| Short term | 0.90 | 0.90 | |
| Momentary | 1.10 | 1.10 | |
| Chipboards (PN-EN 312-4:2000, PN-EN 312-5:2000), Oriented Stress Boards class 2 (PN-EN 300:2000), hard fibreboards (PN-EN 662-2:2000) | | | |
| Permanent | 0.30 | 0.20 | |
| Long term | 0.45 | 0.30 | |
| Medium term | 0.65 | 0.45 | |
| Short term | 0.85 | 0.60 | |
| Momentary | 1.10 | 0.80 | |
| Hard and medium fibreboards (PN-EN 622-3:2000) | | | |
| Permanent | 0.20 | | |
| Long term | 0.40 | | |
| Medium term | 0.60 | | |
| Short term | 0.80 | | |
| Momentary | 1.10 | | |

**Table 6.6** Values of the coefficient $\gamma_M$ (PN-B-03150:2000)

| Border states | Value of the coefficient |
|---|---|
| Border states of load-carrying capacity | |
| Basic combinations of loads | 1.3 |
| Wood and wood-based materials | 1.1 |
| Exceptional states | 1.0 |
| Border state of usability | 1.0 |

When connecting two elements using nails, the satisfactory length of the insert (without taking into account the length of the sharp blade) should amount to 8 times nail diameters (Fig. 6.82). Joining three elements requires the nail to pierce one or two pieces and stick in the second middle or third external at a depth of eight times diameters (Fig. 6.83a, b).

**Fig. 6.82** Single-shear nail joint

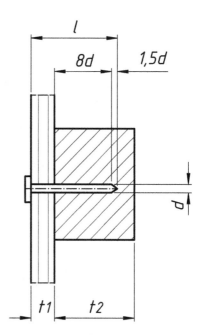

**Fig. 6.83** Nail joints: **a** double shear hammered bilaterally, **b** single shear hammered bilaterally

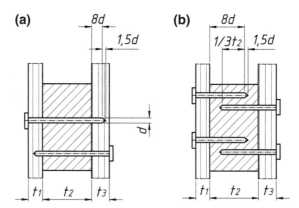

Figure 6.84 shows methods of connecting wooden elements with wood-based materials: chipboards, fibreboards and plywood. It should be noted that the length of the insert of the nail should in these cases amount to a minimum of 14 of its diameters.

When connecting wooden elements, the appropriate order and optimal spacing of the connectors should be used (Fig. 6.85). Minimal spacing and distances between nails are specified in the norm PN-B-03150:2000 (Table 6.7).

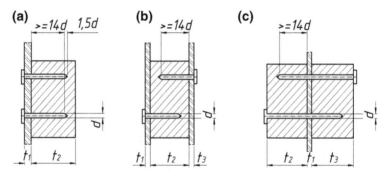

**Fig. 6.84** Methods of connecting wood and board wood-based materials using nails: **a** single shear hammered unilaterally, **b** single shear hammered bilaterally, **c** double shear hammered bilaterally

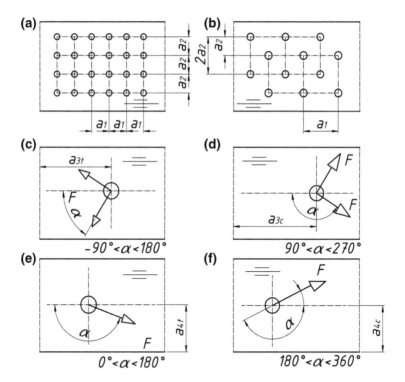

**Fig. 6.85** Hammering patterns, distances and spacing of connectors (nails and bolts) according to PN-B-03150:2000: **a** rectangular layout, **b** alternate layout, **c** distance between the connector and loaded end, **d** distance between the connector and unloaded end, **e** distance between the connector and loaded edge, **f** distance between the connector and unloaded edge

**Table 6.7**  Minimal spacing and distances between nails according to PN-B-03150:2000

| Distance signature | Minimal distances of connectors with drilled holes | Minimal distances of connectors without drilled holes | |
|---|---|---|---|
| | | $\rho_k \leq 420$ kg/m$^3$ | $420 < \rho_k < 500$ kg/m$^3$ |
| $a_1$ | $(4 + 3|\cos\alpha|)d$ | $d < 5$ mm: $(5 + 5|\cos\alpha|)d$ $d \geq 5$ mm: $(5 + 7|\cos\alpha|)d$ | $(7 + 6|\cos\alpha|)d$ |
| $a_2$ | $(3 + 1|\sin\alpha|)d$ | $5d$ | $5d$ |
| $a_{3t}$ | $(7 + 5\cos\alpha)d$ | $(10 + 5\cos\alpha)d$ | $(15 + 5\cos\alpha)d$ |
| $a_{3c}$ | $7d$ | $10d$ | $15d$ |
| $a_{4t}$ | $(3 + 4\sin\alpha)d$ | $(5 + 5\sin\alpha)d$ | $(5 + 5\sin\alpha)d$ |
| $a_{4c}$ | $3d$ | $5d$ | $7d$ |

Strength of Bolt Joints

Bolt joints are formed by drilling holes in the connected elements, inserting the screws, putting on washers and tightening the nuts (Fig. 6.86). The pressure caused by the tightening generates friction between connectors, which decreases or increases along with the changes of equivalent moisture of wood or moisture of wood-based materials.

Minimal spacing and distances between bolts are specified in the norm PN-B-03150:2000 (Table 6.8).

Pressure strength of a bolt joint with wooden connectors can be calculated using the equation (PN-B-03150:2000):

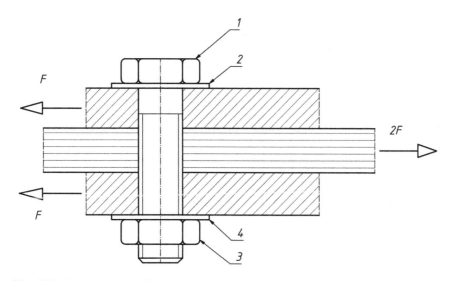

**Fig. 6.86**  Example of a bolt joint: *1* bolt head, *2, 4* washer, *3* nut

**Table 6.8** Minimal spacing and distances between bolts according to PN-B-03150:2000

| Distance signature | Orientation | Minimal distances of connectors |
|---|---|---|
| $a_1$ | ∥ to fibres | $(4 + 3\|\cos\alpha\|)d$ |
| $a_2$ | ⊥ to fibres | $4d$ |
| $a_{3t}$ | $-90° \leq \alpha \leq 90°$ | $7d$ min 80 mm |
| $a_{3c}$ | $150° \leq \alpha \leq 210°$ | $4d$ |
|  | $90° \leq \alpha \leq 150°$ | $(1 + 6\|\sin\alpha\|)d$ min $4d$ |
|  | $210° \leq \alpha \leq 2700°$ | $(1 + 6\|\sin\alpha\|)d$ min $4d$ |
| $a_{4t}$ | $0° \leq \alpha \leq 180°$ | $(2 + 2\|\sin\alpha\|)d$ min $3d$ |
| $a_{4c}$ | Other values of the angle $\alpha$ | $3d$ |

$$f_{h,\alpha,k} = \frac{f_{h,0,k}}{k_{90} \sin^2 \alpha + \cos^2 \alpha}, \tag{6.228}$$

where
$f_{h,0,k} = 0.082(1 - 0.01d\rho_k)$,
$k_{90} = 1.35 + 0.015d$  for coniferous wood species,
$k_{90} = 0.90 + 0.015d$  for deciduous wood species,
$d$                              diameter of bolt core,
$\rho_k$                          density of wood.

### Strength of Screw Joints

Screw joints are a common way of joining elements of furniture construction. They can be found in skeletons of chairs and tables, in frames of armchairs and sofas and in bodies of case furniture. As a general rule, screws are inserted into previously drilled holes, whose diameter must be adapted to the appropriate, characteristic part of the screw. In the part without thread, the hole diameter should be equal to the screw diameter $d_2$, and in the threaded part, it amounts to $d_1 = 0.7d_2$ (Fig. 6.87). And the depth of insert should amount to a minimum of $4d_2$.

In some load states of furniture, screw connectors can work on pulling and pressure (Fig. 6.88). For screws with a diameter of $d_2 < 8$ mm, the same calculation methods as for the nails apply, and in the case of screws with a diameter of $d_2 \geq 8$ mm, we proceed in accordance with the rules that apply when calculating bolts.

For axial loads of screws in wooden constructions, in which connectors were placed perpendicular to the wooden fibres, the constructor should set load-bearing capacities. In this case, we can use the formula given in the norm PN-B-03150:2000:

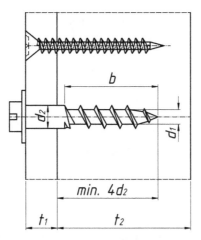

**Fig. 6.87** Example of a screw joint

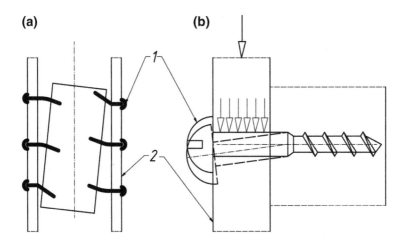

**Fig. 6.88** Scheme of work of connectors in a joint: **a** pulling and pressure, **b** bending and pressure: *1* connector, *2* veneer

$$R_d = f_{3,d}\left(l_{ef} - d_2\right),\tag{6.229}$$

where

$$f_{3,d} = \frac{k_{\mathrm{mod},2}f_{3,k}}{\gamma_M},\tag{6.230}$$

$$f_{3,k} = (1.5 + 0.6d_2)\sqrt{\rho_k},\tag{6.231}$$

$d_2$      diameter of the smooth part of the core,
$l_{ef}$     length of the screwed part of the core from the sharp side,
$\rho_k$     density of wood,
$k_{mod}$  partial modification coefficient (Table 6.5),
$\gamma_M$     partial safety coefficient (Table 6.6).

### 6.2.4  Stiffness and Stability of Chairs and Stools

#### 6.2.4.1  Stiffness of Chairs

Stiffness of construction of furniture is determined by both stiffness of joints and stiffness of components. Determining the dominant element or structural node, which decides about the stiffness of the product is not simple, because in furniture several parts at the same time are actively involved. We can, however, in the course of typical validation tests (Fig. 6.89), assess deformations, on the basis of which it is possible to formulate the criterion of stiffness of furniture correctly enough. In this case, coefficients $k$ may be the measures of the stiffness of construction, calculated as follows:

$$k = \frac{F}{\delta_{iF}}, \tag{6.232}$$

or

$$k = \frac{Fl}{\varphi_i} = \frac{M}{\varphi_i}, \tag{6.233}$$

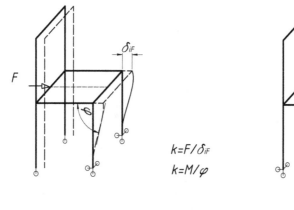

$$k = F/\delta_{iF}$$
$$k = M/\varphi$$

**Fig. 6.89** Scheme of determining stiffness of the skeleton

where

$F$    concentrated force corresponding to operational loads,
$M$    bending moment,
$\delta_{iF}$  displacement of the point i in the direction of the force $F$,
$\varphi_i$    rotation of the node caused by the bending moment $M$.

When calculating stiffness of furniture, we assume that each cross section of each of the construction elements of the skeleton (beams) remains perpendicular to the deflected axis. So, it is tangential to the bending line and with the X-axis form the angle $\varphi$, equal to the angle of rotation of the cross section (without taking into account the effect of lateral forces on the bending). Assuming also constant stiffness of the beams $(EJ)$ and structural nodes $(E_Z J_Z)$ between deflection of beams $y(x)$, angle of rotation of cross sections $\varphi(x)$, bending moment $M(x)$, transverse force $T(x)$ and load $q(x)$, these are the following relations:

$$\frac{d}{dx}(EJy) = EJ\varphi, \tag{6.234}$$

$$\frac{d^2}{dx^2}(EJy) = \frac{d}{dx}EJ\varphi = M, \tag{6.235}$$

$$\frac{d^3}{dx^3}(EJy) = \frac{d^2}{dx^2}(EJ\varphi) = \frac{dM}{dx} = T, \tag{6.236}$$

$$\frac{d^4}{dx^4}(EJy) = \frac{d^3}{dx^3}(EJ\varphi) = \frac{d^2M}{dx^2} = \frac{dT}{dx} = -q. \tag{6.237}$$

Displacements of bent beams can be specified with the use of several methods (Dyląg et al. 1986; Nowacki 1976; Zielnica 1996): integration of differential equation of the bending line, virtual loads, energy method. Below, the energy method is presented, which assumes that elastic energy of linear elastic system under load of the forces $F_1$, $F_2$, $F_3$, ..., $F_n$, equals:

$$U = \frac{1}{2}\sum_{i=1}^{n} F_i \delta_{iF}, \tag{6.238}$$

where

$\delta_{iF}$  displacement of the point and applying the force $F$ in the direction of this force (Clapeyron equation).

Elastic energy of the bent beam expressed as a function of internal forces amounts to:

$$U = \frac{1}{2}\sum \int_x \frac{M^2}{EJ} dx. \tag{6.239}$$

By comparing both sides of the above equations, we obtain a relation which allows us to determine the displacement $\delta_i$, if it constitutes the only unknown in this equation. In the cases of arbitrary load of beams, Castigliano's theorem is used:

$$\frac{\partial U}{\partial F_i} = \delta_{iF} = \sum \frac{1}{2} \int_x \frac{M}{EJ} \frac{\partial M}{\partial F_i} \mathrm{d}x. \tag{6.240}$$

Partial derivative of elastic energy of the system in relation to one of the independent loading forces is equal to the displacement of the point of application of this force in its direction. In the case of searching for displacement of any point $k$, in which there is no force or there are forces $F_k$ with directions different from the direction of the sought displacement $\delta_k$, the beam should be loaded with a fictitious force $1_k$ acting in point k in the direction of the sought displacement. When calculating derivative of elastic energy of the system in relation to $1_k$, after integrating, we compare $F_k$ to zero. In this way, we determine the value of $\delta_k$, which is the sought displacement of any point k of the construction loaded with the system of forces $F_1, F_2, F_3, \ldots, F_n$.

For example, let us determine the vertical displacements and rotation of the cross section $C$ of the frame of a chair from Fig. 6.90.

Horizontal displacement is calculated from the Clapeyron equation:

$$U = \frac{1}{2} F \delta_F = \int_0^a \frac{M_{CB}^2}{2EJ} \mathrm{d}x + \int_0^l \frac{M_{BA}^2}{2EJ} \mathrm{d}x, \tag{6.241}$$

$$F \delta_F = \frac{1}{EJ} \left( \int_0^a (Fx)^2 \mathrm{d}x + \int_0^l (Fa)^2 \mathrm{d}x \right) = \frac{1}{EJ} \left( \frac{F^2 a^3}{3} + F^2 a^2 l \right), \tag{6.242}$$

**Fig. 6.90** Static scheme of a side frame of a chair: A, B, C chosen cross sections

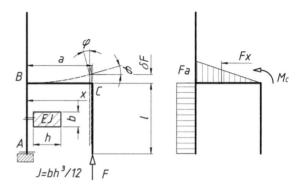

hence,

$$\delta_F = \frac{Fa^2}{3EJ}(a + 3l),\tag{6.243}$$

or from the Castigliano's theorem:

$$\delta_F = \int_0^a \frac{M_{CB}}{2EJ}\frac{\partial M_{CB}}{\partial F}\,dx + \int_0^l \frac{M_{BA}}{2EJ}\frac{\partial M_{BA}}{\partial F}\,dx,\tag{6.244}$$

$$\delta_F = \frac{1}{EJ}\left(\int_0^a Fx \cdot x\,dx + \int_0^l Fa \cdot a\,dx\right) = \frac{1}{EJ}\left(\frac{F^2 a^3}{3} + F^2 a^2 l\right).\tag{6.245}$$

Determining rotation $\varphi$ requires loading the cross section $C$ with the moment $M_c$. Bending moments and their derivatives in relation to $M_c$ in individual cross sections amount to:

$$M_{CB} = Fx + M_C, \quad \frac{\partial M_{CB}}{\partial M_C} = 1,\tag{6.246}$$

$$M_{BA} = Fa + M_C, \quad \frac{\partial M_{BA}}{\partial M_C} = 1.\tag{6.247}$$

Rotation of the cross section is therefore equal to:

$$\phi = \int_0^a \frac{Fx + M_C}{EJ}\,1\,dx + \int_0^l \frac{Pa + M_C}{EJ}\,1\,dx,\tag{6.248}$$

$$\varphi = \frac{1}{EJ}\left((0.5Fx^2 + M_Cx)\big|_0^a + (Fax + M_Cx)\big|_0^l\right).\tag{6.249}$$

By substituting $M_C = 0$, we obtain:

$$\varphi = \frac{Fa}{EJ}(a + 2l).\tag{6.250}$$

Stiffness of the examined construction will therefore amount to:

$$k = \frac{F}{\delta_F} = \frac{3EJ}{a^2(a + 3l)},\tag{6.251}$$

or

$$k = \frac{M_{CB}}{\varphi} = \frac{EJ}{a + 2l}. \tag{6.252}$$

### 6.2.4.2   Stability of Chairs and Armchairs

Determining stability of side frames of skeletal furniture requires knowledge of values of both vertical load $F$ and horizontal load $H$ (Fig. 6.91). Value of the force $F$ can be expressed in the form:

$$F = \frac{(a - x)}{e} \sum_{i=1}^{n} Q_i \leq F_{kr}, \tag{6.253}$$

and the value of the force $H$:

$$H = \frac{(a - x)}{h} \sum_{i=1}^{n} Q_i \leq F_{kr}, \tag{6.254}$$

where
$a$      width of the furniture base,
$e$      arm of acting of vertical force in relation to the edge of the base,
$x$      location of the furniture's centre of gravity,
$h$      height of the backrest,
$\Sigma Q$   sum of the mass loads,
$F_{kr}$   critical load.

**Fig. 6.91** Calculation scheme of stability of a chair in the system: **a** symmetrical, **b** asymmetrical

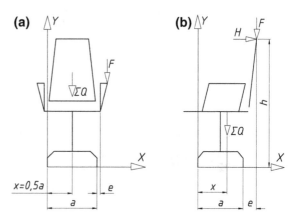

The most common cause of the loss of stability of furniture, by moving the centre of gravity outside of the base area, is the erroneous distribution of elements. For asymmetrical constructions (Fig. 6.91b), it is necessary to check the position of the centre of weight $x$ and determine the permissible loads $H$ and $F$, causing the chair or armchair to tip over.

## 6.2.5 Optimisation of Skeletal Furniture

### 6.2.5.1 Aspects of Optimisation

The issues of strength analysis of furniture presented in the previous chapter are used for analysis or synthesis of interesting issues. In many cases, constructing furniture means creating a new whole from more or less known elements and joints. It is obvious that every subsequent decision made during designing should be the best decision. If the issue is simple, then it is correct, and even the best decisions can be taken on the basis of a general analysis of the object, designing experience and intuition of the designer. However, if the issue is complex, and the proposed construction is too complicated or the designer lacks the proper experience in designing, then it is appropriate to clarify the process of making the best decisions and searching for the best construction solution. Optimisation models can then be used.

When designing furniture, the selection of appropriate parameters takes place, for example destination, functional dimensions, material, elements, type and shape of joints. Some of the properties of furniture or its components may be undesirable—for instance high material consumption, high production costs, low strength of joints and low stiffness of the construction and large deflection of elements—or desirable, such as high durability and stiffness, low weight and overall comfort of use.

The construction being a result of the proceedings, the purpose of which is to obtain the minimum adverse effects or maximum desired effect is called an optimal construction. In such proceedings, we use mathematical models of optimisation and use mathematical techniques of optimisation. To build a good optimisation model of a complex construction task, with which designers deal with, is much more difficult and often requires much more experience than simply making structural decisions, particularly in the simple issues. Regardless of the development of mathematical methods and their more and more widespread use, construction remains an art, and optimisation methods are and will be the only effective accessory tool.

### 6.2.5.2 Mathematical Model of Optimisation

Each construction is described explicitly by the parameters of shape, type of material, exchange, stress, deformation, load, functionality, colour, etc. If any of these features can be attributed to a certain value, then the whole construction can be described by a set of n numbers, where $n$ is a number of characteristics that describe the proposed construction. Therefore, construction can be treated as a certain point $x$ in an $N$-dimensional space $R$. This point is a mathematical model of the construction, which we can write as follows:

$$x = (x_1, x_2, \ldots x_N),$$
$$x \in R^N, \tag{6.255}$$

where
$x_n$   decision variables.

Construction parameters written in the form of a vector can be divided into three groups:

-  determined parameters, e.g. by the designer or recipient,
-  imposed parameters, e.g. by the technological requirements,
-  parameters chosen by the manufacturer.

Construction parameters chosen by the manufacturer are called decision variables:

$$x = (x_1, x_2, \ldots, x_n, x_{n+1}, \ldots, x_N),$$
$$P = x_N - x_n, \tag{6.256}$$

where
$x$   decision variables,
$P$   parameters.

There can be one or many decision variables. They can be expressed by numbers or constitute functions of other variables. Depending on the type and number of decision variables, different optimisation procedures are used. Mathematical relations in the model of optimisation can take the form of inequality that can be written in the structured form:

$$\varphi_i(x_n, P) \leq 0 \quad i = 1, \ldots, k \tag{6.257}$$

and equations

$$\psi_j(x_n, P) = 0 \quad j - 1, \ldots, s. \tag{6.258}$$

They constitute certain conditions imposed on decision variables, having the character of restrictions. Example of restrictions of inequality type are strength conditions. So there are restrictions of maximum stresses to the values not exceeding destructive stresses, or conditions limiting maximum deflection/displacements to the values specified in the applicable normative acts. Examples of equalities are obvious. When building a mathematical model, the number of restrictions should be taken into consideration. The number of inequality restrictions can be arbitrarily large, with one reservation that this set is not empty. On the other hand, the number of equality restrictions must be smaller than the number of decision variables, since these restrictions reduce the dimensionality of the permissible area $\Phi$ created in an N-dimensional space by a set of points fulfilling the construction conditions. When searching for the optimal construction solution, such an element from the set of permissible decisions should be selected, which corresponds to this optimal solution. This choice depends on the optimisation criterion. In the literature devoted to optimisation (Brdyś and Ruszczyński 1985; Dziuba 1990; Findeisen et al. 1980; Goliński 1974; Leśniak 1970; Ostwald 1987; Pogorzelski 1978; Smardzewski 1989, 1992), there are three groups of criteria. These are as follows:

– criteria concerning minimum construction costs,
– criteria of the maximum stiffness or smallest pliability,
– criteria for evening the stresses.

The criterion for minimum construction costs results from the natural human aspiration to ensure the maximum effects for minimum cost. There are two most commonly used optimisation criteria resulting in minimum volume and minimum weight. If we assume that construction costs are proportional to the volume of material, the minimum cost criterion is equivalent to the criterion of minimum volume or minimum weight for all furniture constructions. Criteria for the maximum stiffness or smallest pliability concern elastic deformations of pliable systems. These criteria express the belief of the dominant importance of pliability of construction in the assessment of its value. It mainly concerns case furniture. According to the criterion of evening the stresses in an optimal construction, the stress is low under a specified load in all points where this is possible. Meeting the condition of evening the stresses in all directions is possible only in certain constructions, e.g. in three articulated side frames of chairs. In other constructions, meeting the condition of evening the stresses is possible only in certain areas of construction or only on certain surfaces.

Optimisation criterion is expressed as a function of purpose, which is a function of decision variables:

$$Q = f(x_1, x_2, \ldots, x_n) = \text{optimum}. \qquad (6.259)$$

Decision variables should be selected in such a way that the function of purpose reaches the maximum value. The optimal solution is therefore a point at which the value of the function of purpose is the optimal value. This point, representing the optimal construction, should be written as follows:

$$X_{\text{opt.}} = \left(x_1^{\text{opt.}}, x_2^{\text{opt.}}, \ldots, x_n^{\text{opt.}}\right) \in \varphi, \tag{6.260}$$

and for the problem of maximisation:

$$\left(X_{\text{opt.}} \in \phi\right) : \left\{ \bigwedge_{X_{\text{opt.}} \in \phi} Q(X) \leq Q(X_{\text{opt.}}) \right\}. \tag{6.261}$$

When building a mathematical model, the manufacturer should formulate the optimisation problem in such a way that its mathematical model is adequate to the actual solution to the problem. This is particularly important in the case of the use of computers to solve optimisation tasks. They allow us to solve virtually insoluble problems through traditional calculation methods. But computer is only a machine, and when solving the intended problem, it must be presented to it in an understandable way.

Mathematical models of construction due to the nature of parameters can be classified as follows (Goliński 1974; Ostwald 1987):

- deterministic models, where all the parameters are known and constant, i.e. every possible decision corresponds to one and only one value of the function of purpose,
- probabilistic models, where one or more parameters are random variables with known distribution of probability,
- statistical models, where one or more parameters are random variables with unknown distribution of probability or the distribution of parameter in the function of time is known (stochastic process),
- discrete models, where one or more parameters can assume one of many possible values, while the set of these possible values is in most cases known.

The construction of a mathematical model of optimisation of construction should include the following:

- determining decision variables,
- determining the permissible area (N-dimensional cube of decision variables in Euclidean space),
- determining the criterion and function of purpose.

Below, we will discuss some of the deterministic and random optimisation methods.

### 6.2.5.3 Deterministic Methods of Optimisation

Using this method, one looks for the optimum in the sense of the function extreme by its differentiation. This method leads to the exact determination of the global extreme and all local extremes, which can be treated as the solution of the task if it is located in the area of permissible decisions and if there is no need to examine the edge of the area.

**Fig. 6.92** Scheme of
construction with a
reinforcing rib

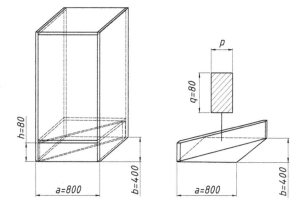

*Example*

The dimensions of the transverse cross section of the socle rib shown in Fig. 6.92
should be chosen so that the weight of the length of its section is the smallest. The
construction before the introduction of rib has the stiffness of 20,000 N/m, and after
the improvement, it should have the rigidity of 30,000 N/m. The rib is made of
pinewood, for which the elasticity modulus amounts to $E = 12,000$ MPa.

We will begin solving these tasks from building an optimisation model. For this
purpose, we determine decision variables. In the discussed case, the cross section of
the rib must be described by the dimensions $q \times p$. Restrictions in this task are the
conditions for the stiffness of the entire construction $k = 20,000$ N/m and stiffness of
the rib itself $k_z \geq 30,000 - 20,000 = 10,000$ N/m,
    whereby

$$k_z = \frac{3}{4}E \cdot \frac{p \cdot q^3}{\sqrt{(a^2 + b^2)}} \geq 10,000 \quad \text{N/m}, \qquad (6.262)$$

and also dimensions of the socle:

$$\begin{aligned} 0 < y = p \leq 80 \text{ mm} \\ 0 < x = q \leq 80 \text{ mm}. \end{aligned} \qquad (6.263)$$

The function of purpose should correspond to the desired minimisation of the
weight of the rib. With constant cross-sectional dimensions, along the length of the
rib, the condition of the minimum weight corresponds to the condition of the
minimum area of cross section. Function of purpose can be therefore formulated as
follows:

$$f(p,q) = p \cdot q. \qquad (6.264)$$

**Fig. 6.93** Graph of the area
of permissible decisions

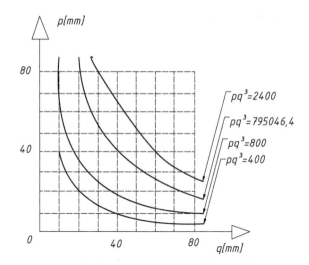

When using the above relations, the minimum value of the function of purpose is obtained, with the accuracy of the variable $y$, that is,

$$f(q_{\text{opt.}}) = \frac{40,000 \cdot \sqrt{(a^2 + b^2)}}{3 \cdot E \cdot p}. \qquad (6.265)$$

Because the function of purpose does not have an extreme, the searched optimum must be placed on the edge of the area of permissible decisions (Fig. 6.93). This task is best solved using the graph of the area of permissible decisions, with the values of purpose $f(x,y)$ = const for:

$$p \cdot q \geq \frac{40}{3E} \cdot \sqrt{(a^2 + b^2)^3} = 795,046.4 \text{ mm}^4. \qquad (6.266)$$

In Fig. 6.93, it can be seen that the sought optimum lies at the intersection of the edge determined by the inequalities. Assuming, therefore, that $q = 80$ mm, we obtain the optimal result $p = 1.55$ and $q = 80$ mm, for which the function of purpose has the value $f(p, q) = 124$ mm$^2$. Of course, a question must be asked, whether such a construction can be made? Probably very hard, due to technological problems related to mounting of the rib. Therefore, if we want to include technology in the mathematical model, additional technological conditions must be taken into account, in this case:

$$20 \leq p \leq 80. \qquad (6.267)$$

The obtained solution $q = 80$ and $p = 20$ is then the optimum for both in terms of construction and in terms of technology.

#### 6.2.5.4   Random Methods of Optimisation

Rapid progress in the field of data processing and electronic calculation means creating not only new numerical methods of designing constructions, but also methods for assessing the strength of the products already made and subject to validation. In order to develop reasonable methods of constructing furniture, fulfilling the function of purpose to minimise the consumption of wood raw material and maximise the strength of the joints, one can use the optimisation theory for selection of the minimum permissible cross section of the components and structural joints.

In engineering calculations, it is recommended to use statistical methods being the component of numerical methods of static optimisation (Dziuba 1990; Goliński 1974; Ostwald 1987; Smardzewski 1989, 1992). The methods of systematic searches, random walk and Monte Carlo are worth discussing. They consist in systematic or random search of the permissible area and the evaluation of the optimal value on the basis of the results.

Systematic Search Method

Let us assume that the subject of optimisation is a chair with a bar, whose side frame construction is shown in Fig. 6.94.

The cost of manufacturing is considered a natural optimisation criterion. This often leads to adopting the dimensions of cross sections of the construction components as the function of purpose. Achieving the optimal solution should be preceded by a conjunctive fulfilment of a number of restriction conditions. To illustrate the way of discretisation of the decision variables cube and the build of the appropriate restrictions, Fig. 6.95 shows the scheme of internal forces in four nodes of the side frame.

**Fig. 6.94** Construction of side frame of a chair with a bar: $A$ flat T-type joints, $B$ flat L-type joints

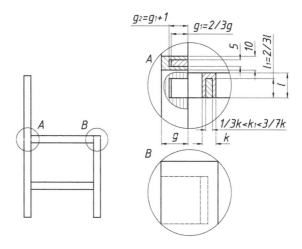

**Fig. 6.95** Internal forces in
nodes of side frame of a chair

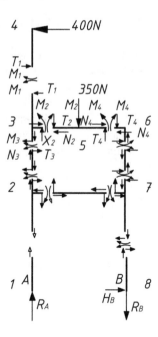

Mathematical model of optimisation of the side frame of a chair includes the
following:

(a) Decision variables, for which the cube of variables $K_z$, have the form:

$$K_z = \{\bar{x} = (x_1 \ldots x_4) : x_{i(min)} \le x_i \le x_{i(max)} : i = 1, \ldots, 4\}, \qquad (6.268)$$

where

$i$                          number of decision variables for one construction node,

$x_2 = g$, $x_1 = h$    dimensions of cross sections of vertical elements,

$x_4 = k$, $x_3 = l$    dimensions of cross sections of horizontal elements.

(b) Parameters, which constitute internal forces in the construction (Fig. 6.96) and
border strength of the material and glue-line in joints, that is,

$M$   bending moment in the $j$th node,
$T$   cutting force in the $j$th node,
$N$   normal force in the $j$th node,
$k_g$   bending strength of the wood,
$k_t$   shearing strength of the wood,
$k_s$   shearing strength of the glue-line.

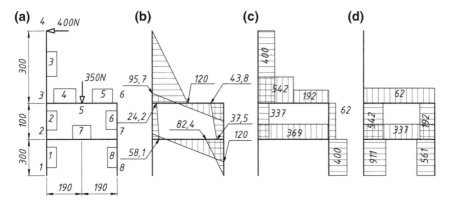

**Fig. 6.96** Static scheme of the load of side frame of a chair: **a** distribution of internal forces, **b** bending moments, **c** cutting forces, **d** normal forces

(c) Permissible set $\varphi$ is made of inequality restrictions, $\varphi_i = (x) \geq 0$, that is,

$$\varphi = \{\bar{x} = (x_1, \ldots, x_4) : \varphi_i(\bar{x}) \geq 0 : \quad i = 1, \ldots, 4\}. \tag{6.269}$$

(d) Strength restrictions, $\varphi_i = (\bar{x}) \geq 0$, shearing strength of the tenon in the $j$th node

$$(k_1 l_1) \geq \frac{T_2 n}{k_t}, \tag{6.270}$$

where
$n$ safety coefficient,
bending strength of the tenon on the $j$th joint

$$(k_1 l_1^2) \geq \frac{6 M_2 n}{k_g}, \tag{6.271}$$

shearing strength of the element in the cross section of mortise for loads:

$$(gh - g_1 k_1 + 2) \geq \frac{T_{\max} n}{k_t}, \tag{6.272}$$

bending strength of the element in the cross section of mortise, centre of gravity of the examined cross section according to Fig. 6.97:

$$y_e = \frac{\frac{1}{2}(g^2 h - g_2^2 k_1)}{(gh - g_2 k_1)}, \tag{6.273}$$

**Fig. 6.97** Geometry of the cross section at the place of implementation of the mortise and tenon joints

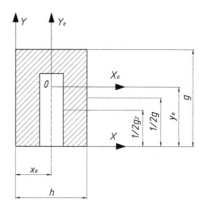

moment of inertia of cross section in relation to $x_e$-axis:

$$J_x = g^3 h \left[ \frac{1}{3} + \left( \frac{y_e}{g} \right)^2 - \frac{y_e}{g} \right] - g_2^3 k_1 \left[ \frac{1}{3} + \left( \frac{y_e}{g_2} \right)^2 - \frac{y_e}{g_2} \right], \qquad (6.274)$$

bending strength of the examined cross section for loads:

$$\sigma_x \geq \frac{M_{max} n}{k_g} y_e, \qquad (6.275)$$

shearing strength of the glue-line at the torsion of the moment $M_{max}$:

$$\tau_s = (3g_1 + 1.8 \cdot l_1) \frac{M_{max}}{2(g_1^2 l_1^2)}, \qquad (6.276)$$

maximum shearing stress in the direction of normal force $N_2$:

$$\tau_{s(max)} = \left( \frac{N_2}{2(g_1 l_1)} \right) n \leq k_s, \qquad (6.277)$$

maximum shearing stress in the direction of cutting force $T_2$:

$$\tau_{s(max)} = \left( \frac{T_2}{2(g_1 l_1)} \right) + \tau_s \Big) n \leq k_s. \qquad (6.278)$$

(e) Construction conditions resulting from the provisions of BN-76/7140-02:

- length of the tenon $g_1 \leq \frac{2}{3} g$,
- depth of the mortise for the tenon $g_2 = g_1 + 2$ mm
- thickness of the tenon $\frac{1}{3} k \leq k_1 \leq \frac{3}{7} k$,
- height of the tenon $l_1 \leq \frac{2}{3} l$.

(f)  Aesthetic conditions for individual horizontal and vertical elements: $h \geq k$.
Minimisation of the area of cross section of components of the side frame of a
chair at the place of their connection has been assumed as the function of
purpose. Therefore, the minimum area of cross section of the horizontal element with a tenon is as follows:

$$A_1 = k \cdot l \rightarrow \min, \tag{6.279}$$

minimum area of cross section of the vertical element with a mortise is as
follows:

$$A_2 = g \cdot h \rightarrow \min. \tag{6.280}$$

When attempting to optimise a chosen construction of a chair, all the optimisation parameters had to be established. To this end, the geometric parameters
discussed previously, and presented in Fig. 6.94, were additionally complemented
with the values of internal forces caused by the standard loads (Fig. 6.96).
A detailed description of each of the planned static optimisation methods can be
found in the rich literature of the subject, and therefore, only the way of determining
the number of samplings $N_o$ and the choice of the optimisation step $t$ and size of the
cell $T$ have been discussed below.

Systematic search method (Fig. 6.98) consists in a gradual search of the entire
permissible area, counting the function value at each point, choosing the point in
which the value of the function of purpose is the smallest. In this case, discretisation
points can be selected so that the optimisation step is equal to 0.5 mm. Then, the
total number of discretisation points for a single structural node is equal to:

$$N_o = \prod_{i=1}^{4} \left( \frac{x_{1(\max)} - x_{i(\max)}}{t_1} \right) + 1. \tag{6.281}$$

The following values can be used as initial data for the process of optimisation:

- internal forces $M_1, M_2, M_3, T_1, T_2, T_3, N_1, N_2, N_3$, as in Figs. 6.94 and 6.96b–d,
- minimum dimensions—assumed arbitrarily—$g_{\min} = 15$ mm, $h_{\min} = 15$ mm,
  $l_{\min} = 15$ mm, $k_{\min} = 15$ mm,
- maximum dimensions—assumed arbitrarily—$g_{\max} = 60$ mm, $h_{\max} = 60$ mm,
  $l_{\max} = 60$ mm, $k_{\max} = 60$ mm,
- bending strength of the pinewood along the fibres $k_g = 105$ MPa,
- shearing strength of the pinewood across the fibres $k_t = 8$ MPa,
- shearing strength of the polyvinyl acetate glue $k_s = 12$ MPa,
- safety coefficient $n = 1.5$,
- optimisation step $t = 0.5$ mm,
- number of samplings $N_o = 68{,}610$.

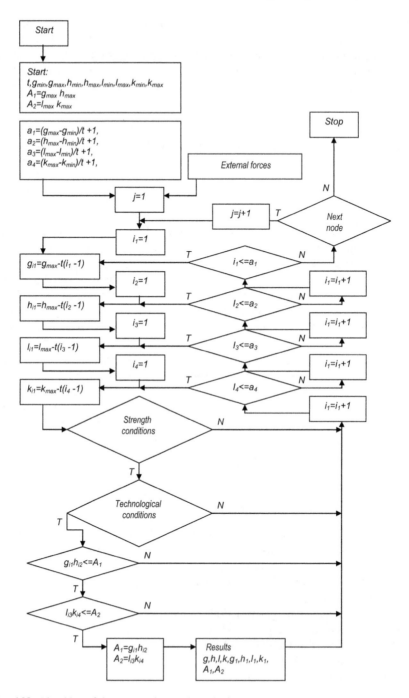

**Fig. 6.98** Algorithm of the systematic search method

Monte Carlo Method

Construction parameters selected by the manufacturer as a result of using optimisation must meet the basic criterion, which in this case is the minimum area of cross section of rails. And the given optimum dimensions must meet numerous additional restricting conditions. For example, let us consider the construction of an external wing of a reinforced single window with the dimensions given in Fig. 6.99.

Structural joints of window wings should shift loads of 100 N on the arm 250 mm and of internal wings—330 N on the same arm (Fig. 6.100a). For the wing of the given standard dimensions, the load state corresponds to the distribution of bending moments caused by concentrated force $P$ of 500 N, as it is shown in Fig. 6.100b. Stiffness and strength of such a loaded window wing depend on the strength of the joints and on the bending strength of rails.

Figure 6.100a shows the scheme of a structural node of the external window wing, whose strength will be the main source of restrictions in the process of optimisation. Cross section of rails shown in Fig. 6.100a results from adopting for the analysis only active part of cross section of rails, in which connections have been made. The remaining part of the cross section, situated outside the hatched

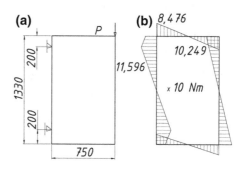

**Fig. 6.99** Window: **a** basic state, **b** distribution of bending moments

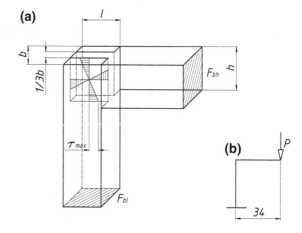

**Fig. 6.100** Structural node as an object of optimisation: **a** geometric scheme, **b** method of standard loading

**Fig. 6.101** Lower corner
connector of window's wing

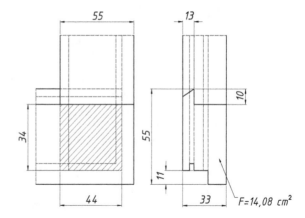

outline of bridle joint (Fig. 6.101), meets only specific technological requirements,
without the effect on strength of the joint.

Mathematical model of optimisation of construction is as follows:

-   selected construction parameters—according to Fig. 6.100a—constitute trans-
    verse dimensions of rails: vertical $l_i \cdot b_i$ and horizontal $h_i \cdot b_i$, accepted in the
    subsequent $n$-steps of optimisation for i = (1, 2, ..., n),
-   restricting conditions resulting from the construction of the window wing
    constitute:

(a)  bending strength of the tenon loaded with bending moment $Mg_1$ (Fig. 6.99b),
     described by the inequality:

$$b_i h_i^2 \geq \frac{18 M g_1}{k_g},  \qquad (6.282)$$

    where
    $k_g$  bending strength of the wood,

(b)  shearing strength of the tenon loaded with the force $P$, expressed as follows:

$$b_i h_i \geq \frac{3P}{k_t},  \qquad (6.283)$$

    where
    $k_t$  shearing strength of the wood,

(c)  shearing strength of the glue-line, which has been presented as follows:

$$l_i h_i^2 \geq \frac{Mg_1}{n_1 \cdot \alpha \cdot \tau_{max}},$$

$$\tau_{max} = k_S,$$

(6.284)

where

$k_S$    shearing strength of the glue-line,
$n_1$    number of glue-line,
$\alpha$    coefficient dependent from the relation $h_i/l_i$,
$\tau_{max}$    maximum static stress.

(d)  bending strength of the loaded vertical rail, at the point of mounting of hinges, maximum bending moment $Mg_2$ (Fig. 6.99b), which has been illustrated below:

$$b_i l_i^2 \geq \frac{6 Mg_2}{k_g},$$

(6.285)

as well as construction conditions tested in practice, which have been presented by the following relations:

$$4 \geq \frac{h_i}{b_i} \geq 1,$$

(6.286)

$$4 \geq \frac{l_i}{b_i} \geq 1.$$

(6.287)

A natural optimisation criterion, as previously, is the cost of manufacturing. Assuming that these costs are proportional to the amount of used materials, the cross sections of rails should be minimised. Therefore, we should try to determine:

$$F_{bh_{min}} = b_i \cdot h_i,$$

(6.288)

$$F_{bl_{min}} = b_i \cdot l_i.$$

(6.289)

Mathematical model of optimisation of construction of a window wing, on the example of the window, therefore includes the following:

- decision variables $l_i$, $b_i$, $h_i$,
- parameters $M_{g1}$, $M_{g2}$, $P$, $k_g$, $k_t$, $k_s$, $n_1$, $\alpha$,
- permissible area determined by restricting conditions,
- function of purpose.

Before solving the model, the 3-D variability cube presented in Fig. 6.102 should be described in detail. Therefore, we have to give minimum—intuitively

**Fig. 6.102** Decision
variables cube

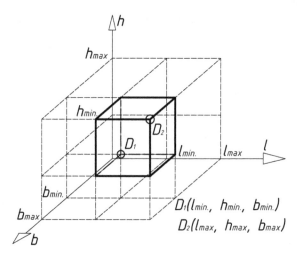

$D_1(l_{min.}, h_{min.}, b_{min.})$
$D_2(l_{max}, h_{max}, b_{max})$

estimated—values of dimensions of the cross sections of rails $l_{min}$, $b_{min}$, $h_{min}$, and maximum values of dimensions $l_{max}$, $b_{max}$, $h_{max}$, which we intend to optimise. In the decision variables cube, these values are described by extreme points $D_1$, $D_2$. In the optimisation process, computer randomly chooses points from the given cube $D_i$, $l_i$, $h_i$, $b_i$ and remembers only those which fulfil all the restricting conditions. The point, which describes the function of purpose best, represents optimal dimensions of cross sections.

For the external wing of the examined window construction with the specified dimensional characteristics and given—according to Fig. 6.99b—distribution of bending moments, optimisation of construction using Monte Carlo method was carried out, according to the algorithm presented in Fig. 6.103.

The following values have been used as initial data for the process of optimisation using Monte Carlo method:

- maximum bending moment lowering the joint $Mg_1 = 10.249$ daNm,
- maximum bending moment lowering the vertical rail $Mg_2 = 11.586$ daNm,
- concentrated force (cutting the tenon) $P = 50$ daN,
- bending strength of the pinewood $k_g = 87.0$ MPa,
- shearing strength of the pinewood $k_g = 10.0$ MPa,
- shearing strength of the urea glue $k_s = 8.0$ MPa,
- minimum dimensions (freely assumed) $l_{min} = 20$ mm, $h_{min} = 20$ mm, $b_{min} = 10$ mm,
- maximum dimensions of the cross sections of rails, according to Fig. 6.101, $l_{max} = 44$ mm, $h_{max} = 34$ mm, $b_{max} = 33$ mm.

Entering the number of samplings $N_o = 50$, which the computer should perform in order to find the optimal solution, is followed by generation of random numbers and selection from the cube of decision variables of any points $D_i(l_i, h_i, b_i)$. The point that specifies the minimum value of the function of purpose after performing

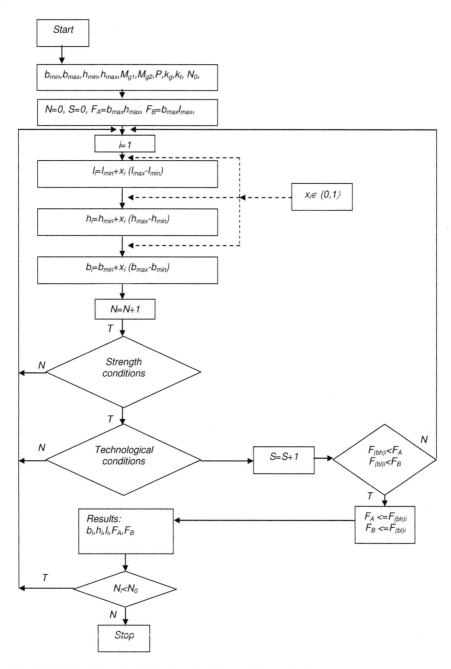

**Fig. 6.103** Algorithm of optimisation using Monte Carlo method

**Fig. 6.104** Cross sections of
rails after optimisation:
**a** vertical rail, **b** horizontal rail

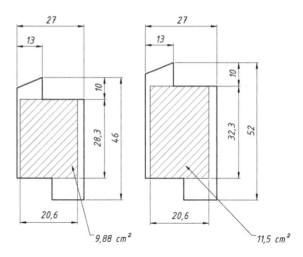

$N_o$ samplings presents the optimal dimensions of cross sections of rails. And this
process is considered completed, when the result after a great number of samplings
has not improved. For the data presented above, we obtain the following results:

– cross section of the vertical rail $L = 26.3$ mm, $b = 20.6$ mm,
– cross section of the horizontal rail $h = 32.3$ mm, $b = 20.6$ mm.

The obtained optimal dimensions—as it has been stated earlier—refer to the
dimensions of the bridle joints. Therefore, taking into account the necessary technological profiles, according to Fig. 6.100, a new, optimal cross section of rails is
obtained with dimensions and surface as shown in Fig. 6.104.

Random Walk Method

Random walk method (Fig. 6.105) is based on the claim that the optimum point is
most often located on the edge of the permissible area $\Phi$. The size of the cell
$T = 10$ mm, the length of the favourable series $t_o = 100$ and the length of the
unfavourable series $p_o = 100$ were assumed arbitrarily, similar to the multiplexing
of increase or reduction of the size of the cell $s = 2$. The above selection of the
values $t_o$ and $p_o$ provides the same search for both the interior and the edge of the
permissible area $\Phi$.

Results of optimisation of the construction of a chair with a bar are presented in
Tables 6.9 and 6.10 and in Fig. 6.106. Table 6.9 presents the criteria allowing for
the assessment of the efficiency of each of the applied methods of numerical
optimisation. This table shows that the fastest are random methods, consisting in
selective searching of the permissible area $\Phi$. The optimal solution is achieved
already after a few minutes of the computer work. The time required to determine
the optimal solution in the systematic search method is, however, a few dozen times

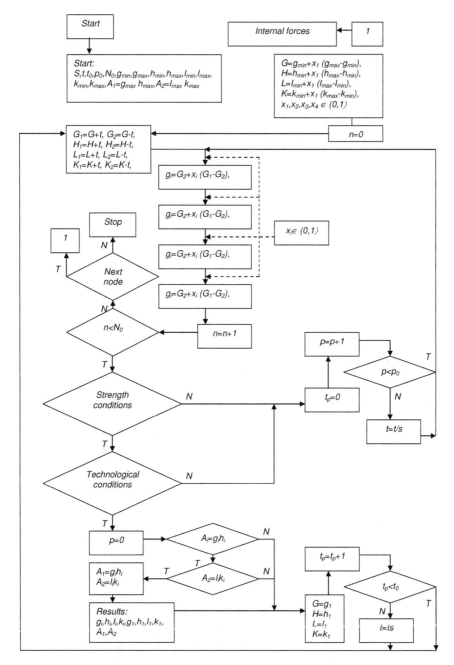

**Fig. 6.105** Algorithm of optimisation using random walk method

**Table 6.9** Efficiency of numerical methods of optimisation of construction of a chair with a muntin

| Method | Node | Cross section of the element [mm²] | | Number of drawings | | Time of optimisation [min] |
|---|---|---|---|---|---|---|
| | | Vertical $A_1$ | Horizontal $A_2$ | General | Successful | |
| Systematic search | 2 | 522.0 | 616.3 | $11 \times 10^7$ | 248 | 245.0 |
| | 3 | 806.0 | 922.2 | $52 \times 10^5$ | 186 | 130.0 |
| | 6 | 525.0 | 547.5 | $27 \times 10^5$ | 88 | 70.0 |
| | 7 | 643.3 | 922.3 | $10 \times 10^6$ | 270 | 210.0 |
| Monte Carlo | 2 | 549.3 | 531.2 | 73,589 | 8 | 4.0 |
| | 3 | 853.2 | 921.3 | 88,289 | 8 | 5.0 |
| | 6 | 589.3 | 650.9 | 384 | 3 | 0.5 |
| | 7 | 912.5 | 978.5 | 5600 | 6 | 1.5 |
| Random walk | 2 | 617.1 | 644.3 | 83,892 | 10 | 4.5 |
| | 3 | 795.9 | 906.1 | 63,148 | 8 | 3.5 |
| | 6 | 770.7 | 436.7 | 5275 | 8 | 1.5 |
| | 7 | 788.3 | 601.5 | 83,892 | 7 | 4.5 |

**Table 6.10** Optimised dimensions of joints

| Method of optimisation | Node | Dimensions of mortise and tenon joints [mm] | | | | | | |
|---|---|---|---|---|---|---|---|---|
| | | $g$ | $g_1$ | $h$ | $l$ | $l_2$ | $k$ | $k_3$ |
| Systematic search | 2 | 36.00 | 24.00 | 14.50 | 42.50 | 28.33 | 14.50 | 6.21 |
| | 3 | 52.00 | 34.67 | 15.50 | 59.50 | 38.67 | 15.50 | 6.64 |
| | 6 | 35.00 | 23.33 | 15.00 | 36.50 | 24.33 | 15.00 | 6.43 |
| | 7 | 41.50 | 27.67 | 15.55 | 59.50 | 39.67 | 15.50 | 6.64 |
| Monte Carlo | 2 | 30.82 | 20.55 | 17.82 | 50.97 | 33.98 | 10.42 | 4.46 |
| | 3 | 54.83 | 36.56 | 15.56 | 59.70 | 39.80 | 15.43 | 6.61 |
| | 6 | 32.73 | 21.82 | 18.01 | 40.17 | 26.78 | 16.20 | 6.94 |
| | 7 | 46.79 | 31.19 | 19.50 | 59.55 | 39.70 | 16.43 | 7.04 |
| Random walk | 2 | 39.85 | 26.56 | 15.49 | 42.03 | 28.02 | 15.33 | 6.57 |
| | 3 | 43.41 | 28.91 | 18.34 | 59.62 | 39.75 | 15.20 | 6.51 |
| | 6 | 50.29 | 33.53 | 15.32 | 28.53 | 19.02 | 15.31 | 6.56 |
| | 7 | 52.19 | 34.80 | 15.10 | 40.09 | 26.07 | 15.01 | 6.43 |

longer. The obtained result is, however, most beneficial in comparison with the results obtained using the Monte Carlo or random walk method. This does not mean, however, that these two methods have doubtful application in the process of optimisation of wooden structures. On the contrary, their solutions are completely satisfactory for the engineering practice, and the argument for the advisability of their use is the extremely short time of calculations.

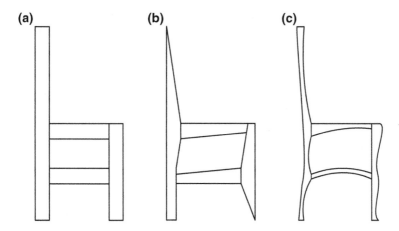

**Fig. 6.106** Change of construction of the side frame of a chair: **a** state before optimisation, **b** state after optimisation, **c** state after taking into account aesthetic elements

Optimised dimensions of mortise and tenon joints of a construction of a chair are presented in Table 6.10. These dimensions allow for rational designing of the furniture construction. The furniture can have the shape as in Fig. 6.106b. It is a theoretical model, matched only for the given loading scheme. It can be appropriately shaped, taking into account aesthetic qualities, which is shown in Fig. 6.106c.

### 6.2.5.5   Mixed Methods of Optimisation

Mixed methods of optimisation are combination of random methods and deterministic methods. Below, we will discuss one of them, the complex method. This method assumes that in an $N$-dimensional area R, at least one starting point is known. We will divide the restrictions of the R area into two groups:

- functional restrictions:

$$\varphi_i(x_n, \rho \leq 0) \quad i = 1, \ldots, m, \tag{6.290}$$

- known restrictions:

$$h_j \leq x_j \leq g_j \quad j = 1, \ldots, r \leq n. \tag{6.291}$$

In the first stage, to a known starting point, we sample more, until the total number of points amounts to $k \geq n + 1$. The sampling is carried out in the area of known restrictions, and then, we check whether the sampled point meets the functional restrictions. If yes, the point is considered as favourable, and if not, the sampled point is moved along the straight line passing through the sampled point and the last favourable point in the direction of a favourable point for half the length of the section between them. This procedure is repeated as long as certain conditions are met. Figure 6.107 illustrates this procedure. Let us assume that as a result of the sampling, we obtain subsequently the points $x_1^0, x_2^0, \ i \ x_3^0$. The first two points are favourable and the third not; therefore, we move it along the straight line passing through the points $x_2^0, \ i \ x_3^0$ in the direction of the point $x_2^0$ by half the distance between points $x_2^0, \ i \ x_3^0$. This way, we obtain the point $x_4^0$. In this example, already one move was enough, because the point $x_4^0$ turned out to be favourable. A set of points obtained in this way has been named "a complex" by the author of the method. In the discussed example, three starting points are enough, because the area of decision variables is two-dimensional.

In the second and subsequent stages, the procedure is as follows: the centre of gravity of $x_3^0$ points of the complex is determined, as well as the value of the function of purpose in each of these points. The most unfavourable point is selected and replaced with another point, whereas the coefficient of homotheticity $\alpha$ should be greater than 1. If a new point does not depend on the area R, then it is moved in the direction of the centre of gravity by half the length of the section between them, this procedure is carried out as long as the new point is found in the area R. After completing this procedure, a new complex is obtained, on which identical operations as on the starting complex are made. In the example shown in Fig. 6.107, the new complex is points $x_1^0, x_4^0, \ i \ x_2^1$. The criterion to stop searching for the optimum in this method is to achieve the equality of the function of purpose, at all points in the complex, with the precision set by the user.

**Fig. 6.107** Illustration of the complex method

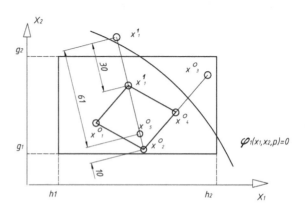

## 6.3 Operational Loads on Tables

One of the basic quality tests of the project and the implementation of the construction of a table is to determine its stability. The basic static scheme here is loading the table worktop with a vertical force with the value of 600 N at a distance of 50 mm from the front edge of the top and a horizontal force with the value from 300 to 400 N (Fig. 6.108a). Durability of tables is tested by loading the surface of the worktop with a weight of 50 kg and applying a horizontal load with the value from 200 to 300 N, at a distance of 50 mm from the side edges of the table top (Fig. 6.108b). Another way of loading, which checks the durability of the construction of the table, is applying a vertical concentrated force with the value of 400–600 N, applied at a distance of 50 mm from the side and front edge of the worktop (Fig. 6.108c).

Testing durability of tables is aimed at, among others, to confirm or exclude the ability of the construction to ensure that, at the time of using, it can endure moving with no apparent deformations of the worktop. During laboratory tests, the worktop is evenly loaded with a maximum weight of up to 100 kg, while the weights should not: protrude beyond the edges and be able to move during the test. Then, a horizontal force of 300 N should be applied at right angle to the edge of the surface of the worktop of the table, at a point 50 mm distant from the corners. This force shall be applied alternately (Fig. 6.109a).

Another type of test is checking the strength of the worktop and the frame of the table to occasional, short vertical loads. In these tests, a downward vertical force of 1000 N, pointing down, should be 10 times applied to the worktop. The acting of the force shall be maintained for 10 s ± 2 s. The test is carried out at four points shown in Fig. 6.109b or elsewhere on the worktop, where damage may occur.

An important load scheme is also a test leading to determine the stiffness indicator of the construction of tables. This indicator is included in the evaluation of vibrations

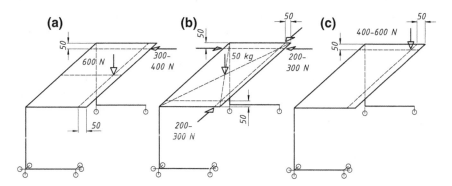

**Fig. 6.108** Table loads according to PN-EN 1729-2:2006, PN-EN 1730:2002: **a** determination of stability with acting of horizontal and vertical forces, **b** determination of durability under load with horizontal and **c** vertical forces (mm)

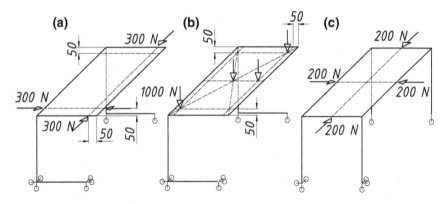

**Fig. 6.109** Determining strength and stiffness of office tables and desks according to PN-EN 527-3:2004. Testing durability under load: **a** horizontal, **b** vertical and **c** testing stiffnesses of the construction (mm)

of the frames activated during use and causes unpleasant sensations to the user. These tests are conducted by applying a force of 200 N, at the height of the worktop and along its longitudinal axis, towards the centre of the table (Fig. 6.109c).

### 6.3.1  Stresses in Construction Elements

A large part of the furniture of a skeletal as well as case structure is based on load-bearing solutions called frames. They are more often exposed to loads than other construction components of the product. Therefore, the construction solution for this part of the product should ensure its sufficient strength. Stiffness of the frame and stresses that occur in its construction elements are dependent on the positioning of the elements in relation to the base, on the positioning of the legs and bars, as well as on the geometry of elements, i.e. variability in their cross section. With a certain linear variable stiffness of the construction components of the frame (legs), it is necessary to change the smallest cross section of the loaded element and the method and location of mounting the bar. It is also easy to identify dangerous places in these elements with the changing dimensions of their cross sections. To this end, we will load the front edge of the table, as it is shown in Fig. 6.110. For a construction without a muntin, the calculation scheme is the statically determinate frame, loaded as in Fig. 6.110a. The distribution of bending moments, being the result of the load, is shown in Fig. 6.111. Assuming that the chair leg is similar to a roller, i.e. it is characterised by a constant stiffness, and then, the biggest bending stresses will occur at the place of its connection to the rail, which can be written as

$$\sigma_g = \frac{M_g}{W},$$                    (6.292)

**Fig. 6.110** Calculation
scheme of a table with a
construction: **a** without rail,
**b** with rail

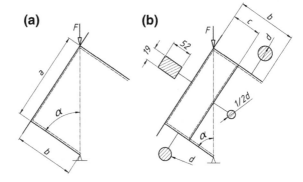

**Fig. 6.111** Distribution of
bending moments in the frame
of a table without rail

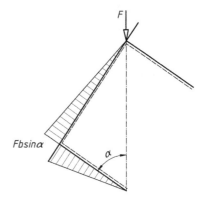

whereby

$$M_g = Fb\sin\alpha, \quad W = \frac{\pi d^3}{32} = 0.1d^3, \tag{6.293}$$

$$\sigma_g = \frac{10}{d^3}Fb\sin\alpha, \tag{6.294}$$

where
$\sigma_g$   normal stresses from bending,
$M_g$   bending moment,
$F$   concentrated force.

The value of these stresses in cross sections $\beta$-$\beta$ at any height $b_{\beta-\beta}$ depends
rectilinearly on the length $b$ (Fig. 6.112a). By selecting at the same time the loaded
joint, the value of the bending moment shifted by it can be reduced, using the side
bars as shown in Fig. 6.110b.

In many constructions of case furniture and frames of case furniture, legs with
variable cross-sectional geometry are designed, both in the form of slant cones and

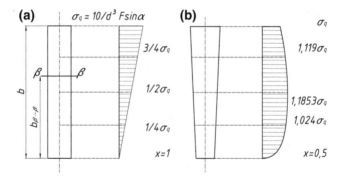

**Fig. 6.112** Course of changes of bending moments along the length of the leg: **a** with constant cross section, **b** with variable cross section

**Fig. 6.113** Calculation scheme for a leg with variable cross section

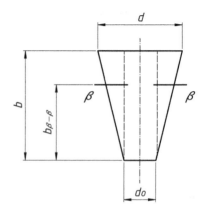

legs turned with profile, whose cross sections change from top to bottom. Therefore, for the scheme shown in Fig. 6.112 and with the leg convergence above 6°, for which the hypothesis of permanent cross sections has been rejected, bending stresses along the length of the leg change in a disproportionate manner. By considering the following geometric scheme (Fig. 6.113) and assuming proportions determining the positioning of the examined cross section, we introduce the relations:

The value of these stresses in cross sections $\beta$-$\beta$ at any height $b_{\beta-\beta}$ depends rectilinearly on the length b (Fig. 6.112a). By selecting at the same time the loaded joint, the value of the bending moment shifted by it can be reduced, using the side bars as shown in Fig. 6.110b.

In many constructions of case furniture and frames of case furniture, legs with variable cross-sectional geometry are designed, both in the form of slant cones, as well as legs turned with profile, which cross sections change from top to bottom. Therefore, for the scheme shown in Fig. 6.112 and with the leg convergence above 6°, for which the hypothesis of permanent cross sections has been rejected, bending stresses along the length of the leg change in a disproportionate manner. By considering the following geometric scheme (Fig. 6.113) and assuming proportions determining the positioning of the examined cross section, we introduce the relations:

$$\frac{b_{\beta-\beta}}{b} = y, \tag{6.295}$$

$$\frac{d_o}{d} = x, \tag{6.296}$$

where
$d$    diameter of the cross section at the leg base,
$d_o$   diameter of the cross section at the leg bottom.

By introducing the following geometric dependencies,

$$\frac{1}{2}\frac{d - d_o}{b} = \frac{1}{2}\frac{d_{\beta-\beta} - d_o}{b_{\beta-\beta}}, \tag{6.297}$$

diameter of the examined cross section at the height $b_{\beta-\beta}$ amounts to:

$$d_{\beta-\beta} = d(y(1 - x) + x). \tag{6.298}$$

For the examined cross section $\beta$-$\beta$ and the appropriate size of legs $b_{\beta-\beta}$, the value of bending stresses is expressed by the equation:

$$\sigma_{\beta-\beta} = \frac{10Fb\sin\alpha}{[d(y(1 - x) + x)]^3}. \tag{6.299}$$

Assuming (for example) that the bottom diameter of the chair leg is two times smaller than that of the top, i.e. $x = 0.5$, then the change in bending stresses on its length (Fig. 6.112b) will amount to:
for

$$
\begin{aligned}
y &= 0.25, \sigma_{\beta-\beta} = 1.024 \cdot (10/3)Fb \sin\alpha = 1.024 \cdot \sigma_g, \\
y &= 0.50, \sigma_{\beta-\beta} = 1.185 \cdot \sigma_g, \\
y &= 0.75, \sigma_{\beta-\beta} = 1.119 \cdot \sigma_g, \\
y &= 1.00, \sigma_{\beta-\beta} = 1.0 \cdot \sigma_g.
\end{aligned}
\tag{6.300}
$$

The most important bending stresses appear in this case halfway up the leg ($y = 0.5$), rather than in the upper part, that is at the point of the connection with the rail of the chair. The discussed dependencies, for the full variability of dimensions of the leg and its two diameters, are shown in Fig. 6.114. A significant effect of the

**Fig. 6.114** Changes of
maximum bending stresses in
legs of tables of linear
variable stiffness

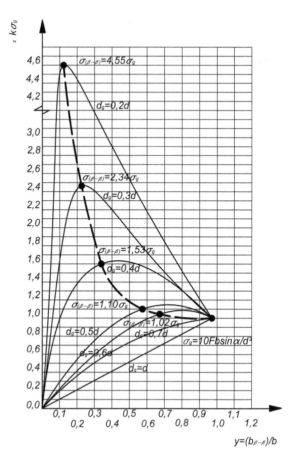

$$y=(b_{\beta-\beta})/b$$

change in the geometry of the chair leg on the course of bending stresses must be
noted. Maximum bending stresses move in the direction of the upper cross section
in case of the increment of the diameter of the lower cross section. Therefore, it can
be concluded that at the relation $d_o/d < 0.5$, the greatest bending stresses focus not
at the place of acting of the biggest bending moment, but in the lower cross sections
of the leg and in a way that clearly lowers its strength. With these construction
solutions, mounting the lateral bar at will is questionable. Because when making a
mortise for the tenon or a hole for the dowel in a clearly excessively loaded place,
the strength of the already weak cross section is reduced.

Therefore, when considering the scheme from Fig. 6.110, the chair construction
can be narrowed down to an overstiff three-time internally statically indeterminate
frame, in which the distribution of bending moments can be determined on the basis
of energetic methods known in the construction theory. By using for this purpose
the Menabrea's theorem:

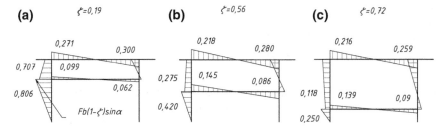

**Fig. 6.115** Distribution of bending moments in the frame of a table with a variable location of the bar: **a** for $\xi = 0.19$, **b** for $\xi = 0.56$, **c** for $\xi = 0.72$

$$\frac{\partial U}{\partial x_i} = \sum \int\limits_x \frac{M_g}{EJ} \frac{\partial M_g}{\partial x_i} dx = 0, \qquad (6.301)$$

where
$x_i$    overvalues,
$U$    elastic energy of the entire system,
$M_g$    bending moment,
$E$    linear elasticity module,
$J$    moment of inertia of the cross section.

the values of the subsequent overvalues xi are determined, and graphs of bending moments are made for constructions with various locations of the side bar at the height of the leg ($c/b = \zeta = 0.19, 0.56, 0.72$), as in Fig. 6.115. It can be noticed that for the scheme from Fig. 6.110a, the bending moment acting on a connection in a chair without bar with standard load depends only on its geometry. From Fig. 6.115, it results that if this bar is located lower than the seat, it causes a reduction of the value of the bending moment acting on both the leg connections with the rail and with the bar itself. It is accompanied by a reduction of the values of bending stresses. Therefore, the question arises: Will this relation be confirmed for schemes of linear variable stiffness? Therefore, another considerations must be made.

When evaluating bending stress in full cross section of the legs at the points of mounting the bar, as previously, a concurrent profile must be assumed, that is $x = 0.5$. Because maximum bending moments in these locations for each type of construction from Fig. 6.115 amount to, respectively,

$$\sigma_\zeta = \frac{10Fb(1 - \zeta)\sin\alpha}{d_\zeta^3}, \qquad (6.302)$$

therefore,

for $\xi = 0.19$, $\sigma_\xi = 1.092 \ldots \sigma_g$,
for $\xi = 0.56$, $\sigma_\xi = 1.178 \ldots \sigma_g$,

and as it can be seen, they evidently exceed the value of bending stresses at the place of connection of the rear leg with the rail. Therefore, the proposed convergence of the leg, expressed by the proportion $d_o/d = 0.5$, does not provide its rational connection to the bar without compromising the strength of construction. The full relation between the geometry of the leg of a chair and connection of a bar and the value of bending stresses at the place of its connection is shown in Fig. 6.116. It can be seen that the value of bending stresses clearly increases along with the reduction of the bottom diameter of the leg below the proportion $x < 0.6$. Lowering the position of the bar in relation to the seat additionally worsens the working conditions of the construction, causing an increase of stresses at the places of their connections with the legs.

**Fig. 6.116** Changes of bending stresses at the place of connection of the leg of linear variable stiffness with the bar

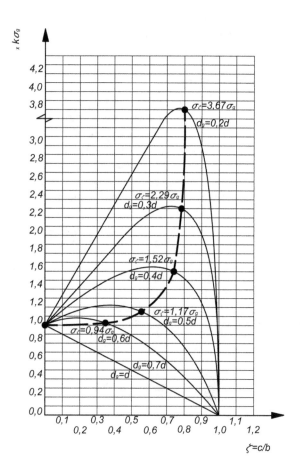

# References

Aasheim E (1993) Glulam trusses for olympic arenas. Struct Eng Int 2:86–87

Ambarcumian SA (1967) Teorija anizotropnykh plastin. Fizmatgiz, Moscow

Apalak KM, Davies R (1993) Analysis and design of adhesively bonded corner joints. Int J Adhes Adhes 139(4):219–235

Apalak KM, Davies R (1994) Analysis and design of adhesively bonded corner joints: fillet effect. Int J Adhes Adhes 14(3):163–174

Ashkenazi EK (1958) Anizotropija miechaniczieskich swoistw driewiesiny i faniery. Gostiechnizdat, Moscow, Leningrad

Biblis EJ, Carino HF (1993) Factors influencing the flexural properties of finger-jointed southern pine LVL. Forest Prod J 43(1):41–46

BN-76/7140-02 Basic furniture connector–requirements

Bodig J, Goodman JG (1973) Prediction of elastic parameters for wood. Wood Sci Tech 4: 249–264

Brdyś H, Ruszczyński A (1985) Metody optymalizacji w zadaniach. WNT, Warsaw

Brüninghoff H (1993) The Essing timber bridge, Germany. Struct Eng Int 2:70–72

Clada W (1965) Über die Fugenelastizität ausgehorteter Leimfugen bei Holzverleimungen. Holz als Roh u. Werkstoff 2:58–67

Dyląg Z, Krzemińska-Niemiec E, Filip F (1986) Mechanika budowli. PWN, Warsaw

Dzięgielewski S, Smardzewski J (1995) Meblarstwo. Projekt i konstrukcja. Państwowe Wydawnictwo Rolnicze i Leśne, Poznań

Dzięgielewski S, Wilczyński A (1990) Sprężystość elementów meblowych sklejanych warstwowo. Zeszyty Problemowe Postępów Nauk Rolniczych 379:89–107

Dziuba T (1974) Obliczanie odkształceń w ustrojach prętowych z węzłami podatnymi. In: II Sesja Młodych Pracowników Nauki i Techniki Drzewnictwa, Poznań, pp 193–198

Dziuba T (1990) Optimierung der Konstruktion von Stuhlseitenteilen. Holztechnologie 30(6): 303–306

Eckelman CA (1970) Codoff, computer design of furniture. User's manual. Research Bulletin, Wood Research Laboratory, No. 357

Findeisen W, Szymanowski J, Wierzbicki A (1980) Teoria i metody optymalizacji. PWN, Warsaw

Francis EC, Gutierrez-Lemini D (1984) Effect of scrim cloth on adhesively-bonded joints. Adhesive joints. Formation, characteristics and testing. Plenum Press, New York. pp 679–685

Godzimirski J (1985) Określenie naprężeń w spoinach klejowych metodą elementów skończonych. Biuletyn WAT 11:77–82

Godzimirski J (2002) Wytrzymałość doraźna konstrukcyjnych połączeń klejowych. Wydawnictwo Naukowo-Techniczne, Warsaw

Goliński J (1974) Metody optymalizacyjne w projektowaniu technicznym. Wydawnictwo Naukowo-Techniczne, Warsaw

Groth HL, Nordlund P (1991) Shape optimization of bonded joints. Int J Adhes Adhes 11(4):204–212

Haberzak A (1975) Analiza rozkładu naprężeń w spoinie klejowej w połączeniu na czopy elementów drewnianych. Przemysł Drzewny 10:11–12

Hearmon RFS (1948) Elasticity of wood and plywood. Forest Products Researsch, Department of Scientific and Industrial Research, London, Special Raport No. 7

Hill MD, Eckelman CA (1973) Flexibility and bending strength of mortise and tenon joints. In: Reprint from the Jan. & Feb. 1973 Issues of Furniture Design and Manufacturing

Ieandrau JP (1991) Analysis and design data for adhesively bonded joints. Int J Adhes Adhes 11(2):71–79

Janowiak JJ (1993) Finger joint strength evamation of three Northeastern hardwoods. Forest Prod J 43(9):23–28

Kline RA (1984) Stress analysis of adhesively bonded joints. Adhesive joints. Formation, characteristics and testing, vol 1. Plenum Press, New York, pp 587–610

Korolew W (1973) Osnovy racionalnovo proiektirovania miebeli. Lesnaja Promyszliennost, Moscow

Kuczmaszewski J (1995) Podstawy konstrukcyjne i technologiczne oceny wytrzymałości adhezyjnych połączeń metali. Wydawnictwo Uczelniane, Politechnika Lubelska, Lublin

Lekhnickij SG (1977) Teorija upugosti anizotropnovo tela. Izdatielstwo Nauka, Glavnaja Redakcja Fizyko-Matiematicheskojj Literatury, Moscow

Leśniak Z (1970) Metody optymalizacji konstrukcji przy zastosowaniu maszyn matematycznych. Arkady, Warsaw

Leontiev NL (1952) Uprugie deformaci drewiesiny. Gostekhnizdat, Moscow

Lindemann ZR, Zimmerman J (1996) Wyznaczenie wytrzymałości warstwy kleju. CADCAM. Forum 1:69–70

Łączkowski R (1988) Katastroficzny model nieliniowej mechaniki pękania. Przegl. Mech. 12:8–15

Matsui K (1990a) Effects of size on nominal ultimate tensile stresses of adhesive bonded circular or rectangular joints under bending or peeling load. Int J Adhes Adhes 2:90–98

Matsui K (1990b) Size effects on average ultimate shear stresses of adhesive bonded rectangular or tubular lap joints under tension-shear. Int J Adhes Adhes 2:81–98

Matsui K (1991) Size effects on nominal ultimate shear stresses of adhesive bonded circular or rectangular joints under torsion. Int J Adhes Adhes 2:59–64

Misztal F (1956) Rozkład naprężeń ścinających w spoinach połączeń klejowych. Archiwum Budowy Maszyn 3(1):35–63

Mitinskij AN (1948) Uprugie postojannye drewiesiny kak ortotropnovo materiała. Leningrad, Trudy Lesotiekhniczieskojj Akademii, p 63

Nakai T, Takemura T (1995) Torsional properties of tenon joints with ellipsoid like tenons and mortises. J Jap Wood Res Soc 4:387–392

Nakai T, Takemura T (1996a) Stress analysis of the through-tenon joint of wood under torsion I. Measurements of shear stresses in the male by using rosette gauges. J Jap Wood Res Soc 4:354–360

Nakai T, Takemura T (1996b) Stress analysis of the through-tenon joint of wood under torsion II. Shear stress analysis of the male using the finite element method. J Jap Wood Res Soc 4:361–368

Niskanen E (1955) On the distribution of shear stress in a glued specimen of isotropic or anisotropic material. State Inst Technol Res (Finland, Publ. 30, Helsinki)

Niskanen E (1957) On the distribution of shear stress in a glued single shear test specimen of Finnish birch timber. State Inst Technol Res (Finland, Publ. 36, Helsinki)

Norris CB (1942) Technique of plywood. Laucks, USA

Nowacki W (1970) Teoria sprężystości. PWN, Warsaw

Nowacki W (1976) Mechanika budowli. PWN, Warsaw

Ostwald H (1987) Optymalizacja konstrukcji. Wydawnictwo Politechniki Poznańskiej, Poznań

Pellicane PJ (1994) Finite element analysis of finger-joints in lumber with dissimilar laminate stiffnesses. Forest Prod J 3:17–22

Pellicane PJ, Gutkowski RM, Jaustin C (1994) Effect of glueline voids on the tensile strength of finger-jointed wood. Forest Prod J 6:61–64

PN-B-03150:2000 Wood constructions—static calculations and designing

PN-EN 300:2000 Oriented Strand Boards (OSB)—definition, classification and specifications

PN-EN 312-4:2000 Chipboards—technical requirements—requirements for boards bearing operational loads in dry conditions

PN-EN 312-5:2000 Chipboards—technical requirements—requirements for boards bearing operational loads in humid conditions

PN-EN 312-6:2000 Chipboards—technical requirements—requirements for heavy duty boards bearing operational loads in humid conditions

PN-EN 527-3:2004 Office furniture. Work tables and desks. Methods of test for the determination of the stability and the mechanical strength of the structure

PN-EN 622-2:2000 Fibreboards technical requirements—requirements for hardboards

PN-EN 622-3:2000 Fibreboards technical requirements—requirements for medium boards

PN-EN 1335-2:2009 Office furniture. Office chair for work. Part 2: Safety requirements

PN-EN 1335-3:2009 Office furniture. Office chair for work. Part 3: Methods of testing safety
PN-EN 1728:2012 Furniture. Chairs and stools. Determining strength and durability
PN-EN 1729-2:2006 Furniture. Chairs and tables for educational institutions. Part 2: Safety requirements and methods of testing
PN-EN 1730:2002 Furniture. Tables. Determining stability
Pogorzelski W (1978) Optymalizacja układów technicznych w przykładach. WNT, Warsaw
Proszyk S, Krystofiak K, Wnnik A (1997) Studies on adhesion to the wood of two Component PVAC adhesives hardened with aluminum chloride. Folia For. Pol. PWN, B. 28:99–106
Rabinowicz AN (1946) Ob uprugikh postojannych i procznosti aviacionnykh materialov. Trudy CAGI 582:1–56
Rónai F (1969) Untersuchungen zur Festigkeit von Fensterflugel-Eckeferbindungen. Holz als Roh u. Werkstoff 3:103–110
Smardzewski J (1989) Optymalizacja konstrukcji skrzydeł okiennych. Przemysł Drzewny 4:21–24
Smardzewski J (1990) Numeryczna analiza konstrukcji mebli metodą elementów skończonych. Przemysł Drzewny 7:1–5
Smardzewski J (1992) Numeryczna optymalizacja konstrukcji krzeseł. Przemysł Drzewny 1:1–6
Smardzewski J (1994) Model matematyczny i analiza numeryczna rozkładu naprężeń stycznych w prostokątnych spoinach połączeń poddanych skręcaniu. In: Badania dla Meblarstwa, Wydawnictwo Akademii Rolniczej w Poznaniu, pp 31–45
Smardzewski J (1995) Rozkład naprężeń stycznych w prostokątnych spoinach połączeń kątowych. In: Badania dla Meblarstwa, Wydawnictwo Akademii Rolniczej w Poznaniu, pp 45–55
Smardzewski J (1996) Distribution of stresses in finger joints. Wood Sci Technol 6:477–489
Smardzewski J (1998a) Numerical analysis of furniture constructions. Wood Sci Technol 32:273–286
Smardzewski J (1998b) Wpływ niejednorodności drewna i spoiny klejowej na rozkład naprężeń stycznych w połączeniach meblowych. Rocznik Akademii Rolniczej w Poznaniu, p 282
Smardzewski J, Dzięgielewski S (1994) Rozkład naprężeń stycznych w spoinie połączenia czopowego. In: Materiały Sesji Naukowej Badania dla meblarstwa, Wydawnictwo Akademii Rolniczej w Poznaniu, pp 62–73
Timoshenko S, Goodier JN (1951) Theory of elasticity. McGraw-Hill Book Company Inc., New York
Tomusiak A (1988) Analyse der Spannungen in Klebfugen bigebelasteter Keilzinkenverbindungen. Holztechnologie 1(29):25–26
Wnuk MP (1981) Podstawy mechaniki pękania. AGH, Kraków
Wilczyński A (1988) Badania naprężeń ścinających w spoinie klejowej w drewnie. Wydawnictwo Uczelniane WSP, Bydgoszcz
Zielnica J (1996) Wytrzymałość materiałów. Wydawnictwo Politechniki Poznańskiej, Poznań

# Chapter 7
# Stiffness and Strength Analysis of Case Furniture

## 7.1 Properties of Case Furniture

Unlike craft furniture, industrially manufactured furniture should be created according to a specific, repetitive technology, in the conditions of mechanised serial production. Also a number of important structural, technical, organisational and commercial requirements have to be fulfilled, unknown to crafts factories and furniture workshops. These requirements have become an essential stimulus to the overall mechanisation and technification of furniture production, which has also resulted in the need to prepare their design documentation. Engineering design methods, without which it is hard to imagine building, aviation and machine construction, have never been systematically or on a large scale introduced into furniture production. The implementation of engineering methods of construction requires the gathering of detailed information concerning:

- the function of the piece of furniture and maximum operational loads resulting from it, or also in extreme cases, unusual loads, especially when it concerns children's and teenager's furniture;
- elastic properties of materials used, taking into account the orthotropic characteristics of wood and wood materials, together with indicating maximum stresses; and
- elastic and durability properties of normalised glued and separable furniture joints and connectors, hinges and accessories carrying operational loads.

The structural development, ensuring durable and safe use, is one of the three components of the design process. Next to strength requirements, aesthetic and functional requirements are a part of them. Functional design is of the greatest practical importance, as it ensures the most effective possible fulfilment of the assumed operational functions. The purpose of aesthetic activities is shaping the

© Springer International Publishing Switzerland 2015
J. Smardzewski, *Furniture Design*,
DOI 10.1007/978-3-319-19533-9_7

proportions and spatial forms of the piece of furniture and the choice of surface colour, texture and drawing to the satisfaction of the most demands of the user. This part of the project in many cases clearly dominates the whole design works, shifting functional and durability features further away. In this situation, the functional requirements should have strong preference, sometimes at the expense of aesthetic or durability values.

In a properly planned furniture production process, the assessment of the stiffness of individual products should be begun already at the constructional design stage. This helps eliminate any furniture construction errors by setting the correct parameters for individual components, subassemblies and assemblies in accordance with the prescribed criteria of stiffness and durability. Formulating the engineering process of designing furniture in such a way will enable to limit destructive tests of finished products and shorten the cycle of implementing the piece of furniture to production, limit the number of complaints in continuous production and save a significant amount of time and material. Stiffness and strength of furniture can be assessed with the use of:

- theoretical analysis and the guideline details resulting from it for constructors and technologists of furniture and
- damage tests of the prototype or finished product, as well as checking calculations in order to justify the constructional errors found.

The first works concerning the stiffness of case furniture appeared in the year 1957–58 (Kotaś 1957, 1958a, b, c). It demonstrates the deformations of furniture bodies and analyses examples of increasing stiffness of the entire construction and strength of joints. Static and strength problems were later extended in national and foreign literature (Chia-Lin and Eckelman 1994; Dzięgielewski and Smardzewski 1989, 1990, 1992; Eckelman and Resheidat 1984; Eckelman and Rabiej 1985; Eckelman 1967; Ganowicz and Kwiatkowski 1978; Ganowicz et al. 1978; Joscak 1986; Joscak and Vacek 1989; Kuhne and Kroppelin 1978; Korolew 1970, 1973; Lapszyn 1968; Smardzewski and Dzięgielewski 1994), focusing both on the global stiffness of the entire construction, as well as stiffness and strength of elements and joints.

Within the framework of the stiffness analysis, deformations of the furniture body and bends of bars, shelves, partitions, bottoms of drawers and containers, as well as deformations of wall joints are examined. While strength calculations primarily concern the strength of wall angular joints and suspending cupboards and bottoms, attempts to assess the stability of designed furniture constitute a separate group of analytical studies. The stability of furniture is the most essential part of quality studies on a finished product, and its loss may be a direct threat to the life or health of the user.

## 7.2   Operational Loads on Furniture During Their Usage

Studying furniture in validation stations is to simulate the forces causing damage to the furniture piece or its components, the loss of balance or deterioration of durability and safety, which may occur during normal use and during improper use, which can be expected with great probability. Before commencing any type of study, the furniture piece must be appropriately seasoned, ensuring it with complete strength.

By studying the components of furniture for storage, the capacity of extended elements is assumed as the product of the surfaces of the bottom of the extended element and the height of the span. All parts intended for storage should be loaded equally, in accordance with the values provided in Table 7.1, unless the design of the manufacturer says otherwise.

Drawers and other moving elements should be extended up to the blocking stops. If the element is not equipped with opening stops and is intended to be removable, then it is opened until the point in which one-third of the internal length remains inside the furniture piece. Then, a vertical force of 250 N must be applied, equal to the total mass of the extended element, in accordance with the scheme provided in Fig. 7.1.

**Table 7.1** Load of part for storage

| Part | Load | |
|---|---|---|
| Shelves | $kg/dm^2$ | 1.5 |
| Clothes bars | kg/dm | 5.0 |
| Extended elements | $kg/dm^3$ | 0.5 |
| Hooks for files | $kg/dm^a$ | 4.0 |

[a]Measured perpendicular to the file hooks

**Fig. 7.1** Study of durability of extended elements

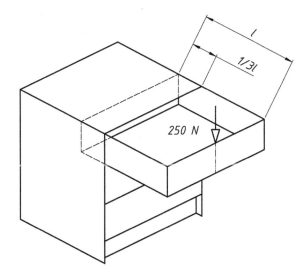

Door with a vertical rotational axis should be loaded by a force of 300 N placed at a distance of 100 mm from their outer edge. Then, the door needs to be moved, performing 10 full cycles from the position of 45° to the position of 10°, measured from the position of a complete opening, but to a maximum of 135° (Fig. 7.2a). Loading the door with horizontal force takes place using a load with a value of 80 N applied perpendicular to the plane of the door on its horizontal central line, 100 mm from the outer edge, in the direction of opening (Fig. 7.2b).

The durability of a hinged door and revolving door is tested by loading the edges with vertical force with a value of 20 N on the central line (Fig. 7.3). The quality of

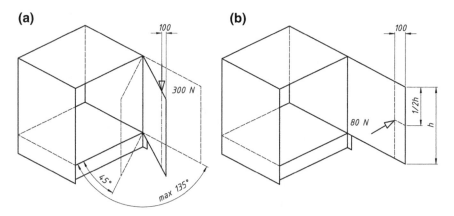
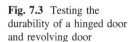

**Fig. 7.2** Loading the door with static force: **a** vertical, **b** horizontal (mm)

**Fig. 7.3** Testing the durability of a hinged door and revolving door

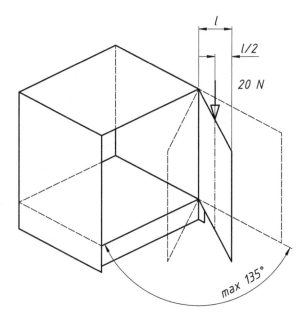

suspension is assessed by examining the appearance and functioning of an unloaded door after the applied force subsides.

Testing the durability of sliding doors and horizontal louvered fronts consists in performing 40,000 cycles of opening and closing the door, and in the case of louvered doors, 20,000 cycles (Fig. 7.4). Before and after the test, the appearance and functioning of sliding doors or louvered fronts should be checked, and if necessary, also the force of opening and closing.

Sliding doors are also tested in terms of resistance to dynamic closing and opening, by suspending on a block a force with a value of 40 N, along with mass $W$ (Fig. 7.5).

A component element of a furniture piece that causes many problems to constructors is the flap of a bar, buffet or dresser. Therefore, in validation stations of furniture, there are tests that are to assess the durability of this element. For this purpose, at flaps that are positioned in full opening/closing, a static force with a value of 250 N must be applied (Fig. 7.6). Before commencing tests and after their completion, the appearance and functioning of the flaps, hinges, etc., need to be checked and then adjusted again, paying attention to whether the friction strut will allow the flap to open only under its own weight.

Within the framework of general safety requirements of office furniture, it is recommended that the edges and corners were without burrs and rounded or chamfered, not to leave open ends of tubes, all the moving parts, available during normal use, should be—during movement in any position—at a safe distance

**Fig. 7.4** Testing the durability of sliding doors and horizontal louvered fronts (mm)

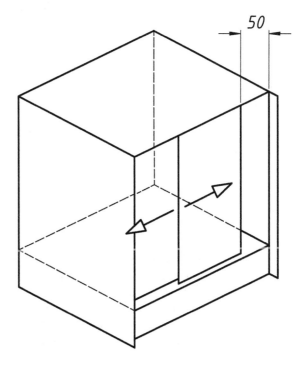

**Fig. 7.5** Dynamic
closing/opening of sliding
doors and horizontal louvered
fronts (mm)

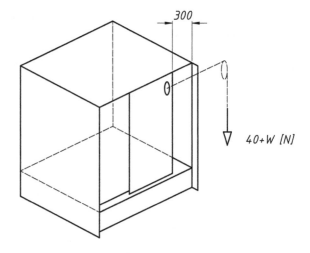

**Fig. 7.6** Scheme of testing
the strength of flaps (mm)

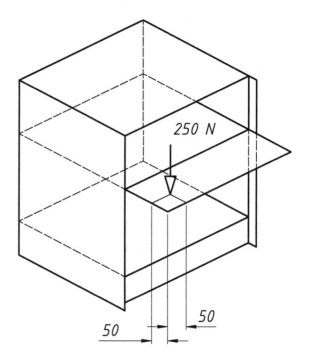

<8 mm or >25 mm. This applies to any two elements moving relative to each other, with the exception of doors (including the hinges), flaps (including the hinges) and extended elements (including the runners). Safe distances should be maintained also between handles and other parts. Adjustable parts need to be constructed in such a way, as to prevent unintended operation or release. No louvered doors sliding vertically should close on their own from a height of more than 200 mm from the

closed position, as this may cause injury. Extended elements should have effective opening stops, which endure pulling out the case with a horizontal force of at least 200 N, exerted on the handle of the loaded pulled out element.

Safety tests provided in Table 7.2 must be carried out in accordance with PN-EN 14073-3:2006 and PN-EN 14074:2006. However, it is accepted that safety tests according to EN 14073-3 and EN 14074 constitute a part of a series of tests, when all essential tests are carried out according to EN 14073-3 and EN 14074.

For furniture hanging on a partition or wall, the order of their tests has been listed in Table 7.3.

Movable furniture should be tested in accordance with PN-EN 14074:2006. If the manufacturer did not establish otherwise, all parts of the furniture intended for storage, according to Table 7.1, must be subjected to load. Close the extended elements, flaps, louvered fronts and doors and apply a static horizontal force of 350 N at point A (Fig. 7.7), located on the vertical central line of the furniture's construction, 50 mm below the highest point of this line, but not higher than at a distance of 1600 mm from the floor. If the furniture piece has a tendency to fall in one direction, then the point of applying force is lowered to a height at which tilting only in that direction is prevented.

An important test, from the operational point of view, is the assessment of strengths of shelf supports. If the shelves and their supports are identical, it is enough to examine only one shelf. If the shelves and their supports are not identical, then each combination thereof needs to be examined. During the test, the weight should be evenly distributed on the shelf, outside the section—with a length of about 220 mm, counting from one support—on which a blow bar should be knocked over 10 times at a point that lies as close as possible to this support (Fig. 7.8).

**Table 7.2** The order of tests of furniture standing on the floor—free-standing or mounted to the building

| Test no. | Test | Reference |
|---|---|---|
| 1 | Removing shelves | PN-EN 14073-3:2006 |
| 2 | Strength of shelf supports | PN-EN 14073-3:2006 |
| 3 | Strength of the top surface | PN-EN 14073-3:2006 |
| 4 | Strength of extended elements | PN-EN 14074:2006 |
| 5 | Dynamic opening of extended elements | PN-EN 14074:2006 |
| 6 | Examination of stops blocking the opening | PN-EN 14074:2006 |
| 7 | Vertical load on revolving doors | PN-EN 14074:2006 |
| 8 | Dynamic closing/opening of sliding doors and horizontal louvered fronts | PN-EN 14074:2006 |
| 9 | Strength of flaps | PN-EN 14074:2006 |
| 10 | Furniture standing on the floor mounted to the wall of the building | PN-EN 14073-3:2006 |
| 11 | Stability[a] | PN-EN 14073-3:2006 |

[a]In the case of furniture which might not meet stability requirements, appropriate stability tests can be performed before commencing a sequence of tests listed in the table

**Table 7.3** The order of tests of furniture hanging on a partition and wall

| Test no. | Test | Reference |
|---|---|---|
| 1 | Detachment of furniture mounted on a partition or wall | PN-EN 14073-3:2006 |
| 2 | Removing shelves | PN-EN 14073-3:2006 |
| 3 | Strength of shelf supports | PN-EN 14073-3:2006 |
| 4 | Strength of the top surface | PN-EN 14073-3:2006 |
| 5 | Strength of extended elements | PN-EN 14074:2006 |
| 6 | Dynamic opening of extended elements | PN-EN 14074:2006 |
| 7 | Examining the opening stop | PN-EN 14074:2006 |
| 8 | Vertical load on revolving doors | PN-EN 14074:2006 |
| 9 | Dynamic closing/opening of sliding doors and horizontal louvered fronts | PN-EN 14074:2006 |
| 10 | Strength of flaps | PN-EN 14074:2006 |
| 11 | Strength of partition and mounting elements | PN-EN 14073-3:2006 |

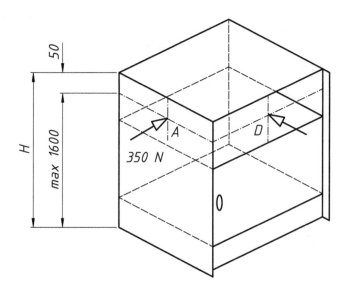

**Fig. 7.7** Scheme of testing the strength of a furniture piece (mm)

## 7.2.1 Methods of Determining Computational Loads

Constructions of case furniture are mainly made from board elements or frame-board elements. Due to the high shearing stiffness of boards in comparison with their torsional stiffness, case furniture subjected to torsion forces has a regular, characteristic way of deforming, which can be caused by various force systems (Fig. 7.9).

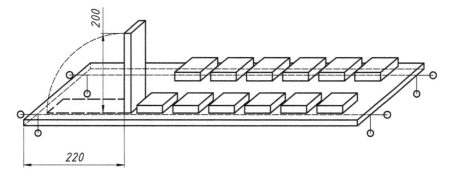

**Fig. 7.8** Strength of the shelf supports (mm)

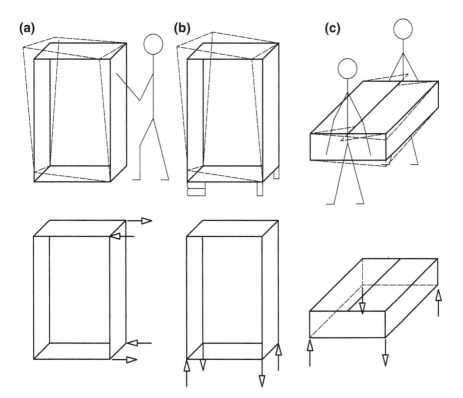

**Fig. 7.9** Equivalent systems of forces causing torsional deformation of the body: **a** shifting, **b** unevenness of the base, **c** moving

With regard to operational requirements, the quality of the furniture is determined by a stiffness and strength test both of the whole structure and on its individual elements and joints, as well as impact on mass and operational loads (Fig. 7.10).

**Fig. 7.10** Subjecting the
body of the furniture piece to
mass and operational loads

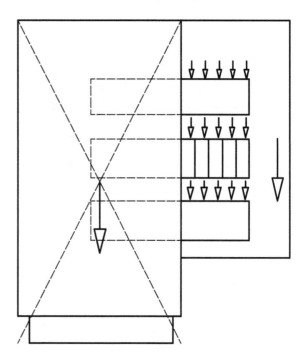

   The stiffness of case furniture can be defined as the quotient of the values of
external load $P_z$ applied to the body's side wall, at a height of the top (Fig. 7.11) to
the value of the displacement $\Delta P_z$ measured in the direction of the effect of this
load:

$$k = \frac{P_z}{\Delta P_z},\qquad(7.1)$$

where
$k$      stiffness of the furniture body,
$P_z$     external load and
$\Delta P_z$ displacement in the direction of the load $P_z$.

   The experiments of the testing and validation station of furniture show that the
value of this coefficient should not be smaller than 10,000 N/m. In engineering
practice, however, it is worth taking into account two values:

- $k \geq 10{,}000$ N/m for house furniture and
- $k \geq 20{,}000$ N/m for library bookcases, kitchen furniture, office furniture and
  other heavily loaded structures.

**Fig. 7.11** Scheme of support
and load of the furniture body
during tests of global stiffness

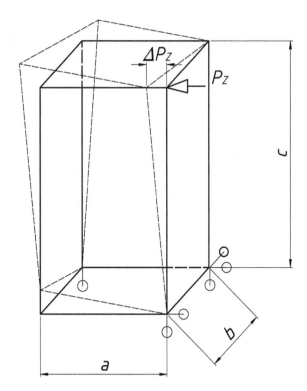

**Fig. 7.11** Scheme of support
and load of the furniture body
during tests of global stiffness

The support and load scheme of the body contained in Fig. 7.11 differ, however, from laboratory tests, in that the external load $P_z$ is not applied 50 mm below the top or at the height of 1600 mm, but exactly at the height of the top, in the plane of the front case. Therefore, as an external operational load $P_z$ it should be assumed:

$$P_z = \frac{C - 0.05}{C} P_U, \qquad (7.2)$$

for furniture of height $C \leq 1.65$ m,

$$P_z = \frac{1.6}{C} P_U, \qquad (7.3)$$

for furniture of height $C > 1.65$ m,

where
$C$   furniture height and
$P_U$   operational load.

The operational load $P_U$ constitutes a reduced sum of operational loads and weight of the furniture piece, expressed in the form of the equation:

$$P_U = \frac{a}{2H}(Q_m + Q_u),\tag{7.4}$$

where
$a$    width of the furniture body,
$H$    height of applying force, where $H = 1.6$ m for $C \geq 1.65$ m, $H = C - 0.05$ m for $C < 1.65$ m.

And the operational load $Q_u$ represents the sum of fixed evenly distributed loads, acting on the board of the bottom, partition, shelf, drawer and bars of the furniture piece, calculated according to Table 7.1, depending on the surface of the board element, the length of the bar or the volume of the container, from the equation:

$$Q_u = \sum_{i=1}^{n}(q_{Vi}V_i + q_{Ai}A_i + q_iL_i),\tag{7.5}$$

where
$q_{Vi}$    volume load,
$q_{Ai}$    surface load,
$q_i$    linear load,
$V_i$    container or drawer capacity,
$A_i$    surface of bottom, shelves, partitions and
$L_i$    length of bars.

Mass load $Q_m$ is mostly determined by weighing the furniture piece. In the absence of relevant scales, the value of this load can be determined as follows:

$$Q_m = 1.05 \sum_{i=1}^{n} \rho_i V_i g,\tag{7.6}$$

where
$g$    gravity acceleration,
$\rho_i$    density of material and
$V_i$    volume of element,

by summing up the weight of all structural elements and adding the weight of fittings constituting around 5 % of $Q_m$.

## 7.2.2 Stiffness of Case Furniture

### 7.2.2.1 Stiffness of Shelves and Horizontal Elements

Engineering practice requires the selection of the appropriate thickness for boards intended for shelves or partitions of case furniture. The general differential equation of the bending of an elastic thin isotropic board in a rectangular system of coordinates $X$, $Y$ and $Z$ (Fig. 7.12a) has a known form of Legrange's equation:

$$\frac{\partial^4 w}{\partial x^4} + \frac{\partial^4 w}{\partial x^2 \partial y^2} + \frac{\partial^4 w}{\partial y^4} = -\frac{q(x, y)}{D},\qquad (7.7)$$

where

| | |
|---|---|
| $q(x, y)$ | operating surface load, |
| $D = \frac{Ed^3}{12(1+v^2)}$ | bending stiffness of the board, |
| $w$ | function of bending the board, |
| $v$ | Poisson's ratio, |
| $E$ | linear elasticity module and |
| $d$ | thickness of the board. |

Expenditures of bending moments $M_x$, $M_y$ and torsion moments $M_{xy}$, $M_{yx}$ and expenditures of transverse forces $F_x$, $F_y$, acting on the sides of the elementary section of the board $dA = dxdy$ (Fig. 7.12b), are determined by the following equations:

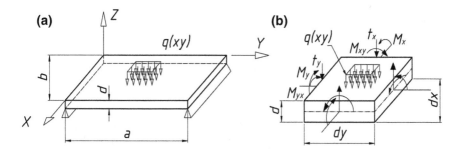

**Fig. 7.12** Analysis of operation of boards in rectangular coordinates: **a** external load of the board, **b** internal forces

$$\left\{ \begin{array}{l} M_x = D\left(\dfrac{\partial^2 w}{\partial x^2} + v\dfrac{\partial^2 w}{\partial y^2}\right) \\[2ex] M_y = D\left(\dfrac{\partial^2 w}{\partial y^2} + v\dfrac{\partial^2 w}{\partial x^2}\right) \\[2ex] M_{xy} = -M_{yx} = D(1-v)\dfrac{\partial^2 w}{\partial y \partial x} \\[2ex] F_x = -D\dfrac{\partial}{\partial x}\left(\dfrac{\partial^2 w}{\partial x^2} + \dfrac{\partial^2 w}{\partial y^2}\right) \\[2ex] F_y = D\dfrac{\partial}{\partial y}\left(\dfrac{\partial^2 w}{\partial x^2} + \dfrac{\partial^2 w}{\partial y^2}\right). \end{array} \right. \tag{7.8}$$

The above-mentioned equation is solved, taking into account the border conditions of support of rectangular boards: of a fixed edge, freely supported and free edge. If we place these conditions at the edge of the board $y = 0$, then they have the form:

for a fixed edge (horizontal partitioning)

$$(w)_{y=0} = 0, \quad \left(\frac{\partial w}{\partial y}\right)_{y=0} = 0, \tag{7.9}$$

which means that the deflection and angle of deflection on this edge are equal to zero, for free support (shelf)

$$(w)_{y=0} = 0, \quad (M_y)_{y=0} = 0, \quad \left(\frac{\partial^2 w}{\partial y^2}\right)_{y=0} = 0, \tag{7.10}$$

for free edge

$$(M_y)_{y=0} = 0, \quad (M_{xy})_{y=0} = 0, \quad (F_y)_{y=0} = 0. \tag{7.11}$$

In the last two cases, we assume that the edges are not loaded by bending moments and transverse forces. The solution to the above equation is obtained by entering a rapidly convergent triple trigonometric series for the bending function. The exception here is the cylindrical bending of shelves and partitions supported or fixed on the two opposite sides (Fig. 7.13a, b). This occurs when a board of the dimensions $b < a$, supported freely along the edges with a length of $x = 0$ and $x = a$, is loaded in any way in the direction of $x$ and in a constant way in the direction of $y$. In such case,

$$\left(\frac{\partial w}{\partial y}\right) = 0, \quad \left(\frac{\partial^2 w}{\partial y^2}\right) = 0, \tag{7.12}$$

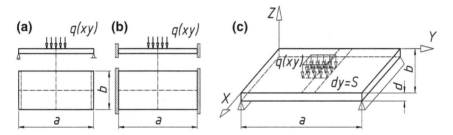

**Fig. 7.13** Bending rectangular boards: **a** bending freely supported shelves, **b** bending fixed supported partitions, **c** cylindrical bending of a board

and according to these equations,

$$
\begin{cases}
M_x = D\dfrac{\partial^2 w}{\partial x^2}, \\[2mm]
M_y = Dv\dfrac{\partial^2 w}{\partial x^2}, \\[2mm]
M_{xy} = 0, \\[2mm]
F_x = -D\dfrac{\partial^3 w}{\partial x^3}, \\[2mm]
F_y = 0.
\end{cases}
\tag{7.13}
$$

Therefore, eventually, we obtain a differential equation of bending an elastic, thin isotropic board in the form as follows:

$$
\frac{\partial^4 w}{\partial w^4} = -\frac{q(x,y)}{D}.
\tag{7.14}
$$

Therefore, cutting an element from a board with a width dy = S and calculating the course of the bending moment using known methods:

$$
M_x = \frac{\partial^2 w}{\partial w^2}D,
\tag{7.15}
$$

we obtain all the necessary information in order to determine the deflection and the $M_y$ moment in a simple way. For support and load schemes from Fig. 7.13a, b we write an equation of moments for the left half of the beam as a function of length and insert it to the differential equation of the deflection line. The bending value is calculated by integrating the equation:

$$
w = \frac{1}{D}\iint M_x dx,
\tag{7.16}
$$

and determining constant integrations with the appropriate border conditions.

Solutions for the discussed boards have the form:

maximum deflection of a freely supported shelf (Fig. 7.14a)

$$w = \frac{5}{384} \frac{q_A a^4}{D} \leq w_p, \quad w_p = 0.004a, \tag{7.17}$$

maximum deflection of a partition fixed on both sides (Fig. 7.14b)

$$w = \frac{1}{384} \frac{q_A a^4}{D} \leq w_p, \quad w_p = 0.002a, \tag{7.18}$$

where
$q_A$  surface load of the board element according to Table 7.1,
$a$   length of the free side of the board,
$w_p$  acceptable deflection in the middle of the shelf's span, which should not
    exceed 4 mm/m in length, acceptable deflection in the middle of the partition's
    span, which should not exceed 2 mm/m in length.

In many structural solutions of furniture, the boards can additionally be supported on a third edge from the side of the rear wall (Fig. 7.15). Strict solutions of these cases come down to the integration of the general equation of a board subject to bending and torsion. A similar solution can also be obtained on the basis of the elementary strength of materials.

A shelf that is additionally supported can be regarded as a rod subject to bending or torsion. External load can then be broken down into two states (Eckelman and Resheidat 1984):

- balanced load of a freely supported shelf (Fig. 7.16a),
- concentrated or continuous load (Fig. 7.16b, c) in the location of support.

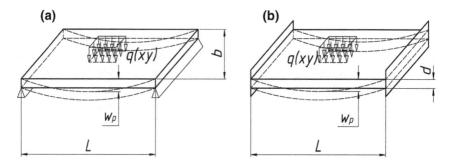

**Fig. 7.14** Acceptable deflection of board furniture elements: **a** shelves, **b** partitions

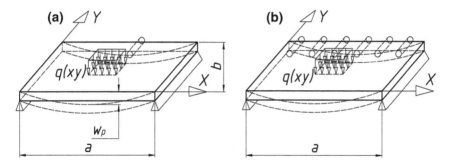

**Fig. 7.15** The deflections of shelves supported along three edges: **a** by point from the side of the rear wall, **b** discreetly from the side of the rear wall

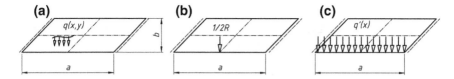

**Fig. 7.16** Load states of the shelf: **a** board bending, **b**, **c** board tension

Deflection in the first load state is the same as for cylindrical bending $w_p = (5/384) \cdot (q_A a^4/D)$. The load of the second state can be broken down into two components, symmetrical (Fig. 7.17a) and asymmetrical (Fig. 7.17b). Symmetrical load gives cylindrical bending of a maximum ordinate:

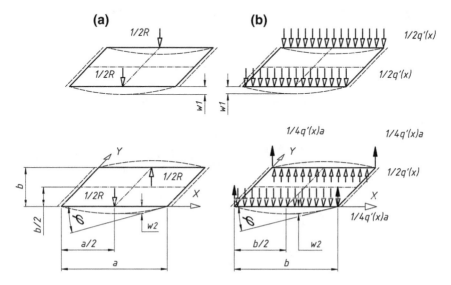

**Fig. 7.17** The state of edge load of boards: **a** symmetrical, **b** asymmetrical

in the case of concentrated load $R$

$$w_1 = \frac{1}{48}\frac{Ra^3}{EJ}. \tag{7.19}$$

in the case of distributed load along the free edge

$$w_1 = \frac{5}{384}\frac{q'_{(x)}a^4}{EJ}. \tag{7.20}$$

Asymmetrical load causes the torsion moment of the board:

$$M = \frac{1}{2}Rb \quad \text{or} \quad M = \frac{1}{2}q'_{(x)}ab. \tag{7.21}$$

This moment is applied in the middle of the shelf's span. Taking into account the dependence of calculating the torsion angle $\varphi$ of the board:

$$\varphi = \frac{M_{(x)}x}{GJ} = 24(1+v)\frac{M_{(x)}x}{Ebd^3}, \tag{7.22}$$

where

$M_{(x)}$            rod torsion moment,
$x$              torsion length,
$G = 2E(1+v)$    shear modulus,
$J_s = bd^3/12$     moment of inertia of rectangular cross section of the board.

Deflection of the board at mid-length $x = a/2$ loaded by the moment $M_{(0.5b)} = Rb/4$ amounts to

$$w_2 = x\phi = \frac{3}{8}\frac{Rab}{Ed^3} \tag{7.23}$$

whereas for a board evenly loaded along its edge, this deflection is expressed by the formula:

$$w_z = \varphi\frac{b}{2} = \frac{1}{GJ}\left(\frac{1}{4}q'_{(x)}ab - \int\limits_0^x \frac{1}{2}q'_{(x)}bdx\right)x, \tag{7.24}$$

where

$$\frac{1}{4}q'_{(x)}ab - \int\limits_0^x \frac{1}{2}q'_{(x)}bdx = M_{(x)}, \tag{7.25}$$

is the moment causing the torsion of the board. In solving this equation, we obtain the value of deflection of the board caused by torsion in the form of the equation:

$$w_z = \frac{3}{16}\frac{q'_{(x)}a^2b}{Ed^3} = \frac{6}{384}\frac{q'_{(x)}a^4b^2}{EJa^2}. \tag{7.26}$$

Because bending along a supported edge is equal to zero, the value of reaction $R$ and $q'_{(x)}$ can be calculated from the equations:

for a shelf supported at one point

$$\frac{5}{384}\frac{q_Aa}{D} + \frac{1}{48}\frac{Ra^3}{EJ} + \frac{3}{8}\frac{Rab}{Ed^3} = 0, \tag{7.27}$$

which gives

$$R = \frac{5}{8}q_Aab\left(\frac{2a^2}{2a^2+3b^2}\right), \tag{7.28}$$

for shelves supported continuously from the side of the rear wall

$$\frac{5}{384}\frac{q_Aa}{D} + \frac{5}{384}\frac{q'_{(x)}a^4}{EJ} + \frac{6}{384}\frac{q'_{(x)}a^4b^2}{EJa^2} = 0, \tag{7.29}$$

hence,

$$q'_{(x)} = q_Ab\left(\frac{5a^2}{5a^2+6b^2}\right). \tag{7.30}$$

The maximum deflection can be determined from the appropriate sum of component deformations:

$$w = w_o + w_1 - w_2, \tag{7.31}$$

thus for a shelf supported locally:

$$w = \frac{5}{384}\frac{q_Aa^4}{D}\left(1 + \frac{3b^2-2a^2}{2a^2+3b^2}\right) \le f_{dop}a, \tag{7.32}$$

and for a shelf supported on the entire length of the rear edge:

$$w = \frac{5}{384}\frac{q_Aa^4}{D}\left(1 - \frac{5a^2-6b^2}{5a^2+6b^2}\right) \le f_{dop}a. \tag{7.33}$$

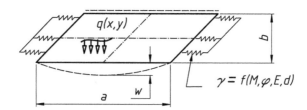

**Fig. 7.18** A board mounted to the walls of the body using semi-rigid joints of known stiffness

A separate example of a furniture board is a partition supported elastically from the side of the rear wall and attached to the side walls using joints of the known stiffness $\gamma$ (Fig. 7.18). According to Eckelman (1967), the maximum deflection of free edges is determined by the equation:

$$w = \frac{5}{384} \frac{q_A a^4}{D} \left(1 - \frac{5a^2\gamma - 6b^2}{5a^2\gamma + 6b^2}\right) \le f_{\text{dop}} a, \qquad (7.34)$$

where

$$\gamma = \frac{10 + \delta}{10 + 5\delta}, \qquad (7.35)$$

$$\delta = \frac{12\gamma}{Ed^3}, \qquad (7.36)$$

$$\gamma = \frac{M_{(x)}}{\varphi}. \qquad (7.37)$$

Bending stiffness $D$, present in the formulas, is a certain substitute stiffness, which value should be determined experimentally in the test of static bending. This size, for multi-layer boards (Fig. 7.26), can also be calculated analytically on the basis of tabular data and the equation:

$$D_z = \frac{1}{12} \left(\frac{E_r d_r^3}{1 - v_r^2} + \sum_{i=2}^{n} \frac{E_i \left(d_i^3 - d_{i-1}^3\right)}{1 - v_i^2}\right), \qquad (7.38)$$

whereby

$$\sum_{i=2}^{n} d_i - d_{i=1} = \frac{1}{2} d_z. \qquad (7.39)$$

For a board that is symmetrically veneered on both sides (Fig. 7.27), stiffness is expressed in the form as follows:

$$D_z = \frac{1}{12}\left(\frac{E_w d_w^3}{1 - v_w^2} + \frac{E_o\left(d_z^3 - d_w^3\right)}{1 - v_o^2}\right), \tag{7.40}$$

where

$d_z$   thickness of the board after veneering,
$d_w$   thickness of the core of the board (particle board),
$E_w$   linear elasticity module of the particle board,
$E_o$   linear elasticity module of the veneer,
$v_w$   Poisson's ratio of the particle board and
$v_o$   Poisson's ratio of the veneer.

### 7.2.2.2 Stiffness of Bottoms of Drawers and Containers

In the most common construction solutions, the bottom of a drawer is inserted into a groove on the entire perimeter without additional fastenings or inserted into grooves on three sides and fixed to the back of the drawer with staples. The first method of mounting the bottom corresponds to the theoretical scheme of the board supported on the entire perimeter in an articulated manner. The second is the scheme of a board of three sides supported elastically and one in a discontinuous manner. The deflection of such a board (Fig. 7.19), made from isotropic material (e.g. a thin particle board or fibreboard), can be expressed in the form of a double trigonometric series:

$$w(x, y) = \sum_{m-1}\sum_{n=1} A_{mn} \sin\frac{m\pi x}{a}\sin\frac{n\pi y}{b}, \tag{7.41}$$

where
$a$ and $b$   dimensions of the board.

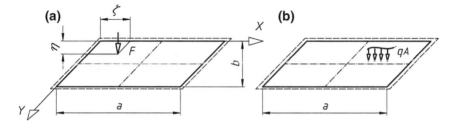

**Fig. 7.19** Rectangular board supported articulately and loaded with: **a** concentrated force, **b** evenly on the surface

By applying energy methods known in the theory of elasticity and using the equation for energy of bending an isotropic board,

$$U = \frac{1}{2}D \int\limits_0^a \int\limits_0^b \left( \left( \frac{\partial^2 w}{\partial x^2} \frac{\partial^2 w}{\partial y^2} \right)^2 - 2(1-v) \left( \frac{\partial^2 w}{\partial x^2} \frac{\partial^2 w}{\partial y^2} - \frac{\partial^2 w}{\partial x \partial y} \right)^2 \right) dx dy, \quad (7.42)$$

We obtain the equation for the total energy of a bent board in the form

$$U = \frac{1}{2}abD \sum_{m-1} \sum_{n=1} A_{mn}^2 \left( \frac{m^2\pi^2}{a^2} + \frac{n^2\pi^2}{b^2} \right)^2. \quad (7.43)$$

By providing the coefficient $A_{mn}$ with a small increment $\partial A_{mn}$, we obtain

$$\frac{\partial U}{\partial A_{mn}} = \partial A_{mn} = P \partial w(x, y). \quad (7.44)$$

By entering the unit load 1 for P, we obtain the equation of the deflection of the board at a point of the coordinates $x_0 = \xi$ and $y_0 = \eta$:

$$K(x, y, \xi, \eta) = \frac{4}{abD\pi^4} \sum_{m-1} \sum_{n=1} \frac{\sin \frac{m\pi}{a} \sin \frac{n\pi}{b}}{\left( \frac{m^2}{a^2} + \frac{n^2}{b^2} \right)^2} \sin \frac{m\pi x}{a} \sin \frac{n\pi y}{b}. \quad (7.45)$$

If a board (bottom of a drawer) is subjected to concentrated load (Fig. 7.19a) $P(\xi, \eta)$ at the point of coordinates $(\xi, \eta)$, then the deflection of this board (in accordance with the principle of superposition) will amount to

$$w(x, y) = PK(x, y, \xi, \eta). \quad (7.46)$$

However, when an evenly distributed load works on the bottom of a drawer (Fig. 7.19b), then the deflection of that bottom at the point of the coordinates $x$ and $y$ can be presented in the following function:

$$w(x, y) = \frac{18q_A}{D\pi^4} \sum_{m-1} \sum_{n=1} \frac{\sin \frac{m\pi x}{a} \sin \frac{n\pi y}{b}}{mn \left( \frac{m^2}{a^2} + \frac{n^2}{b^2} \right)^2}, \quad (7.47)$$

where in all equations $m$ and $n$ are odd numbers.

A solution to this problem on the basis of elemental strength of materials requires determining the deflection of the board caused by cylindrical bending along every axis of symmetry of the board (Fig. 7.20). Presenting unit load in the

**Fig. 7.20** The calculation
scheme of the deflection of
the bottom of the drawer

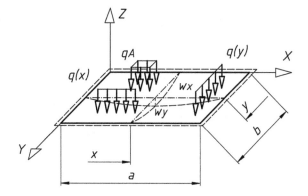

direction of the $x$ axis as $q(x)$ and the load in the direction of the $y$ axis as $q(y)$, the
total load on both axes of symmetry of the board can be written in the form:

$$q_A = q_{(x)} + q_{(y)}, \tag{7.48}$$

while the deformation at the intersection point of these axes amounts to

$$w = w_{(x)} = w_{(y)}, \tag{7.49}$$

where

$$
\begin{cases}
w_x = \dfrac{5}{384} \dfrac{q_{(x)} a^4 b}{EJ_x}, \\[2mm]
J_x = \dfrac{bd^3}{12}, \\[2mm]
w_y = \dfrac{5}{384} \dfrac{q_{(y)} ab^4}{EJ_y}, \\[2mm]
J_y = \dfrac{ad^3}{12}.
\end{cases} \tag{7.50}
$$

By comparing the deflections of the boards $w_x = w_y$, we obtain the equation

$$q_{(y)} = q_{(x)} \frac{a^4}{b^4}, \tag{7.51}$$

and transform it further

$$q_{(x)} = q_A \frac{b^4}{a^4 + b^4}. \tag{7.52}$$

The deflection of the bottom of the drawer of a depth $h$ at the point of the coordinates $(x = a/2, y = b/2)$ shows the equation:

$$w_{\left(\frac{a}{2},\frac{b}{2}\right)} = \frac{5}{384} \frac{q_v a^4 bh}{EJ_x} \frac{b^4}{a^4 + b^4},$$ (7.53)

where

$q_v$  volume load of the drawer,
$a$   dimension of the bottom of the drawer in the direction of the $x$ axis,
$b$   dimension of the depth of the drawer,
$E$   linear deformations module and
$J_x$  moment of inertia of the cross section.

### 7.2.2.3  Stiffness of Closet Bars

Bars for hanging clothing, from the mechanical point of view, can be treated as freely supported beams and subjected to even load (in accordance with Table 7.1) (Fig. 7.21).

The maximum deflection of such a bar is described by the equation:

$$w = \frac{5}{384} \frac{qL^4}{EJ} \leq f_{dop}L,$$ (7.54)

where

$q$    continuous load on the length of the beam,
$L$    length of the beam,
$E$    linear elasticity module,
$J$    moment of inertia of the cross section (see Table 7.4) and
$f_{dop}$  8 mm/m—acceptable deflection of the bar.

**Fig. 7.21** Bar of a wardrobe evenly loaded

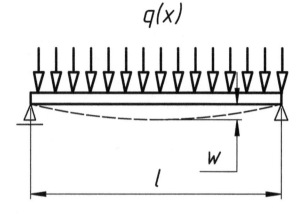

**Table 7.4**  Some of the more important characteristics of cross sections

| Cross section of the element | Field of cross section | Moment of inertia | | |
|---|---|---|---|---|
| | | Bending $J_y$ | Bending $J_x$ | Torsion $J_o$ |
| (circle, radius $R$) | $\pi R^2$ | $\dfrac{\pi R^4}{4}$ | $\dfrac{\pi R^4}{4}$ | $\dfrac{\pi R^4}{2}$ |
| (hollow circle, outer $R$, inner $r$) | $\pi(R^2 - r^2)$ | $\dfrac{\pi(R^4 - r^4)}{4}$ | $\dfrac{\pi(R^4 - r^4)}{4}$ | $\dfrac{\pi(R^4 - r^4)}{2}$ |
| (ellipse, semi-axes $a$, $b$) | $\pi ab$ | $\dfrac{\pi ab^3}{4}$ | $\dfrac{\pi a^3 b}{4}$ | $\dfrac{\pi a^3 b^3}{a^2 + b^2}$ |
| (triangle, base $2a$, height $2a$) | $a^2\dfrac{\sqrt{3}}{3}$ | $\dfrac{bh^3}{18}$ | $\dfrac{b^3 h}{6}$ | $\dfrac{a^4\sqrt{3}}{5}$ |
| (rectangle, width $b$, height $h$) | $bh$ | $\dfrac{b^3 h}{12}$ | $\dfrac{bh^3}{12}$ | $\dfrac{b^4}{4}\left(\dfrac{h}{b} - 0.63 + \dfrac{0.052}{\left(\frac{h}{b}\right)^2}\right)$ |

#### 7.2.2.4  Stiffness of Bodies of Furniture

By analysing the next stages of assembling the body of the furniture piece, starting from the side walls and bottom, it is easy to notice that the side walls, due to low stiffness of joints, are not able to shift the side load $P$ (Fig. 7.22). This structure can be improved by adding a rear wall, which will begin to shift side load $P$ in its plane. By subsequently introducing load on free corners of the side walls by forces $Q$, a clearly curved deformation of boards is caused. And because the torsional stiffness is many times smaller than the shearing stiffness, in order to improve the quality of the structure, load $Q$ should be shifted in the top plane, in this case, shield-loaded. Another loading of the top with force $V$ vertical to its surface causes torsional deformation of the entire body of the furniture piece. In this situation, the only treatment, leading to the elimination of figural deformations of boards, is to

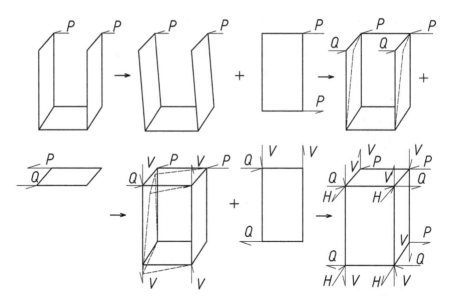

**Fig. 7.22** Example of eliminating torsional deformations of the boards of the body of case furniture

mount doors inserted in the body in such a way, that the bottom, top and sides of the wardrobe meet with their outline. Forces $V$ will then be shield-displaced by the doors, and the entire construction of the furniture piece will remain almost perfectly stiff.

Situating board elements in such a manner enables to obtain a high stiffness of the body and strength of joints, only when the doors of the furniture piece are closed. In the construction of case furniture without doors or with lift-off doors, during operational loads, the form of torsional deformation dominates.

The stiffness of furniture's body depends on the value of displacement $\Delta P_z$, measured at the direction of the operation of the force $P_z$. This displacement is caused by torsional deformation of individual boards. Therefore, without neglecting essential construction requirements of furniture in many studies (Ganowicz et al. 1977, 1978; Ganowicz and Kwiatkowski 1978), an important assumption was adopted that the elements of a wardrobe are connected together articulately and only in the corners. This enables to use energy methods, known in the theory of construction and express the value $\Delta P_z$ in the course of analytical calculations. Knowing that the work of external forces $P_z$ on external displacement $\Delta P_z$ must be equal to the sum of the work of internal forces $P_i$ on internal displacements $\Delta_i$, it can be written as

$$\frac{1}{2} P_z \Delta P_z = \sum_{i=1}^{n} \frac{1}{2} P_i \Delta_i, \tag{7.55}$$

where

$n$    number of boards (excluding shelves and doors) forming the body of furniture,
$P_i$   internal load of the board and
$\Delta_i$   internal displacement of the board.

As it has already been mentioned, these displacements are the result of torsions that the sides, partitions, rear walls and other fixtures on permanent board elements of furniture are subject to. It should be added here that shelves, as elements that are unrelated to the body of the wardrobe, are not subject to torsions, and thus in no way does their number and location determine the stiffness of the body, that is the values of sought displacements $\Delta P_z$.

The angle of torsion of the $i$th element in the state of torsional loads $P_i$, according to Fig. 7.23, can be determined from equations binding for the rod of a rectangular cross section subjected to torsion. Therefore,

$$\varphi_i = \frac{M_s l_{1i}}{\beta G_i d_i^3 l_{2i}} = \frac{\Delta_i}{l_{2i}}, \tag{7.56}$$

where

$M_s = P_i(l_2)_i$   torsion moment (N m),
$P_i$           internal load (N),
$l_{(1.2)i}$       dimensions of the $i$th board (m),
$d_i$           thickness of the $i$th board (m),
$\beta$           coefficient dependent on the relation $(l_2/d)_i$, for $(l_2/d)_i > 10\beta = 1/3$ and
$G_i$         shear modulus of the $i$th board.

By solving the system of equations:

$$\begin{cases} \varphi_i = \dfrac{\Delta_i}{l_{2i}}, \\[2mm] \varphi_i = \dfrac{3P_i l_{1i}}{G_i d_i^3}, \\[2mm] P_z \Delta P_z = \displaystyle\sum_{i=1}^{n} P_i \Delta_i, \end{cases} \tag{7.57}$$

**Fig. 7.23** Torsional deformation of a board

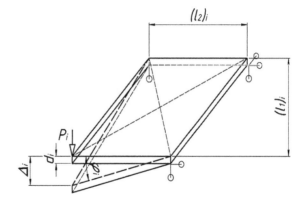

we obtain the expression connecting external displacements $\Delta P_z$ with internal displacements $\Delta_i$ in the form as follows:

$$P_z \Delta P_z = \sum_{i=1}^{n} \frac{G_i d_i^3}{3(l_1 l_2)_i} \Delta_i^2. \qquad (7.58)$$

From the geometry of deformations of the wardrobe (Fig. 7.24) and on the assumption of the perfect stiffness of board elements in their plane, particular relations result between the displacement $\Delta P_z$ of vertical external board elements and the displacement $\Delta_c$ of horizontal elements and $\Delta_b$ rear walls in the form of the equations:

$$\Delta_z = \frac{b_y a_x}{bc} \Delta P,$$

$$\Delta_x = \frac{b_y c_z}{bc} \Delta P,$$

$$\Delta_y = \frac{a_x c_z}{bc} \Delta P, \qquad (7.59)$$

**Fig. 7.24** The state of torsional deformation of a multi-chamber body

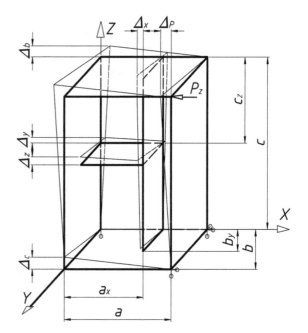

where

a, b and c          appropriate overall dimensions of furniture in the directions $x$, $y$ and
                    $z$ of the local coordinate system,
$a_x$, $b_y$ and $c_z$  dimensions of board elements corresponding to the directions of
                    overall dimensions of the furniture piece

By marking the displacement of the $i$th board in the general form as

$$\Delta_i = \alpha_i \Delta P \tag{7.60}$$

where

$\alpha_i$  the coefficient determining the geometric dependency of the element and the
     body of furniture according to the equations:

$$\alpha_z = \frac{b_y a_x}{bc}, \quad \alpha_x = \frac{b_y c_z}{bc}, \quad \alpha_y = \frac{a_x c_z}{bc}, \tag{7.61}$$

external deformation $\Delta P_z$ can be written down as

$$\Delta P = \frac{P_z}{\sum_{i=1}^{n} \frac{G_i d_i^3}{3(l_1 l_2)_i} \alpha_i^2}. \tag{7.62}$$

For practical purposes, the concept of stiffness of the furniture's body is used
more often, proving the quality of approval tests based on the value of the stiffness
coefficient $k$ and visual inspection of the state of damage. By comparing the
obtained equations, it is easy to demonstrate that the global stiffness of any
multi-chamber furniture body is the sum of torsional stiffness of its particular
elements and is expressed in the form of:

$$k = \sum_{i=1}^{n} \frac{G_i d_i^3}{3(l_1 l_2)_i} \alpha_i^2. \tag{7.63}$$

A natural characteristic of wood-based board, used to construct case furniture, is
the module of figural deformation $G$. For the purposes of designing furniture, this
module must be specified, because it is one of the main stiffness parameters of the
body. The method of determining the module of transverse elasticity for boards has
been shown in Fig. 7.25. Experimental studies are carried out on rhombus and
preferably square samples of boards of the dimensional proportion:

$$25d \leq l \leq 40d \tag{7.64}$$

where

d  thickness of the board and
l  length of the side of the sample

**Fig. 7.25** Scheme of how to measure the figural elasticity module

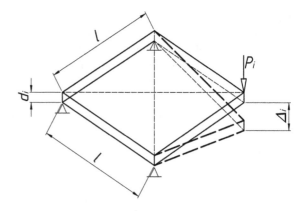

Load value $P_i$ in material tests should be adopted in such a way that deformation $\Delta_i$ measured in the direction of the force was smaller than the triple thickness of the board $d$ ($\Delta_i \leq 3d$). And the value of the module is established from the already known equation:

$$G = 3\frac{P_i l_i^2}{d^3 \Delta_i}\alpha_i^2 . \qquad (7.65)$$

For boards with a complex construction, e.g. veneered or strengthened by socle skirts and fins, in an analogous experiment, only a certain substitute of their stiffness $D_z$ can be determined. This stiffness can also be established in a theoretical way, if the geometric and physical characteristics of individual components of the system are known. The substitute torsion stiffness of the board after veneering (Fig. 7.26) is expressed in the form

$$D_z = \frac{G_z d_z^3}{12}, \qquad (7.66)$$

where
$D_z$   substitute torsion stiffness,
$G_z$   substitute figural deformations module and
$d_z$   substitute thickness of the board after veneering.

Taking into account the properties of the board constituting the core of the layer system and properties of the veneer, the substitute figural deformations module $G_z$ of furniture board built from n layers can be written by the following equation:

$$G_z = G_r \left(\frac{d_r}{d_z}\right)^3 + \sum_{j=1}^{n} G_i \left(\left(\frac{d_j}{d_z}\right)^3 - \frac{d_{j-1}}{d_z}\right), \qquad (7.67)$$

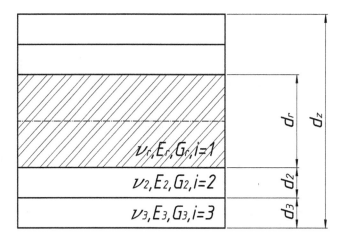

**Fig. 7.26** Three-layer board with a symmetrical system of veneers

where

$$d_z = 2\sum_{j=1}^{n} d_{j-1} \quad d_{j=1},$$

| | |
|---|---|
| $n$ | number of layers, |
| $d_{j-1}$ | thickness of veneers, |
| $d_{j=1} = d_r$ | thickness of the core (of a particle board, MDF, etc.), |
| $G_i$ | shear modulus of the veneer and |
| $G_r$ | shear modulus of the core, |

And for a symmetrical board built from two layers of veneer (Fig. 7.27), the substitute module of figural deformations amounts to

$$G_z = G_w\left(\frac{d_w}{d_z}\right)^3 + G_o\left(1 - \left(\frac{d_w}{d_z}\right)^3\right), \qquad (7.68)$$

where

| | |
|---|---|
| $d_z = d_w + 2d_o$ | thickness of the system, |
| $d_w$ | thickness of the core (particle board), |
| $G_o$ | shear modulus of the veneer and |
| $G_w$ | shear modulus of the particle board |

In order to determine $G_z$ of a veneered board, the shear modules of boards and veneers must be known. By analysing the stiffness model of case furniture, where the body of the furniture piece constitutes an open case, supported by three corners and loaded in such a way that causes its torsional deformation, the doors are only extra mass load and do not improve the global stiffness of the body. Only when they are placed inside the construction in such a way as to fit tightly to the sides, the

**Fig. 7.27** Particle board
veneered symmetrically

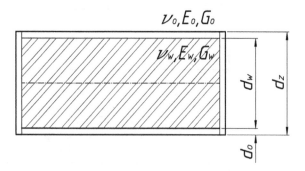

bottoms and tops, then they cause that instead of torsional deformations, all components of the furniture piece shift shield loads.

Sealed single-chamber case furniture, assuming that they are made from linear-elastic boards interconnected articulately, in accordance with the theory of torsion of thin-walled profiles can be treated as a construction of a cohesive closed structure, in which the rear wall and door fulfil the function of membranes closing streams of tangential stresses. Therefore, in the calculation scheme, it was adopted that the construction was torsionally loaded, as it has been shown in Fig. 7.28a. The deformation, which the closed thin-walled case will be subject to under the influence of the torsion moment in relation to the bending centre 00′, is explicitly described by the values of the rotation angle $\varphi$ and displacements $\Delta_{1P}$, $\Delta_{2P}$, $\Delta_{3P}$ (Fig. 7.28b).

Due to the symmetry of the construction of the case subjected to torsion in theoretical calculations, it was decided to assume that the bending centre of the case is located in the geometric centre overlapping with the centre of the gravity of the block. Moments of inertia of the cross section, as it has been shown in Fig. 7.28c, therefore amount to

$$I_x = \frac{1}{6} db^2 (b + 3a),$$

$$I_y = \frac{1}{6} da^2 (a + 3b). \tag{7.69}$$

Determining the streams of tangential intensities $q_p$:

$$q_p = \bar{q} + q_o, \tag{7.70}$$

whereby

$$q_o = \frac{M_s}{\Omega} - \frac{1}{\Omega} \oint \bar{q} h ds, \tag{7.71}$$

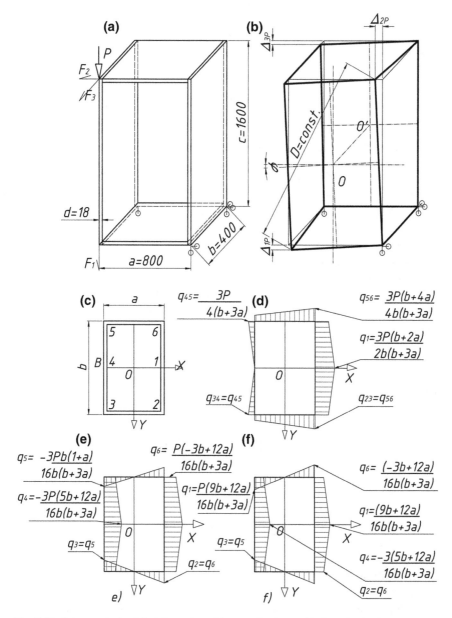

**Fig. 7.28** Scheme of **a** load, **b** deformation of the case furniture with doors inserted to the interior of the body, as well as the characteristics of the streams of intensities in the thin-walled profile: **c** cross-sectional geometry, **d** streams of intensities of tangents relative to pole B, **e** the basic state of streams of tangential intensities, **f** virtual state of streams of tangential intensities

$$\overline{q} = \frac{PS_x}{J_x},$$ (7.72)

$$\Omega = 2ab,$$ (7.73)

where
$M_s$   torsion moment,
$P$     load,
$S_x$   static moment of the considered part of the cross section,
$I_x$   moment of inertia in relation to $x$ axis,
$I_y$   moment of inertia in relation to $y$ axis,
$\Omega$   double surface of the profile,
$h$     distance of the centre of gravity from the pole,
$a$     width of the cabinet,
$b$     height of the cabinet,
$c$     depth of the cabinet and
$d$     thickness of the board,

and assuming the pole in point B (Fig. 7.28c), for which $M_j = 0$, the streams of intensities $q$ of values provided in Fig. 7.28d are obtained. For a coherent closed profile, streams of intensities of tangents $q_p$ relative to the bending centre in given points of the profile have been provided in Fig. 7.28e, while the graph of streams of intensities of tangents $q$, at the virtual state of loads have been provided in Fig. 7.28f. The displacement of a closed case in the direction of force impact can be determined from the general equation:

$$\delta_{ip} = \oint \int_0^c \frac{\sigma_i \sigma_p}{E} d\,dzds + \oint \int_0^c \frac{q_i q_p}{Gd} dzds,$$ (7.74)

where

$$\sigma_i = \frac{q_i}{d}, \quad \sigma_p = \frac{q_p}{d}.$$ (7.75)

### 7.2.2.5  Stiffness of Socle

An open body of a case furniture piece can be stiffened to torsion by increasing the thickness of one or a few of its boards. The use of thick boards for all elements of the furniture piece improves the efficiency of the construction, but at the same time increases its weight, material consumption and the total cost of the product. The low effectiveness of this solution causes that designing works should be concentrated more on focusing the material in one board (Fig. 7.29). The stiffness of a wardrobe made from the same boards of the thickness $d$ amounts to

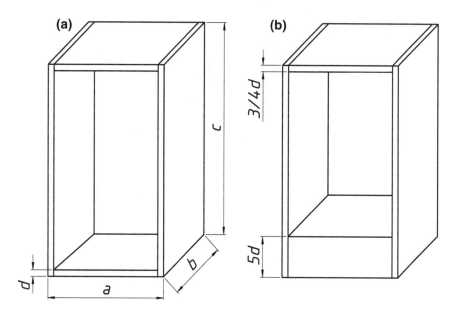

**Fig. 7.29** Stiffening the body of the furniture piece by applying one thick board and reducing the thickness of other boards: **a** all boards of the same thickness, **b** bottom of increased thickness

$$k_d = \sum_{i=1}^{5} \frac{G_i d_i^3}{3(l_1 l_2)_i} \alpha_i^2 = \frac{2a}{3bc} Gd^3 \left( \frac{1}{a} + \frac{1}{c} + \frac{1}{2b} \right), \tag{7.76}$$

whereas the stiffness of the body, in which the thickness of the bottom has been increased and the thickness of remaining elements have been decreased, has the value

$$k_{5d} = \frac{2a}{3bc} Gd^3 \left( 0.844 \frac{1}{a} + 62.7 \frac{1}{c} + 0.21 \frac{1}{b} \right), \tag{7.77}$$

Assuming the dimensional proportions $c = 2a = 4b$, we obtain that

$$k_{5d} = 13 k_d. \tag{7.78}$$

The stiffness of the body of a furniture piece with one horizontal board thickened is thirteen times greater, while 12.5 units of stiffness comes from a board with a thickness $5d$, and the remaining 1/2 units from boards with the thickness $3/4d$. Stiffening the body of the furniture piece with a thick-walled board is expensive due to the high content of material. This cost can be significantly reduced by using reinforcement in the form of plinths and socles with fins (Dzięgielewski and Smardzewski 1995).

**Fig. 7.30** Deformation of the furniture body along with the socle

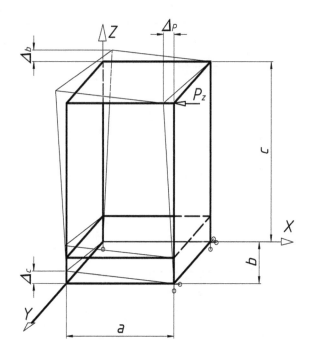

Let us consider the deformation of a single-chamber wardrobe with a socle (Fig. 7.30). Like other boards, the socle is also subject to torsional deformation. Its stiffness can be determined, similarly to the stiffness of a five-element body of a furniture piece, by placing the socle in a local Cartesian system of coordinates $X$, $Y$ and $Z$ (Fig. 7.31). The value of the coefficient of stiffness for this furniture element is expressed by the equation:

$$k_c = \frac{P_i}{\Delta_i} = \sum_{i=1}^{n} \frac{G_i g_i^3}{3(l_1 l_2)_i} \alpha_i^2. \tag{7.79}$$

For the socle of a furniture piece of rack structure, the stiffness coefficient $k_c$ is

$$k_c = \frac{G_i g_i^3}{3(a - 2g_i)b} \left( \frac{a - 2g_i}{a} \right)^2 + 2\frac{G_i g_i^3}{3hb} \left( \frac{h}{a} \right)^2 + S\frac{G_i g_i^3}{3(h - g_i)(a - 2g_i)} \left( \frac{h - g_i}{b} \right)^2, \tag{7.80}$$

while for the flange structure

$$k_c = \frac{G_i g_i^3}{3ab} + 2\frac{G_i g_i^3}{3(h - g_i)(b - 2g_i)} \left( \frac{h - g_i}{a} \right)^2 + S\frac{G_i g_i^3}{3(h - g_i)a} \left( \frac{h - g_i}{b} \right)^2, \tag{7.81}$$

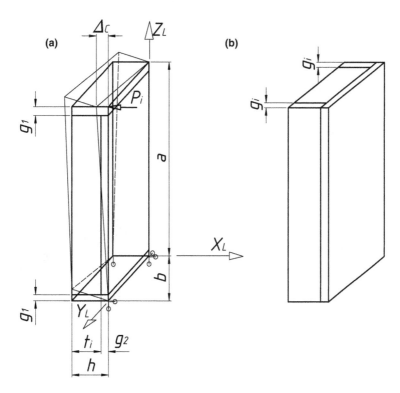

**Fig. 7.31** Deformation of the socle: **a** of rack structure, **b** of flange structure

where

| | |
|---|---|
| $k_c$ | stiffness of the socle, |
| $g$ | thickness of the $i$th element, |
| $h, b, a$ | appropriate overall dimensions, |
| $G$ | shear modulus of the element and |
| $S$ | the number of elements corresponding to the dimensions of the frontal socle skirt, $S = 2$ for a four-element socle, $S = 1$ for a three-element socle (no skirt from the side of the rear wall). |

In the considered case, the stiffness of the body depends on the torsional stiffness of the socle and that on the torsional stiffness of components. It is already known, however, that the torsional stiffness of boards depends, among others, on the module of figural deformations $G$, which value in relation to the module of linear deformations for isotropic materials remains in the familiar equation:

$$G = \frac{E}{2(1 + v)},$$

(7.82)

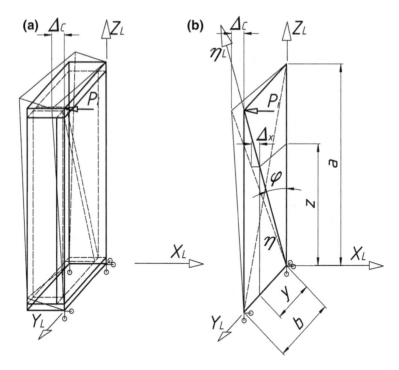

**Fig. 7.32** Deformation of the socle along with the fin: **a** general scheme of deformation, **b** calculation scheme

where

$v$   Poisson's ratio (for particle boards $v = 0.28$).

By taking advantage of the bending work of wood or wood-based materials, the stiffness of the socle, thus the entire body, can be easily increased. This can be achieved by applying fins in socle boards (Fig. 7.32). A finned board subjected to clean torsion deforms similarly to a board without fins. The diagonal of the board is bent according to the parabola, and the fin is in a state of clear bending (Dzięgielewski and Smardzewski 1995). This kind of deformation can be written, according to the indications in Fig. 7.32, as follows:

$$\Delta_x = \frac{\Delta_c}{ba} zy, \qquad (7.83)$$

where

$$y = bz/a, \tag{7.84}$$

$$z = \eta \cos \varphi = \eta \frac{a}{\sqrt{a^2 + b^2}}. \tag{7.85}$$

The function of deflection in the direction $\eta$ has the form:

$$f(\eta) = \frac{\Delta_c}{a^2 + b^2} \eta^2. \tag{7.86}$$

Because the work of external forces performed over the system must be equal to the sum of internal forces, then in the case of fin bending, we shall obtain

$$\frac{1}{2} P_i \Delta_c = \frac{1}{2EJ} \int_\eta W^2 d\eta, \tag{7.87}$$

where

$$W = EJ \frac{\partial^2 f(\eta)}{\partial \eta^2} = 2EJ \frac{\Delta_c}{a^2 + b^2} \tag{7.88}$$

$$J = \frac{gh^3}{12} \tag{7.89}$$

which gives

$$P_i \Delta_c = EJ \frac{4\Delta_c^2}{(a^2 + b^2)^2} \int_0^{\sqrt{a^2+b^2}} d\eta = 4EJ \frac{\Delta_c^2}{\sqrt{(a^2 + b^2)^3}} \tag{7.90}$$

The stiffness of a fin can, therefore, be written by the equations:

$$k_z = \frac{P_i}{\Delta_c} = 4EJ \frac{1}{\sqrt{(a^2 + b^2)^3}} \tag{7.91}$$

or:

$$k_z = Egh^3 \frac{1}{3\sqrt{(a^2 + b^2)^3}} \tag{7.92}$$

By determining the stiffness of a single-chamber wardrobe, we demonstrated that the doors, if they are not inserted into the case, only constitute additional mass load

**Fig. 7.33** Closed socle
loaded by torsional forces

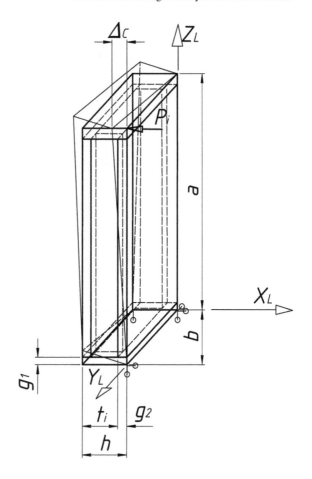

and do not improve the stiffness of the furniture piece. By inserting them into the body, it makes the furniture a coherent closed profile, which stiffness increases by dozens. Closing a socle with a thin board from the side of the socle (Fig. 7.33) makes it a stiff, six-sided box torsionally loaded, similarly to the entire construction of the furniture piece.

The presented ways of stiffening the furniture body refer to one element, which is the bottom in the form of a thick, single board or socle constituting a subassemblage. By inserting this element into the furniture body, the stiffness of other components of the furniture piece should be established, disregarding the bottom (or socle) excluded in thought. In this case, the stiffness of the body is expressed by the equation:

$$k = \sum_{i=1}^{n-1} \frac{G_i d_i^3}{3(l_1 l_2)_i} \alpha_i^2 + k_n \left(\frac{a}{c}\right)^2 \qquad (7.93)$$

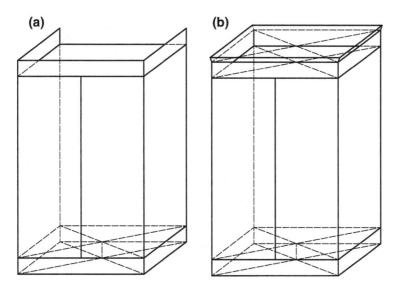

**Fig. 7.34** Construction of furniture with top in the form of: **a** socle, **b** crown

where

$$\sum_{i=1}^{n-1} \frac{G_i d_i^3}{3(l_1 l_2)_i} \alpha_i^2$$ 　the stiffness of all elements of the furniture body with a socle excluded in thought and

$k_n$ 　stiffness of the socle as the $n$th element of the furniture piece.

The stiffness of the complex socle constitutes the sum of the stiffness of components, which is why for the n-element socle with a fin, we can write as

$$k_n = k_c + k_z. \qquad (7.94)$$

The previously presented methods of designing stiffening of the construction of case furniture concern mainly the bottom and the socle. However, the same reasoning can be applied to the top boards, especially since furniture having decorative masking skirts over the top board in the shape of a socle or crown is becoming increasingly common on the market (Fig. 7.34). These decorative elements are perfect for hiding the designed reinforcements, also on the top surface. Cooperation of fins and skirts with the board, however, requires the constructional insurance of a fixed angle between the board and the fin. In the case of a skirt with a free end (Fig. 7.35a), this end should be connected with the board using an extra element, e.g. a block. Using two intersecting fins (Fig. 7.35b) requires a constructional solution of the middle node. Another solution to this problem is to mount one fin under the board of the bottom, and the other over the surface top of the furniture (Fig. 5.35c).

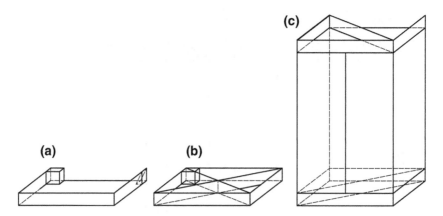

**Fig. 7.35** The method of connecting elements of the socle between one another and the top board: **a** three-element socle, **b** four-element socle with a fin, **c** alternating spacing of fins on the *top* and *bottom*

### 7.2.2.6 Stiffness of Eccentric Joints

The load-carrying capacity of eccentric joints with eccentric connectors to a great extent determines the strength of structural nodes of case furniture. The distribution of internal forces generated while forcing torsional deformations of the furniture body results in the fact that in structural nodes that join the side walls with the bottom and top, the biggest loads on metal connectors of wall angular joints occur (Fig. 7.36). Such a state of loads makes it necessary to check the strength of the connection due to the shearing strength, splitting strength and compression strength of the particle board. Therefore, strength calculations should be carried out on models, for which initial data must derive from studies of elementary properties of elastic materials used to make joints and connectors.

The stiffness of joints is determined by experimental tests of models of corner joints (Fig. 7.37). On the basis of these studies, the coefficient of stiffness $\gamma$ is determined from the equations:

$$\gamma = \frac{M}{\varphi} \, \mathrm{N\,m/rad}, \tag{7.95}$$

where
$M$    bending moment and
$\varphi$    rotation angle

Figure 7.38 illustrates the stiffness of joints established on the basis of clenching tests. Then, the stiffness coefficients were determined as the value of the derivative of function $M = f(\varphi)$ in a point (Table 7.5). The provided illustrations and tables

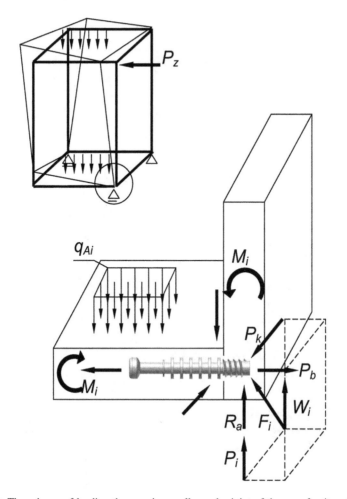

**Fig. 7.36** The scheme of loading the core in a wall angular joint of the case furniture body

show that the tested joints belong to the group of semi-rigid or flexibility connection structural nodes of furniture.

For example, let us consider the horizontal partition loaded by the force $q_{xy}$ causing its bending (Fig. 7.39). In this case, as long as internal forces are balanced,

$$T = \mu 2F = q_{xy}y\partial x, \tag{7.96}$$

where
$T$     friction force,
$\mu$     friction coefficient for particle boards,
$F$     operational tension forces,
$q_{xy}$     operational forces of the partition,

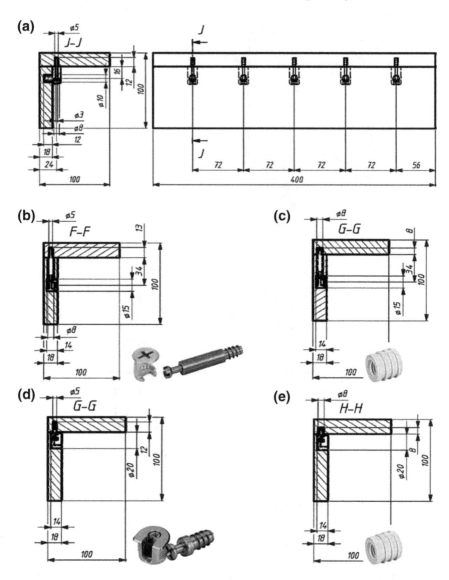

**Fig. 7.37** Models of eccentric joints: **a** trapezoid, **b** Rastex 15, **c** Rastex 15 with sleeve, **d** VB35, **e** VB35 with sleeve

$y$     current coordinate (for the partition) and

$\partial x$     elementary depth of the partition,

then the pressures $q'_{zx} > 0$ are distributed evenly ($q'_{zx} = q''_{zx}$). Increasing the value of operational load $q_{xy}$ will cause a change in the form of these pressures gradually to a trapezoid ($0 < q'_{zx} < q''_{zx}$), triangular ($0 = q'_{zx} < q''_{zx}$) and finally a one-point.

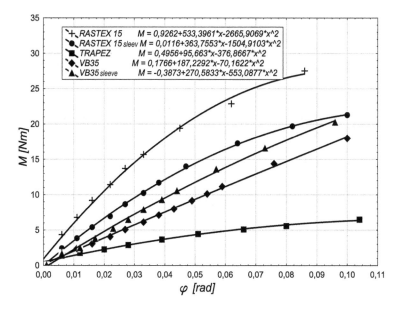

**Fig. 7.38** Stiffness of joints subject to closing

**Table 7.5** Stiffness of joints in the range of deformations $0 < f < 0.07$ (rad)

| Joint | $y = M'(x)$ | $\gamma = M'(x)\|_{x=\varphi=0.07}$ (N m/rad) | $\gamma = \frac{M}{\varphi}$ (N m/rad) |
|---|---|---|---|
| Rastex 15 | $y = -3583.8x + 819.7$ | 568.8 | 743.8 |
| Rastex 15 with sleeve | $y = -4173.4x + 825.0$ | 532.9 | 556.3 |
| Trapezoidal | $y = -633.8x + 235.8$ | 192.4 | 168.8 |
| VB35 | $y = -1512.8x + 466.0$ | 360.1 | 381.3 |
| VB35 with sleeve | $y = -263.8x + 332.8$ | 314.3 | 393.8 |

As it can be seen from Fig. 7.39, the increase of the value of operational load $q_{xy}$ causes that the resultant $Q_n$ of pressures $q'_{zx}$ moves downwards, away from the axis of the core of the value $z$. The location from the resultant vector $Q_n$ of forces of mutual pressures of board elements can be calculated from the equation:

$$z = \sum_{i=1}^{n} A_i z_i \bigg/ \sum_{i=1}^{n} A_i, \qquad (7.97)$$

where
$A_i$   surfaces of pressures,

**Fig. 7.39** The state of
operational loads

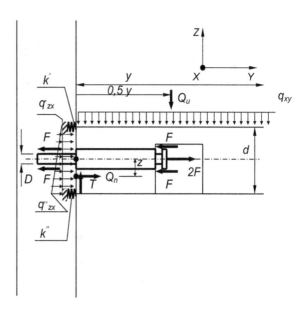

and after substituting values like in Fig. 7.39, we shall get

$$z = \frac{1}{6} d \alpha_q,$$  (7.98)

where

$$\alpha_q = \left( \frac{q''_{zx} + 2q'_{zx}}{q''_{zx} + q'_{zx}} \right),$$  (7.99)

$q''_{zx}$   surface pressures at the lower edge of the board,
$q'_{zx}$   surface pressures at the top edge of the board and
$d$     thickness of the board.

There are such working conditions for which pressures $q_{zx}$ assume the form of even loads, growing linearly or concentrated forces. Because by assuming,

for $Q_u = T$, that is $q''_{zx} = q'_{zx}$, we obtain that $z = 0$,
for $Q_u > T$, that is $q''_{zx} > q'_{zx}$, we obtain that

$$z = \frac{1}{6} d \left( 3 - 2 \frac{q''_{zx} + 2q'_{zx}}{q''_{zx} + q'_{zx}} \right)$$  (7.100)

for $Q_u \gg T$, that is $q''_{zx} > 0$ and $q'_{zx} = 0$, we obtain that $z = 1/6\ d$,
for $Q_u \gg T$, that is $q''_{zx} = 0$ and $q'_{zx} = 0$, we obtain that $z = 1/2\ d$,

where
$Q_u$  resultant operational load.

When considering the cases,

$$Q_u > Tiq''_{zx} > q'_{zx}, \tag{7.101}$$

$$Q_u \gg T \text{ and } q''_{zx} > 0 \text{ and } q'_{zx} = 0, \tag{7.102}$$

in the joint a balance of moments must occur deriving from external forces $Q_u$ and internal forces $Q_n$ in the form:

$$\frac{1}{2}Q_u y = \frac{1}{2}q_{xy}y^2 \partial x = Q_n z, \tag{7.103}$$

where

$$Q_n = \frac{1}{2}(q''_{zx} + q'_{zx})d\partial x. \tag{7.104}$$

By ensuring the connection made from particle boards and eccentric connectors sufficient stiffness and strength, it needs to be made sure that the size of stresses $q''_{zx}$ and $q'_{zx}$ does not exceed the acceptable compression strength for particle boards $k_t^w \approx 4$ MPa. By using the scheme presented in Fig. 7.40, the value of these stresses can be determined, resulting from the equations in effect for passive forces $R_A$ and $R_B$ caused at stress points A and B by the resultant force $Q_n$.

By writing the equations of balance for any 2-D system of forces relative to points $A$ and $B$, we obtain that

$$R_B = \frac{7}{12}Q_n(3 - 2\alpha_q), \quad R_A = Q_n\left(1 - \frac{7}{12}(3 - 2\alpha_q)\right). \tag{7.105}$$

While the balance of power in the entire joint shows that

$$q_{xy}y^2 \partial x = \frac{1}{6}d^2 \partial x(q''_{zx} + q'_{zx})\left(3 - 2\frac{q''_{zx} + 2q'_{zx}}{q''_{zx} + q'_{zx}}\right), \tag{7.106}$$

which gives

$$q_{xy} = \frac{1}{6}\frac{d^2}{y^2}(q''_{zx} - q'_{zx}). \tag{7.107}$$

**Fig. 7.40** Reactions of board
elements

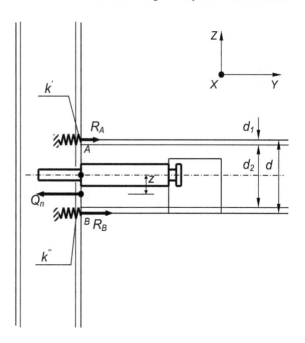

Assuming also that reactions $R_A$ and $R_B$ determined on the edges of the boards in points A and B, attributed to elementary sections of surfaces of pressures $\partial A = \partial z \partial x$, should correspond to the pressures $q''_{zx}$ and $q'_{zx}$, the equations can be formulated

$$\frac{R_A}{\partial z \partial x} = q''_{zx} = E\varepsilon''_y, \quad \frac{R_B}{\partial z \partial x} = q''_{zx} = E\varepsilon''_y, \tag{7.108}$$

where

$E$      linear elasticity module of the particle board in the direction of $y$ axis,

$\varepsilon'_y\varepsilon''_y$    relative normal strain of the board in point $A$ and $B$,

$\partial z \partial x$   elementary sections of the surface of pressure.

Knowing the value of the linear elasticity module $E$ for the board and assuming acceptable relative normal strain of the board $\varepsilon'_y\varepsilon''_y$ at points A and B, the values $q_{xy}$ in the function of strains can be estimated as follows:

$$q_{xy} = \frac{1}{6}\frac{d^2}{y^2}E\left(\varepsilon''_{zx}, -\varepsilon'_{zx}\right). \tag{7.109}$$

**Fig. 7.41** Resultant forces of edge pressures caused by operational load

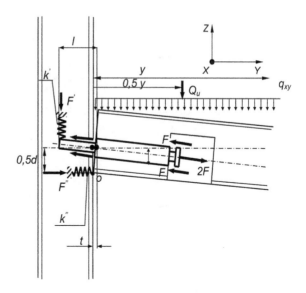

Acceptable reactions at points of pressure should not exceed the values calculated from the following equation:

$$R_B = \frac{7}{24} Ed\partial x \left( \varepsilon_y'' - \varepsilon_y' \right), \quad R_A = \frac{1}{24} Ed\partial x \left( 5\varepsilon_y'' + 19\varepsilon_y' \right). \tag{7.110}$$

If acceptable values of operational loads are exceeded, for which $Q_u \gg T$ and $q_{zx}'$ = 0 and $q_{zx}'$ = 0, we obtain that $z = 1/2\ d$. Therefore, at the junction of board elements surface pressures do not appear, but edge pressures do (Fig. 7.41).

From the balance of forces in this node, it results that

$$\frac{1}{2} q_{xy} y^2 \partial x = \frac{1}{2} F'' d + F'(l - t), \tag{7.111}$$

where
F' and F"   resultant forces of pressures and
l                  arm or the force F'

and that the value of squeezes at the point of forces F" and F' operating amount to, respectively:

$$\Delta l'' = \frac{F''}{k''}, \quad \Delta l' = \frac{F'}{k'}, \tag{7.112}$$

$$k = \gamma/a^2, \tag{7.113}$$

where $a = l$ or $0.5\ d$.

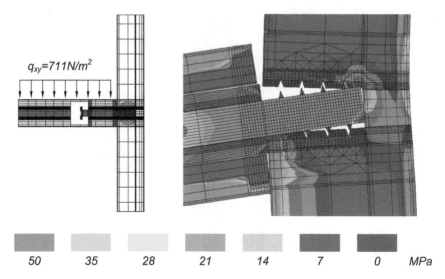

$q_{xy}=711N/m^2$

| 50 | 35 | 28 | 21 | 14 | 7 | 0 | MPa |

**Fig. 7.42** Reduced stresses according to Mises caused by the load on an eccentric joint with the stress $q_{xy} = 711$ N/m$^2$

Assuming a controlled size of compression for the particle board, resulting from its strength to compression, the acceptable operational load should be as follows:

$$q_{xy} = \frac{1}{y^2 \partial x}[\Delta l'' k'' d + 2\Delta l' k' (l - t)]. \tag{7.114}$$

The analytical determination of acceptable values of operational loads requires the use of the stiffness coefficient $y$, which values for specific types of connectors have been provided in Table 7.5. Adopting from this table the stiffness of the joint Rastex 15 for the elastic range $y = 568.8$ N m/rad, and using the above formula, for the horizontal partition with dimensions 0.4 × 0.8 m, we can determine the acceptable operational load $q_{xy}$ at the level 711 N/m$^2$. For the trapezoid joint, in which $y = 192.4$ N m/rad, $q_{xy}$ cannot exceed the value 240.5 N/m$^2$. The results of these analyses can be verified by numerical calculations. Figure 7.42 shows the model of a semi-cross-joint built from a mesh of finite elements. 20-node ortho-tropic block elements and gap-type contact elements were used to create it. Together with this, elastic characteristics of the board were used according to data provided in Table 7.6.

As it can be seen in Fig. 7.42, as a result of the rotation of the core, contact stresses appear on the particle board which contribute to decalibrating the diameter of the hole, in which an outline of the nut thread was made. The form of defor-mations of the node determined on the basis of numerical calculations indicates that the assumptions as to the location of the rotating points of the core for the math-ematical model are correct.

**Table 7.6** Mechanical properties of particle board (based on Bachmann 1983; Smardzewski 2004b, c)

| Property | Unit | Value |
|---|---|---|
| $E_1$—Young's modulus of external layers of the board | MPa | 4656 |
| $E_2$—Young's modulus of the internal layer of the board | | 1080 |
| $E_s$—Young's modulus of steel | | 200,000 |
| $v$—Poisson's coefficient for all materials | – | 0.3 |
| $k_r^w$—delamination resistance of the board | MPa | 0.71 |
| $k_t^w$—shearing strength of the board | | 5.35 |
| $E_g$—stiffness of the gap-type element | | 1080 |
| $h_p$—thickness of the board | mm | 18 |
| $h_1$—thickness of the external layer of the board | | 3 |
| $h_2$—thickness of the internal layer of the board | | 12 |
| $L_z$—length of the screw | | 50 |
| $l_i$—height of the $i$th surface of the cone of impact | | |
| $D_o = d$—external diameter of the thread | | 7 |
| $D_i$—diameter of the $i$th surface of the cone of impact | | |
| $D_z$—external diameter of the screw head | | 10 |
| $d_o$—screw core diameter | | 4 |
| $\beta$—angle of elementary section of the friction surface | deg | |
| $\gamma$—angle of cone of the screw head | | 21 |
| $A_i$—surface of the $i$th part of the cone of impact | mm$^2$ | |
| $A$—friction surface of cone of the screw head | | |
| $S$—skip of the thread | mm | 3 |
| $dz$—width of elementary section of the friction surface | | |
| $M_T$—friction moment | N mm | |
| $M_G$—moment on the thread | | |
| $N$—pressure force | N | |
| $T$—friction force | | |
| $\mu$—coefficient of friction of the board against metal | – | |

#### 7.2.2.7   Stiffness of Screw Joints

Screw joints of case furniture belong to the group of susceptible connections, for which the total stiffness of the system depends on the stiffness of all its components. In engineering calculations, using models of these connections as perfectly stiff nodes it seems too optimistic, while articulated models are too unfavourable. Only the susceptible models allow the correct mathematical description and numerical modelling of the stiffness of construction of case furniture assembled with screws.

Previous studies on the load-carrying capacity of confirmat-type connectors (Bachmann 1983) show that the average strength of the screw, set at a depth of 30 mm, for pulling in the parallel direction to the wide surfaces of the board amounts to 1250 N (Fig. 7.43). By adopting this value as force in the core of the screw, and

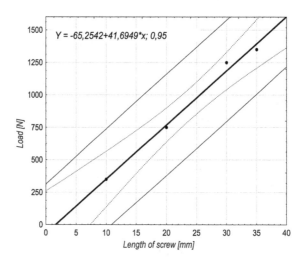

**Fig. 7.43** Strength of the screw for pulling in the parallel direction to the width of the board surface (own development based on Bachmann 1983)

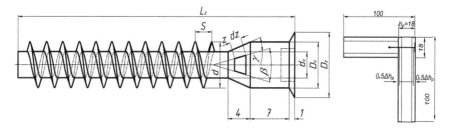

**Fig. 7.44** The dimensions of the screw and angular confirmat-type joint

choosing the geometric characteristics of the screw and material constants of a three-layer particle board (Table 7.6), in order to determine the stiffness of angular joints, as shown in Fig. 7.44, the cones of impact must be determined, taking into account only stresses of the screw head (Fig. 7.45) and stresses from the screw cone (Fig. 7.46). Then, the following needs to be calculated in the order:

- stiffness coefficient of screw $c_s$,
- stiffness coefficient of board (sleeve) $c_k$,
- load coefficient $\xi$,
- initial stress of joint $Q_o$,
- maximum stress of joint $Q_{max}$,
- residual stress of joint $Q_r$,
- tightening moment of screw $M$,
- acceptable compression of the board caused by initial stress of the screw $\Delta h_p$ and
- maximum load $P_r$ of the screw causing a contraction in the board equal to $\Delta h_p$.

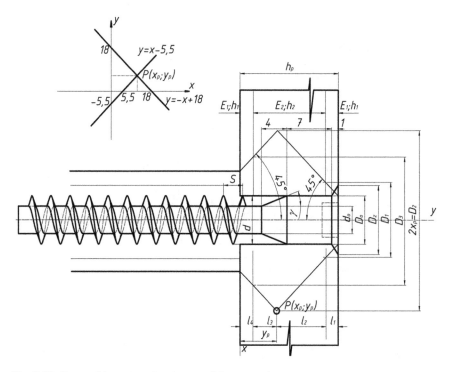

**Fig. 7.45** Cones of impact on the stresses of the screw sleeve

For the scheme from Fig. 7.45, we get

$$c_x = \int_{x_1}^{x_2} \frac{1}{E_x A_x} dx, \tag{7.115}$$

$$\frac{1}{c_s} = \frac{4}{\pi E_s} \left( \frac{7}{D_o} + \frac{4 \cdot 4}{(D_o + d_o)^2} + \frac{18 - 7 - 4}{d_o^2} \right) = 45,364,796 \times 10^{-6} \left( \frac{mm}{N} \right), \tag{7.116}$$

$$\frac{1}{c_k} = \frac{l_1}{E_1 A_1} + \frac{l_2}{E_2 A_2} + \frac{l_3}{E_2 A_3} + \frac{l_4}{E_1 A_4}, \tag{7.117}$$

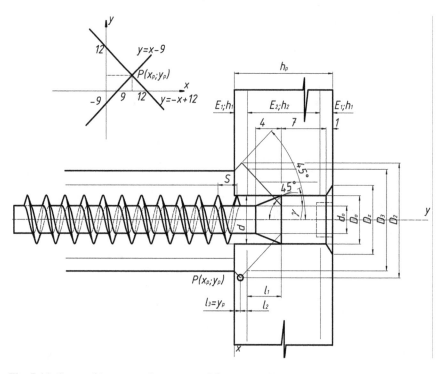

**Fig. 7.46** Cones of impact on the stresses of the screw cone

$$
\frac{1}{c_k} = \frac{4h_1}{\pi E_1\left((D_o + h_1)^2 - D_o^2\right)} + \frac{4(h_p - h_1 - y_p)}{\pi E_2\left((D_o + h_1 + x_p)^2 - D_o^2\right)} +
$$

$$
+ \frac{4(y_p - h_1)}{\pi E_2\left(\left(\frac{D_o + 2x_p + h_p + 2h_1}{2}\right)^2 - D_o^2\right)} + \frac{4h_1}{\pi E_1\left((h_p + h_1)^2 - D_o^2\right)} \tag{7.118}
$$

$$
= 4,803,085 \times 10^{-5} \left(\frac{\mathrm{mm}}{\mathrm{N}}\right),
$$

$$
\zeta = \frac{c_s}{c_s + c_k} = 0.913, \tag{7.119}
$$

$$
Q_o = Q(1 - \zeta)k_z = 129.36\,\mathrm{N}, \tag{7.120}
$$

$$
Q_{\max} = Q_o + \zeta Q = 1270.6\,\mathrm{N}; \tag{7.121}
$$

$$
Q_r = Q_{\max} - Q = 20.6\,N. \tag{7.122}
$$

The value of shortening $\Delta h_p$ of the vertical connection element, caused by maximum force $Q_{max}$ deriving from initial installation, was obtained from the equation:

$$\Delta h_p = Q_{max}\left(\frac{l_1}{E_1 A_1} + \frac{l_2}{E_2 A_2} + \frac{l_3}{E_2 A_3} + \frac{l_4}{E_1 A_4}\right) = 0.061 \text{ mm}. \tag{7.123}$$

And for the scheme from Fig. 7.46, the following was obtained:

$$\frac{1}{c_s} = \frac{4}{\pi E_s}\left(\frac{4}{\left(\frac{D_o+d_o}{2}\right)^2} + \frac{6}{d_o^2}\right) = 3229 \times 10^{-6}\left(\frac{\text{mm}}{N}\right), \tag{7.124}$$

$$\frac{1}{c_k} = \frac{l_1}{E_2 A_1} + \frac{l_2}{E_1 A_2} + \frac{l_3}{E_1 A_3} = 5.9 \times 10^{-5}\left[\frac{\text{mm}}{N}\right], \tag{7.125}$$

$$\zeta = 0.9484, \tag{7.126}$$

$$Q_o = 77.25\,N; \quad Q_{max} = 1262.86\,N; \quad Q_r = 12.86\,N, \tag{7.127}$$

$$\Delta h_p = 0.0885 \text{ mm}. \tag{7.128}$$

This result can be easily verified using numerical modelling (Smardzewski and Ożarska-Bergandy 2005). As the calculation scheme, the model shown in Fig. 7.46 was selected with a force value $Q_{max} = 1262.89$ N. Additionally, taking into account the friction force on the cone and thread of the screw (Fig. 7.47), the tightening moment $M$ has been determined from the equation:

$$M = M_T + M_G, \tag{7.129}$$

where

$$M_T = \mu\frac{4\sin\gamma Q_{max}}{(\mu\cos\gamma + \sin\gamma)(D_o^2 - d_o^2)}\sin\gamma \int_{\frac{d_o}{2\sin\gamma}}^{\frac{D_o}{2\sin\gamma}} z^2 \int_0^{2\pi} dz d\beta, \tag{7.130}$$

$$M_G = \frac{1}{2}Q_{max}D_o \text{tg}\left(\text{atg}\frac{S}{\pi D_o} + \text{atg}\,\mu\right), \tag{7.131}$$

$$M = 18646.9\,[\text{N mm}]. \tag{7.132}$$

Building a numerical model of a susceptible confirmat-type joint, the effect of compression of a vertical board was simulated by the compression of the

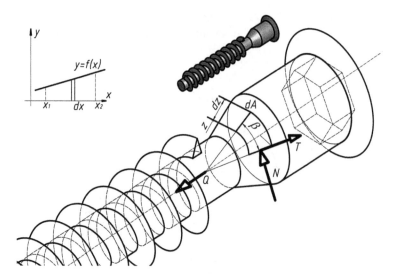

**Fig. 7.47** Internal forces for a screw

non-threaded part of the screw, set on the section $h_p$ with the force $P_r = 24{,}714.76$ N, causing a contraction of the core $\Delta h_p = 0.0885$ mm.

The distribution of reduced stresses according to Mises demonstrated that the biggest strains of the particle board occur in the part tightened by the screw head and horizontal element. Along the threaded part of the screw, the stresses are minor and do not have an impact on damaging the material of the middle layer of the board. In addition, exerting installation stresses only with the cone part of the screw head worsens the conditions of work of the joint through the increase of linear shortenings in the direction of the force of initial stress. The maximum tightening moment of the screw does not cause stresses that are destructive to the board at the length of the thread; however, it does cause stresses that are destructive in the board tightened by the head of the confirmat.

The solutions presented above included the case of loading a connector with axial installation forces $Q_{max}$. In practice, in the construction of case furniture, more complex states of loads act on the nodes. The most dangerous include the bending moments, which cause mutual stresses of connection elements. Deformations of the connector and deformations of board elements cause that the stiffness of the structural node depends on the geometry of elements after deformation and material susceptibility. The deformation of a joint loaded by the bending moment takes place gradually. Along with the growth of the value of the bending moment $M$, the values of stress force $P$, the character of surface stresses $q_{yz}$ and the value of the angle $\varphi$ change (Fig. 7.48).

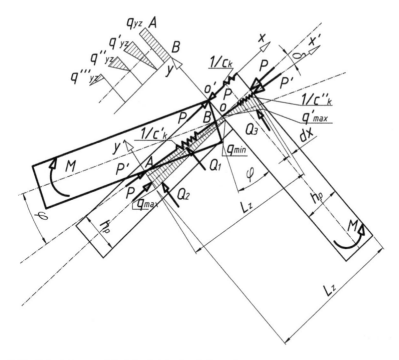

**Fig. 7.48** Calculation model of the joint after deformation

Assuming the indicators as in Table 7.6, it can be written that for $\varphi = 0$ and $P = 0$ (Figs. 7.48 and 7.49):

$$q_{yz} = \frac{Q_o}{h_p \int dz} \tag{7.133}$$

where

$$Q_o \le k_t^w \pi D_o sn \tag{7.134}$$

or

$$Q_o \le \frac{k_r^w \pi (D_o^2 - d_o^2) n}{4} \tag{7.135}$$

or

$$Q_o \le 2k_t^w \sin\gamma \int_{\frac{d_o}{2\sin\varpi}}^{\frac{D_o}{2\sin\varpi}} \int_0^{2\pi} z \, dz \, d\beta. \tag{7.136}$$

**Fig. 7.49** Calculation model of the joint before deformation

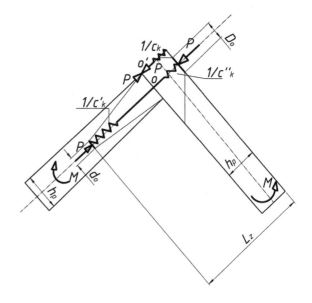

For $\varphi = 0$ and $P > 0$ (Fig. 7.49), the stresses change the value from $q_{yz}$ to $q'_{yz}$. The stiffness of the joint and value of the bending moment is written as follows:

for the part of screw with the thread,

$$\frac{1}{c'_k} = \frac{4\left(L_z - h_p\right)}{E_2 \pi \left(h_p^2 - d_o^2\right)} = \frac{\Delta l}{P},\tag{7.137}$$

$$M = \frac{1}{8} h_p \varepsilon_2 E_2 \pi \left(h_p^2 - d_0^2\right),\tag{7.138}$$

for the part of screw with the head,

$$\frac{1}{c'_k} = \frac{4h_p}{E_1 \pi \left(\left(\frac{h_p + D_o}{2}\right)^2 - D_o^2\right)} = \frac{\Delta l}{P},\tag{7.139}$$

$$M = \frac{1}{8} h_p \varepsilon_1 E_1 \pi \left(\left(\frac{h_p + D_0}{2}\right)^2 - D_o^2\right).\tag{7.140}$$

For $\varphi > 0$ and $P > 0$ (Fig. 7.48), the stresses change the value from $q'_{yz}$, to $q'''_{yz''}$ depending on the value $M$. Therefore, assuming the stiffness of the joint in the following form:

$$k = \frac{M}{\varphi}, \qquad (7.141)$$

for the threaded part of the screw, we obtain

$$\varepsilon_i = \frac{\Delta l}{l} = 0.02, \qquad (7.142)$$

$$q_{min} = q_{max} \frac{\sin\left(\frac{\varphi}{2}\right)h_p}{L_z - h_p}, \qquad (7.143)$$

$$Q_1 = q_{max} \sin\left(\frac{\varphi}{2}\right)d_o = \frac{1}{2}E_2\varepsilon_2 d_o, \qquad (7.144)$$

$$Q_2 = \frac{1}{2}q_{max}d_o \int_0^{L_z-h_p} dx = \frac{1}{2}E_2\varepsilon_2 d_o \int_0^{L_z-h_p} dx, \qquad (7.145)$$

$$P = E_1\varepsilon_1 \mathrm{tg}(\alpha)d_o \int_0^{h_p} xdx, \qquad (7.146)$$

$$M = \frac{1}{2}\left(\frac{1}{2}E_1\varepsilon_1 h_p^3 + E_2\varepsilon_2\left(\frac{2}{3}\left(L_z - h_p\right)^2 d_o + \sin\left(\frac{\varphi}{2}\right)h_p d_o\left(L_z - h_p\right)\right)\right), \qquad (7.147)$$

and for the part with the head:

$$q'_{max} = \frac{Q_3}{\frac{D_o - d_o}{2}x} = E_2\varepsilon_2, \qquad (7.148)$$

$$Q_3 = \frac{1}{2}E_2\varepsilon_2 d_o \int_0^{h_p} dx, \qquad (7.149)$$

$$P = E_1\varepsilon_1 \mathrm{tg}(\alpha)d_o \int_0^{h_p} xdx, \qquad (7.150)$$

$$M = \frac{1}{12}h_p^2\left(3E_1\varepsilon_1 h_p + 4E_2\varepsilon_2 d_o\right). \qquad (7.151)$$

As it can be seen, the stiffness of the structural node depends on the suscepti-bility of the connector and type of board materials used.

### 7.2.2.8  Stability of Side Walls

Operational load of the top, acting along the side wall of the furniture body, depending on its intensity, can cause a change in the shape which is the result of the loss of their stability. From a mechanical point of view, the side wall constitutes a board in which one longitudinal edge is free, the second one parallel to it and two transverse ones are supported in a discreet way. Therefore, the form of buckling of side walls depends on the number of horizontal partitions related to them (Figs. 7.50 and 7.51).

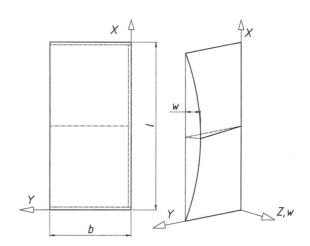

**Fig. 7.50**  The buckling of the side wall not connected with partitions

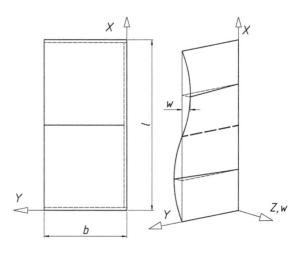

**Fig. 7.51**  The buckling of the side wall connected with partitions

Critical loads of an orthotropic side wall can be determined by solving the equation:

$$D_x \frac{\partial^4 w}{\partial x^4} + 2D \frac{\partial^4 w}{\partial x^2 \partial y^2} + D_y \frac{\partial^4 w}{\partial y^4} = N_{xy} \frac{\partial^2 w}{\partial x \partial y}, \tag{7.152}$$

where

$$D_x = \frac{E_x h^3}{12(1 - \upsilon_x \upsilon_y)} \tag{7.153}$$

$$D_y = \frac{E_y h^3}{12(1 - \upsilon_y \upsilon_x)} \tag{7.154}$$

$$D = D_{xy} + 2D_z, \tag{7.155}$$

$$D_{xy} = D_x \nu_y = D_y \nu_x \tag{7.156}$$

$$D_z = \frac{G h^3}{12} \tag{7.157}$$

$D_i$    bending stiffness of the board ($i = x, y, z,$),
$E_x$   linear elasticity module in the direction of $x$,
$E_y$   linear elasticity module in the direction of $y$,
$G$     shear modulus,
$h$     thickness of the board and
$w$    function describing the surface of the board bending,

substituting the following expression for the deflection of the board:

$$w(x, y) = f(y) \sin \frac{m \pi x}{a}. \tag{7.158}$$

Hence, we obtain the equation describing the value of critical force for the side wall loaded longitudinally:

$$N_{xy} = D_y \theta^2(m) + \frac{\left( D_x D_y - D_{xy}^2 \right)}{D_z} \frac{m^2 \pi^2}{a^2} \tag{7.159}$$

where

$m = 1, 2, 3, \ldots$

### 7.2.2.9  Stability of Rear Wall

In addition to torsional loads, the board elements also shift shield loads, that is forces lying in their plane. Hence, for flaccid elements, in this case the rear walls, the possibility of the loss of stability occurs (Smardzewski 1991). Therefore, it is necessary to check the stability of this element for the most unfavourable load. Such a load occurs in the case of supporting the furniture body in four corners (Fig. 7.52).

To determine the value of critical loads in boards, the method proposed by Southwell (1954) is commonly used. This method in laboratory works on wood-based boards was also used by Ożarska-Bergandy (1983). It consists in measuring deflections of the board and the loads corresponding to these deflections. By drawing up a graph of the load-deformation dependencies for the central points of the surface of the shield, the value $F_{kr}$ can be read from it as an ordinate of asymptote, to which the deformation curve strives (Fig. 7.53).

The nature of work of rear walls, being in a state of clear shearing, requires a broader discussion. The primary task is determining the value of critical loads in which the rear walls, as isotropic and orthotropic boards attached to the body in various ways, lose their stability. The values of loads and critical stresses are calculated on the basis of the linear theory. For boards subjected to shearing like in Fig. 7.54, the differential equation of the bending surface takes the following form:

$$D\left(\frac{\partial^4 w}{\partial x^4} + 2\frac{\partial^4 w}{\partial x^2 \partial y^2} + \frac{\partial^4 w}{\partial y^4}\right) = 2N_{xy}\frac{\partial^2 w}{\partial x \partial y}, \qquad (7.160)$$

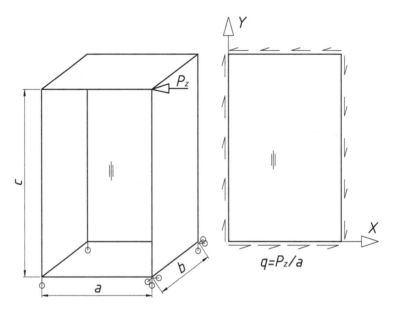

**Fig. 7.52** Shield loads of the rear wall caused by supporting the body in four corners

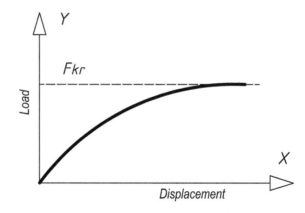

**Fig. 7.53** The relationship between the deformation and the value of critical forces (Skan and Southwell 1954)

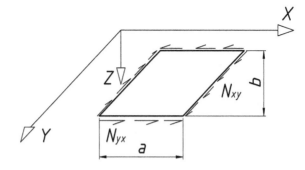

**Fig. 7.54** Scheme of a sheared isotropic board

where

$N_{xy}$   edge tangential forces,

$w$   function describing the surface of the board bending, $D = Eh^3/(12(1 - v^2))$,

$E$   linear elasticity module of the board,

$h$   thickness of the board and

$v$   Poisson's ratio.

By applying the energy methods commonly used in the theory of elasticity, critical values of edge contact loads are sought. By determining the work of external forces by $\Delta W$, the energy of the bent board by $\Delta U$, we determine the value of critical forces from the equation:

$$\Delta W = \Delta U \tag{7.161}$$

whereby

$$\Delta W = \int_0^a \int_0^b N_{xy} \frac{\partial w}{\partial x} \frac{\partial w}{\partial y} \, dx dy, \tag{7.162}$$

and

$$\Delta U = \frac{1}{2}D \int\limits_0^a \int\limits_0^b \left( \left( \frac{\partial^2 w}{\partial x^2} + \frac{\partial^2 w}{\partial y^2} \right)^2 - 2(1-v)\left( \frac{\partial^2 w}{\partial x^2}\frac{\partial^2 w}{\partial y^2} - \left( \frac{\partial^2 w}{\partial x \partial y} \right)^2 \right) \right) dxdy$$

$$(7.163)$$

The orthotropic board is a board of orthogonal anisotropy. This type of anisotropy occurs when the structural elements of the board are mutually perpendicular, e.g. the perpendicular-fibre plywood, particle board veneered on both sides or a glued wooden board. And both directions of the constructions are the main axes of elasticity of an orthotropic board. By determining the main directions of anisotropy for a board loaded like in Fig. 7.55, the differential equation of the bent surface takes the form:

$$D_x \frac{\partial^4 w}{\partial x^4} + 2H \frac{\partial^4 w}{\partial x^2 \partial y^2} + D_y \frac{\partial^4 w}{\partial y^4} = 2N_{xy}\frac{\partial^2 w}{\partial x \partial y} \qquad (7.164)$$

where

$$H = \frac{Gh^3}{6} + \frac{1}{2}\left( D_x v_y + D_y v_x \right) \qquad (7.165)$$

$$D_x = \frac{E_x h^3}{12\left(1 - v_x v_y\right)} \qquad (7.166)$$

$$D_y = \frac{E_y h^3}{12\left(1 - v_y v_x\right)} \qquad (7.167)$$

$E_x$  linear elasticity module in the direction of $x$,
$E_y$  linear elasticity module in the direction of $y$,
$G$   shear modulus and
$h$   thickness of the board,

**Fig. 7.55** Scheme of a sheared orthotropic board

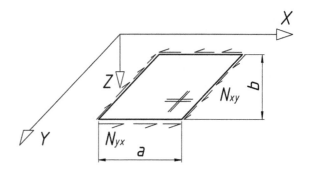

whereby

$$D_x v_y = D_y v_x. \tag{7.168}$$

Solutions of critical stresses of rectangularly non-unidirectional (orthotropic) boards loaded by contact forces should be sought by integrating the differential equation or on the basis of the previously presented energy methods.

Stability of Board with Articulated Support on Perimeter

The buckling surface for a free support of edges of the isotropic board can be adopted in the form of a double trigonometric series fulfilling all the border conditions:

$$w = \sum_{m=1}^{\infty} \sum_{n=1}^{\infty} A_{mn} \sin \frac{m\pi x}{a} \sin \frac{n\pi y}{b}. \tag{7.169}$$

By using one of the presented solving methods, we obtain an expression for the value of critical edge forces in the following form:

$$N_{xy} = k \frac{\pi^2}{b^2} D \geq \frac{P_z}{b} \tag{7.170}$$

where
b  smaller dimension of the rear wall,
k  coefficient dependent on the relation $c/a$, determined on the basis of Table 7.7.

A freely supported sheared orthotropic board is described by Dutko (1976). He demonstrates how to calculate critical loads according to the following equation:

$$N_{kr} = k \frac{4\pi^2}{b^2} \sqrt{4 D_x D_y^3}. \tag{7.171}$$

In this equation, the value of the coefficient $k$ is determined on the basis of chart (Fig. 7.56), for the following characteristics:

Table 7.7 Value of the coefficient $k$ for isotropic boards

| b/a | 1.0 | 1.2 | 1.4 | 1.5 | 1.6 | 1.8 | 2.0 | 2.5 | 3.0 |
|-----|-----|-----|-----|-----|-----|-----|-----|-----|-----|
| k | 9.34 | 8.0 | 7.3 | 7.1 | 7.0 | 6.8 | 6.6 | 6.1 | 5.9 |

**Fig. 7.56** Dependence of the
coefficient $k$ from the
characteristics $\alpha$ and $\eta$ (Dutko
1976)

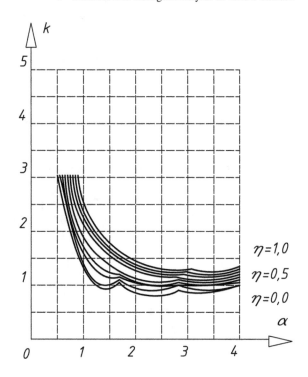

stiffness characteristic of the board,

$$\eta = \frac{H}{\sqrt{D_x D_y}} \qquad (7.172)$$

reduced ratio of sides,

$$\alpha = \frac{a}{b} \sqrt[4]{\frac{D_y}{D_x}}. \qquad (7.173)$$

where
$H, D_x, D_y$  as previously.

Stability of Board with Fixed Support on Perimeter

The method of completely attaching the rear walls of furniture in the body by
gluing, due to the inability to disassemble, has never been used in industrial
practice. From the cognitive point of view, to check the theoretical assumptions,

this example should be regarded as necessary. Timoshenko and Gere (1963), when presenting the approximate method for solving the case of a rectangular shield mounted on four sides, subject to shearing, introduced an approximate form of a curved sheet described by the function:

$$w = \frac{A}{4}\left(1 - \cos\frac{2\pi x}{a}\right)\left(1 - \cos\frac{2\pi y}{b}\right). \tag{7.174}$$

Huber (1922) suggests assuming a similar function of buckling for unidirectional boards loaded symmetrically. Further analysis of sheared boards, in solving cases of fixing around the whole perimeter, was developed by Skan and Southwell (1954) and also Budiansky and Connor (1948). Studies of critical states of rectangular boards freely supported on the perimeter, and applied to surface girders, were conducted by Girkmann (1957). Studies related to the stability of shields for various support conditions were also conducted by Cox (1933). A convenient and simple method for determining critical stresses for boards of various border conditions was also suggested by Wolmir (1956). He recommended calculating values of the coefficient $k$, according to the specified method of supporting a board. Therefore, the value of critical edge forces for the case of fixing edges can be expressed by the following equation:

$$N_{xy} = k\frac{\pi^2}{b^2}D, \tag{7.175}$$

where

$$k = 8.98 + \frac{5.5}{\beta^2}. \tag{7.176}$$

For an orthotropic board, the critical values of contact forces are calculated according to the equation:

$$N_{kr} = 4k\frac{\sqrt{4D_x D_y^3}}{b^2}, \tag{7.177}$$

where
$k$  coefficient dependent on the relation $1/\eta$, determined on the basis of Table 7.8

Table 7.8  Value of the coefficient $k$ for isotropic boards

| $1/\eta$ | 0.0 | 0.2 | 0.5 | 1.0 | 2.0 | 3.0 | 5.0 | $\infty$ |
|---|---|---|---|---|---|---|---|---|
| $k$ | 18.6 | 18.9 | 19.9 | 22.12 | 18.8 | 17.6 | 16.6 | 15.1 |

Stability of Board with Discontinuous Support on Perimeter

Huber (1923) was the first to analyse the scheme partially fixed elastically on the perimeter of an isotropic board, subject to shearing (Huber 1923). He proposed to adopt the function of a curved shaped board in the following form:

$$w = A \sin \frac{\pi x}{b} \sin \frac{\pi y}{b} \sin \frac{\pi}{a} \left( x - \frac{a}{b} y \right). \tag{7.178}$$

This equation does not satisfy the border conditions for free support and gives a finite value of the moment on the perimeter. Therefore, when calculating the energy of buckling board $\Delta U$ and work of external forces $\Delta W$, the sought critical edge load is obtained, expressed in the form

$$N_{kr} = 4\pi^2 \frac{D}{b^2} \left( \beta + \frac{1}{\beta} + \frac{1}{\beta^2} \right). \tag{7.179}$$

The application of an orthotropic board in a discontinuous manner (discreet) corresponds to the scheme of fixing the rear wall to the body using stables or bolts. Korolew (1970) dealt with the stability of such a board, at the mentioned support conditions, noting that, during the course of normal load of the furniture body, the protuberances of the rear wall assume the form of oblique waves. In this case, for the equation of a buckling surface, fulfilling all the border conditions, the following function should be adopted:

$$w = A \left( 1 - \cos 2 \left( \frac{m\pi x}{a} + \frac{n\pi y}{b} \right) \right). \tag{7.180}$$

By substituting this expression to the differential equation of a sheared shield and specifying the number of oblique buckling waves for conditions of minimum kinetic energy loading the furniture body, we get the following equation:

$$N_{kr} = \frac{15.2}{b^2} (k + 2) \sqrt{k - 1} \sqrt{\frac{\left( 3D_x v_y + Gh^3 \right)^3}{27 D_x}}, \tag{7.181}$$

where

$$k = \sqrt{1 + \frac{27 D_x D_y}{\left( 3D_x v_y + Gh^3 \right)^2}}. \tag{7.182}$$

### 7.2.2.10  Stability of Case Furniture

Due to the safety of the user, the stability of furniture is probably the most important characteristic. As far as fractures of elements or cracks in the joints cause a gradual loss of stiffness in the construction, depending on the size of these defects, the loss of stability of the furniture, especially of a large weight, may suddenly and immediately endanger the health or life of the user. This especially concerns furniture for children and infants, who cannot respond to states of imminent danger. Current methods for assessing construction stability come down to laboratory measurements of horizontal or vertical load values, for which the furniture is subject to displacement. Below are solutions that enable to closely assess, through analysis, the stability of the designed piece of furniture.

In furniture of a case construction, first the coordinates of the centre of gravity of the construction in a state of operational load must be determined. To this end, the transverse cross section of a furniture piece in a side view should be considered (Fig. 7.57), taking into account both the mass loads and operational loads. An interesting value of the $x$ coordinate of the centre of gravity is determined from the equation:

$$x = \frac{\sum Q_i x_i}{\sum Q_i},\qquad(7.183)$$

**Fig. 7.57** The calculation scheme of the location of the centre of gravity of a case furniture body

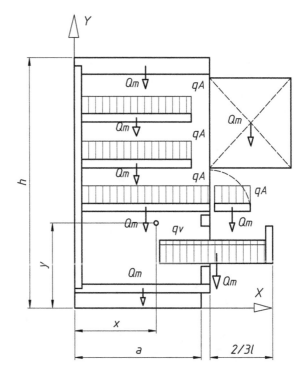

in which

$$\sum Q_i x_i = \sum_{i=1}^{n} \rho_i x_i + \sum_{i=1}^{m} A_i q_{Ai} x_i + \sum_{i=1}^{k} V_{si} q_{Vi} x_i + \sum_{i=1}^{p} A_{\rho i} q_{\rho i} x_i, \qquad (7.184)$$

$$\sum Q_i = \sum_{i=1}^{n} V_i q_i + \sum_{i=1}^{m} A_i q_{Ai} + \sum_{i=1}^{k} V_{si} q_{Vi} + \sum_{i=1}^{m} A_{\rho i} q_{\rho i}, \qquad (7.185)$$

where

$V_i$    volume of element,
$\rho_i$    density of element,
$A_i$    area of shelves, horizontal partitions, bottom and top,
$q_{Ai}$    surface load of shelves, horizontal partitions of the bottom,
$V_{si}$    volume of the drawer,
$q_{Vi}$    volume load of the drawer,
$A_{\rho i}$    area of the door of horizontal rotation axis,
$q_{\rho i}$    surface load of the door of horizontal rotation axis and
$x_i$    abscissa of coordinates of location of the centre of gravity of element or load, in relation to the beginning of the system.

Knowing the location of the centre of gravity $x$ of the furniture block, the balance state of the body based on known dependencies can be established, and when

- $x > a$ the body loses balance on its own (falls over without the use of external force),
- $x = a$ the body maintains in shaky balance, that is in a state when any small horizontal force $P$ causes a loss of its stability,
- $x < a$ the body maintains fixed balance and a certain horizontal force is needed to throw the furniture piece off this state. The value of this load can be written in the form

$$P = \frac{a}{h} \sum Q_i \left(1 - \frac{x}{a}\right) \geq P_{kr}, \qquad (7.186)$$

where
$a$ and $h$    dimensions of the side cross section of the body and
$P_{kr}$    acceptable critical load.

Dressers composed of a bottom part and an upper part set on top pose a particular danger to the user (Fig. 7.58). No connection of both parts and a change in the location of the centre of gravity of the body by opening a door cause the risk of the upper part tipping over. In order to answer the question of whether a piece of furniture can lose stability on its own and/or what minimum force can lead to a loss

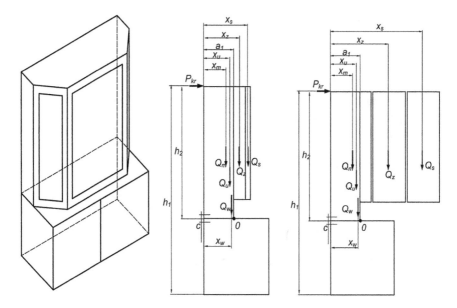

**Fig. 7.58** Calculation scheme of the loss of stability of a dresser's top part before and after opening the doors

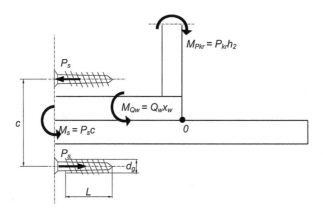

**Fig. 7.59** The calculation scheme of the strength of the screw joint connecting the top part with the base

of stability, the static schemes set out in Fig. 7.58 should be used. This drawing shows the construction of a furniture piece filled with operational load, in which the doors of the extension are closed. The same drawing, on a scheme next to it, illustrates an identically loaded construction, in which the doors of the extension are open. Fig. 7.59 shows the solution of the structural node, which is to prevent the extension from tipping over as a result of external loads. For all calculation schemes, the following indicators have been adopted:

$a_1$     width of the side wall of top extension,
$c$      spacing of screw connectors,
$d_g$    thread diameter of screw connector,
$h_2$    height of the side wall of top part,
$h_1$    total height of the segment (with the top part),
$k_t^w$  shearing strength of the wood-based material (particle board),
$L$      thread length of screw,
$M_{Pkr}$ moment of force $P_{kr}$,
$M_{Qw}$ moment of force $Q_w$,
$M_s$    moment of force $P_s$,
$P_{kr}$ external critical force causing loss of stability of the furniture top part,
$P_s$    force pulling out the screw connector from the body of the top part or bottom cabinet,
$Q_m$    resultant mass load of the top part,
$Q_s$    resultant of central load of the door wing,
$Q_u$    resultant operational load of the top part,
$Q_w$    resultant force deriving from the sum of all mass and operational forces,
$Q_z$    resultant of external load of the door wing,
$x_m$    location of the resultant vector of the mass load of top part $Q_m$,
$x_s$    location of the resultant vector of the central load of the door wing $Q_s$,
$x_u$    location of the resultant vector of the operational load of top part $Q_u$,
$x_w$    location of the vector of the resultant force $Q_w$ and
$x_z$    location of the resultant vector of the external load of the door wing $Q_z$

The calculation methodology presented below allows to establish the following:

- whether the furniture piece is stable without any external load at the most favourable and least favourable scheme of usage,
- when the furniture piece will lose its stability under the influence of external load and
- what the minimum value of an external critical load is, when a self-stable furniture piece can lose stability and tip over.

For the provided constructions, the location of the vector of resultant force and the value of the resultant force should be calculated according to the following formulas:

$$x_w = \frac{Q_m x_m + Q_u x_u + 2Q_z x_x + Q_s x_s}{Q_m + Q_u + 2Q_z + Q_s}, \qquad (7.187)$$

$$Q_w = Q_m + Q_u + 2Q_z + Q_s. \qquad (7.188)$$

The furniture loses its stability under the influence of external load if the value of the external force is greater than that determined on the basis of the following formula of critical force values:

$$P_{kr} = Q_w \frac{(a_1 - x_w)}{h_2}.$$ (7.189)

In securing the top part against tipping over, by additionally fixed screws, the value of external force causing the loss of stability of the top part can be increased; thus, the safety of the structure can be increased. After fastening a metal connector, the value of external critical force can be calculated from the equation:

$$P_{kr} = \frac{Q_w x_w + c\pi d_g L k_t^w}{h_2}.$$ (7.190)

### 7.2.3 Strength of Case Furniture

#### 7.2.3.1 State of Internal Forces in Corner Joints

The primary determinant of the strength of a furniture body is the strength of its joints. Therefore, the forces acting on the nodes of a loaded construction must be established and the strength of joints specified. The previously assumed reservations that the boards of the furniture body are connected articulately and only in the corners allow to conclude that the impacts between the boards are also focused in the corners. Such assumptions correspond well to the constructional solutions using separable joints like eccentric joints, minifix or confirmat, distributed in ones in the corners (Fig. 7.60a). In the case of joints distributed in more than two, on the length of the edge of the board, e.g. dowels or more confirmants, we recognise that calculation angular corner forces are displaced by connectors distributed on half the length of the edge (Fig. 7.60b). From precise solutions, it is known that in joints

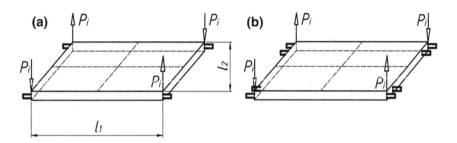

**Fig. 7.60** Corner forces shifted by: **a** one connector in the corner, **b** a few connectors in the middle of the length of the edge

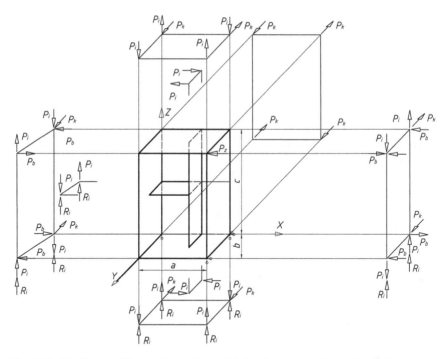

**Fig. 7.61** Distribution of internal forces in a multi-chamber construction torsionally loaded

between boards, forces $P$ occur perpendicular to each of the joined boards (Fig. 7.61). The force $P_i$, causing twisting of the ith board of the dimensions $(l_1 l_2)_i$, amounts to

$$P_i = \frac{G_i d_i^3}{3(l_1 l_2)_i} \alpha_i \frac{P_z}{\sum_{i=1}^{n} \frac{G_i d_i^3}{3(l_1 l_2)_i} \alpha_i^2}. \tag{7.191}$$

Figure 7.61 also shows that connectors of external elements (bottom, top and side walls) are loaded by edge forces $P_k$, originating from an external load $P_z$, with a value of

$$P_k = \frac{a}{b} P_z. \tag{7.192}$$

For corners not loaded by external forces $P_z$, the force $P_i$ is a force shifted by the connector of the construction's node. And in the case of a corner loaded by external forces, the values of internal torsional force of the board are the difference between external load and force $P_i$:

**Fig. 7.62** Load of horizontal partition

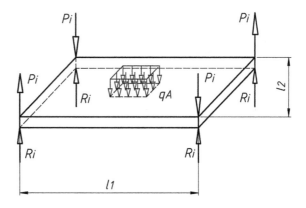

$$P_b = P_z - P_i. \tag{7.193}$$

Connectors of vertical partitions of course only shift $P_i$ loads perpendicular to its plane, while a group of connectors of corner horizontal partitions, in addition to torsional loads $P_i$, also shifts the fourth part of surface loads attributed to this partition (Fig. 7.62) according to the following equation:

$$P_A = P_i + R_i = P_i + \frac{1}{4} q_{A_i} (l_1 l_2)_i. \tag{7.194}$$

### 7.2.3.2 Strength of Inseparable Joints

Knowing the distribution and values of external forces between board elements of the furniture body, the strength of appropriate connectors can be determined and the number of joints attributed to the length of the board's edge can be therefore designed. Table 7.9 provides geometric examples, as well as the strength criteria for designing dowel joints connecting the side wall with the bottom (bottom flange) of the body of rack and flange structure. Considering this example seems to be the most effective, because in the whole construction of the furniture, this node is the most heavily loaded with all kinds of internal forces.

For design calculations, the following loads should be adopted:

- for internal vertical partitions only forces $P_i$,
- for horizontal partitions forces $P_i$ and $R$,
- for top board forces $P_i$ and $P_k$,
- for bottom board forces $P_i$, $R$ and $P_k$ and
- for side walls $P_b$.

**Table 7.9** Strength conditions for the dowel joint connecting the side wall with the bottom (bottom flange)

| Type | Scheme of distribution of forces | Strength conditions | Border conditions |
|---|---|---|---|
| Rack construction | | $$n \geq \frac{4m\sqrt{(P_i + R)^2 + (0.5P_k)^2}}{\pi d^2 k_t^b}$$ $g_1 = 3/4g, \ g_2 = l - g_1$ | $k_t^b$—shearing strength of connector material, for beech wood $k_t^b = 6.5{-}18$ MPa |
| | | $$n \geq \frac{m(P_i + R)}{(2eg_1 + ed)k_t^w}$$ | $k_t^w$—shearing strength of the board, for particle board $k_t^w = 4{-}6$ MPa |
| | | $$n \geq \frac{2m(P_i + R)}{\pi g_2^2 k_r^w}$$ | $k_r^w$—splitting strength of the slot material, for particle board $k_r^w = 0.4{-}0.5$ MPa |
| | | $$n \geq \frac{mP_b}{\pi d g_{\min} k_t^k}$$ $g_1 < g_2, \ g_{\min} = g_1$ | $k_t^k$—shearing strength of the glue, $k_t^k = 2.5{-}7$ MPa |
| Flange construction | | $$n \geq \frac{4m\sqrt{P_b^2 + (0.5P_k)^2}}{\pi d^2 k_t^b}$$ | $k_t^b$—shearing strength of connector material, for beech wood $k_t^b = 6.5{-}18$ MPa |
| | | $$n \geq \frac{mP_b}{(2eg_1 + ed)k_t^w}$$ | $k_t^w$—shearing strength of the board, for particle board $k_t^w = 4{-}6$ MPa |
| | | $$n \geq \frac{2mP_b}{\pi g_2^2 k_r^w}$$ | $k_r^w$—splitting strength of the slot material, for particle board $k_r^w = 0.4 - 0.5$ MPa |
| | | $$n \geq \frac{m(P_i + R)}{\pi d g_{\min} k_t^k}$$ | $k_t^k$—shearing strength of the glue, $k_t^k = 2.5{-}7$ MPa |

In the formulas: $n$ means the number of individual connectors on half the length of the edge, $m$ coefficient of safety determined experimentally, can be assumed $m = 1.1{-}2$

### 7.2.3.3 Strength of Separable Joints

Strength of Screw Joints

As an example of the strength of screw joints, let us consider the widely used method of fixing the rear wall to the side walls, bottoms and top (Fig. 7.63). On the number and spacing of connectors $b_x$, $a_y$, depends not only the stability of the rear wall, but also on stiffness and strength of the furniture body. In the construction loaded by horizontal force $Q$ on the length of the edges of the rear wall, there are streams of contact intensities $q_{yx} = Q/b$ causing shearing of connectors. Depending on the number of connectors distributed on the length of edge $a$ and $b$, the forces attributed to one connector can be calculated as

$$F_H = \frac{q_{yx}b}{m}, \quad F_V = \frac{q_{xy}a}{n}, \tag{7.195}$$

where
$F_H$          force acting on the connector located in horizontal rows,
$F_V$          force acting on the connector located in vertical rows,
$q_{yx}$ and $q_{xy}$   appropriate streams of contact intensities,
$a$           height of the rear wall,

**Fig. 7.63** The distribution of screws fixing the rear wall of the body

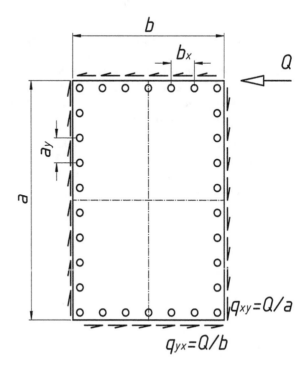

**Fig. 7.64** Calculation
scheme of a screw

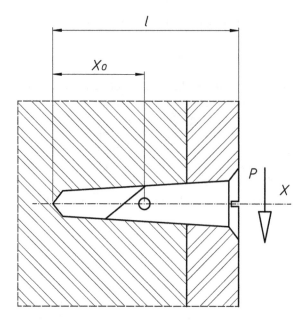

b                    width of the rear wall,
n                    number of connectors on the edge of the length $a$ and
m                    number of connectors on the edge of the length $b$.

In order for the constructor to provide the number and type of screws, it is
necessary to specify the acceptable stresses on the screw and the greatest normal
stresses caused by its bending. The static scheme of the connector with loading by
shearing forces has been shown in Fig. 7.64.

In order to improve the calculation formulas, Korolew (1973) suggested the
following simplifying assumptions:

- the intensity of the load in the plane of bending the screw changes according to
  the curve of the third degree $q_x = q_o x^3$. The beginning of the system of coordinates have been placed at a point of balance, which location is also the subject
  of calculation (Fig. 7.65),
- on the transverse cross section of the screw, the contact stresses are
  cosine-distributed and disappear only on side surfaces, in contact with the
  surface of the board (Fig. 7.66) where

$$-\frac{\pi}{2} \leq \varphi \leq \frac{\pi}{2},$$                           (7.196)

$$p(\varphi) = p_o \cos \varphi,$$                                          (7.197)

**Fig. 7.65** The distribution of stresses on the length of the screw at shield load of the rear wall

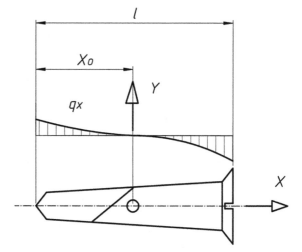

**Fig. 7.66** The distribution of stresses at the cross section of the screw at shield load of the rear wall

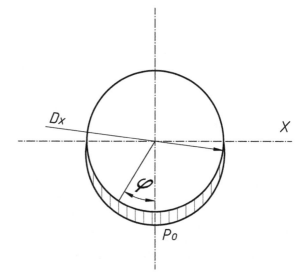

$$|\varphi| \geq \frac{\pi}{2}, \tag{7.198}$$

$$p(\varphi) = 0. \tag{7.199}$$

The dependence of the maximum values of contact stresses in the transverse cross section of screw $x$ from the intensity of the attributed load is calculated as

$$\int\limits_{0}^{\pi/2} p_o(x)D_x \cos^2 \varphi d\varphi = q(x), \qquad (7.200)$$

from which we can obtain

$$p_o(x) = \frac{4q(x)}{\pi D_x}. \qquad (7.201)$$

For the adopted system of coordinates, the diameter of the screw in the chosen cross section $x$ amounts to

$$D_x = D(\rho + (1 - \rho)(\xi + \xi_o)), \qquad (7.202)$$

where
$D$          largest screw diameter,
$P = d/D$    reference coefficient of screw diameters,
$\xi = x/l$      reference coefficient of the location of the transverse cross section of the screw,
$\xi_o = x_o/l$   reference coefficient of the location of the beginning of the system of coordinates and
$d$          smallest screw diameter

In order to determine unknown values $q_o$ and $\xi_o$, the conditions of balance of the screw are used:

$$\begin{cases} q_o \int\limits_{-x_o}^{l-x_o} x^3 dx = P \\ q_o = \int\limits_{-x_o}^{l-x_o} x^4 dx = P(l - x_o). \end{cases} \qquad (7.203)$$

After integrating equations and elementary transformations in order to determine $q_o$ and $\xi_o$, we obtain a nonlinear system of algebraic equations:

$$\begin{cases} q_o\left(1 - 4\xi_o + 6\xi_o^2 - \xi_o^3\right) = \frac{4P}{l^4}, \\ q_o\left(1 - 5\xi_o + 10\xi_o^2 - 10\xi_o^3 + 5\xi_o^4\right) = \frac{5P}{l^4}(1 - \xi_o). \end{cases} \qquad (7.204)$$

By calculating $q_o$, from the above equation, the location of the origin of the system of coordinates is determined in the form:

$$10\xi_o^3 - 10\xi_o^2 + 2\xi_o - 1 = 0. \qquad (7.205)$$

According to Korolew's (1973) calculations, $\xi_o = 0.38$ determines the origin of the system of coordinates measured from the left end of the screw. Therefore, by solving the system of equations,

$$\begin{cases} q_o\left(1 - 4\zeta_o + 6\zeta_o^2 - \zeta_o^3\right) = \frac{4P}{l^4}, \\ \zeta_o = 0.38 \end{cases}$$

(7.206)

we obtain that

$$q_o = \frac{31.4P}{l^4}.$$

(7.207)

The greatest intensity of load distribution on the length of the screw is formed in its largest cross section at $\xi = 1 - \xi_o$ and can be expressed by the equation:

$$q_{max} = q_o(l - x)^3 = 7.6\frac{P}{l}.$$

(7.208)

And the biggest value of stress force can be determined by

$$p_{max} = \frac{4q_{max}}{\pi D} = 9.65\frac{P}{Dl}.$$

(7.209)

Strength calculations for the screw usually come down to determine the cross section, in which most normal stresses disappear. The bending moment forming in the cross section of the screw specified by the parameter $\xi$ is determined from the following equations:

$$\begin{cases} -\xi_o \leq \xi \leq 0 \\ M(\xi) = 6.3Pl\left(\xi^5 + \xi_o^5\right) \\ 0 \leq \xi \leq 1 - \xi_o \\ M(\xi) = 6.3Pl\left[(1 - \xi_o)^5 - \xi^5\right] \end{cases}$$

(7.210)

Hence, the normal stresses arising in the transverse cross sections of the screw during its shearing are as follows:
for $-\xi_o \leq \xi \leq 0$

$$\sigma_{max} = 64.2\frac{Pl}{D^3}\frac{\left(\xi^5 + \xi_o^5\right)}{(\rho + (1 - \rho)(\xi + \xi_o))l^3},$$

(7.211)

for $0 \leq \xi \leq 1 - \xi_o$

$$\sigma_{max} = 64.2\frac{Pl}{D^3}\frac{\left[(1 - \xi_o)^5 - \xi^5\right]}{(\rho + (1 - \rho)(\xi + \xi_o))l^3}.$$

(7.212)

The main cross section of the screw is the cross section located at the distance $\xi = 0$, so for this cross section, the largest normal stresses are as follows:

$$\sigma_{\max} = 64.2 \frac{Pl}{D^3} \frac{(1 - \xi_o)^5}{(\rho + (1 - \rho)\xi_o)^3}. \tag{7.213}$$

Eventually, the strength conditions of fixing a rear wall to the body using screws can be written in the form:

$$\begin{cases} \frac{5.8Pl}{D^3(0.38+0.62p)^3} \leq k_g^r \\ \frac{9.6P}{Dl} \leq k_c^w \end{cases}, \tag{7.214}$$

where
$k_g^r$   bending strength of the screw material and
$k_c^w$   compression strength of the board

## Strength of Eccentric Joints

The load-carrying capacity of eccentric joints with eccentric connectors to a great extent determines the strength of structural nodes of case furniture. The distribution of internal forces generated while forcing torsional deformations of the furniture body (Dzięgielewski and Smardzewski 1995) results in the fact that in structural nodes that join the side walls with the bottom, the biggest loads on wooden or metal connectors of wall angular joints occur (Fig. 7.67).

Such a state of loads makes it necessary to check in the joint the shearing strength, splitting strength and compression strength of the particle board. Therefore, strength calculations should be carried out on in situ models, for which initial data must derive from elementary studies of elastic properties of materials used to make joints. Most often, bolt connections in eccentric joints are mounted directly in the particle board, which in this case constitutes a type of nut. Its elastic features have been given in the literature of the subject and in the vast majority these results can be used directly in engineering studies. However, in order to get the most reliable results of engineering calculations for particle board, an orthotropic model of the elastic properties is proposed, taking into account various properties of the surface layers (with microchip layer no. 1) and middle layer of the board (layer no. 2).

For orthotropic material, characterised by the appropriate dependencies between strains $\varepsilon_L$, $\varepsilon_T$, $\varepsilon_R$ and stresses $\sigma_L$, $\sigma_T$, $\sigma_R$, we use known dependencies for stress tensors:

$$\sigma_i = E_i \cdot \varepsilon_i, \tag{7.215}$$

$$\tau_{ij} = G_{ij} \cdot \gamma_{ij}, \tag{7.216}$$

**Fig. 7.67** Distribution of internal forces in a wall angular joint

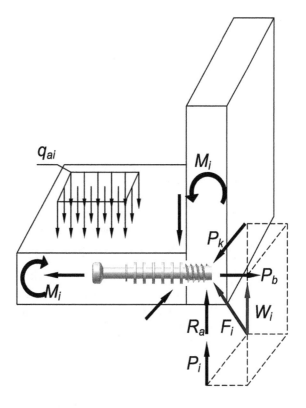

whereby

$$\frac{v_{xz}}{E_x} = \frac{v_{zx}}{E_z}; \frac{v_{xy}}{E_x} = \frac{v_{yx}}{E_y}; \frac{v_{zy}}{E_z} = \frac{v_{yz}}{E_y}. \tag{7.217}$$

For a three-layer particle board with the thickness of 18 mm, where the thickness of individual layers amounts to $h_1 = 3$ mm, $h_2 = 12$ mm, it can be assumed that (Kociszewski et al. 2002) $E_{x1} = 3850$ MPa, $E_{x2} = 1030$ MPa, and calculate substitute Young's modulus, using the equations:

$$E_{zast}J_{zast} = \sum_{i=1}^{n} E_i J_i, \tag{7.218}$$

$$E_{zast}h_{zast}^3 = 2E_1\left(4h_1^3 + 3h_2h_1h_{zast}\right) + E_2h_2^3, \tag{7.219}$$

$$E_{x-subst} = 2523.2\,MPa. \tag{7.220}$$

**Table 7.10** Elastic properties of a three-layer particle board

| Physical quantity | Layer $h_1$ = 3 mm | Layer $h_2$ = 12 mm |
|---|---|---|
| $E_x$ (MPa) | 3850 | 1030 |
| $E_y$ (MPa) | 3301 | 883 |
| $E_z$ (MPa) | 123.9 | 33.2 |
| $v_{xz}$ | 0.368 | 0.098 |
| $v_{xy}$ | 0.270 | 0.072 |
| $v_{zy}$ | 0.059 | 0.016 |

For the most of the examined particle boards $E_{x-subst}$ usually amounts to 2950 MPa, and the remaining modules, respectively, $E_{y-subst}$ = 2530 MPa, and $E_{z-subst}$ = 95 MPa, whereas Poisson's ratios $v_{xz-subst}$ = 0.282, $v_{xy-subst}$ = 0.207 and $v_{zy-subst}$ = 0.045.

Bearing in mind the assumed proportions and determined values of substitute linear elasticity modules and Poisson's ratios, other elastic values for individual layers of the particle board can be specified (Table 7.10).

For practical reasons, a particle board is mostly treated as a homogeneous isotropic material, assuming that the connection of the core with the particle board should shift only post-axial forces $P_b$ (Fig. 7.67) and rotational moments $M_s$ caused by mounting operations. Acceptable loads for the examined joint are transferred through the adhesion surface of the thread and nut according to the scheme shown in Fig. 7.68.

At a given point of the loaded threaded surface, unit forces $\sigma_i$ of normal impact act, with the versor compatible with the normal at the given point $i$ to the surface of the thread. Also unit forces $\tau_i$ will occur, originating from the friction impact of

**Fig. 7.68** Load scheme of bolt and nut coils

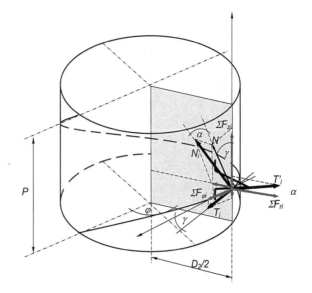

**Fig. 7.69** Load scheme of one bolt coil

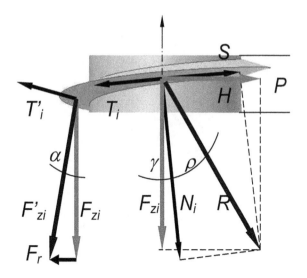

tangentials to the bolt line and the line creating the loaded side of the thread outline. Counterparts of these stresses are elementary forces $dN_i$, $dT_i$ and $dT_i'$.

When solving the load scheme of one thread coil (Fig. 7.69), we check the conditions of self-suppression of the thread and values of resultant forces.

$$\frac{T_i}{N_i} = \mu = \mathrm{tg}\,\rho, \tag{7.221}$$

$$H = F_{zi}\mathrm{tg}(\gamma + \rho), \tag{7.222}$$

$$F_{zi}' = \frac{1}{\cos \alpha} F_{zi}, \tag{7.223}$$

$$T_i = F_{zi}'\mu = F_{zi}\frac{\mu}{\cos \alpha}, \tag{7.224}$$

$$\mu' = \frac{\mu}{\cos \alpha} = \frac{\mathrm{tg}\,\rho}{\cos \alpha}. \tag{7.225}$$

For the examined connector of the eccentric joint, geometric characteristics in accordance with the values indicated in Fig. 7.70 can be assumed.

In order to describe the distribution of stresses along the length of the thread core or in the nut body (particle board) of a semi-cross-wall joint (Fig. 7.71), the distribution of loads of the thread has to be described, whereas

$$E_s > E_p, \quad E_p = E_n = E_{x-\mathrm{subst}}, \tag{7.226}$$

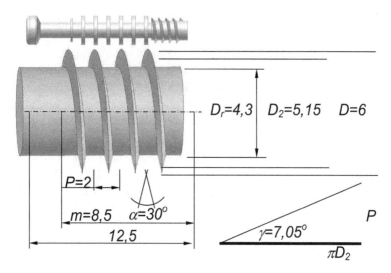

**Fig. 7.70** Geometric parameters of the core thread coils

**Fig. 7.71** Model of a
semi-cross-wall joint

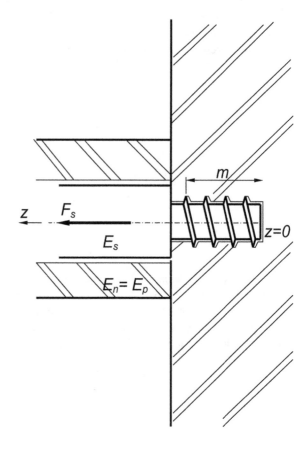

**Fig. 7.72** Scheme of deformations of the thread coils

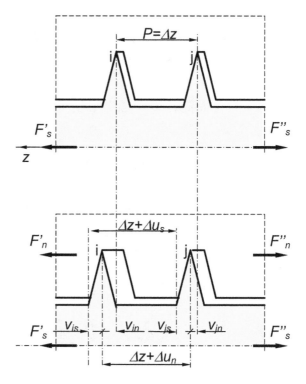

where

$E_s$  linear elasticity module of the bolt core,
$E_p$  linear elasticity module of the board and
$E_n$  linear elasticity module of the nut,

Detailed model of the load of the thread has been given in the work (Dietrich 2008), explaining reasons and character of various loads of the thread coils (Fig. 7.72).

When considering total relative displacements of chosen coils $i$ and $j$ distant from each other by $\Delta_z$ in the post-axial direction, we will find identity connection of elongations or shortenings of the bolt and nut body with the deflections of observed coils,

$$\Delta u_s - \Delta u_n = (v_{is} + v_{in}) - (v_{js} + v_{jn}), \qquad (7.227)$$

where

$\Delta u_s$  change of distance of the bolt thread coils as a result of elastic elongation or shortening of the bolt core,

$\Delta u_n$  change of distance of the nut thread coils as a result of elastic shortening of the nut core,

$v_{is}$ and $v_{js}$   elastic deflections of the $i$th or $j$th coil of the bolt thread, respectively, measured on the average thread diameter,

$v_{in}$, $v_{jn}$   elastic deflections of the $i$th or $j$th coil of the nut thread, respectively, cooperating with the bolt coils.

Degree of differentiation (concentration) of the distribution of expenditures or pressures for loaded parts of the thread can be written by the equations:

$$q(z) = \frac{1}{k \sinh(km)} [q'(m) \cosh(kz) - q'(0) \cosh(k(m - z))], \qquad (7.228)$$

$$\sigma(z) = \frac{1}{k \sinh(km)} [\sigma(m) \cosh(kz) - \sigma(0) \cosh(k(m - z))], \qquad (7.229)$$

where

$$q'(0) = \frac{1}{C} \left( -\frac{F'_s}{E_s A_s} + \frac{F''_n}{E_n A_n} \right), \qquad (7.230)$$

$$q'(m) = \frac{1}{C} \left( -\frac{F'''_s}{E_s A_s} + \frac{F'_n}{E_n A_n} \right), \qquad (7.231)$$

$$k^2 = \frac{e}{C}, \qquad (7.232)$$

$$e = \frac{1}{E_s A_s} + \frac{1}{E_n A_n}, \qquad (7.233)$$

$$C = \frac{P^2}{A_{0r}} \left( \frac{C_s}{E_s} + \frac{C_n}{E_n} \right). \qquad (7.234)$$

We will consider cases of operational load of the core with forces caused during mounting or during exploitation of the furniture body.

The initial stress caused by screwing in the core or loading the thread with the force from the eccentric joint (Fig. 7.73) allows to assume the following operating conditions of the joint: the bolt core is extended and the particle board (nut body) is compressed. For such assumptions, border conditions can be written in the following form:

for $z = 0$

$$F''_s = F, \qquad (7.235)$$

$$F''_n = -F, \qquad (7.236)$$

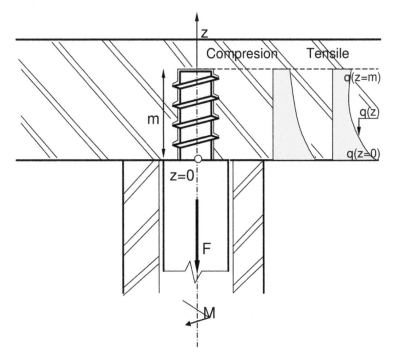

**Fig. 7.73** Distribution of loads of thread coils, caused by screwing in of the core or loading of the thread with the force from the eccentric joint

for $z = m$

$$F'_s = F'_n = 0, \qquad (7.237)$$

$$q'(0) = -k^2 F, \qquad (7.238)$$

$$q'(m) = 0, \qquad (7.239)$$

therefore,

$$q(z) = \frac{kF}{\sinh(km)}\cosh(k(m - z)). \qquad (7.240)$$

The force necessary for the initial stress (before mounting the joint) is determined from the condition of compression strength of the particle board $k_c^w = 4$ MPa and by receiving a value equal to

$$F \leq n \frac{\pi (D - D_r)^2}{4} k_c^w, \qquad (7.241)$$

Therefore, $F \leq 36.30$ N.

The moment $M$ on the clutch screwing the core in the particle board should therefore amount to

$$M \leq \frac{FD_r \left( \text{tg} \gamma + \frac{\mu}{\cos \alpha} \right)^2}{2 \left( 1 + \frac{\mu}{\cos \alpha} \text{tg} \gamma \right)}, \qquad (7.242)$$

therefore $M = 88.42$ N mm.

More unfavourable, however, is the work of the joint associated with the forces caused by mounting of connections and mounting of the furniture body. Stress of the core caused by the eccentric joint, however, should not be greater than the value of the force possible to be shifted, due to the shearing strength of the particle board. Using this condition, the value of forces can be determined:

of the operating stress from the condition of compression strength of the particle board $k_t^w = 3.5$ MPa:

$$F \leq \pi D m k_t^w, \qquad (7.243)$$

therefore, $F \leq 560.5$ N,

the moment $M$ on the clutch of the screwdriver:

$$M \leq \frac{FD_r \left( \text{tg} \gamma + \frac{\mu}{\cos \alpha} \right)^2}{2 \left( 1 + \frac{\mu}{\cos \alpha} \text{tg} \gamma \right)}, \qquad (7.244)$$

equal to $M = 1365.38$ N mm.

For such conditions of use of the furniture, both the bolt core (shaft) and particle board (nut body) shall be subject to stretching. Operational conditions for this joint are as follows:

for $z = 0$

$$F_s'' = F, \qquad (7.245)$$

$$F_n' = 0, \qquad (7.246)$$

for $z = m$

$$F'_s = 0, \tag{7.247}$$

$$F''_n = F, \tag{7.248}$$

$$q'(0) = -\frac{1}{C}\frac{F}{E_s A_s}, \tag{7.249}$$

$$q'(m) = -\frac{1}{C}\frac{F}{E_n A_n}, \tag{7.250}$$

$$q(z) = \frac{F}{Ck}\frac{1}{\sinh(km)}\left[\frac{\cosh(kz)}{E_n A_n} + \frac{\cosh(k(m-z))}{E_s A_s}\right], \tag{7.251}$$

where

$$k^2 = \frac{e}{C}, \tag{7.252}$$

$$e = \frac{1}{E_s A_s} + \frac{1}{E_n A_n}, \tag{7.253}$$

$$C = \frac{P^2}{A_{0r}}\left(\frac{C_s}{E_s} + \frac{C_n}{E_n}\right), \tag{7.254}$$

$$C_s = 0.86 + 0.108\frac{D_2}{P}, \tag{7.255}$$

$$C_n = 1 + 0.234\frac{D_2}{P}. \tag{7.256}$$

On this basis, assuming the following numerical data:
M    8.5 mm,
P    2 mm,
$A_{0r}$   $\pi(D - D_r)^2/4$,
$E_n$    1800 MPa,
$E_s$    200,000 MPa,
$A_n$    $\pi(R_p)^2$,
$A_s$    $\pi(D_r)^2/4$,
$D_2$    5.15 mm,
$D_r$    4.3 mm, and
D    6 mm,
expenditures can be calculated, which have been presented in Fig. 7.74.

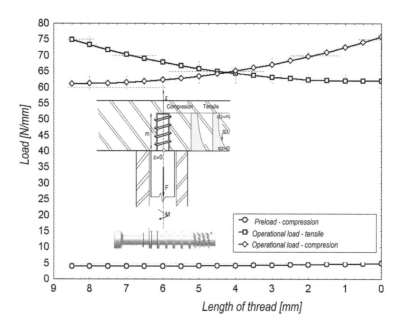

**Fig. 7.74** Expenditure of forces for the thread core in an eccentric joint

On the basis of values of the expenditures illustrated in Fig. 7.74, it can be noted that the initial mounting stresses related to mounting of the connector in the board do not contribute to a significant loading of the thread. Only operational loads dependent on the nature of the work of the core and nut significantly increase the level of load up to a value of 75 N/mm.

### 7.2.3.4  Strength of Joints of Wall Cupboards

Wall cupboards, especially bathroom and kitchen cupboards, are attached to the walls with the use of furniture handles screwed to the body, top or simultaneously to the top and side wall using screws (Fig. 7.75). Embedded in the wood, they should be screwed into the properly drilled holes with a diameter 2 mm smaller than the diameter of the screw core. Drillings should be made on a section of about 80 % of the length of the screw. In structures made of particle boards, drillings are not used.

Because screws of handles of metal wall cabinets work on bending and shearing, the number $t$ of single-sheared screws should be determined from the equations:

$$t \geq \frac{4}{5}\frac{1}{d^2}\left[\sum_{i=1}^{n}(V_i\rho_i g + A_i q_i) + \sum_{i=1}^{m}V\rho_i g\right], \qquad (7.257)$$

**Fig. 7.75** Methods of
mounting cupboards

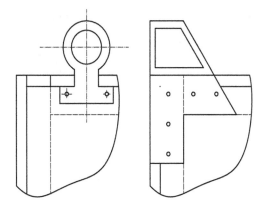

where
$t$    number of single-sheared screws,
$d$    screw diameter,
$V_i$    volume of element,
$\rho_i$    density of element material,
$A_i$    usable area of horizontal elements,
$q_i$    surface load of horizontal elements,
$g$    9.81 m/s$^2$,
$n$    number of horizontal elements and
$m$    number of vertical elements

The above relation applies when the depth of embedding the screw $L$ amounts to

$$L \geq 8d. \tag{7.258}$$

In other cases, i.e. when

$$4d \leq L \leq 8d, \tag{7.259}$$

the calculated number of screws must be corrected by the coefficient $k$:

$$k = L/8d. \tag{7.260}$$

### 7.2.3.5   Strength of Door of Vertical Rotation Axis

Criterion of strength for hinged doors of vertical rotation axis is the value of force
resulting in breaking out of the hinge form the slot made in the particle board
constituting side wall of the body. Distribution of internal forces acting on the hinge

**Fig. 7.76** Scheme of
examining suspension of door
of vertical rotation axis

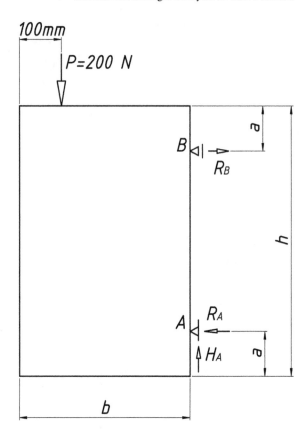

has been shown in Fig. 7.76. Taking into account the weight of the door, the
resultant $W$ of the forces $R_A$ and $H_A$ acting on the lower hinge must be determined:

$$W = \left( (P + V\rho g)^2 + \left( \frac{P(b - 100) + 0.5V\rho gb}{h - a} \right) \right)^{0.5}, \qquad (7.261)$$

where
$W$  resultant force,
$P$  operational load,
$V$  volume of the door board,
$\rho$  density of element material,
$g$  9.81 m/s$^2$,
$h$  height of the door,
$a$  height of the setting of hinges and
$b$  width of the door.

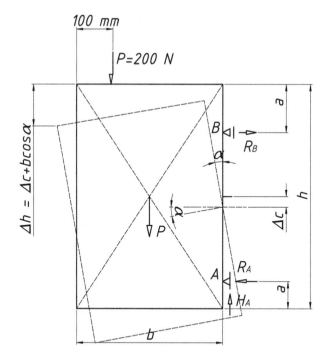

**Fig. 7.77** Scheme of door deformation

Results of studies on the strength of box hinges mounted in particle boards of various densities indicate that the criterion of strength for hinges refers to the plasticity limit, $R_e$, designated from the experimental graph of the load-deformation function $P = f(\Delta h)$. Theoretically established value of the force $W$ should not be greater than the experimentally designated plasticity limit $R_e$:

$$W \leq R_e. \tag{7.262}$$

The value $R_e$ for box hinges, mounted in a board with a density from 650 to 700 kg/m$^3$, is at a medium level of 365 N. An additional criterion for the evaluation of the quality of hinged joints of vertical rotation axis is the stiffness of the joints. For aesthetic reasons, the value of the door hanging down during use should be limited to 2 mm (Fig. 7.77). This value results from the possibility of improving the geometric fault through adjusting the position of the screws on the assembly plate of box hinges.

A necessary and sufficient condition is, therefore, that the displacement $\Delta h$ fulfils the condition of inequality:

$$\Delta h = \frac{(P(b-100) + 0.5V\rho gb)b}{(h-2a)^2}\left(\frac{1}{k_1} + \frac{1}{k_2}\right) \leq 2\,\text{mm} \qquad (7.263)$$

where
$k_1, k_2$  stiffness of the upper and lower hinge determined on the basis of the results of laboratory tests (N/mm).

### 7.2.3.6  Strength of Door of Horizontal Rotation Axis

Doors of horizontal rotation axis are used in furniture fulfilling the function of bars and davenports. Their work tops in addition to mass and concentrated loads also shift the load equally distributed on the surface of the board. The strength of the door suspension is determined by the hinges mounted in horizontal partition of the body. A tangible indicator of this strength is the value of the force aiming to break out the hinge box from the partition. Taking into account the actual conditions of the furniture use, two calculation schemes can be applied. The first, shown in Fig. 7.78, describes a condition in which a correctly operating guide shifts

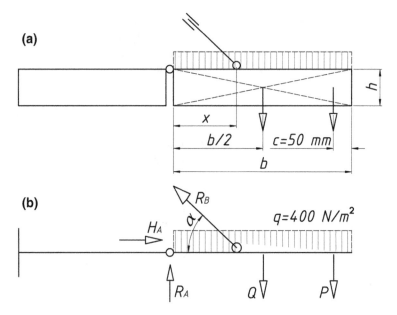

**Fig. 7.78** Door of horizontal rotation axis with correctly working guides: **a** geometric scheme, **b** calculation scheme

operational loads, at the same time relieving the box hinge. Rules for first class lever, which are valid for this calculation scheme, clearly indicate that the durability of the door suspension will be greater when the distance $x$ of mounting the guide from the hinge will be similar to the width of the door $b$. By making calculations for half of the value of the operational load (per one hinge), and taking into account the weight of the door, the value of the resultant force $W$ acting on the hinge, must be written in the form:

$$W = R_w = R_A = \frac{1}{2x}\left(\frac{1}{2}b - x\right)\left(V\rho g + 2P\left(1 - \frac{1}{2}c\right) + q_A bl\right), \qquad (7.264)$$

where
$W$   resultant force,
$b$   width of the door,
$c$   50 mm,
$g$   9.81 m/s$^2$,
$l$   length of the door,
$\rho$   density of element material,
$q_A$   surface load,
$V$   volume of the door board and
$x$   distance of mounting the guide.

Calculation scheme in Fig. 7.79 shows the door suspension in which the guides work faultily, not shifting the relevant operational loads. As a consequence, the result is that the edge of the door and horizontal partition, by pushing each other, trigger a particularly high concentration of forces gathered in the hinge. Value of this force can be written down as

$$W = H_A = \frac{1}{4h}b\left(V\rho g - 2P\left(1 - \frac{c}{b}\right) - q_A bl\right), \qquad (7.265)$$

where
$h$   thickness of the door board,

## 7.3  Durability of Usage of Case Furniture

### 7.3.1  Reliability of Case Furniture

The European Union Directive No. 2001/95/EC, relating to general product safety, includes also furniture. The appropriate European norms define safety requirements for the use of furniture; however, they do not apply to the evaluation of the

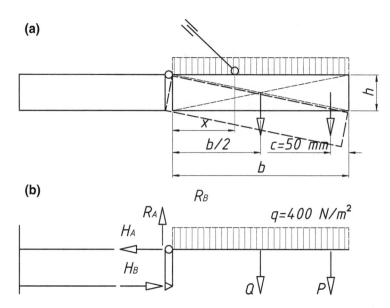

**Fig. 7.79** Door of horizontal rotation axis with faultily working guides: **a** geometric scheme, **b** calculation scheme

construction reliability at an extended period of time of their use. The use of traditional methods of designing and evaluating the strength requirements for furniture, based on subjectively accepted safety coefficients and certainty reserve, does not allow to judge the time and probability of damage to the furniture or its element. It is therefore obvious that a deterministic approach to furniture design is not reliable and makes it difficult for furniture manufacturers to establish warranty conditions beneficial for themselves and the customer, including the period of free warranty repairs. Therefore, it is appropriate to introduce new methods of designing, which would allow for the random nature of construction parameters, so that the reliability of furniture could be determined at the stage of construction. In most cases, durability of the furniture is determined by the strength of the structural nodes (Gozdecki and Smardzewski 2005; Smardzewski 2002a, b; Smardzewski and Gozdecki 2007). All construction parameters of systems needed to analyse the reliability of the furniture construction are determined by the relevant distribution of stresses and strength or loads and load-carrying capacity, above all in relation to the joints and elements. If both of these distributions are established, it will be possible to determine the probability of damage to the joint, and then the probability of damage to the furniture. The problem of reliability of the furniture construction has been addressed in a few publications (Smardzewski 2005; Smardzewski and Ożarska-Bergandy 2005), where the issue of testing stiffness of dowel or bolt joints in furniture for storage was discussed.

**Fig. 7.80** Permeation of
distribution of stress $f_\sigma(\sigma)$ and
strength $f_Z(Z)$

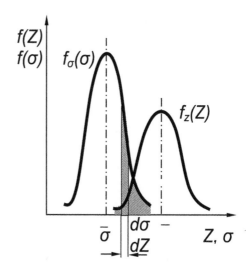

**Fig. 7.80** Permeation of distribution of stress $f_\sigma(\sigma)$ and strength $f_Z(Z)$

The basis for evaluation of reliability of furniture joints should be the probability of exceeding the border level of their strength. From the point of view of reliability, the calculation of strength amounts to determining the probability of exceeding a given level of border stresses $Z$, with the specified dispersion area, by random loading, which causes working load $\sigma$ at a given time. When using the furniture, one can observe complex cases, requiring to consider both the reduction of the connection strength, as well as the increase of internal stresses. Because the values $Z$ and $\sigma$ are random variables, on the basis of their characteristics the probability of structural damage $\Phi(u)$ in the planned period of use should be determined. For random variables of strength $Z$ and stresses $\sigma$, characteristic for furniture joints, the form of distributions $f(Z)$ and $f(\sigma)$ can be established, and then, on the basis of the size of the area of surface permeation (Fig. 7.80), determine the probability of damage:

$$F(Z<\sigma).\tag{7.266}$$

According to Murzewski (1989), the probability that a certain value of strength $Z$ is located in a narrow range $dZ$ (Fig. 7.81) and that stress $\sigma$ does not exceed the strength $Z_0$ is equal to

$$R(\sigma \leq Z) = f_Z(Z_0)dZ \int_{-\infty}^{Z_0} f_\sigma(\sigma)d\sigma.\tag{7.267}$$

**Fig. 7.81** Determining probability of failure-free work

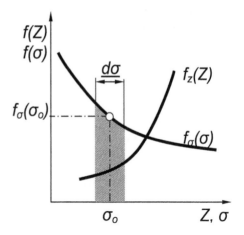

Unreliability that is the probability of structural damage will therefore amount to

$$F = F(Z \leq \sigma) = 1 - \int_{-\infty}^{\infty} f_\sigma(\sigma)(1 - F_Z(\sigma))\mathrm{d}\sigma = \int_{-\infty}^{\infty} F_Z(\sigma)f_\sigma(\sigma)\mathrm{d}\sigma \quad (7.268)$$

or for

$$R = \int_{-\infty}^{\infty} f_Z(Z)\left(\int_{-\infty}^{Z} f_\sigma(\sigma)\mathrm{d}\sigma\right)\mathrm{d}Z, \quad (7.269)$$

$$F = F(Z \leq \sigma) = 1 - \int_{-\infty}^{\infty} f_Z(Z)F_\sigma(Z)\mathrm{d}Z = \int_{-\infty}^{\infty} (1 - F_\sigma(Z))f_Z(Z)\mathrm{d}Z. \quad (7.270)$$

In order to assess the reliability of joints and case furniture, at the Department of Furniture of the University of Life Sciences in Poznań, a study on the strength of three populations of angular joints was carried out, 10 pieces each, in which two connectors were used: confirmat screw Ø5 × 50 mm, beech dowels with the dimensions of Ø6 × 32 mm and dowels Ø8 × 32 mm. The joints were made of unveneered particle board with the thickness of $h_p$ = 18 mm, density $\rho$ = 660 kg m$^{-3}$, bending strength $k_G$ = 16 MPa, splitting strength $k_R$ = 0,35 MPa, absolute moisture content 8 %, as well as the shearing strength of glue $k_S$ = 9 MPa and shearing strength of beech wood $k_B$ = 17 MPa. Compression loads were distributed by the strength machine ZWICK 1445, at the same time registering the value of the load $P$ and displacement $\Delta P$ at the point of application of the force (Fig. 7.82). Based on these results, values of destructive force and the most important indicators of strength of the joints were determined.

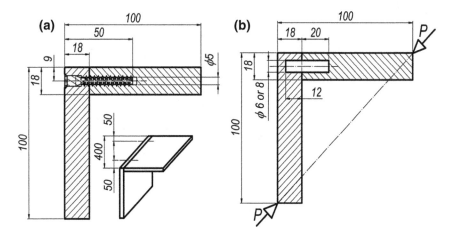

**Fig. 7.82** Construction and method of loading the joints: **a** confirmat screw, **b** dowel

The following were accepted as indicators of strength of the joints:

shearing strength of the connector

$$Z_B = \frac{4P\sin(45°)}{\pi d^2} \leq k_B, \tag{7.271}$$

shearing strength of the adhesive bond

$$Z_S = \frac{3\left(P\cos(45°) + 2M_i/h_p\right)}{2\pi d h_p} \leq k_S, \tag{7.272}$$

splitting strength of the particle board

$$Z_R = \frac{P\sin(45°)}{\left(L - 2/3h_p\right)^2} \leq k_R \tag{7.273}$$

while the value of the bending moment $M_i$ in the joint has been determined on the basis of the equation,

$$M_i = P\cos(\Delta\varepsilon_2)\left(\left(L - h_p\right)^2 + h_p^2\right)^{0.5}, \tag{7.274}$$

where

$$\Delta\varepsilon_2 = 45° + \arcsin\frac{h_p}{\left((L-h_p)^2+h_p^2\right)^{0.5}} - \varphi_2, \tag{7.275}$$

$$\varphi_2 = \gamma_3 - \beta_2, \tag{7.276}$$

$$\gamma_3 = \arccos\frac{\frac{1}{2}(a-\delta_p)\left((L-h_p)^2+h_p^2\right)^{0,5}}{(L-h_p)^2+h_p^2}, \tag{7.277}$$

$$\beta_2 = \arccos\frac{\frac{1}{2}a\left((L-h_p)^2+h_p^2\right)^{0.5}}{(L-h_p)^2+h_p^2}, \tag{7.278}$$

established on the basis of geometric dependencies in the deformed joint (Fig. 7.83).

Values of the coefficients of the strength of joints have been presented in Table 7.11. According to them, only low splitting strength of particle boards could pose a serious threat to the failure-free work of the furniture. For these reasons,

**Fig. 7.83** Geometric dependencies in the deformed joint

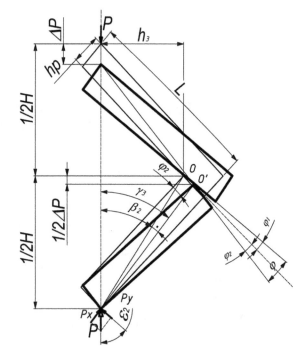

**Table 7.11**  Indicators of the strength of joints

| Type of joint | Symbol | Strength (MPa) | | | | | | |
|---|---|---|---|---|---|---|---|---|
| | | Shearing strength of the connector $Z_B$ | | Shearing strength of the adhesive bond $Z_S$ | | Splitting strength of the board $Z_R$ | | |
| | | Average | Std. deviation | Average | Std. deviation | Average | Std. deviation | |
| Dowel 6 | Z6 | 7.03 | 0.59 | 8.64 | 0.72 | 0.501 | 0.044 | |
| Dowel 8 | Z8 | 5.22 | 0.27 | 8.87 | 0.41 | 0.662 | 0.033 | |
| Confirmat screw | ZK | 19.86 | 1.84 | | | 0.272 | 0.025 | |

for further research on the nature of the random variable of strength $f(Z)$ and stresses $f(\sigma)$, only the results of the determination of splitting strength of particle boards were selected.

Bearing in mind the distribution of internal forces acting on nodes of the furniture body (Fig. 7.84a, b), and also taking into account only the criterion of splitting strength of particle boards, stresses caused by operational load were calculated from the equation:

$$\sigma_R = \frac{P_i + R_A}{\left(L - 2/3h_p\right)^2} \leq k_R, \tag{7.279}$$

whereby

$$P_i = \frac{G_i h_{pi}^3}{3(l_1 l_2)_i} \zeta_i \cdot \frac{P_Z}{\displaystyle\sum_{i=1}^{5} \frac{G_i h_{pi}^3}{3(l_1 l_2)_i} \zeta_i^2}, \tag{7.280}$$

$$R_A = \frac{1}{4} q_{Ai}(l_1 l_2)_i, \tag{7.281}$$

$$l_1, l_2 = a, b, c, \tag{7.282}$$

$$\zeta_i = a/c; a/b. \tag{7.283}$$

For the calculations, it was assumed that the furniture may be loaded by the user with an external force $P_Z$ of a variable value (600, 490, 480, 420, 360, 300 N) and standard deviation 105.9 N. Stresses established for these loads have been presented in Table 7.12.

On the basis of the information from Tables 7.11 and 7.12, presenting results of calculations of strength Z and stresses $\sigma$ of selected angular joints, characteristics of probability have been developed (Fig. 7.85) on the basis of which it can be seen that the permeation surfaces of both distributions are dependent on the type of the

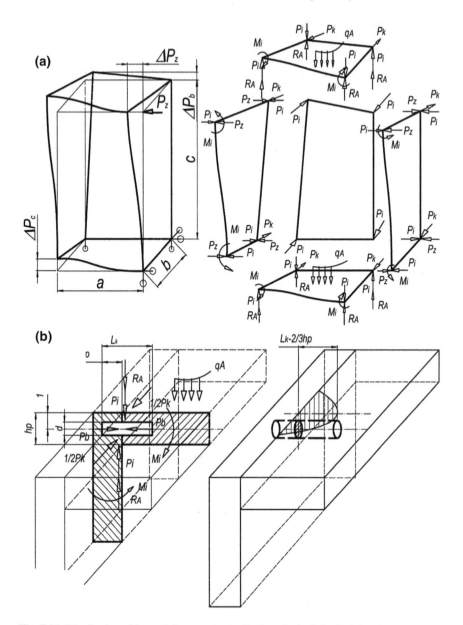

**Fig. 7.84** Distribution of internal forces: **a** in the furniture body, **b** in the joint

examined joints. Greater surface area of the charts imposition suggests a higher probability of damage to the joint. The lack of imposition, however, informs that the joint is fail-safe within a given range of given loads.

**Table 7.12** Values of stresses in joints according to the criterion of splitting strength of boards

| Type of joint | Symbol | Stresses $\sigma_R$ (MPa) | |
|---|---|---|---|
| | | Average | Std. deviation |
| Dowel 6 | N6 | 0.368 | 0.088 |
| Dowel 8 | N8 | 0.368 | 0.088 |
| Confirmat screw | NK | 0.105 | 0.024 |

In order to determine numeric values of the probability of damage to the joint, a new random variable should be considered:

$$Y = Z - \sigma. \tag{7.284}$$

The condition for the security of the construction is then the assumption that

$$R = F(Y > 0), \tag{7.285}$$

hence, the probability of damage is

$$F = \int_{-\infty}^{0} f_Y(Y)\mathrm{d}Y = \int_{-\infty}^{0} \int_{-Y}^{\infty} f_Z(Y + \sigma)f_\sigma(\sigma)\mathrm{d}\sigma\mathrm{d}Y. \tag{7.286}$$

For the discussed case furniture joints, it has been assumed that both random variables $Z$ and $\sigma$, as is shown in Fig. 7.85, are described by a normal distribution:

$$f_\sigma(\sigma) = \frac{1}{S_N\sqrt{2\pi}}\exp\left(-\frac{1}{2}\left(\frac{\sigma - \overline{\sigma}}{S_N}\right)^2\right), \tag{7.287}$$

and

$$f_Z(Z) = \frac{1}{S_Z\sqrt{2\pi}}\exp\left(-\frac{1}{2}\left(\frac{Z - \overline{Z}}{S_Z}\right)^2\right), \tag{7.288}$$

where
$\overline{\sigma} = N$ and $\overline{Z} = Z$ average values of stress and strength and
$S_N$ and $S_Z$ standard deviations of stress and strength;

Hence, the standard deviation of the new random variable $Y$ is described by the following equation:

$$S_Y = \sqrt{S_N^2 + S_Z^2}. \tag{7.289}$$

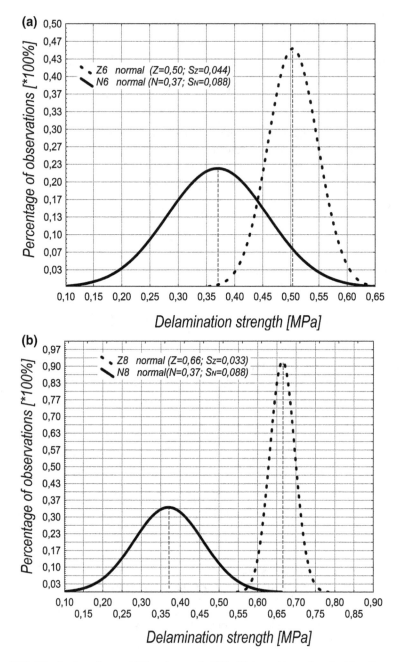

**Fig. 7.85** Distribution of stress $f(N)$ and strength $f(Z)$ of joints with **a** a dowel with a diameter of 6 mm, **b** a dowel with a diameter of 8 mm, **c** a confirmat screw

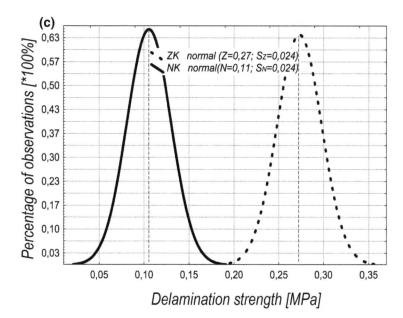

**Fig. 7.85**  (continued)

**Fig. 7.86** Characteristics of the probability of distribution of values $Y = Z - \sigma$

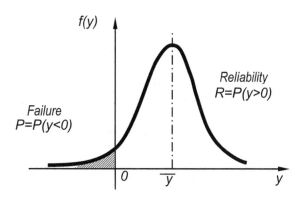

The distribution of the random variable $Y$ has therefore the form (Fig. 7.86)

$$f_Y(Y) = \frac{1}{S_Y\sqrt{2\pi}}\exp\left(-\frac{1}{2}\left(\frac{Y - \overline{Y}}{S_Y}\right)^2\right).$$

(7.290)

The probability of fail-safe work is therefore described by the equation:

$$R = P(Y > 0) = \int_0^\infty f_Y(Y)\mathrm{d}Y = \int_0^\infty \frac{1}{S_Y\sqrt{2\pi}}\exp\left(-\frac{1}{2}\left(\frac{Y - \overline{Y}}{S_Y}\right)^2\right)\mathrm{d}Y.$$

(7.291)

By assuming the marking:

$$u = \frac{Y - \overline{Y}}{S_Y}, \tag{7.292}$$

$$dY = S_Y du, \tag{7.293}$$

with

$$u = \begin{cases} \frac{0 - \overline{Y}}{S_Y} = \frac{\overline{Z} - \overline{\sigma}}{\sqrt{S_Z^2 + S_N^2}}; & \text{dla } Y = 0, \\ \longrightarrow \infty & \text{dla } y \longrightarrow \infty \end{cases} \tag{7.294}$$

thereby

$$R = \frac{1}{2\pi} \int\limits_u^\infty e^{-\left(\frac{u}{2}\right)^2} du = 1 - \Phi(u). \tag{7.295}$$

The value of $\Phi(u)$ corresponds to the value of probability of the damage occurring and amounts to

$$\Phi(u) = F(u) = 0.5^{\left(\frac{u}{2}+1\right)^{2.46}} \quad \text{dla } u \geq 0. \tag{7.296}$$

On this basis, the probabilities of damage to the various types of joints have been determined (Table 7.13). This table shows that the most unreliable joints are those, where dowels with a diameter of 6 mm have been used as a connector. In this case, the probability of fail-safe work in the given load conditions amounts to 0.927149. For dowel joints with a diameter of 8 mm, this probability amounts to 0.999178, while for confirmat screw joints—0.999999.

Reliability of furniture as a system consisting of many unreliable joints can be calculated on the basis of the component reliabilities of the joints. In this case, case furniture should be treated as a system of serially connected structural nodes (Fig. 7.87). It is characterised by the fact that damage to one structural node leads to the damage to the whole construction.

**Table 7.13** Values of the probability of damage to the various types of joints

| Type of joint | Probability of damage | Probability of fail-safe work |
| --- | --- | --- |
| Dowel 6 | 0.078251 | 0.927149 |
| Dowel 8 | 0.000822 | 0.999178 |
| Confirmat screw | 0.000001 | 0.999999 |

**Fig. 7.87** Case furniture as
serial structure of joints

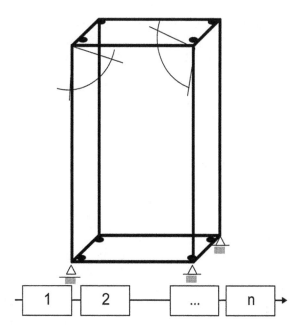

Probability of fail-safe operation of this construction is expressed by the product
of the probabilities of independent events of the 8 component items:

$$R(t) = \prod_{j}^{8} R_j(t).$$   (7.297)

In the discussed case $R_1(t) = R_2(t) = \ldots R_8(t)$; therefore, the probability of fail-safe
work of a furniture body with dowel joints with a diameter of 6 mm amounts to

$$R(t) = R_1^8(t) = 0.927149^8 = 0.546004.$$   (7.298)

Probability of fail-safe work of a furniture body with a construction using the
other joints has been presented in Table 7.14.

From summaries given in Table 7.14, it appears that the reliability of the furniture construction is the lower, the more it contains elements joined with unreliable
construction nodes. Systems built with many structural elements must contain a

**Table 7.14** Probability of fail-safe work of a furniture body

| Type of joint | Probability of damage | Probability of fail-safe work |
|---|---|---|
| Dowel 6 | 0.453996 | 0.546004 |
| Dowel 8 | 0.006558 | 0.993442 |
| Confirmat screw | 0.000008 | 0.999992 |

very high reliability joints, such as confirmat screw joints. If a high reliability of the system consisting of a large number of elements is required, then reserves should be used, which allow to greatly reduce the chance of damage, and at the same time connections with a lower probability of fail-safe operation should be used. In constructions of case furniture, and in particular in multi-chamber systems, double or triple reinforcement systems should therefore be used.

## 7.3.2  Warranty Services

The document obliging the producer to provide periodic, free service repairs is a warranty card. This document should normally include:

- the name of the product or its components covered by the warranty,
- the warranty period for the entire piece of furniture or its components,
- starting date of the warranty validity,
- the territorial area of the warranty validity,
- the warranty conditions, that is the requirements, which fulfilment maintains the warranty services of the guarantor or which unfulfilment causes the loss of these services,
- a description of the complaint procedure, which is a precise description of the acceptance and processing of the complaint and the manner of the implementation of furniture repair,
- description of the furniture construction,
- description of the raw materials and components used, including the characteristics of the materials used for making the furniture given in accessible form, including the specifics of the natural origin of components and the resulting conditions, and
- terms of use and maintenance of the furniture, including a description of the allowed and prohibited uses of the furniture and an indication of the ways, means and tools intended for maintenance and care of the furniture.

*Example*
**Warranty for furniture xxx.**
The guarantor AAA Ltd. provides good quality and functioning of the furniture, provided that the appropriate rules for their use are followed. The warranty period is 12 months from the date when the furniture has been released to the buyer. The buyer has the possibility of extending the warranty period:

- up to 2 years for the wooden bearing structure of the furniture and
- up to 5 years for the metal frame in products with a reclining function and for the polyurethane foam fillings.

**Warranty conditions**:

- The condition for the extension of the warranty period is purchasing a service package,
- In order for the warranty to gain validity, it is necessary for the seller to fill in the data relating to the transaction and product,
- The warranty is valid, provided that the buyer observes the instructions attached to the furniture, concerning use, regular cleaning and maintenance,
- The warranty covers only residential furniture used in home conditions,
- The warranty is valid on the territory of Poland,
- The warranty does not cover changes of softness of cushions and shrinking of the cover material, being the result of normal use of the furniture,
- Characteristic features of furniture made entirely from natural leather or in combination with other materials, such as faux leather and upholstery fabrics, are the differences in the texture of hides of leather, traces of scars, natural marks, stretch marks or skin smell; they are not furniture defects and are not subject to complaint.

**The loss of warranty services occurs in the case of the following:**

- Wrong use of the furniture, not in accordance with the instructions for use;
- Exposing the furniture to very intense sunlight;
- Spilling water or other liquid on furniture (chemically active solutions can be especially dangerous for furniture, e.g. acids, dyes and body care lotions);
- Performing repairs or corrections by the buyer on his own; and
- Mechanical damage to the furniture.
- Complaint procedure:
- The buyer declares the complaint in the commercial unit, in which he purchased the piece of furniture, and in the case of its liquidation—to the guarantor;
- The buyer provides documentation concerning the purchase of the furniture (invoice, bill or receipt) and the warranty card;
- The buyer participates in the completion of the complaint by the seller;
- The complaint containing the warranty card number is transmitted via fax to the guarantor from the commercial unit;
- After accepting the complaint, the guarantor shall establish with the buyer the date of inspection of the furniture at the place indicated by the buyer;
- The guarantor through his representative—service specialist—will inspect the furniture and make a decision about the complaint;
- If the complaint is justified, then the guarantor will remove its defects or deliver furniture free from defects within 14 days from the date of inspection of the furniture, whereas

  - the guarantor chooses the warranty service (removal of defects or replacement) and
  - replacement of the furniture is possible in the case of two prior repairs of the same element of the furniture system;

- When carrying out repair of the furniture, the guarantor shall reserve the right to replace the materials used in the manufacture of the furniture with others, of comparable quality and use value;
- In the case of introducing construction changes, it is possible to make the repair in accordance with new technology; and
- If the complaint is found to be unjustified, then the guarantor will present his position in writing and forward it to the buyer and the seller along with the reasons for his decision within 14 days from the date of receipt of the complaint.

### Terms of use and maintenance of the furniture:

- Complying with the requirements given in this instruction will allow the long-term and fail-safe use of the furniture;
- When moving the furniture, all upholstered elements must be secured, so that they are not subject to mechanical damage;
- When unpacking the furniture, do not use sharp tools, as there is a danger of mechanical damage of the upholstery material;
- When moving the furniture, always hold them underneath at stiff elements of the bearing structure;
- It is not allowed to drag or lift furniture holding it at the lining or cushions;
- Furniture should be used in accordance with their intended purpose;
- Furniture should be used in dry, enclosed spaces, protected from adverse weather conditions and direct sunlight;
- Upholstered furniture must be placed within more than 1 m from effective heat sources;
- It is not allowed to sit on backrests and armrests of upholstered furniture. Large loads in places not intended for them can expose the furniture to damage to the cover, breaking the bearing structure or deformation of the soft elements;
- It is not allowed to expose upholstered surfaces to strong, point tensions, for example standing on the seat; and
- It is recommended to use the upholstered furniture in rooms with a temperature between 15 and 30 °C and air humidity between 40 and 70 %.

### Protection and care of leather upholstered furniture:

- Natural and faux leather have a protective layer, but the use of inappropriate cleaning agents may cause damage to the cover. Therefore, under the threat of losing the warranty, the leathers have to be protected against solutions such as alkaline, acidic, etc. Instead, use solutions recommended by: (name of a company).
- The frequency of cleaning the leather depends on the intensity of use of the furniture. Fresh contaminations can be removed immediately using a clean, dry cloth or tissue. If these methods prove ineffective, one should use solutions recommended by (name of a company).

**Protection and care of fabrics in upholstered furniture:**

- It should be regularly cleaned, vacuumed, and in the case of fabrics with hair, the furniture cover should be brushed;
- Colourants used to dye fabrics are sensitive to sunlight; therefore, you should avoid exposing the furniture to direct sunlight through windows without curtains;
- In the case when on the surface of the fabrics a liquid is spilled, e.g. coffee, first the stain must be drained using a paper napkin or cleaned using easily absorbing cotton cloth, and then, the contamination must be removed, e.g. coffee grounds, and only then the stain can be removed;
- In the course of removing the stain with perchloroethylene-based stain removers and aqueous detergents, one should avoid soaking the fabrics too much;
- Before using a product for cleaning upholstery, make sure to check its effect on not exposed parts of the furniture, in order to avoid damaging the fabric, e.g. discolouration due to too aggressive effects of the solution;
- Pets may significantly damage the furniture cover, so appropriate bedspreads must be used; however, they should be arranged in a way that prevents contact of the fabrics hair with the bedspread hair; and
- Recommended stain removers and products for cleaning upholstery are available commercially.
- Description of raw materials and components used:
- Bearing structure,
- Skeleton internal construction of the furniture is built with wooden elements and plywood. The seat is reinforced with sinusoidal springs and upholstery belts and
- In some furniture with a reclining function, metal frames equipped with elastic skirts and mattresses are applied.

**Filling:**

- Comfort of furniture is achieved through the use of appropriately selected fillings from highly flexible polyurethane foams. They provide the right softness at the key points of furniture—on the backrests, seat cushions, armrests. To increase the durability of the seat, special bonnell-type springs have been used. Properly filled cushions ensure comfort for the furniture users and ensure a high level of ergonomics. To achieve the effect of fluffiness, in selected series of upholstered furniture granulated, eco-friendly silicone fibres have been used;
- Leather is a natural material and each hide of leather is different; therefore, the differences in colour and texture are a common and natural phenomenon; and
- Upholstery fabrics meet the requirements in terms of durability, resistance to abrasion and contamination, elasticity and shrinkage.

# References

Bachmann G (1983) Festigkeit eingedrehter Holzschrauben in Abhängigkeit von Eindrehtiefe, Schraubendurchmesser und Festigkeit der Verwendeten Spanplatten. Möbel und Wohnraum 36(1):21–25

Budiansky B, Connor R (1948) Buckling stresses of clamped rectangular. Flat plates in shear. NASA technical note 1559

Chia-Lin H, Eckelman CA (1994) The use of performance test in evaluating joint and fastener strength in case furniture. For Prod J 44(9):47–53

Cox HL (1933) Summary of the present state of knowledge regarding sheet metal. Reports and memoranda, no 1553

Dietrich M (2008) Podstawy konstrukcji maszyn. PWN, Warsaw

Dutko P (1976) Drevne konstrukcie. Alfa, Bratislava

Dzięgielewski S, Smardzewski J (1989) Jakość połączeń a sztywność mebli skrzyniowych. Przemysł drzewny 11(12):4–8

Dzięgielewski S, Smardzewski J (1990) Der Einflußder Befestigungsart von Ruckwand und Tur auf die Steifigkeit von Behaltnismobeln. Holztechnologie 3:136–139

Dzięgielewski S, Smardzewski J (1992) Metody analityczne oceny sztywności korpusów mebli skrzyniowych. Prace Instytutu Technologii Drewna, Poznań Y XXXVI Bulletin 1(2):53–76

Dzięgielewski S, Smardzewski J (1995) Meblarstwo. Projekt i konstrukcja. Państwowe Wydawnictwo Rolnicze i Leśne, Poznań

Eckelman CA (1967) Furniture mechanics, the analysis of paneled case and carcass furniture. Purdue University agricultural experiment station progress report no 274, West Lafayette, Indiana

Eckelman CA, Resheidat M (1984) Deflection analysis of shelves and case top end bottoms. For Prod J 6:55–60

Eckelman CA, Rabiej R (1985) A comprehensive method of analysis of case furniture. Forest Prod. J. 4:62–68

Ganowicz R, Kwiatkowski K (1978) Experimentelle Pruffung der Theorie der Verformungen von Schrankkonstruktionen. Holztechnologie 4:202–206

Ganowicz R, Dziuba T, Ożarska-Bergandy B (1978) Theorie der Verformungen von Schrankkonstruktionen. Holztechnologie 2:100–104

Ganowicz R, Dziuba T, Kwiatkowski K, Ożarska-Bergandy B (1977) Badania zmierzające do ustalenia zasad konstruowania mebli skrzyniowych. Typescript Department of Mechanical Engineering and Heat Technology, Agricultural Academy of Poznań

Girkmann K (1957) Dźwigary powierzchniowe. Arkady, Warsaw

Gozdecki C, Smardzewski J (2005) Detection of failures of adhesively bonded joints using the acoustic emission method. Holzforschung 59:219–229

Huber MT (1922) Teorya płyt prostokątnie różnokierunkowych wraz z technicznymi zastosowaniami do płyt betonowych, krat balkonowych itp. Archive of Towarzystwo Naukowe in Lviv, Department III, vol 1, bulletin 4

Huber MT (1923) Studia nad belkami o przekroju dwuteowym. Wyd. Warszawskiego Towarzystwa Politechnicznego. Sprawozdania i Prace. vol 2, March–June, bulletin 1 and 2

Joscak P (1986) Vypocet nosnych casti skrinioveho nabytku metodu konecnych prvkov. Drevo 8:227–229

Joscak P, Vacek V (1989) Kontrola priehybuvodorovnych nosnych casti skrinioveho nabytku statickym vypoctom. Drevo 44:15–17

Kociszewski M, Warmbier K, Wilczyński M (2002) Właściwości mechaniczne przy zginaniu warstw płyty wiórowej. Przemysł Drzewny 2:12–15

Korolew W (1970) K rasczotu na procznost niekatorych uzlow i elementow korpusnoj mebeli. Sbornik Rabot MLTI 35:104–129

Korolew W (1973) Osnovy racionalnovo proiektirovania miebeli. Lesnaja Promyszliennost, Moscow

Kotaś T (1957) The theoretical and experimental analysis of cabinet structures. Furniture development council research report no. 6, London

Kotaś T (1958a) Design Manual for Cabinet Furniture. Furniture Development Council Pergamon Press, New York

Kotaś T (1958b) Sztywność mebli skrzyniowych cz.1. Przemysł drzewny 10:15–18

Kotaś T (1958c) Sztywność mebli skrzyniowych cz.2. Przemysł drzewny 11:10–14

Kuhne G, Kroppelin U (1978) Untersuchungen zum Beanspruchungsverhalten von Eckeverbindungen durch Dubel. Holztechnologie 19:95–99

Lapszyn YG (1968) Raszciet procznosti ugłowych soiedinienji korpusnoj miebieli. Naucznyje Trudy MLTI 30:60–68

Murzewski J (1989) Niezawodność konstrukcji inżynierskich. Arkady, Warsaw

Ożarska-Bergandy B (1983) Badanie w stanie nadkrytycznym płyt pilśniowych i sklejki pod działaniem naprężeń stycznych. PhD dissertation, Typescript, Department of Mechanical Engineering and Heat Technology, Agricultural Academy of Poznań

PN-EN 14074:2006 Office furniture. Tables and desks and storage furniture. Test methods for the determination of strength and durability of moving parts

PN-EN 14073-3:2006 Office furniture. Storage furniture. Test methods for the determination of stability and strength of the structure

Smardzewski J (1991) Analiza odkształceń połączeń drewnianych. Roczniki Akademii Rolniczej w Poznaniu CCXVI:83–103

Smardzewski J (2002a) Strength of profile-adhesive joints. Wood Sci Technol 36:173–183

Smardzewski J (2002b) Technological heterogenity of adhesive bonds in wood joints. Wood Sci Technol 36:213–227

Smardzewski J (2004a) MebelCAD, Parametryczna nakładka dla systemu AutoCAD. Meble Materiały Akcesoria 12(55):40–42

Smardzewski J (2004b) Stereomechanika połączeń mimośrodowych. Modelowanie półsztywnych węzłów konstrukcyjnych mebli. In: Branowski B, Pohl P (eds) Wydawnictwo Akademii Rolniczej im. A. Cieszkowskiego w Poznaniu

Smardzewski J (2004c) Modeling of semi–rigid joints of the confirmate type. AnnWarsaw Agric Univ For Wood Technol 55:486–490

Smardzewski J (2005) Niezawodność konstrukcji mebli skrzyniowych. Przemysł Drzewny 6:24–27

Smardzewski J, Dzięgielewski S (1993) Stability of cabinet furniture backing board. Wood Sci Technol 28:35–44

Smardzewski J, Dzięgielewski S (1994) Rozkład naprężeń stycznych w spoinie połączenia czopowego. In: Materiały Sesji Naukowej Badania dla meblarstwa, Wydawnictwo Akademii Rolniczej w Poznaniu, pp 62–73

Smardzewski J, Gozdecki C (2007) Decohesion of glue bonds in wood connections. Holzforschung 61(3):291–322

Smardzewski J, Ożarska-Bergandy B (2005) Rigidity of cabinet furniture with semi-rigid joints of the confirmat type. Electron J Pol Agr Univ, Wood Technol 8(2). http://www.ejpau.media.pl

Skan S, Southwell R (1954) On the stability under shearing forces of flat elastic strip. Proc R Soc Lond A 105

Timoshenko S, Gere J (1963) Teoria stateczności sprężystej. Arkady, Warsaw

Wolmir AS (1956) Gibkije płastinki i obołoczki. Gostiechnizdat, Moscow

# Chapter 8
# Stiffness and Strength Analysis of Upholstered Furniture

## 8.1 Stiffness and Strength of Upholstery Frames

Frames of upholstery furniture belong to 3D structures, made of rails not lying in one plane, or 2D frames, but loaded with forces not lying in the system plane. (Fig. 8.1).

The general principles and proceedings when solving frame 3D systems with the method of forces are the same as with solving 2D frames (see calculations on the side frames of chairs). First, the degree of static indeterminacy $n_s$ should be determined:

$$n_s = p + 6t - 6, \qquad (8.1)$$

where

$p$   number of support nodes,
$t$   number of necessary cuts of closed contours and
6   number of equilibrium equations for any 3D force system

Therefore, the frame presented in Fig. 8.2 is a 19-fold statically indeterminate system $n_s = 19$.

Then, the frame of references x, y, z should be assumed, so that as many rods lay in the xy plane as possible. The loads of frames should be the most disadvantageous for them (Fig. 8.3). The distribution of evenly distributed load $q$ shown in Fig. 8.3a is caused by the tension of belts or springs. The value of this load, dependent on the strength of the belt tension, number of belts and their length, can be determined from the equation:

$$q = \frac{iP_n}{l}, \qquad (8.2)$$

© Springer International Publishing Switzerland 2015
J. Smardzewski, *Furniture Design*,
DOI 10.1007/978-3-319-19533-9_8

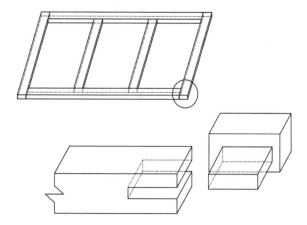

**Fig. 8.1** Examples of upholstery frame constructions: **a** covered mortise and tenon joints, **b** bridle joints

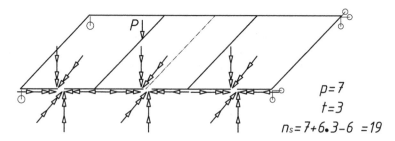

**Fig. 8.2** Calculation diagram of static indeterminacy of the system

where
$q$   even distribution on the length of external rails,
$i$   number of belts or springs,
$l$   length of the rail and
$P_n$   strength of tension of one belt or one spring.

Values of concentrated loads shall be assumed on the basis of standardised data. By discarding the extra nodes and replacing them with the appropriate forces, we obtain the basic system. This system must be unchanging and statically determinate, and at the same time as easy as possible to solve. If a given frame is symmetrical, then in order to simplify the calculations also a symmetrical basic system should be assumed. Symmetrical and asymmetrical overvalues will cause symmetrical and asymmetrical graphs of moments.

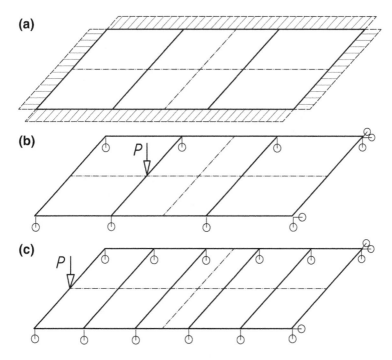

**Fig. 8.3** Methods of loading upholstery frames: **a** transverse with upholstery belts, **b** with force focused on internal rail, **c** with force focused on external rail

Below are the results of the numerical calculations determining the impact of the construction of upholstery frame on its stiffness and strength. During the calculations, the following assumptions were made as:

- upholstery frame is a flat grid statically loaded,
- operational load in the form of concentrated force $P = 1000$ N is applied,
- in the middle of the length of the longitudinal external element (rail),
- support of the grid results from the way the base of the case of the sofa is placed and mounted,
- in the calculations, only bending moments Mg and cutting forces $T$ will be considered,
- cross section of the element (rail) is $32 \times 50$ mm,
- frame elements are made of flat-pressed particle board of module $E = 3500$ MPa and Poisson's ratio $v = 0.3$. The stiffness of these elements does not change on their length and
- construction (carpentry) joints are perfectly stiff.

Calculations were made using the finite elements method, while the basic static schemes, i.e. the support state and frame loads are shown in Fig. 8.4. The main purpose of this analysis was to identify the optimal solution for the frame

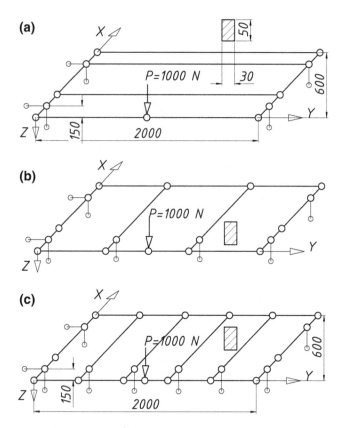

**Fig. 8.4** Static schemes of frames, along with division into calculation elements: **a** frame with longitudinal system of internal rails, **b** frame with transverse system of two internal rails, **c** frame with transverse system of four internal rails (mm)

construction, by assuming the maximum strength of components and significant material saving as the optimisation criterion. Based on the static schemes of the frames assumed for consideration, it was concluded that the values of reaction forces of the lifter locks are equal. This obviousness results from the fixed dimensional proportions and fixed location of mounting of the fittings.

However, depending on the number of support points for transverse rails on the case of the sofa, the value of reaction at these points is reduced proportionally to the number of supports. This has a significant impact on the distribution of internal forces in elements (Fig. 8.5).

From the given schemes of distribution of internal forces, it is easy to note that the maximum bending moment of the longitudinal front rail of the construction from Fig. 8.4a is 3 to 4 times greater than the corresponding bending moments in constructions as in Fig. 8.4b, c. Therefore, a more correct constructions are structures containing a transverse system of internal rails. Moreover, in all the

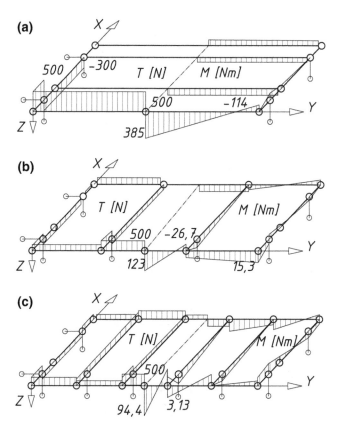

**Fig. 8.5** Distribution of bending moments and cutting forces in rails: **a** frame with longitudinal system of internal rails, **b** frame with transverse system of two internal rails, **c** frame with transverse system of four internal rails

considered construction variants, the maximum bending moment occurs under concentrated force $P$ that is at half the length of the longitudinal front rail. The difference between the values of maximum moments for frames with a transverse system of internal rails was about 30 % in favour of the system with four internal rails. However, for this reason, one should not expect significant material savings or the possibility of reduction of the coefficient of the cross-sectional strength resulting from them. These savings will not compensate for an increase in expenditure arising from the application of two additional internal rails in the system as shown in Fig. 8.4c. Besides, the construction stiffens slightly, which is shown by the values of displacement of individual nodes illustrated in Fig. 8.6.

The results shows that frame of upholstery furniture with a transverse rails is less prone to damage than a frame with a longitudinal system.

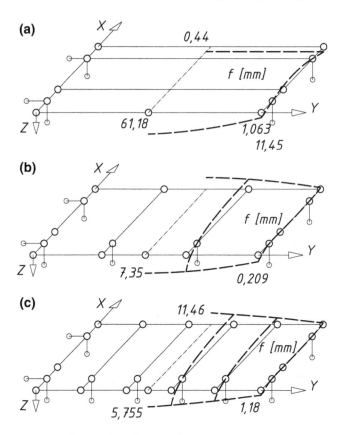

**Fig. 8.6** Displacement of nodes of the frames: **a** frame with longitudinal system of internal rails, **b** frame with transverse system of two internal rails, **c** frame with transverse system of four internal rails

## 8.2 Properties of Polyurethane Foams

### 8.2.1 Properties of Hyperelastic Polyurethane Foams

Hyperelastic polyurethane foams are produced in the process of foaming polyure-thanes, with the use of cross-linkers, foaming agents and catalysts. In the produc-tion of these types of foams, it is important that after seasoning the block, a mechanical opening of closed pores takes place. Properly opened pores provide the level of flexibility of the foam.

Foams are one of the basic materials used in the construction of subassemblages of upholstered furniture, both with springs and without springs. In particular in the latter, polyurethane foams form an essential part of the upholstery construction. However, due to the variety and availability of these raw materials, one should specify the rules for their selection, depending on the size and nature of the operational loads.

From the point of view of human physiology, maintaining sitting position for a couple of hours is not beneficial for the nervous and musculoskeletal system. In addition, the wrong distribution of weight on a seat can cause point loads on the cardiovascular system. As a consequence, long-term positioning of the body in the wrong position, on an improperly fitted base, often causes pains, changes in degenerative arthritis and blood clots, as well as superficial inflammation of the venous system of lower limbs (Smardzewski et al. 2007).

In production practice, highly flexible K-type foams, standard T-type foams and flame resistant foams are commonly used. Elastic properties of some of them are summarised in Table 8.1. Foam manufacturer's signatures contain a letter and numeric part. The letter part means type of foams, the first two digits of the numeric part inform of the foam density in kg/m$^3$ and the last two of the foam stiffness in kPa.

A characteristic feature of highly flexible polyurethane foams is their degressive–progressive, nonlinear stiffness. The division of the foam stiffness curve into three characteristic parts allows you to determine the stiffness coefficients and elasticity modules for each of the characteristic stages of deformation (Fig. 8.7). Stiffness coefficient $k$ is expressed as the quotient of the load value $P$ to the displacement $\Delta l$:

$$k_i = \frac{P_i - P_{i-1}}{\Delta l_i - \Delta l_{i-1}}, \tag{8.3}$$

while the linear elasticity modules $E_1$, $E_2$, $E_3$ are determined for each part of the function $P = f(\Delta l)$ located between the points of inflection of the function and the beginning and end points of the curve, according to the scheme shown in Fig. 8.7.

Figure 8.8 shows the stiffness characteristics of highly flexible foams, and Fig. 8.9 shows the stiffness characteristics of standard-type foams.

From the stiffness analysis of polyurethane foams for the furniture industry, it appears that they are materials of nonlinear stiffness characteristics. This means that in the process of designing upholstered furniture, elastic materials should be chosen based on the selection of the specific foam stiffness and not the foam density, as it usually takes place in practice.

## 8.2.2 Mathematical Models of Foams as Hyperelastic Bodies

Elastomers are a class of polymers having the following characteristics:

- They include natural and synthetic rubbers; they are amorphous and consist of long molecular chains (Fig. 8.10);
- The molecular chains are strongly twisted, spiral and randomly oriented in undeformed form; and
- The molecular chains during stretching get partially straightened; however, when the load stops, they go back to their original form.

**Table 8.1** Types of hyperelastic $K$ and standard $T$ foams and their characteristics specified by the manufacturer

| No. | Producer's signature | Density acc. to PN-77/C-05012.03 (kg/m$^3$) | Stiffness acc. to DIN EN ISO 3386 (kPa) | Permanent deformation acc. to PN-77/C05012.10 no more than (%) | Resiliency acc. to ISO 8307:2007 no less than (%) |
|---|---|---|---|---|---|
| 1 | K-2313 | 20.0–23.5 | 1.0–1.6 | 8 | 47 |
| 2 | K-2518 | 23.5–25.5 | 1.3–2.2 | 12 | 45 |
| 3 | K-2525 | 22.0–25.0 | 1.6–3.0 | 6 | 45 |
| 4 | K-3028 | 27.0–31.0 | 2.3–3.0 | 5 | 50 |
| 5 | K-3037 | 27.0–31.5 | 3.3–4.3 | 5 | 50 |
| 6 | K-3530 | 31.5–35.0 | 2.4–3.2 | 5 | 56 |
| 7 | K-3536 | 32.0–35.5 | 3.3–4.3 | 5 | 55 |
| 8 | K-4036 | 37.0–41.0 | 3.1–3.8 | 6 | 59 |
| 9 | K-4040 | 36.0–40.0 | 3.4–4.6 | 6 | 80 |
| 10 | K-4542 | 41.0–46.0 | 3.3–4.8 | 4 | 60 |
| 11 | T-1619 | 14.5–17.5 | 1.6–2.3 | 6 | 39 |
| 12 | T-1828 | 16.0–20.0 | 2.3–3.4 | 8 | 37 |
| 13 | T-2121 | 19.5–22.5 | 1.8–2.6 | 9 | 38 |
| 14 | T-2130 | 19.5–22.5 | 2.7–3.6 | 7 | 38 |
| 15 | T-2237 | 20.5–23.5 | 3.2–4.4 | 7 | 38 |
| 16 | T-2516 | 23.0–25.5 | 1.3–1.7 | 5 | 45 |
| 17 | T-2520 | 23.0–27.0 | 1.8–2.5 | 7 | 40 |
| 18 | T-2538 | 22.0–26.0 | 3.3–4.6 | 5 | 40 |
| 19 | T-2544 | 22.0–26.0 | 3.8–5.0 | 5 | 38 |
| 20 | T-2550 | 22.0–26.0 | 4.4–5.8 | 5 | 40 |
| 21 | T-2838 | 24.0–28.0 | 3.5–4.5 | 5 | 45 |
| 22 | T-3030 | 27.5–30.0 | 2.4–3.4 | 5 | 50 |
| 23 | T-3038 | 27.0–31.0 | 3.5–4.5 | 5 | 45 |
| 24 | T-3050 | 27.0–31.0 | 4.4–5.8 | 5 | 45 |
| 25 | T-3530 | 32.0–36.0 | 2.6–3.5 | 4 | 55 |
| 26 | T-3543 | 32.0–36.0 | 3.8–5.0 | 5 | 45 |
| 27 | T-3550 | 36.0–40.0 | 4.4–5.8 | 5 | 45 |
| 28 | T-4040 | 36.0–40.0 | 3.3–4.4 | 5 | 50 |
| 29 | T-4060 | 36.0–40.0 | 5.1–6.9 | 5 | 50 |

At the macroscopic level, the elastomer shows the following characteristics:

- may be subject to large deformations from 100 to 700 % of the initial dimension, depending on the degree of twisting of the molecular chains;
- to a minor extent, it changes the volume under load during deformation, which is why elastomers are almost incompressible; and
- the stress–strain relationship, as shown previously, is highly nonlinear.

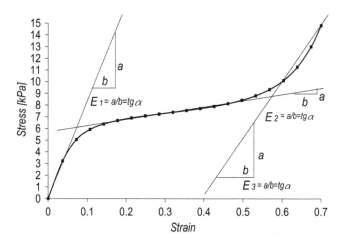

**Fig. 8.7** Scheme of determining linear elasticity modules of foams

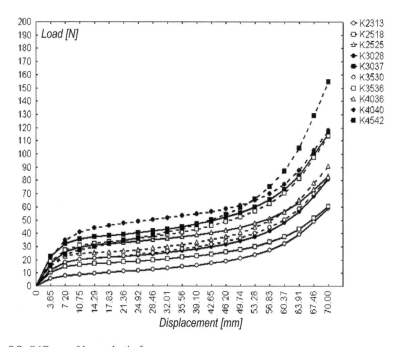

**Fig. 8.8** Stiffness of hyperelastic foams

An attempt of compression of polyurethane foams shows that load and unload curves do not overlap, creating a significant hysteresis of strains and absorbing a significant part of energy in the stress–strain cycle (Fig. 8.11). In the case of bodies of linear viscoelasticity, the shape of hysteresis residue is independent of the size of

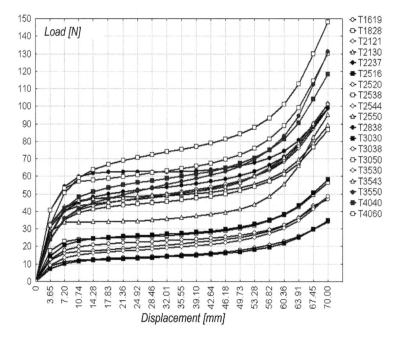

**Fig. 8.9** Stiffness of standard-type foams

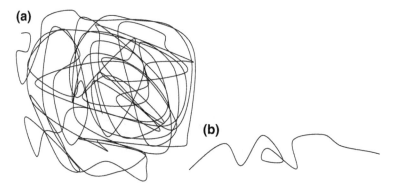

**Fig. 8.10** Example of a molecular chain: **a** unstretched, **b** stretched

the deformation, so the time of experiment has no impact on the properties of such materials being determined. The amount of energy absorbed by materials of non-linear viscoelasticity, such as polyurethane foams, depends on the size of the strains, and therefore, mechanical properties of these bodies will vary depending on the speed of the pressure and the duration of the experiment. Figure 8.12 shows hysteresis loop with a grey field of total load energy, and the light grey colour shows the amount of total unload energy. The difference between these fields is the

**Fig. 8.11** Hysteresis of
viscoelastic material in the
stress–strain cycle

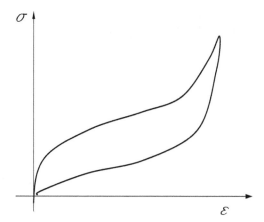

amount of absorbed energy. If the size of the strains is smaller and load and offload
curves are approaching each other, then the amount of distributed energy will be
minimal (Fig. 8.12b).

Foams, like other hyperelastic materials, can be described using the theory of
energy of potential strains $U$, which determines the size of the energy accumulated
in a volume unit of a material, as a function of strain at a point (Anonim 2000a, b;
Hill 1978; Mills and Gilchrist 2000; Ogden 1972; Renz 1977, 1978; Storakers
1986). Before a detailed analysis of different forms of density functions of strains
energy, basic concepts will be defined, such aselongation coefficient

$$\lambda = \frac{L}{L_0} = \frac{L - \Delta u}{L_0} = 1 + \varepsilon_E, \qquad (8.4)$$

where
$\varepsilon_E$   unit strain

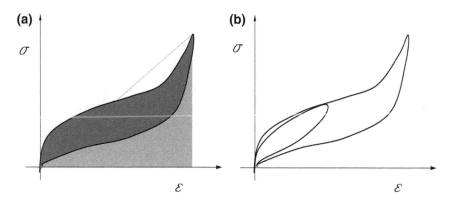

**Fig. 8.12** The amount of distributed energy depending on the size of the strains

**Fig. 8.13** Strains of the
sample in a flat biaxial state of
loads

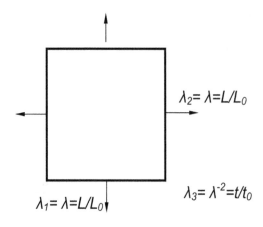

In addition, we also have three elongation coefficients $\lambda_1, \lambda_2, \lambda_3$, which allow the measurement of strain and are used to determine the density functions of strains energy. The following example illustrates strains of a rectangular element in the state of biaxial compression (Fig. 8.13). The main strains coefficients $\lambda_1, \lambda_2$, show the strain in a plane. In the case of the sample thickness, the coefficient $\lambda_3$ presents the change of material thickness $(t/t_0)$, and for an incompressible material, $\lambda_3 = \lambda^{-2}$.

Three parameters of strains are used to define the density functions of strains energy:

$$I_1 = \lambda_1^2 + \lambda_2^2 + \lambda_3^2, \tag{8.5}$$

$$I_2 = \lambda_1^2 \lambda_2^2 + \lambda_2^2 \lambda_3^2 + \lambda_3^2 \lambda_1^2, \tag{8.6}$$

$$I_3 = \lambda_1^2 \lambda_2^2 \lambda_3^2. \tag{8.7}$$

For an incompressible material, $I_3 = 1$.
Volume coefficient $J$ is defined as:

$$J = \lambda_1 \lambda_2 \lambda_3 = \frac{V}{V_0}. \tag{8.8}$$

Density function of strains energy can be most often defined as $W$. The function of strains energy can also be the function of major strain coefficients or the function of strain parameters

$$W = W(I_1 I_2 I_3) \quad \text{or} \quad W = W(\lambda_1 \lambda_2 \lambda_3). \tag{8.9}$$

Based on $W$, the second Piola-Kirchhoff stresses (as well as Green-Lagrange strains) can be written in the form:

$$S_{ij} = \frac{\partial W}{\partial E_{ij}}.$$  (8.10)

Because the material is incompressible, we can separate the components of the function of strain energy into differentiable (with index $d$) and volume (with index $b$). As a result, volume strains are only a function of the volume coefficient $J$:

$$W = W_d(\bar{I}_1, \bar{I}_2) + W_b(J),$$  (8.11)

$$W = W_d(\bar{\lambda}_1, \bar{\lambda}_2, \bar{\lambda}_3) + W_b(J),$$  (8.12)

where the differential of major extensions and the differential of parameters $I$ are expressed in the form:

$$\bar{\lambda}_p = J^{-\frac{1}{3}}\lambda_p \quad \text{for } p = 1, 2, 3,$$  (8.13)

$$\bar{I}_p = J^{-\frac{2}{3}}I_p \quad \text{for } p = 1, 2, 3.$$  (8.14)

By writing $I_3 = J^2$, $I_3$ is not used in the definition of $W$.

There are many forms of functions that describe the value of potential energy of strains (Anonim 2000a, b), e.g. the equations: Arruda–Boyce, Marlow, Mooney–Rivlin, neo-Hookean, Ogden, polynomial, reduced polynomial, Yeoh and van der Waals. Each of them is presented below.

**Mooney–Rivlin model**

There are 2-, 3-, 5- and 9-parametric Mooney–Rivlin models known.

Mooney–Rivlin model with two parameters:

$$W = C_{10}(\bar{I}_1 - 3) + C_{01}(\bar{I}_2 - 3) + \frac{1}{D}(J_{\text{el}} - 1)^2.$$  (8.15)

Mooney–Rivlin model with three parameters:

$$W = C_{10}(\bar{I}_1 - 3) + C_{01}(\bar{I}_2 - 3) + C_{11}(\bar{I}_1 - 3)(\bar{I}_2 - 3) + \frac{1}{D}(J_{\text{el}} - 1)^2.$$  (8.16)

Mooney–Rivlin model with five parameters, where $N = 2$:

$$W = \sum_{i+j=1}^{2} C_{ij}(\bar{I}_1 - 3)^i(\bar{I}_2 - 3)^j \frac{1}{D}(J_{\text{el}} - 1)^2.$$  (8.17)

Mooney–Rivlin model with nine parameters, where $N = 3$:

$$W = \sum_{i+j=1}^{3} C_{ij}(\bar{I}_1 - 3)^i(\bar{I}_2 - 3)^j \frac{1}{D}(J_{el} - 1)^2. \qquad (8.18)$$

For all forms of the Mooney–Rivlin function, the initial value of the shear modulus we define as:

$$\mu_0 = 2(C_{10} + C_{01}), \qquad (8.19)$$

where

$C_{10}$ and $C_{01}$   are the coefficients determined in experimental studies
and module $\kappa$:

$$\kappa = \frac{2}{D}. \qquad (8.20)$$

By choosing the right type of function for the tested kind of foam, one can follow the characteristics of the material stiffness $\sigma = f(\varepsilon)$, and so:

- for a curve without point of inflection (Fig. 8.14), an equation with two parameters can be used,
- for a curve with one point of inflection (Fig. 8.15), an equation with five parameters can be used, and
- for a curve with two points of inflection (Fig. 8.16), an equation with nine parameters can be used.

**Fig. 8.14** Characteristics of stiffness of foam without point of inflection

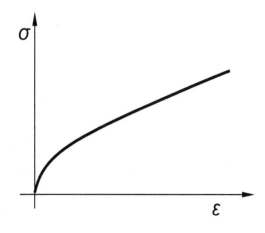

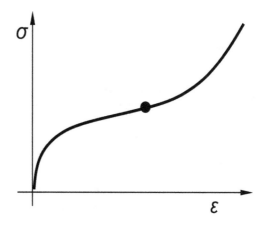

**Fig. 8.15** Characteristics of stiffness of foam with one point of inflection

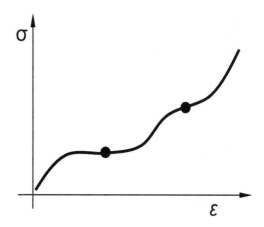

**Fig. 8.16** Characteristics of stiffness of foam with two points of inflection

**Ogden's model**

General Ogden's model is a model based on coefficients of main extensions, whereby

$$W = \sum_{i=1}^{N} \frac{\mu_i}{\alpha_i} \left( \bar{\lambda}_1^{\alpha_i} + \bar{\lambda}_2^{\alpha_i} + \bar{\lambda}_3^{\alpha_i} - 3 \right) + \sum_{i=1}^{N} \frac{1}{D_i} (J_{\mathrm{el}} - 1)^{2i}, \qquad (8.21)$$

where the initial shear modulus has the form:

$$\mu_o = \frac{\sum\limits_{i=1}^{N} \mu_i \alpha_i}{2}, \qquad (8.22)$$

and module $\kappa_o$:

$$\kappa_o = \frac{2}{D_i}. \tag{8.23}$$

**Ogden's model for hyperelastic foams**

This model is very similar to the incompressible material model:

$$W = \sum_{i=1}^{N} \frac{\mu_i}{\alpha_i} \left( J_{el}^{\frac{\alpha_i}{3}} (\bar{\lambda}_1^{\alpha_i} + \bar{\lambda}_2^{\alpha_i} + \bar{\lambda}_3^{\alpha_i}) - 3 \right) + \sum_{i=1}^{N} \frac{\mu_i}{\alpha_i \beta_i} \left( J_{el}^{-\alpha_i \beta} - 1 \right), \tag{8.24}$$

where the initial shear modulus has the form:

$$\mu_o = \frac{\sum_{i=1}^{N} \mu_i \alpha_i}{2}, \tag{8.25}$$

and module $\kappa_o$:

$$\kappa_o = \sum_{i=1}^{N} \mu_i \alpha_i \left( \frac{1}{3} + \beta_i \right). \tag{8.26}$$

Ogden's models are mainly used for modelling foams of deformations above 700 %.

**Arruda–Boyce equation**

This equation has the form:

$$U = \mu \left\{ \frac{1}{2} (\bar{I}_1 - 3) + \frac{1}{20\lambda^2_{,,,}} \left( \bar{I}_1^2 - 9 \right) + \frac{11}{1050\lambda^4_{,,,}} (\bar{I}_1^3 - 27) + \frac{19}{7000\lambda^6_{,,,}} (\bar{I}_1^4 - 81) + \frac{519}{673750\lambda^8_{,,,}} (\bar{I}_1^5 - 243) \right\}$$
$$+ \frac{1}{D} \left( \frac{J_{el}^2 - 1}{2} - \ln J_{el} \right), \tag{8.27}$$

where $U$ is the potential energy of strains per unit volume, $\mu$, $\lambda_{,,,}$ and $D$ temperature-dependent material parameters,

$$\bar{I}_1 = \bar{\lambda}_1^2 + \bar{\lambda}_2^2 + \bar{\lambda}_3^2, \tag{8.28}$$

whereby

$$\bar{\lambda}_i = J^{-\frac{1}{3}} \lambda_i, \tag{8.29}$$

where

$J$    total volume coefficient,

$J_{el}$   elasticity volume coefficient and

$\lambda_i$   physical elongation.

The initial shear modulus $\mu_o$ in relation to $\mu$ is expressed by the equation:

$$\mu_o = \mu\left\{1 + \frac{3}{5\lambda_{,,,}^2} + \frac{99}{175\lambda_{,,,}^4} + \frac{513}{875\lambda_{,,,}^6} + \frac{42,039}{673,375\lambda_{,,,}^8}\right\}. \tag{8.30}$$

Most often, the coefficient $\lambda_{,,,}$ takes the value 7, for which $\lambda_o = 1.0125$, and the initial value of the module $\kappa_o$ is

$$\kappa_o = \frac{2}{D}. \tag{8.31}$$

**Marlow equation**

The equation for potential energy of strains according to Marlow has the form:

$$U = U_{dev}(\bar{I}_1) + U_{vol}(J_{el}), \tag{8.32}$$

where $U$ is the potential energy of strains per volume unit, with $U_{dev}$ as the deformed part and $U_{vol}$ as the volume part—undeformed,

$$\bar{I}_1 = \bar{\lambda}_1^2 + \bar{\lambda}_2^2 + \bar{\lambda}_3^2, \tag{8.33}$$

whereby

$$\bar{\lambda}_i = J^{-\frac{1}{3}}\lambda_i, \tag{8.34}$$

where

$J$    total volume coefficient,

$J_{el}$   elasticity volume coefficient and

$\lambda_i$   main elongation.

**Neo-Hookean equation**

In this case, the equation for potential energy of strains has the form:

$$U = C_{10}(\bar{I}_1 - 3) + \frac{1}{D_1}(J^{el} - 1)^2, \tag{8.35}$$

where $U$ is the potential energy of strains per unit volume, $C_{10}$ and $D_1$ temperature-dependent material coefficients:

$$\bar{I}_1 = \bar{\lambda}_1^2 + \bar{\lambda}_2^2 + \bar{\lambda}_3^2, \tag{8.36}$$

whereby

$$\bar{\lambda}_i = J^{-\frac{1}{3}}\lambda_i, \tag{8.37}$$

where
$J$   total volume coefficient,
$J^{\mathrm{el}}$  elasticity volume coefficient and
$\lambda_i$  main elongation.

The initial value of the figural strains modulus and the module $\kappa_o$ has the form:

$$\mu_o = 2C_{10}, \quad \kappa_o = \frac{2}{D_1}. \tag{8.38}$$

**Polynomial form**
This equation has the form:

$$U = \sum_{\substack{i+j=1}}^{N} C_{ij}(\bar{I}_1 - 3)^i(\bar{I}_2 - 3)^j + \sum_{i=1}^{N} \frac{1}{D_1}(J^{\mathrm{el}} - 1)^{2i}, \tag{8.39}$$

where
$U$              potential energy of strains per volume unit,
$N$              material parameter and
$C_{ij}$ and $D_1$   temperature-dependent material coefficients,

$$\bar{I}_1 = \bar{\lambda}_1^2 + \bar{\lambda}_2^2 + \bar{\lambda}_3^2, \quad \bar{I}_2 = \bar{\lambda}_1^{(2)} + \bar{\lambda}_2^{(2)} + \bar{\lambda}_3^{(2)}, \tag{8.40}$$

whereby

$$\bar{\lambda}_i = J^{-\frac{1}{3}}\lambda_i, \tag{8.41}$$

where
$J$   total volume coefficient,
$J^{\mathrm{el}}$  elasticity volume coefficient and
$\lambda_i$  physical elongation.

The initial value of the figural strains modulus and the module $\kappa$ has the form:

$$\mu_o = 2(C_{10} + C_{01}), \quad \kappa_o = \frac{2}{D_1}. \tag{8.42}$$

When normal stresses are small or moderately large, then the first part of the equation leads to sufficiently correct solutions,

$$U = \sum_{i+j=1}^{N} C_{ij}(\bar{I}_1 - 3)^i(\bar{I}_2 - 3)^j. \tag{8.43}$$

**Reduced polynomial form**

This equation has the form:

$$U = \sum_{i=1}^{N} C_{i0}(\bar{I}_1 - 3)^i + \sum_{i=1}^{N} \frac{1}{D_1}(J^{el} - 1)^{2i}, \tag{8.44}$$

where

| | |
|---|---|
| $U$ | potential energy of strains per volume unit, |
| $N$ | material parameter and |
| $C_{ij}$ and $D_1$ | temperature-dependent material coefficients, |

$$\bar{I}_1 = \bar{\lambda}_1^2 + \bar{\lambda}_2^2 + \bar{\lambda}_3^2, \tag{8.45}$$

whereby

$$\bar{\lambda}_i = J^{-\frac{1}{3}}\lambda_i, \tag{8.46}$$

where

| | |
|---|---|
| $J$ | total volume coefficient, |
| $J^{el}$ | elasticity volume coefficient and |
| $\lambda_i$ | physical elongation. |

The initial value of the figural strains modulus and the module $\kappa_o$ has the form:

$$\mu_o = 2C_{10}, \quad \kappa_o = \frac{2}{D_1}. \tag{8.47}$$

**Van der Waals equation**

The equation for potential energy of strains according to van der Waals has the form:

$$U = \mu \left\{ -\left(\lambda_{,,,}^2 - 3\right)[\ln(1 - \eta) + \eta] - \frac{2}{3}a\left(\frac{I - 3}{2}\right)^{\frac{3}{2}} \right\} + \frac{1}{D}\left(\frac{J_{el}^2 - 1}{2} - \ln J_{el}\right), \tag{8.48}$$

where

$$I = (1 - \beta)\bar{I}_1 + \beta\bar{I}_2 \tag{8.49}$$

and

$$\eta = \sqrt{\frac{I - 3}{\lambda_{,,,}^2 - 3}}. \tag{8.50}$$

Whereby $U$ is the potential energy of strains per unit volume, $\mu$—the initial shear modulus, $\lambda_{,,,}$—observed elongation, $a$—general interaction coefficient, $\beta$—constant coefficient and $D$—parameter influencing the compression. All these parameters are temperature-dependent. Moreover,

$$\bar{I}_1 = \bar{\lambda}_1^2 + \bar{\lambda}_2^2 + \bar{\lambda}_3^2, \quad \text{and} \quad \bar{I}_2 = \bar{\lambda}_1^{(2)} + \bar{\lambda}_2^{(2)} + \bar{\lambda}_3^{(2)}, \tag{8.51}$$

whereby

$$\bar{\lambda}_i = J^{-\frac{1}{3}}\lambda_i, \tag{8.52}$$

where
$J$    total volume coefficient,
$J^{el}$   elasticity volume coefficient and
$\lambda_i$   physical elongation.

The initial value of the figural strains modulus and the module $\kappa_o$ has the form:

$$\mu_o = \mu, \tag{8.53}$$

$$\kappa_o = \frac{2}{D}. \tag{8.54}$$

**Yeoh equation**
This equation can be written as:

$$U = C_{10}(\bar{I}_1 - 3) + C_{20}(\bar{I}_1 - 3)^2 + C_{30}(\bar{I}_1 - 3)^3$$
$$+ \frac{1}{D_1}(J^{el} - 1)^2 + \frac{1}{D_2}(J^{el} - 1)^4 + \frac{1}{D_3}(J^{el} - 1)^6 \tag{8.55}$$

where
$U$   potential energy of strains per volume unit and
$D_i$   temperature-dependent material coefficient,

$$\bar{I}_1 = \bar{\lambda}_1^2 + \bar{\lambda}_2^2 + \bar{\lambda}_3^2, \tag{8.56}$$

whereby

$$\bar{\lambda}_i = J^{-\frac{1}{3}}\lambda_i, \tag{8.57}$$

where
$J$    total volume coefficient,
$J^{el}$    elasticity volume coefficient and
$\lambda_i$    physical elongation.

The initial value of the figural strains modulus and the module $\kappa_o$:

$$\mu_o = 2C_{10}, \tag{8.58}$$

$$\kappa_o = \frac{2}{D_1}. \tag{8.59}$$

If data characterising stiffness of the foams, derived directly from many complex experiments (e.g. from multi-directional compression test), are known, then the most useful models are Ogden's and van der Waals. If limited results of experimental research are available, the models Arruda–Boyce, van der Waals, Yeoh or reduced polynomials should be used. However, if there is only one set of experimental test results (such as axial compression), then the recommended model is Marlow.

## 8.2.3 Correctness of Nonlinear Mathematical Models of Polyurethane Foams

The optimisation of the construction of mattresses and/or seats is very important in the use of furniture for sleeping and relaxation, motor vehicles, aircrafts or rehabilitation medical equipment. Descriptions of the mechanics of hard foams are known on the basis of articles of Renz (1977, 1978). Czysz (1986) described the behaviour of soft polyurethane foams as an elastic Hooke's body. Using the function of strains energy built in the system ABAQUS (Anonim 2000a, b), Mills and Gilchrist (2000) conducted calculations for soft foams under compression. In this research, only the main parameters were compared with the experimental results, without making a thorough comparative analysis of detailed parameters, mainly the parameter $\beta$. The purpose of the research carried out by Schrodt et al. (2005) was the use of the standard function of strains energy "hyperfoam" in the system ABAQUS to describe mechanical properties of the foams. For the research, the authors used polyurethane foams type SAF 6060, with the dimensions $200 \times 200$ mm and a height of 50 mm.

As previous experimental research has shown, soft polyurethane foams behave like viscoelastic materials. Therefore, for their description, constitutive equations of a viscoelastic body are mostly used. In general models, Schrodt et al. (2005) broke up the stress tensor $S$ into the part comprising the equations of stresses in terms of elasticity $S_G$ and $S_{OV}$ part of the stresses representing properties of the material memory. Therefore, the stress tensor was written as:

$$S = S_G + S_{OV}. \tag{8.60}$$

Hyperelastic material, as an example of Cauchy's flexible material, was characterised by the function of strains energy. The stress tensor can be obtained by the differentiation of the function of strains energy, which reflects the strain tensor. Therefore, basing on the mechanical energy equation:

$$\dot{\omega} = JS \cdot D \quad \text{whereby } J = \det F, \tag{8.61}$$

where
$\omega$   function of strains energy,
$F$   strain gradient,
$S$   Cauchy stress tensor and
$D$   gradient of strains tensor

$$D = \frac{1}{2} F^{-T} \dot{C} F^{-1}, \tag{8.62}$$

where
$\dot{C}$   the right Cauchy-Green tensor, the dot above the symbol means differentiation after time.

According to the assumptions of Schrodt et al. (2005), $\omega$ is a scalar, a non-negative function tensor of the right extension of the tensor $U$ or the right Cauchy-Green tensor:

$$\omega = \omega(U) = \omega(C) = \begin{cases} > 0 & \text{for} \quad C \neq I \\ = 0 & \text{for} \quad C = I. \end{cases} \tag{8.63}$$

By entering the above equation to

$$\dot{\omega} = JS \cdot D, \tag{8.64}$$

by reference to

$$D = \frac{1}{2} F^{-T} \dot{C} F^{-1}, \tag{8.65}$$

the general structure of the constitutive equation for nonlinear, hyperelastic and anisotropic body is obtained, in the form:

$$S = 2J^{-1}F\frac{\partial\omega(C)}{\partial C}F^T. \tag{8.66}$$

**The function of strains energy for highly compressible polymers**
For the description of mechanical behaviour of highly compressible polymers, the function of strains energy has the form given by Hill (1978) and Storakes (1986):

$$\omega = \sum_{i=1}^{N}\frac{2\mu_i}{\alpha_i^2}\left[\left(\lambda_1^{\alpha_i} + \lambda_2^{\alpha_i} + \lambda_3^{\alpha_i} - 3\right) + f(J)\right], \tag{8.67}$$

where
$\mu_i, \alpha_i$  material parameters and
$f(J)$   volume function, which fulfils the condition $f(1) = 0$

By using this equation, we obtain the constitutive equation form:

$$S = 2J^{-1}\sum_{j=1}^{3}\sum_{i=1}^{N}\frac{\mu_i}{\alpha_i^2}\left[\lambda_i^{\alpha_i} + \frac{1}{\alpha_i}J\frac{\partial f(J)}{\partial J}\right]n_in_i, \tag{8.68}$$

where
$\lambda_i$  value of the right elongation of the tensor $U$ and
$n_i$  value of the left elongation of the tensor $V$.

The probable form of the volume function $f(J)$ was given by Storakes (1986):

$$f(J) = \frac{1}{\beta_j}\left(J^{\alpha_j\beta_j} - 1\right), \tag{8.69}$$

where
$\beta_j$  additional material parameter.

Hence the number 3 N of material coefficients $\alpha_j$, $\beta_j$ and $\mu_j$ ($j = 1, 2, ...N$) was obtained, which should be determined in the course of experimental research. In addition, the initial value of the shear modulus and compression module has been defined (Anonim 2000a, b):

$$\mu_0 := \sum_{j=1}^{N}\mu_j \quad \text{and} \quad \kappa_0 := \sum_{j=1}^{N}2\left(\frac{1}{3} + \beta_j\right)\mu_j. \tag{8.70}$$

In this way, also the relation between Poisson's ratio $v_j$ and the parameter $\beta_i$ was obtained:

$$v_j = \frac{\beta_j}{1 + 2\beta_j}, \tag{8.71}$$

hence, respectively:

$$\beta_j = \frac{v_j}{1 - 2v_j} \quad \text{for } j = 1, 2 \ldots N. \tag{8.72}$$

For individual cases $\beta_j = : \beta = \text{const}$, $v$ is equal to the standard value of Poisson's ratio. According to Hill (1978) and Storakes (1986), it is justified to use the following equations:

$$\mu_i \alpha_i > 0 (i = 1.2, \ldots N) \quad \text{and} \quad \beta > -\frac{1}{3} \tag{8.73}$$

**Strength–elongation relationship for the axial compression test**
When considering the homogeneity of the strain, the constant strain gradient can be written in the form:

$$F(t) = \lambda_1(t) e_1 e_1 + \lambda_2(t) e_2 e_2 + \lambda_3(t) e_3 e_3, \tag{8.74}$$

where elongations can be expressed as:

$$\lambda_1 = \lambda_2 = \frac{a(t)}{a_0}, \quad \lambda_3 = \frac{h(t)}{h_0}, \quad J = \lambda_1^2 \lambda_3 = \left[\frac{a(t)}{a_0}\right]^2 \frac{h(t)}{h_0}, \tag{8.75}$$

where
$a_0$, $a(t)$   sizes of the angle and
$h_0$, $h(t)$   heights of the examined sample before and during the strain.

If the foam sample is loaded only in the direction of axis 3, then stresses in the direction of 1 and 2 do not occur. Hence, on the basis of the equation

$$S = 2J^{-1} \sum_{j=1}^{3} \sum_{i=1}^{N} \frac{\mu_i}{\alpha_i^2} \left[\lambda_i^{\alpha_i} + \frac{1}{\alpha_i} J \frac{\partial f(J)}{\partial J}\right] n_i n_i \tag{8.76}$$

and using the equations

$$f(J) = \frac{1}{\beta_j} \left(J^{\alpha_j \beta_j} - 1\right) \tag{8.77}$$

and

$$\lambda_1 = \lambda_2 = \frac{a(t)}{a_0}, \tag{8.78}$$

$$\lambda_3 = \frac{h(t)}{h_0}, \tag{8.79}$$

$$J = \lambda_1^2 \lambda_3 = \left[\frac{a(t)}{a_0}\right]^2 \frac{h(t)}{h_0}, \tag{8.80}$$

stresses equation takes the following form:

$$\sigma_{33}(\lambda_1, \lambda_3) = 2\left(\lambda_1^2, \lambda_3\right)^{-1} \sum_{i=1}^{N} \frac{\mu_i}{\alpha_i} \left[\lambda_3^{\alpha_i} - \left(\lambda_1^2, \lambda_3\right)^{-\alpha_i \beta_i}\right], \tag{8.81}$$

$$\sum_{i=1}^{N} \frac{\mu_i}{\alpha_i} \left[\lambda_3^{\alpha_i} - \left(\lambda_1^2, \lambda_3\right)^{-\alpha_i \beta_i}\right] = 0. \tag{8.82}$$

If the sample is loaded with a single load $K$ in the direction of axis 3, then the stress in the direction of 3, after taking into account the conditions of equilibrium, will amount to:

$$\sigma_{33} = -K/(ab) \equiv -K/a^2. \tag{8.83}$$

Hence, the final relation for the axial load has the form:

$$K(\lambda_1, \lambda_3) = 2a^2 \left(\lambda_1^2, \lambda_3\right)^{-1} \sum_{i=1}^{N} \frac{\mu_i}{\alpha_i} \left[\lambda_3^{\alpha_i} - \left(\lambda_1^2, \lambda_3\right)^{-\alpha_i \beta_i}\right], \tag{8.84}$$

and the next resulting relation for extensions in the directions 1 and 2:

$$f(\lambda_1, \lambda_3) = \sum_{i=1}^{N} \frac{\mu_i}{\alpha_i} \left[\lambda_3^{\alpha_i} - \left(\lambda_1^2, \lambda_3\right)^{-\alpha_i \beta_i}\right] = 0. \tag{8.85}$$

For the case $N = 1$ and using $\alpha_1 := \alpha$, $\beta_1 := \beta$, $\mu_1 := \mu$, a dependency between $\lambda_1$ and $\lambda_3$ can be derived:

$$\lambda_1 = f(\lambda_3) = \lambda_3^{-\frac{\beta}{1+2\beta}}, \tag{8.86}$$

and finally,

$$\lambda_1^2 \lambda_3 = \lambda_3^{-\frac{1}{1+2\beta}}. \tag{8.87}$$

The equation $\lambda_1^2 \lambda_3 = \lambda_3^{-\frac{1}{1+2\beta}}$ allows for the separation of the parameter $\beta$, which can be used for a separate analysis of $\beta$. By application $\lambda_1^2 \lambda_3 = \lambda_3^{-\frac{1}{1+2\beta}}$, from the equation,

$$v_j = \frac{\beta_j}{1 + 2\beta_j}, \tag{8.88}$$

or

$$\beta_j = \frac{v_j}{1 - 2v_j} \quad \text{for } j = 1, 2, \ldots N, \tag{8.89}$$

$\lambda_1$ for $N = 1$ could be eliminated. Hence, the final form of the stress-elongation relation, with the assumption that $N = 1$, has the form as:

$$K(h) = 2\frac{\mu}{\alpha}a_0^2 \left[ \left( \frac{h}{h_0} \right)^{-\alpha\frac{1+3\beta}{1+2\beta}} - 1 \right] \left( \frac{h}{h_0} \right)^{\alpha-1} \equiv 2\frac{\mu}{\alpha}a_0^2 \left( \lambda_3^{-\alpha\frac{1+3\beta}{1+2\beta}} - 1 \right) \lambda_3^{\alpha-1}. \tag{8.90}$$

From the above equation, it results that the following conditions are very important:

$$\mu_i\alpha_i > 0 (i = 1, 2, \ldots N) \quad \text{and} \quad \beta > -\frac{1}{3}, \tag{8.91}$$

because for $\beta = -1/3$, the value of $K$ will always equal zero for the elongation $\lambda_3$.

## Calculations using the finite elements method
In order to verify the correctness of the constitutive model describing the mechanics of behaviour of soft foams, Schrodt et al. (2005) conducted numerical calculations using the finite elements method in the environment of the system ABAQUS. The mesh model was built using an 8-node line element of brick type. To the bottom surface of the foam, bonds were assigned to prevent displacement in the direction of the axes x, y and z. Bonds were assigned to the top surface to prevent displacement in the direction of axes x and z. Pressurers were modelled as perfectly stiff bodies. It was also assumed that the coefficient of friction between the surface of the pressurer and the surface of the foam will amount to 0.75. The sum of the actual loads applied to the sample and distribution of this value proportionally on all the nodes of the numeric model was assumed as the model load.

**Table 8.2** Comparison of the values of material parameters of the foams which were subject to axial compression test, calculated numerically or analytically

| Parameters | Analytical calculations (EXP) | Numerical calculations (FEM) | Difference FEM/EXP (%) |
|---|---|---|---|
| $\mu$ [MPa] | $0.831 \times 10^{-2}$ | $0.907 \times 10^{-2}$ | 9.2 |
| $\alpha$ | $0.198 \times 10^{2}$ | $0.213 \times 10^{2}$ | 7.6 |
| $\beta$ | $0.109 \times 10^{-1}$ | $0.849 \times 10^{-2}$ | 0.779 |

**Results of the calculations and their comparison with the results of experimental research**

Table 8.2 compares the values of material parameters of the foams determined in the course of experimental research through axial compression and calculated on the basis of the equation:

$$K(h) = 2\frac{\mu}{\alpha}a_0^2\left[\left(\frac{h}{h_0}\right)^{-\alpha\frac{1+3\beta}{1+2\beta}} - 1\right]\left(\frac{h}{h_0}\right)^{\alpha-1} \equiv 2\frac{\mu}{\alpha}a_0^2\left(\lambda_3^{-\alpha\frac{1+3\beta}{1+2\beta}} - 1\right)\lambda_3^{\alpha-1}, \qquad (8.92)$$

with the results of numerical calculations with the use of the finite elements method.

On the basis of the compiled values, it can be seen that the results of the numerical calculations of parameters $\mu$ and $\alpha$ are about 7–9 % larger in relation to the results of the analytical calculations. In the case of parameter $\beta$, numerical calculations provided a result smaller by about 22 % in relation to the analytical calculations. The presented solutions are therefore valid and can be used for analysis of stiffness of more complex multi-layer sets.

## 8.2.4 Stiffness of Hyperelastic Polyurethane Foams

This chapter presents the results of the axial compression of polyurethane foams study (Fig. 8.17), for which a nonlinear model of Mooney–Rivlin was built and a numerical analysis of contact stresses was conducted.

By building a mathematical model of elastic foam, it was assumed that this is a model

- of isotropic and nonlinear material,
- made up of cells distributed evenly and capable of large deformations,
- capable of large deformations, over 90 % during compression and
- requiring geometrical nonlinearity during subsequent steps of strain analysis.

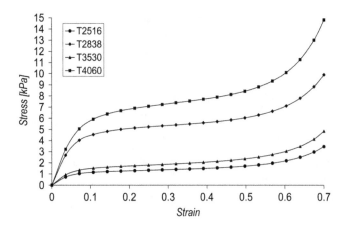

**Fig. 8.17** The stiffness of the foams in the stress–strain system

Stresses in the energy function of stretching for foam have been expressed in the form:

$$\sigma_i = \frac{\partial W}{\partial L_i}. \tag{8.93}$$

Thereby compression energy is expressed by the equation:

$$W = f(I_1, I_2, I_3), \tag{8.94}$$

whereby

$$I_1 = L_1^2 + L_2^2 + L_3^2, \tag{8.95}$$

$$I_2 = L_1^2 L_2^2 + L_2^2 L_3^2 + L_3^2 L_1^2, \tag{8.96}$$

$$I_3 = L_1^2 L_2^2 L_3^2. \tag{8.97}$$

For axial compression, the stress function has the form:

$$\sigma L = \left(L^2 - \frac{1}{L}\right)\left[2\left(\frac{\partial W}{\partial I_2}\right) + \frac{2}{L}\left(\frac{\partial W}{\partial I_1}\right)\right]. \tag{8.98}$$

Therefore, Mooney–Rivlin's equation, appropriate for hyperelastic materials (of large deformations up to 200 %), has been written as:

$$W(I_1, I_2) = C_1(I_1 - 3) + C_2(I_2 - 3). \tag{8.99}$$

For uniaxial compression or stretching, it takes the form:

$$\sigma = 2\left(C_1 + \frac{C_2}{L}\right)\left(L - \frac{1}{L^2}\right). \tag{8.100}$$

Transforming this equation to the form:

$$\frac{\sigma}{2\left(L - \frac{1}{L^2}\right)} = \frac{1}{L}C_2 + C_1, \tag{8.101}$$

the equation of a line was obtained, by means of which the coefficients $C_1$ and $C_2$, were determined, necessary for numerical analysis:

$$y = ax + b, \tag{8.102}$$

where

$$y = \frac{\sigma}{2\left(L - \frac{1}{L^2}\right)}, \quad a = \frac{1}{L}. \tag{8.103}$$

In Fig. 8.17, the dependency stress–strain has been presented for each type of foam. As it can be seen, foams T2838 and T4060 were characterised by the greatest stiffness. The foams T2516 and T3530 were much softer. Therefore, initially it could be concluded that the foams T2516 and T3530 should be used as an outer layer of a mattress, directly in contact with the user's body, while the foams T2838 and T4060 should be used as inner layers, to prevent greater displacements, particularly at a large weight of the user.

Additionally, Table 8.3 shows that at different stages of compression of the foam, they have a variable value of Young's modulus. For foams T2516 and T3530, the ratio $E_3/E_1 = 0.76$–$0.82$, while for foams T2838 and T4060, the ratio $E_3/E_1 = 0.43$–$0.58$. This relevant differentiation allows greater freedom in the selection of the stiffness of foam when modelling complex systems of multi-layer mattresses.

Material constants occurring in Mooney–Rivlin's equation are determined from the dependencies provided in Fig. 8.18. These constants represent the data necessary to build suitable numerical models.

In simulating the stress of the human thigh on the surface of the polyurethane foam mattress, in the system ABAQUS, an appropriate mesh model of the foam was

**Table 8.3** Modules of linear elasticity of foams

| Type of foam | Young's modulus (kPa) | | |
|---|---|---|---|
| | $E_1$ | $E_2$ | $E_3$ |
| T2516 | 14.44 | 1.52 | 11.06 |
| T2838 | 56.12 | 3.75 | 24.30 |
| T3530 | 18.69 | 2.32 | 15.50 |
| T4060 | 70.25 | 6.22 | 40.82 |

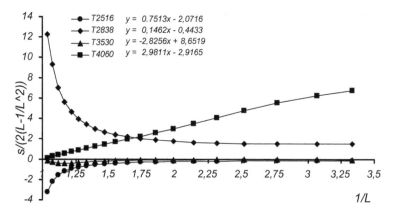

**Fig. 8.18** Functions for determining constants C1 and C2 in Mooney–Rivlin equations

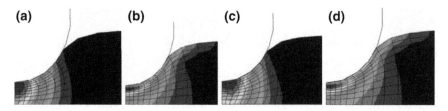

**Fig. 8.19** Distribution of stresses according to Mises in foams caused by operational load: **a** T2516, **b** T2838, **c** T 3530, **d** T4060

made, loading it with an analytical curve of a radius equal to the radius of the thigh of an adult man. The results of these calculations have been shown in Fig. 8.19.

An analysis of the compiled distributions of stresses according to Mises leads to interesting conclusions. The foams T2516 and T3530 are conducive to the concentration of stresses around the sciatica bones and unequally support the user's body, while the foams T2838 and T4060 more evenly move the stresses of the human body and ensure fuller comfort resulting from the reaction of the base.

While modelling contact of the human body with an elastic base, it is also important that in the built calculation models, the elastic properties of soft tissues of a potential user are more or less exactly presented.

## 8.3   Elastic Properties of Human Body Soft Tissues

Numerical modelling of soft tissues requires gathering of experimental data of biomechanical properties of these bodies. In conducting studies on the properties of soft tissue, on large samples of research material (pork liver), Hu and Desai (2005) assumed that the tissue is a material that is incompressible, homogenous and isotropic.

Assuming the load force of the cubic sample as $F$, the elongation coefficient as $\lambda$ and the initial contact surface on the cube $A_o$, stresses $\sigma$ according to Cauchy were written in the form:

$$\sigma = \frac{F}{A_o}\lambda, \qquad (8.104)$$

while strains $\varepsilon$ in the form:

$$\varepsilon = \ln\left(\frac{1}{\lambda}\right). \qquad (8.105)$$

By dividing the experimental load-strain curve (Fig. 8.20) into small subregions, the authors noted that the dependence of force on movement is linear. For each small subregion, $\delta_i$ corresponds to each $F_i$ value; therefore, $\varepsilon_i$ corresponds to each $\sigma_i$ value. Based on the assumption of linear courses of the curve in subregions, it has been assumed that the LEM (Local Elastic Modulus) is expressed by the following equation:

$$E_i = \frac{\sigma_i - \sigma_{i-1}}{\varepsilon_i - \varepsilon_{i-1}} \text{ for } i = 1, 2, 3 \qquad (8.106)$$

For large samples subjected to compression, it was assumed that there were no stresses on the front and side walls of the sample. Piola-Kirchhoff's stresses tensor is connected with the stresses tensor in the chosen point by the formula:

$$P = SF^{-T}. \qquad (8.107)$$

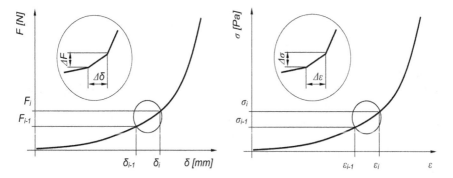

Fig. 8.20 Experimental curves of the stiffness of soft tissue (Hu and Desai 2005)

In this way, Piola-Kirchhoff's stresses tensor has been written in the form:

$$P = -pF^{-T} + 2\omega_1 F - 2\omega_2 B^{-1} F^{-T}, \tag{8.108}$$

where

$p$  Lagrange operator corresponding to limitations of incompressions,

$\text{Det}(F) = 1$, $B = FF^T$—Cauchy-Green tensor,

$$\omega_1 = \frac{\partial U}{\partial I_B}, \quad \omega_2 = \frac{\partial U}{\partial II_B} \tag{8.109}$$

where $I_B = \text{trace }(B)$ and $II_B = (\text{trace}(B)^2 - \text{trace}(B)^2)/2$ are the main constants with respect to $B$.

For incompressible and isotropic materials, the energy of elastic strains $U$ is a function of the main constants with respect to $B$, hence $U = U(I_B, II_B)$. Based on this, Piola-Kirchhoff's stresses tensor $P$ fulfils the equation:

$$\textbf{Div}\,(P) = 0. \tag{8.110}$$

The equations

$$P = -pF^{-T} + 2\omega_1 F - 2\omega_2 B^{-1} F^{-T} \quad \text{and} \quad \textbf{Div}(P) = 0, \tag{8.111}$$

show that the constitutive equation for stresses can be determined when the function of energy strains is known.

To describe the properties of soft tissue, Hu and Desai (2005) used the models of Ogden and Mooney–Rivlin discussed earlier. On this basis, in the system ABAQUS, a numerical model was built by preparing adequate 2D meshes in a flat state of stresses and a flat state of strains. The bone of tissue was modelled as an elastic material consisting of N elements. It was also assumed that the coefficient of friction between the sample and the plate of pressurer will be equal to zero. For modelling, separate flat state of strains and flat state of stresses was used. A four-node CPS4-type element was used in a flat state of stresses, and a four-node CPE4-type element was used in the flat state of strains. The movements of the sample were used as input data, while the reaction forces were used as a comparison with the results of experimental studies. The total strain of the sample (more than 25 % of the initial amount) was divided into 30 subregions (each subregion demonstrated 1 % of elongation increase). For each subregion $j$ ($j = 1, 2\ldots$) of the force–movement curve, the current fragment of the strain was imported in order to calculate the next linear region. Data import was conducted with the use of the standard functionality of the system ABAQUS. Also the value of Poisson's coefficient equal to 0.3 was assumed, as well as the initial LEM value equal to $E_{1,j}$, in order to begin simulation. Then, the experimentally measured values of displacements $\Delta\delta^{\text{EXP}}$ were applied to the nodes of the numerical model. For calculations, the finite elements method was used in order to determine the reaction $\Delta F^{FEM}$. The results of numerical calculations $\Delta F^{FEM}$ were compared with the results of experimental studies $\Delta F^{EXP}$. The value of the linear elasticity modulus was updated based on the equation:

$$E_{i+1,j} = E_{i,j}\left(\frac{\Delta F^{EXP}}{\Delta F^{FEM}}\right) \quad \text{for } i,j = 1,2,3\ldots \tag{8.112}$$

until $\Delta F^{FEM}$ in the new iteration was not similar to the experimentally measured value $\Delta F^{EXP}$. The similarity evaluation criterion was the similarity coefficient expressed by the equation:

$$\frac{\left\|\Delta F_j^{FEM} - \Delta F_j^{EXP}\right\|}{\Delta F_j^{EXP}} \leq 0,02 \quad \text{for } j = 1, 2, 3\ldots \tag{8.113}$$

The corresponding value of the local elasticity modulus has been recorded as $E^{LEM}$. The first index "$i$" describing the linear elasticity modulus $E_{i,j}$ is the number of the iteration in each successive step in the subregion, in which LEM was calculated. The second index "$j$" determines the number of the subregion. The geometry of deformations and mesh for the subregion $(j)$ were imported to the subregion model $(j + 1)$, and then the calculation process was run from the beginning until the entire process of iteration was completed. The procedure of iterations has been shown in Fig. 8.21.

On the basis of the analysis, it was demonstrated that both in the flat state of stresses and in the flat state of strains, identical results were obtained. Therefore, this did not significantly affect the quality of the results of numerical calculations. Slight differences in $E^{LEM}$ values were also obtained by comparing the numerical method with the experimental method. For the obtained dependencies $E^{LEM} = f$ $(\Delta\delta)$, approximation was conducted using polynomials of the 4th degree. Table 8.4 summarises the calculated values of material parameters of Mooney–Rivlin and Ogden's models. It turns out that Ogden's model describes the characteristics of soft tissue much better than Mooney–Rivlin's model. However, this rule is correct only for the quasi-static analysis.

When using furniture, with which the user comes into contact directly, pressure forces, caused by maintaining his position, cannot cause limitations in blood circulation at the surface. By averaging the pressure in the arterial system, we obtain a value of around 100 mmHg (13.32 kPa), in the capillaries around 25 mmHg (3.33 kPa), and in the final part of the venous system, it amounts on average to around 10 mmHg (1.33 kPa) (Guzik 2001). According to Krutul (2004) in places where the bones push onto tissues harder, pressure increases and the lumen of blood vessels is reduced, which leads to the damage of skin tissue. The stresses of the external surface are 3–5 times smaller than internal stresses that occur as a result. Therefore, the limit value of the pressure amounting to 32 mmHg (4.26 kPa) (closing the lumen of capillaries) must be appropriately reduced and range from

**Fig. 8.21** The method of calculating the local elastic modulus (LEM) of soft tissue according to Hu and Desai (2005)

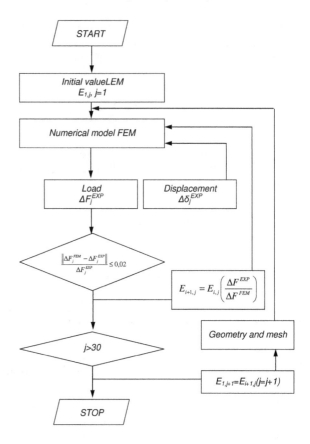

**Table 8.4** Material parameters of different tissues in Mooney–Rivlin and Ogden's model according to Hu and Desai (2005)

| Type of model | Parameter | Tissue A | Tissue B | Tissue C | Tissue D |
|---|---|---|---|---|---|
| Mooney–Rivlin | $C_{10}$ | 0.039 | 0.067 | 0.052 | 0.063 |
|  | $C_{01}$ | −0.041 | −0.067 | −0.052 | −0.066 |
| Ogden | $\mu_1$ | −0.288 | −0.221 | −0.179 | −0.380 |
|  | $\mu_2$ | 0.162 | 0.230 | 0.177 | 0.272 |
|  | $\mu_3$ | 0.067 | −0.002 | 0.007 | 0.109 |
|  | $\alpha_1$ | 11.085 | 7.625 | 6.663 | 13.514 |
|  | $\alpha_2$ | 11.625 | 7.913 | 7.062 | 13.616 |
|  | $\alpha_3$ | 0.996 | −25.000 | 5.846 | 13.465 |

6.4 mmHg (0.85 kPa) to 10.6 mmHg (1.41 kPa). Each pressure greater than these values may result in closing the light of veins, then of arteries, which slows down the flow of blood or stops its circulation, causing local ischaemia. If being in a seated position lasts a long time, it may be the cause of so-called pins and needles of the lower limbs. Another negative effect of the improper choice of seat can be the compression of the spine muscle and gluteal muscle by the sacral vertebra and sciatica, which results in an increase of stiffness of the muscles where the bone meets the muscle. The studies of Gefen et al. (2005) demonstrated that the stiffening of muscle tissue occurs in a living system of muscles exposed to pressure of 35 kPa for 35 min or longer, and in the same muscles that have been exposed to pressure of 70 kPa for 15 min or longer. By simulating using numerical methods immobility in

**Table 8.5** Elastic properties of human body soft tissues

| Type of tissue | Property | | | | | |
|---|---|---|---|---|---|---|
| | Density (kg/m³) | | Linear elasticity module (MPa) | | Poisson's ratio | |
| | Value | Author | Value | Author | Value | Author |
| Muscles | 1000 | Gerard (2004) | 0.50–0.79 | Deuflhard (2003) | 0.45–0.50 | Gerard (2004) |
| | 1040 | Golombeck (1999) | 1.00 | Wang and Lakes (2002) | 0.45–0.50 | William (1993) |
| | 1050 | Douglas (2000) | 0.75 | Linder–Ganz and Gefen (2004) | 0.49 | Wang and Lakes (2002) |
| Bones | 1810 | Golombeck (1999) | 13,100–16,700 | Taylor et al. (1999) | 0.30 | Srinivasan (1999) |
| | 1500 | Morcovescu and Dragulescu (2002) | 10,400–14,800 | Rho et al. (1993) | 0.30 | Hazelwood et al. (1998) |
| | 1900 | Enderle et al. (2000) | 10,900–13,000 | Ashman and Rho (1988) | 0.30 | Guo (2001) |
| Cartilage | 1100 | Prior (2001) | 0.5–3.4 | Ghadiali (2004) | 0.49 | Ghadiali (2004) |
| | – | – | 10 | Patil et al. (1996) | – | – |
| Skin | 1010 | Golombeck (1999) | 0.09–0.50 | Deuflhard (2003) | 0.49 | Chabanas et al. (2002) |
| | 1090 | Douglas (2000) | 0.85 | Qunli (2005) | 0.30–0.50 | Lees et al. (1991) |
| | 1056 | Schneck (1995) | 0.70 | Linder-Ganz and Gefen (2004) | 0.40 | Gefen et al. (1999) |
| Fat tissue | 920 | Golombeck (1999) | 0.001–0.005 | Deuflhard (2003) | 0.50 | Todd and Thacker (1994) |
| | 950 | Douglas (2000) | 0.01 | Qunli (2005) | – | – |
| | 1200 | Mangurian and Donaldson (1990) | 0.08 | Linder-Ganz and Gefen (2004) | – | – |
| Blood vessels | 1060 | Golombeck (1999) | 1.2 | Deuflhard (2003) | 0.40 | Fung (1993) |
| | 1060 | Douglas (2000) | 0.133 | Fung (1993) | – | – |

a seated position on a hard base, it was found that after four hours of maintaining this position, it causes compressive stresses in 50–60 % of the cross section of the muscle at a level of 35 kPa or more. It was also demonstrated that the intensity of injury damaging cells increases during the first 30 min of immobile sitting, which is the cause of the formation of pressure pain.

Therefore, with the numerical simulation of the effect of a base on the human body, it is important to define the mechanical parameters of the human body's tissue. These values, based on the literature, are provided in Table 8.5.

## 8.4  Stiffness of Upholstery Springs

### 8.4.1  Stiffness of Cylindrical Springs

The comfort of the use of upholstered furniture is connected, to a large extent, with the softness of the spring layer. Its quality can be adjusted by choosing or designing the appropriate springs and spring units.

First, let us imagine a cylindrical spring stretched with two forces P acting in its axis (Fig. 8.22.). By cutting this spring in a plane perpendicular to the axis of the wire, which the spring is made of, a balance of cross sections will be conducted (Fig. 8.23). The transverse force $P$ and the resulting torsion moment $M_s = P \cdot R$ must be differentiated by tangential stresses in this cross section, therefore

$$\tau_1 = \frac{P}{\pi r^2}, \tag{8.114}$$

$$\tau_2 = \frac{M_s}{J_o}\rho = \frac{2PR}{\pi r^3}, \tag{8.115}$$

where
$\tau_1$  tangential stresses resulting from transverse forces,
$\tau_2$  tangential stresses resulting from torsion moment,
$P$   load stretching the spring,
$J_o$  moment of inertia of the cross section at axial torsion,
$\rho$   fibres outward in relation to the torsion axis,
$R$   radius of the spring coil and
$r$   radius of the wire cross section.

By adding both stresses $\tau_1$ and $\tau_2$ (Fig. 8.24), we obtain the value of the maximum stress in the cross section of the spring wire in the form:

$$\tau_{max} = \frac{P}{\pi r^2}\left(1 + \frac{2R}{r}\right). \tag{8.116}$$

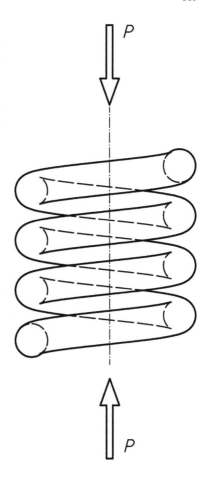

**Fig. 8.22** Cylindrical spring
subjected to stretching

However, since the share of the component $P/(\pi r^2)$ does not exceed 5 % of the $\tau_{max}$, therefore the stresses in the wire section are expressed in a simple form:

$$\tau_{max} = \frac{2PR}{\pi r^3}. \tag{8.117}$$

Knowing the state of stresses in the spring coils, we can proceed to determine its deformations caused by external reasons. To this end, we cut from the spring a certain section of elementary length $ds$, using two planes perpendicular to the axis of the wire and passing through the axis of the cylinder (Fig. 8.25). Let us also assume that the $R$ radiuses running from the axis of the cylinder to the centres of gravity of both cross sections before deformation lie in one plane. After deformation, these cross sections will turn in relation to one another by the angle $d\varphi$ equal to:

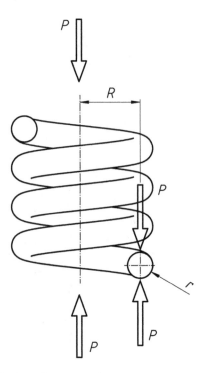

**Fig. 8.23** Forces acting in the rod's cross section

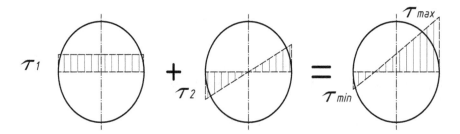

**Fig. 8.24** Distribution of tangential stresses in the cross section of the wire

$$\mathrm{d}\varphi = \frac{M_s}{GJ_o}\mathrm{d}s, \tag{8.118}$$

$$\mathrm{d}\varphi = \frac{\mathrm{d}\lambda}{R}, \tag{8.119}$$

**Fig. 8.25** Section of the
spring coil

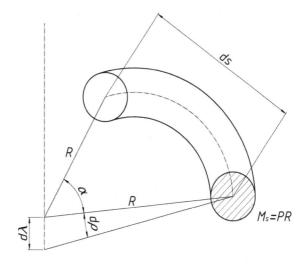

and hence, the value of elementary deformation $d\lambda$ will amount to:

$$d\lambda = \frac{PR^2}{GJ_o}ds,$$ (8.120)

where
$G$   shear modulus of the wire.

Taking into account the fact that all the elements with the length ds will deform identically, total deformation of the cylindrical spring $\lambda$ will be the sum of the elementary deformations $d\lambda$:

$$\lambda = \sum d\lambda,$$ (8.121)

therefore

$$\lambda = \int_0^{2\pi Rn} \frac{PR^2}{GJ_o}ds,$$ (8.122)

which gives, for $J_o = \pi r^4/2$,

$$\lambda = \frac{4PR^3 n}{Gr^4},$$ (8.123)

where
$n$          number of the spring coils and
$2\pi Rn$   length of the spring wire.

A measurable indicator of the quality of the spring is its stiffness $k$, understood as the quotient of the load $P$ to the displacement caused by this load:

$$k = \frac{Gr^4}{4R^3n}.$$
(8.124)

### 8.4.2   Stiffness of Conical Springs

For biconical springs, which are applied in the Bonnell-type spring systems, the torsion moment $M_s$ is not a value dependent only on the load, but a function of the changing length of the coil radius $M_s = f(R)$. This radius depends on the angle of the unstretching of the spring $\alpha$, $R = f(\alpha)$ (Fig. 8.26). The angle $\alpha$ is the angle between the intermediate radius $R$ (variable) and the upper radius of the spring coil $R_1$ depending on the number of the spring coils $n$. Therefore, the increase in the length of the intermediate radius can be written as:

$$y = \frac{R_2 - R_1}{2\pi n}\alpha,$$
(8.125)

**Fig. 8.26** Geometry of conical spring

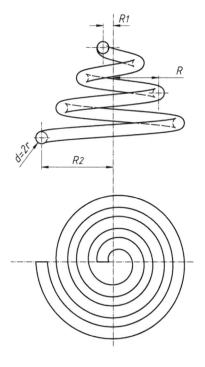

where
$R_1$  the largest radius of the coil,
$R_2$  the smallest radius of the coil,
$\alpha$  the angle of the unstretching of the spring and
$n$  number of the spring coils.

We calculate the strength of the wire in the conical spring by entering into the equation for maximum stresses $\tau_{\max}$ the values of the intermediate radius $R$. When determining the values of the conical spring deflections, the change in the length of the intermediate radius should also be taken into account:

$$R = R_1 + \frac{R_2 - R_1}{2\pi n}\alpha. \tag{8.126}$$

And because the torsion moment $M_s$ is a function of the intermediate radius, the deflection of the spring under the load $P$ can be therefore written as:

$$\lambda = \int_s \frac{P\left(R_1 + \frac{R_2 - R_1}{2\pi n}\alpha\right)^2}{GJ_o}\,\mathrm{d}s, \tag{8.127}$$

where as it can be seen from Fig. 8.25, $\mathrm{d}s = R\mathrm{d}\alpha$.

By assuming the borders of variation $0 < \alpha < 2\pi n$ as the integration boundaries, we eventually obtain the equation for the deformation of the conical spring:

$$\lambda = \frac{Pn}{Gr^4}\left(R_2^2 + R_1^2\right)\left(R_2 + R_1\right). \tag{8.128}$$

For conical springs, the stiffness $k$ can be written as:

$$k = \frac{Gr^4}{n\left(R_2^2 + R_1^2\right)\left(R_2 + R_1\right)}. \tag{8.129}$$

This coefficient has significant importance in the calculation of the stiffness of whole spring systems.

### 8.4.3 Modelling of Stiffness of Conical Springs

The variable stiffness of the spring during operational loads should ensure high softness of the system at surface loads and significant stiffness when exposed to concentrated forces or forces of high intensity. For such exploitation conditions, a construction minimum is a biconical spring consisting of two conical springs differing in stiffness coefficients, but made from a single piece of wire.

The differentiation of the stiffness coefficient value should be forced by the selection of a suitable geometry of each part of the spring, which has been schematically shown in Fig. 8.27.

In the course of designing the shape of such a spring, it should be ensured that the coils of the lower cone are first settled on a hard base, therefore ensuring the exhaustion of the border of the largest soft deformations. At the same time, the upper cone should deform slightly, providing increased stiffness in the second stage of the system operation.

Therefore, let us assume the following: when the lower cone of the spring will be completely compressed $\lambda_1 = H$, the upper cone will deform for only about 10 % of its initial height $H$; that is, $\lambda_2 = 0.1\,H$. The stiffness of the upper cone is therefore defined by the following dependency:

$$k_1 = 10\frac{P}{H}, \tag{8.130}$$

**Fig. 8.27** Calculation scheme of a biconical spring with nonlinear characteristics

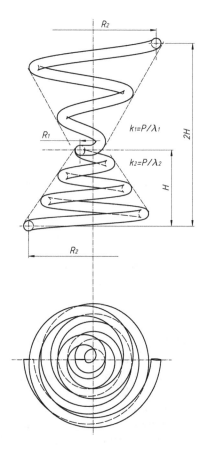

whereas the stiffness of the lower cone

$$k_2 = \frac{P}{H},$$                                                              (8.131)

where
$P$   the force loading the spring and
$H$   half of the height of the biconical spring.

For serially connected conical springs, the total deformation therefore amounts to

$$\lambda_c = \sum_{i=1}^{n} \lambda_i,$$                                           (8.132)

therefore

$$\lambda_c H + 0.1\, H = 1.1\, H,$$                                                (8.133)

whereas the stiffness of the biconical spring as a system of two conical springs with given stiffness coefficients $k_1$ and $k_2$ amounts to:

$$\frac{1}{k_c} = \sum_{i=1}^{n} \frac{1}{k_i},$$                                   (8.134)

which gives

$$k_c = \frac{k_1 k_2}{k_1 + k_2},$$                                                (8.135)

and finally

$$k_c = \frac{10P}{11H}.$$                                                          (8.136)

From this relation, it results that the total stiffness of the $k_c$ system is smaller than the component $k_1$ and $k_2$ stiffnesses. However, it should be noted that this formula ceases to have effect when the deformation reaches a value of 1.1 $H$; that is, when the lower spring cone and 10 % of the upper cone completely settles, then only the upper spring cone will be compressed. Therefore, the stiffness will increase from $k_c$ to $k_2 = 10P/H$ (Fig. 8.28).

Designing the shape of the discussed spring involves the selection of a suitable equation describing the form of the spiral of wound coils (Fig. 8.29). Therefore, knowing the stiffness coefficients of each conical part, the following equation of the spiral was used:

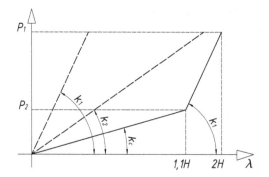

**Fig. 8.28** Two-stage nonlinear characteristics of the asymmetrical biconical spring: P1—force causing total deformation, P2—force causing the deformation of the lower cone

**Fig. 8.29** Projection of the spring spiral

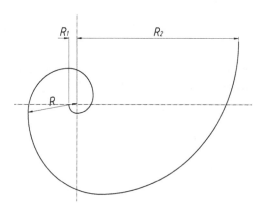

$$2\pi R^{m-1}\partial R = C\partial\lambda, \tag{8.137}$$

where

$C$  constant determined from border conditions,
$R$  the intermediate radius of the spiral determined for different angles between this radius and the smallest radius of the spiral $R_1$,
$m$  the coefficient of the changes in the value of the radius $R$,

$$2\pi \int_R R^{m-1}\partial R = C\partial\lambda, \tag{8.138}$$

$$\frac{2\pi R^m}{m} = C\alpha + C_1, \tag{8.139}$$

and then, taking into account the border conditions, we obtain a system of equations, which allows to determine the constant values $C$ and $C_1$,

$$\frac{R_1^{m-1}}{m} = C_1 \quad \text{for } \alpha = 0 \Rightarrow R = R_1 \text{ and :} \tag{8.140}$$

$$\frac{R_2^m}{m} = C\alpha + C_1 \quad \text{for } \alpha = 2\pi R \Rightarrow R = R_2, \text{ which gives:} \tag{8.141}$$

$$C_1 = \frac{R_1^m}{m}, \quad C = \frac{R_2^m - R_1^m}{mn}, \tag{8.142}$$

where
$n$   the number of coins of a single spring cone

Therefore, the sought equation of the intermediate radius R has the form:

$$R = \frac{R_2^m - R_1^m}{2\pi n}\alpha + R_1^m. \tag{8.143}$$

This formula shows that depending on the selection of the parameter m, the increase of the value of the radius $R$ will be variable for the same angle $\alpha$. This will obviously condition the change of the stiffness of a given conical spring. Because the torsion deformation of the cross section of the wire is expressed by the equation:

$$\frac{d\beta}{ds} = \frac{PR}{GJ_o}, \tag{8.144}$$

where
$J_o = r^4/2$   polar moment of inertia of the cross section, and
$ds = Rd\alpha$   the length of the springs section,

therefore, elementary deflection of the spring is equal to

$$d\lambda = \frac{PR^3}{GJ_o}d\alpha. \tag{8.145}$$

We exclude from this equation the part that corresponds to the stiffness coefficient $k$ and introduce the following auxiliary function $\eta(R)$:

$$\eta(R) = \frac{R^2}{GJ_o}\frac{ds}{dR} = \frac{1}{k}, \tag{8.146}$$

whereby

$$\frac{\partial R}{\partial \alpha} = \frac{C}{2\pi R^{m-1}}, \tag{8.147}$$

therefore, determining the values of parameters $m$ for $k_1$ and $k_2$ requires solving the equation:

$$\lambda = P \int \eta(R) dR, \qquad (8.148)$$

therefore

$$\lambda = P \int_0^{R_2 - R_1} \frac{4mn}{Gr^4 \left(R_2^m - R_1^m\right)} R^{m+2} dR, \qquad (8.149)$$

which gives

$$k = \frac{1}{\dfrac{4mn}{Gr^4 \left(R_2^m - R_1^m\right)} \displaystyle\int_0^{R_2 - R_1} R^{m+2} dR}. \qquad (8.150)$$

Assuming for the analysis the number of coils of the spring equal to $n = 2$, the dimensions of the coils $R_1$ and $R_2$ as well as module $G$ and radius $r$ as for Bonnell springs, the solution of this equation has been shown in Fig. 8.30. On the axis of

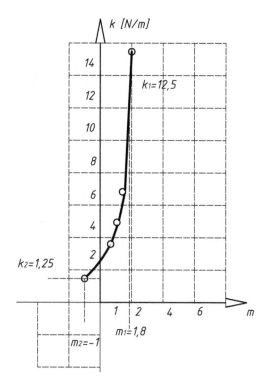

**Fig. 8.30** Stiffness of the conical springs of variable geometry of the spiral: for $m = 0$ the function $k = f(m)$ is not specified

Fig. 8.31 Shape of the spiral: **(a)** **(b)**
a upper, b lower

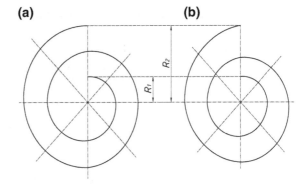

ordinates of the chart, the previously calculated values of the stiffness coefficients $k_2$ and $k_1$ have been marked. On the axis of abscissae, values $m_2 = -1$ and $m_1 = 1.8$ corresponding to these points have been found. In this way, the two basic equations describing the geometry of the spiral of each of the conical springs have been obtained.

By entering to these equations, the arc values of the angle are in the range $0 \Leftarrow \alpha \Leftarrow 4\pi$; the actual shape of the designed springs has been obtained. Their sketches are shown in Fig. 8.31, and the geometry of the whole spring is shown in Fig. 8.32.

Fig. 8.32 Side view of the
designed spring

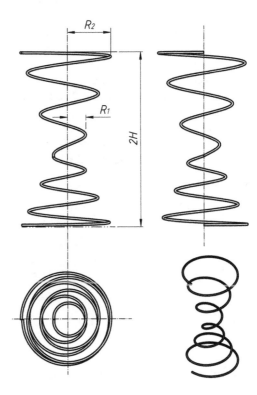

## 8.5  Parallel Systems of Springs of Various Stiffness

In upholstered furniture, there are several construction solutions used for spring units, which constitute the main spring layer. Biconical springs and cylindrical springs (Fig. 8.33a, b) are used in many designs of mattresses. However, the latest solutions tend towards a serial or parallel connection of springs of varying geometry (Fig. 8.33c). Such a compilation of springs enables to obtain nonlinear characteristics of deformations of spring systems of upholstered furniture and also to match the stiffness of the bed to the individual needs of the user.

The analysis of the softness of spring systems in upholstered furniture proves that these layers, from an engineering point of view, should consist of elements of nonlinear and progressive compression characteristics. It is more favourable to use biconical springs for this purpose, in which the progressive characteristics of operation are caused by the settling of coils (Fig. 8.34a).

However, they do not always meet the expectations of the designer, especially when he is looking for an element with a precisely calculated variable of stiffness, which can ensure high softness of the system in places of weak load of the user's body, and at the same time a large hardness in places of strong stresses. The spring, which could meet these requirements, should consist of a minimum of two elementary springs connected in parallel (Fig. 8.34b).

The compression characteristics of the entire system $kz$ will then consist of sections representing the deformation of element $k_1$ until its settling on limiters, a base or another element of $k_2$ stiffness, and joint stiffness of both components $kz = k_1 + k_2$. It is important in this case to determine such stiffnesses $k_1$ and $k_2$, respectively, for the larger and smaller spring, which will correspond to stresses of the human body placed at a given point of the seat or bed surface.

A study of the phenomenon of settling of coils in parallel inserts of cylindrical springs can be carried out using numerical methods. Figure 8.35 shows a mesh model of a fragment of a spring unit, consisting of two cylindrical springs, of the dimensions: diameter 60 mm, height 125 mm, diameter 26 mm, height 100 mm. The model has been made from six- to eight-node isotropic solid elements. In order

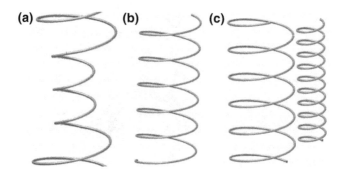

**Fig. 8.33** Upholstery springs: **a** biconical, **b** cylindrical, **c** double system of cylindrical springs

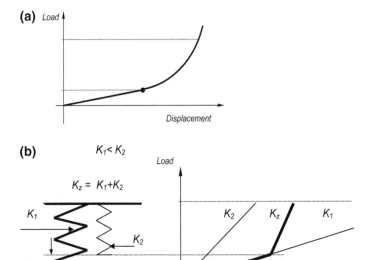

**Fig. 8.34** Characteristics of upholstery springs: **a** biconical, **b** cylindrical in parallel systems

**Fig. 8.35** The mesh model of
the spring system

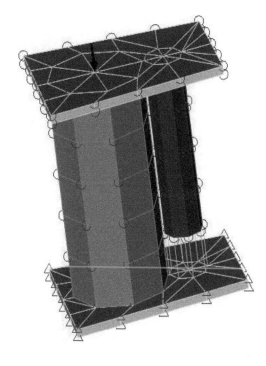

to mark the settling of coils of the shorter spring on a stiff base, it has been supported by gap-type elements. Thanks to them, a case of individual compression has been observed in the larger spring, and after exhausting the height $h_p$ (Fig. 8.34), simultaneous compression of both springs. In the first load scheme, the force 3 N was used. This value was to guarantee that a spring of 60 mm in diameter and $k_1$ stiffness will deform in a controlled manner, and the value of this displacement will amount to $\lambda_1 = P_1/k_1 = 25$ mm. As it can be seen in Fig. 8.36, the simulation brought the intended effect. The displacement amounted to 25 mm, and the spring of 26 mm in diameter set its base on an unmovable support.

In the second calculation scheme, the model was loaded by a force of 38 N. This value derives from the studies of stresses of the human body, lying down in a lateral position, on a mattress of spring construction. When calculating the deformation of the springs system of alternative stiffness $k_z = 0.62$ N/mm, the displacement value that should have been expected was at the level $\lambda_2 = P_2/k_z = 61.1$ mm. As shown in Fig. 8.37, the displacement value marked numerically amounted to almost 61.1 mm. This means that the developed model correctly reflects the nature of work of parallel systems of springs subjected to compression. Therefore, modelling mattresses

**Fig. 8.36** The deformation of the springs system under load 3 N

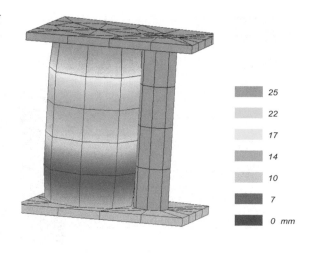

25
22
17
14
10
7
0  mm

**Fig. 8.37** The deformation of the springs system under load 38 N

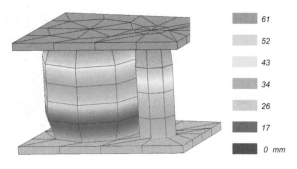

61
52
43
34
26
17
0  mm

systems of furniture for lying down, containing complex spring systems, should not pose too many problems, and the results of the calculations will correspond to real values. In this context, any mattress can be designed which characteristics of stiffness will result not only from applying the right stiffness of springs connected in parallel, but also foams and rubbers combined with springs in serial systems.

## 8.6   Stiffness of Spring Units

When designing furniture for lying down of sitting (with sleep and relaxation functions), anthropometric and physiological rules should be taken into account arising from their use. Mattresses, especially those of orthopaedic character, constitute one of the essential factors of the quality of life for people with musculo-skeletal dysfunction. Rehabilitation of such patients is an ongoing process, and the level of daily activity and wellness determines the behaviour of previously achieved effects or conditions progress in the improvement of abilities. A different perspective of orthopaedics on the cause of health ailments caused by badly designed and manufactured upholstered furniture makes it that most mattresses are designed by intuitively selecting both the spring materials and the shapes and dimensions of beds and seats. Such approach to design, results of a number of inconsistencies between the requirements and expectations of users and normative recommendations. Therefore, engineering design methods of spring methods are looked for in order to determine the most favourable material and construction parameters which improve ergonomics, and thus the user's comfort of sleep and relaxation.

The following analysis concerns states established for objects, which is a system made up of cylindrical upholstery springs with: height $H = 125$ mm, coil diameter $D = 60$ mm, wire diameter $d = 2.1$ mm, the number of active coils $n = 5$ and linear elasticity modulus $E = 2 \times 10^5$ MPa (Fig. 8.38).

In order to determine the deformation of the entire spring system, under the influence of the load of concentrated force, first the behaviour of single springs subjected to compression and deflection is analysed (Fig. 8.39), since these deformations make up the deformation of the entire system of springs connected in parallel.

**Fig. 8.38**  Spring unit of a mattress

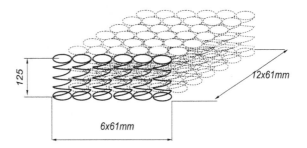

125

12x61mm

6x61mm

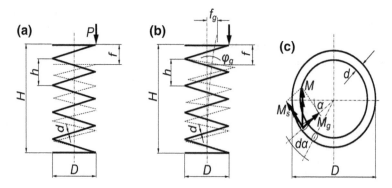

**Fig. 8.39** Scheme of spring load: **a** compression, **b** deflection, **c** deflection and twisting

The stiffness of a compressed spring has been calculated from the known equation:

$$k_1 = \frac{P}{f} = \frac{Gd^4}{8D^3 n}. \tag{8.151}$$

In the case of a spring subjected to deflection, the rotation angle $\varphi_g$ and deflection $f_g$ had to be determined. To this end, the internal energy of the spring $U_w$ was determined:

$$U_w = \sum_{i=1}^{n} \frac{1}{2} \int_s \frac{M^2}{EJ} ds, \tag{8.152}$$

where
$s$    the length of the wire of spring coils and
$M$   torsion moment in the cross section of the wire,

$$M = M_s^2 + M_g^2,\ M_s = M \cos \alpha,\ M_s = M \sin \alpha, \tag{8.153}$$

which gives

$$f_g = \frac{\partial U_w}{\partial P} = \frac{16 P D^3 n (1 + v)}{E d^4}, \tag{8.154}$$

$$\varphi_g = 2 \frac{\partial U_w}{\partial M} = \frac{64 M D n}{G d^4}. \tag{8.155}$$

Finally, the stiffness of the bent spring was written as:

$$k_g = \frac{M}{\varphi_g} = \frac{Gd^4}{64Dn}. \tag{8.156}$$

where

$$G_g = E/2(v+1), \tag{8.157}$$

$v$   Poisson's ratio

The non-axial load of a single spring, however, causes the stiffness of it alone and the system of connected springs may be changed. Figure 8.40 presents the deflections of springs compressed eccentrically. It was assumed that the stiffness $k$ of each spring can be expressed as the stiffness of the parallel system of springs of the stiffness $k = 0.5\,k_1$. As a result of such compression, the deflection of springs under the influence of individual loads has the form:

For the axial compression of the spring

$$F_1' = F_1'' = \frac{1}{2}P, \quad \text{hence } f_1 = \frac{P}{k_1}, \tag{8.158}$$

for the compression of the spring along the peripheral of the cylinder

$$F_1' = 0,\ F_1'' = P, \quad \text{hence } f_1 = \frac{2P}{k_1}, \tag{8.159}$$

for the compression of the spring with force applied at any point on the surface indicated by the passive coil

$$F_1' = P\frac{x_2}{x_1 + x_2}, \tag{8.160}$$

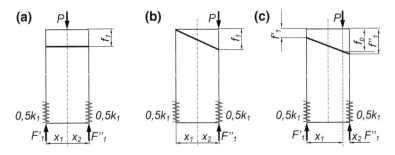

**Fig. 8.40** Scheme of spring compression: **a** axially, **b** along the peripheral, **c** at any point

$$F_1'' = P\frac{x_1}{x_1 + x_2}, \tag{8.161}$$

$$f_1' = \frac{2Px_2}{k_1(x_1 + x_2)}, \tag{8.162}$$

$$f_1'' = \frac{2Px_1}{k_1(x_1 + x_2)}, \tag{8.163}$$

$$f_2 = f_2' + f_2'', \tag{8.164}$$

hence

$$f_p = \left(f_1'' - f_1'\right)\frac{x_1}{(x_1 + x_2)}. \tag{8.165}$$

In the case of a parallel connection of two springs, as it takes place in the spring unit of a mattress, the value of the deflection of the system should depend on the dimensions of springs, the distance between them and their stiffness. Therefore, the load scheme of two springs that were apart from one another and loaded in the middle of the span of the tie connecting them was analysed first (Fig. 8.41).

The maximum deflection of the presented system, assuming an identical stiffness of springs $k_1$, amounts to:

$$f_1 = f_2, \tag{8.166}$$

where

$$f_1 = f_1' + f_1'', \tag{8.167}$$

$$f_2 = f_2' + f_2'', \tag{8.168}$$

**Fig. 8.41** Scheme of load of springs connected with a tie

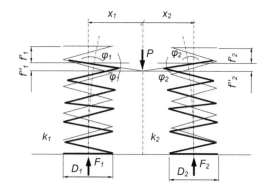

and

$$f_1' = \frac{8PD_1^3 n_1}{G_1 d_1^4} \frac{x_2}{x_1 + x_2},$$  (8.169)

$$f_1'' = \varphi_1 x_1,$$  (8.170)

hence

$$f_1 = f_2 = \frac{8PD_1 n_1 x_2}{G_1 d_1^4 (x_1 + x_2)} \left( D_1^2 + 8x_1^2 \right).$$  (8.171)

For the parallel connection of two springs with their peripherals, the deflection value of the set should depend on the dimensions of the springs and their stiffness. The deformation of the system presented in Fig. 8.42 results from the deflection work of the spring set on the right side.

In this case, it can be written that

$$P = F_1' + F_1'' + F_2' + F_2'',$$  (8.172)

$$\frac{1}{2} F_1' D = F_1'' D + \frac{3}{2} F_2'' D,$$  (8.173)

$$P = 2F_1'' + 3F_2'' + 2F_1'' + F_2'' = 4\left( F_1'' + F_2'' \right),$$  (8.174)

$$\frac{1}{4} P = F_1'' + F_2'',$$  (8.175)

$$\frac{3}{4} P = F_1' + F_2',$$  (8.176)

**Fig. 8.42** Load scheme of springs connected by peripherals

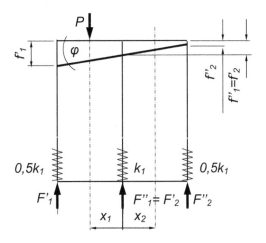

therefore

$$\begin{cases} \dfrac{1}{2}\dfrac{P}{k_1} = f_1'' + f_2'', \\[2mm] \dfrac{3}{2}\dfrac{P}{k_1} = f_1' + f_2', \\[2mm] f_2' = f_1'', \\[2mm] f_1' = 2f_2', \end{cases} \tag{8.177}$$

hence

$$f_1' = \frac{P}{k_1}, \tag{8.178}$$

$$f_1'' = \frac{1}{2}\frac{P}{k_1}, \tag{8.179}$$

$$f_2' = \frac{1}{2}\frac{P}{k_1}, \tag{8.180}$$

$$f_2'' = 0. \tag{8.181}$$

Because comfortable use of a mattress of an upholstered furniture piece should be associated with an even distribution of stresses of the body on the spring layer of the furniture piece, when modelling the stiffness of the spring unit, both the effect of compression and deflection of individual springs should be taken into account. Therefore, the springs have been supported immovably on a stiff base, while on the ends, at the site of mutual contact, they were connected articulately (Fig. 8.43).

At constant parameters $k_1$ and $k_g$, characterising the stiffness of the springs, the stiffness of the considered unit will depend, among others, on the diameter $D$. The reaction of the unit can therefore be expressed as:

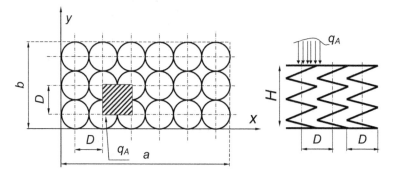

**Fig. 8.43** Geometry of the unit made up of cylindrical springs

$$q_A = \frac{k_1}{D^2} w(x, y), \tag{8.182}$$

$$M_x = \frac{k_g}{D} \frac{\partial w}{\partial x}, \tag{8.183}$$

$$M_y = \frac{k_g}{D} \frac{\partial w}{\partial y}, \tag{8.184}$$

$$w(x, y) = \Delta e^{-\eta(|x-x_P|+|y-y_P|)}, \tag{8.185}$$

where

$M_x$, $M_y$   appropriate bending moments (Fig. 8.44),

$w(x, y)$   a function that describes the deflection of the unit surface caused by concentrated force

By specifying the deflection value of the spring unit, the principle of minimum potential energy was used. During compression with concentrated force $P$ (Fig. 8.45) of the system in point $A(x_p, y_p)$, the work of external forces on the external displacements is equal to the sum of working internal forces and potential energy of the deformed surface. The potential energy $V$ of the whole system has been written in the form:

$$V = U_z - U_w, \tag{8.186}$$

where the work of external forces:

$$U_z = \frac{1}{2} P w(x, y), \tag{8.187}$$

**Fig. 8.44** Spring load by concentrated force and bending moments

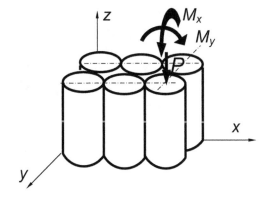

**Fig. 8.45** Loading the mattress surface by concentrated force in point A

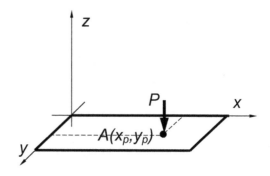

and the work of internal forces:

$$U_w = \frac{1}{2} \int_0^a \int_0^b \left( q_A w(x,y) + M_x \frac{\partial^2 w}{\partial x^2} + M_y \frac{\partial^2 w}{\partial y^2} \right) dxdy, \tag{8.188}$$

hence finally

$$V = \frac{1}{2} \left[ Pw(x,y) - \frac{k_1 + 2Dk_g \eta^3}{D^2} \int_0^a \int_0^b w^2(x,y)dxdy \right]. \tag{8.189}$$

The parameters $\Delta$ i $\eta$ occurring in the equation of deflection of the surface of the unit have been determined from the conditions:

$$\begin{cases} \dfrac{\partial V}{\partial \Delta} = 0, \\ \dfrac{\partial V}{\partial \eta} = 0, \end{cases} \tag{8.190}$$

taking into account that the value of the deflection amplitude will not depend on the coordinates of the load point, the following was established:

$$\Delta = \frac{P\eta^2 D^2}{2k_1 \left( 1 + 2D\eta^3 \frac{k_g}{k_1} \right)}, \tag{8.191}$$

and

$$\eta = \sqrt[3]{2}. \tag{8.192}$$

In this way, the maximum deflection of the spring unit, made up of cylindrical springs connected articulately in the upper coils and loaded by concentrated force in the point, can be written in the form:

$$w_{max} = \frac{2\sqrt[3]{4}PD^2}{(D^2k_1 + 32k_g)} e^{-\sqrt[3]{2}}. \tag{8.193}$$

## 8.7 Experimental Testing of Stiffness of Seats

Most furniture pieces designed for sitting, especially office chairs, cafe chairs, cinema chairs or house chairs, have soft, upholstered seats and/or backrests. Usually these parts of the furniture, on an industrial scale, are made from flexible polyurethane foams, latex foams and spring systems covered by layers of polyurethane foam or coconut mats, as well as using the technology of embedding conical or cylindrical springs in polyurethane or latex foams. The selection of materials, as well as design solutions, in many cases, is coincidental. Designers, based on their own experience and feeling of comfort, often erroneously decide on a design and technology of making a seat. This is how idyllic designs come about, spellbinding with the finesse of the shape, designed for a unique form or colour, but not complying with the essential requirements of ergonomics and functionality. Therefore, it was valuable to gather reliable information about the stiffness of seats depending on the applied design and material solutions.

Experimental studies were conducted on three models of seats of chairs manufactured by one of the reputable factories of house furniture on the polish market at the time of conducting the studies. These models significantly differed among themselves both in terms of the materials used and the design solutions adopted (Fig. 8.46).

The force and movement values recorded during axial compression tests have been provided in Fig. 8.47. On the basis of this chart, it can be seen that the seats selected for the studies have much more different stiffness characteristics. The seat described as model A has a strongly progressive stiffness characteristic, while model S is characterised by an almost linear stiffness. The characteristics of the seats of model B are intermediate between the two previous ones, also with a tendency of progressive increases of forces in relation to movements.

The highest stiffness was shown by type A seats. In linear deformations, the force required to obtain deformations of the seat, the same as in model A, constituted only 86 % of the load value of the seat of model A in model B, while in model S it is 68 % of this load value. In nonlinear deformations, these differences clearly increase. Thus, the force required to obtain deformations of the seat, the same as in model A, constituted 57 % of the load value of the seat of model A in model B, while in model S only 29 % of this load value. This is a result of the fact that for the user, seats of lesser stiffness are more ergonomic and comfortable to use, made from

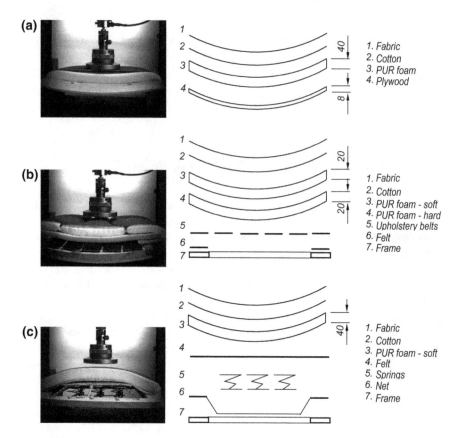

**Fig. 8.46** Structures of seats used for the studies: **a** model A, **b** model B, **c** model S

**Fig. 8.47** The stiffness of
spring systems of an
upholstered chair

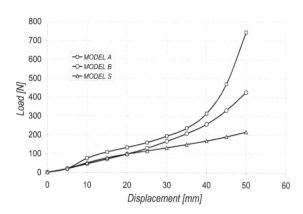

**Fig. 8.48** The dependency of stiffness on the movement of spring systems of an upholstered chair

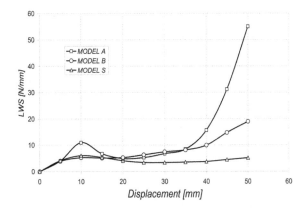

systems of springs and foam, because smaller load values produce the desired deflection of the seat.

It needs to be noted that constant stiffness of the springs is not the decisive factor for full comfort of use of the furniture piece for sitting, but the size of unit pressures and form of their distribution on the contact surface between the human body and the seat.

Figure 8.48 shows that the constant and unfavourable value of the local stiffness coefficient LWS are only featured in seats of model S. Constant stiffness of the seat does not ensure the adjustment of the values of contact stresses to the user's body weight. The proportional deflection of spring layers leads to their overall compression and, consequently, forces the user, with a large weight, to sit on a hard and undeformed base. The system of the seat of model B is the best. The average and gently increasing stiffness of the system provides the ability to model the hardness of the seat and adjust it to the individual needs of the user.

From the point of view of the designer and manufacturer of furniture for sitting, the best characteristics of use, in terms of stiffness of the seat, are characterised by models made using upholstery belts and two layers of polyurethane foams of varying stiffness.

## 8.8 Model of Interaction of the Human-Seat System

Due to human physiology, maintaining sitting position for a couple of hours is not beneficial for the nervous and musculoskeletal systems. Despite the fact that maintaining such a position is physically less tiring compared to the standing position, then with an incorrect position of the body it can cause a much greater (by approx. 40 %) load of the lumbar part of the spine. In addition, the wrong distribution of weight on a seat can cause point loads on the cardiovascular system. As a consequence, long-term positioning of the body in the wrong position, on an improperly fitted base, often causes pains, changes in degenerative arthritis, blood clots, as well as superficial inflammation of the venous system of lower limbs (Kamińska 2001).

Each day, one can see that the sitting position has dominated the human being's contemporary lifestyle. Even passively, he generally relaxes on furniture for sitting and resting, which most frequently cause only mental relaxation for the user, not physiological. This is caused by a poor distribution of stresses on the entire contact surface of the user's body with the flexible layer of the furniture piece. The first step to solve this problem can be the analysis of the human-seat system (Smardzewski et al. 2008), which will enable to assess the accuracy of the human body's position on a selected furniture piece for sitting.

The following parameters for assessing the body's position can be assumed as the most important (Będziński 1997):

- personal information—gender, age, somatic type, body weight and height,
- concerning the sections of the spine and defining their mobility, flexibility and shape of physiological curvatures,
- anthropometric—determining the dimensions and proportions between the basic sections of the body, and
- relating to the musculo-nervous system and musculoskeletal system.

Knowledge associated with being familiar with the mechanical and structural properties of the human body is a fundamental preliminary prerequisite for any theoretical, numerical or experimental approximations in the analysis of adjusting the technical means, which the seat is, to the physiological functions of the body.

Understanding the values of reaction forces at play when the human body is in contact with a technical object requires reducing the human body to the scheme of a multi-joint beam, where each element of the beam between the joints reflects a specific part of the body. (Figure 8.49). Therefore, the centres of gravity of certain parts of the body are defined, like the head, the torso—analysed together with the upper limbs, thighs, shanks and feet. Between these parts, joints are located which reflect the possibility of movement.

The places where reactions occur have been identified based on the measurement of the distribution of stresses using a sensory mat (Fig. 8.50). These studies were carried out on the model of an armchair with geometry that ensures minimal impact of mass forces on the user's body.

By replacing the actual object with a calculation model, simplifications have been introduced not only in the system of the beam, but also in the system of external forces, where the concept of concentrated loads has been applied. External forces were represented by the reaction-supporting system and the mass forces of body parts applied to a construction element (Zielnica 1996).

The presented calculation scheme of elements' reaction forces of the furniture piece on the human body-free load (Fig. 8.51) does not take into account the friction coefficient of each body part on the surface of the furniture piece and changes resulting from the tension of the human muscular system.

**Fig. 8.49** The human-seat system as a model of a joint beam: **a** the centre of gravity for individual parts of the body, **b** beam joint

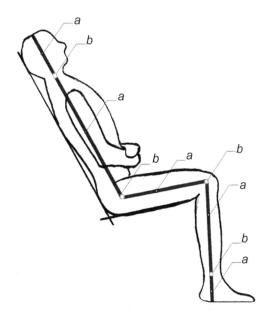

**Fig. 8.50** Distribution map of surface stresses

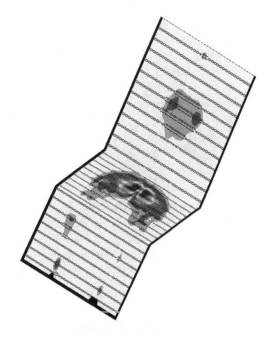

On the basis of the calculations conducted (Smardzewski et al. 2008), it was found that the reactions of supports affecting the beam diminish along with a reduction in the values of mass forces and represent a constant value of share percentage of the sum of all reactions of supports working on the model of the multi-joint beam.

**Fig. 8.51** The calculation
scheme of the base's reaction

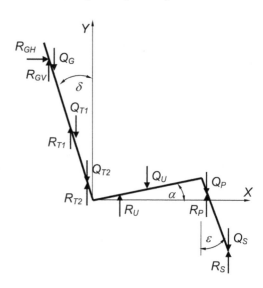

In the analysed structure, a certain regularity was found in the percentage distribution
of forces on individual parts of the body and they are as follows:

- the head—9.60 % of the sum of mass forces and 8.17 % of the sum of reactions
  of supports,
- the torso—57.40 % of the sum of mass forces, as well as 35.27 % for $R_{T1}$ and
  12.94 % for $R_{T2}$ of the sum of reactions of supports,
- the thigh—21.14 % of the sum of mass forces and 33.17 % of the sum of
  reactions of supports,
- the shank—8.96 % of the sum of mass forces and 6.91 % of the sum of reactions
  of supports and
- the feet—2.90 % of the sum of mass forces and 3.56 % of the sum of reactions
  of supports.

Furthermore, it is clear that the multi-point support of the body is beneficial for
the reduction of bending moments. If in the human-seat system, the human body is
supported by a greater number of points, therefore, the contact surface with the
piece of furniture for sitting is greater, then the forces that occur inside the body are
smaller. Limiting the forces affecting the user's body while sitting has a positive
impact on his musculoskeletal system, which is associated with comfort of use.

## 8.9  Numerical Modelling of Human-Seat Systems

Bedsores are a major problem for people who are physically handicapped and
forced to stay lying down or in a reclining position permanently. The practice of
dermatology shows that ulceration begins in the deeper tissues and spreads from

there outwards to the surface of the skin (Krutul 2004). At the same time, cracking of the skin and necrosis of adipose tissue is observed. The finite elements method is an excellent tool to simulate the phenomena in the scope of physical engineering, and with it, one can calculate and present anatomical processes that occur in the human body. Numerical modelling of the human body's tissues or its individual organs consists in discretization, using any flat or spatial elements that are the basic unit of algorithm of the finite elements method. Individual elements are connected with each other in the nodes, forming a flat or spatial grid, where each unit has attributed mathematical formulas describing its stiffness and treatability, as well as specific biomechanical properties. In the final stage of constructing the numerical model, the conditions of support and load are assigned. Hence, various forms of strengths, movements, pressures and other factors, which are necessary to reproduce natural load conditions, are also assigned to the object.

The problem of modelling the phenomenon of bedsores using the finite elements method has been presented in the work of Todd and Thacker (1994). It discusses the effect of the distribution of stresses caused by vertical forces representing gravity loads. However, it is known that horizontal forces appear when the user moves or rotates on a bed or seat. The resultant of vertical and horizontal forces gives the actual use load of the human body in contact with the base. According to Akimoto et al. (2007), such a load is a major cause of the rapid development of bedsores in the deeper layers of tissue. These authors also demonstrated that the application of special pads can reduce the occurrence of horizontal forces and the value of pressures on the resting body.

In the work of Akimoto et al. (2007), results of numerical calculations were presented the effect of the stiffness of a thin cushion on the value of pressures on the human body in standard operational loads. As a comparison model, a system without a cushion was made. The first numerical model developed presented the human body as a cylinder of equal stiffness. Another model consisted of two types of material: soft tissue and hard tissue. Soft tissue corresponded to the layout of the skin, adipose tissue and muscle tissue. Hard tissue constituted the equivalent of a bone. The numerical model, to simplify calculations, contained only the bottom symmetrical half of the cylinder. In cross section, the cylinder consisted of two concentric circles, of which the outer one depicted the soft tissue and the inner one depicted the hard tissue. The cushion was modelled as a layer of elements adjacent to the outer peripheral of the cylinder. The outer diameter of the cylinder, representing the layer of soft tissue, was 200 mm, while the inner diameter of the cylinder was 100 mm. It was also assumed that there is a contact between the cushion and the cylinder.

The actual, biologically living tissue is a nonlinear, anisotropic and viscoelastic material. For calculations, however, the authors adopted that the linear, isotropic body independent of time will correspond to soft tissue. The value of Young's module of soft tissue amounts to 15 kPa and Poisson's ratio 0.49. For the cushion, the value of Young's modulus was calculated using the $T$ coefficient, which figure was defined as the quotient of Young's modulus of cushion $E_c$ to Young's modulus of soft tissue of the user's body $E_{st}$,

$$T = \frac{E_c}{E_{st}}. \tag{8.194}$$

During each calculation cycle, the value of Young's modulus of cushion $E_c$ was changed using the $T$ coefficient equal to 1/1, 1/2, 1/4, 1/8 and 1/16. In clinical conditions, the patient usually does not take a permanent position but during treatments is moved or rotated. In order to represent this state of loads and movements, horizontal and vertical movements were added to the model, each with a value of 10 mm.

In the work of Akimoto et al. (2007), it was also assumed that the patient rests on a horizontal, hard bed, not intended for sleeping (Fig. 8.52). This bed has been reproduced as a horizontal line fixed across all nodes. It was also assumed that there will be contact between the bed and bottom edge of the cushion or bottom edge of the soft tissue layer. The coefficient of friction between the bodies has been established at level 1.

Two places have been specified, in which stress values have been marked reduced according to Misses: the contact border between the hard and soft tissue, and the middle of the layer of soft tissue. In the event of direct contact of the human body with a hard base, the stress value in contact with the hard and soft tissue amounted to 5.83 kPa, while inside the soft tissue it is 4.64 kPa. In systems in which the human body was supported by a flexible cushion, the stresses at the border of the hard and soft tissue and inside the soft tissue decreased along with the decreasing value of the linear flexibility modulus of the cushion.

Chow and Odell (1994) developed an axial symmetrical numerical model of the human buttock. The aim of their work was to determine the distribution of stresses in the soft tissues of the buttocks at various operational loads. The buttocks were modelled by a system consisting of a stiff core and a layer of soft tissue surrounding

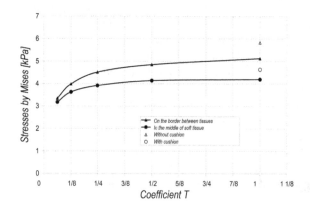

**Fig. 8.52** Dependencies of the stresses reduced according to Misses inside the soft tissue and on the border of contact of soft and hard tissues from the value of Young's modulus of cushion (Akimoto et al. 2007)

it as a hemisphere with properties of a linear elastic and isotropic body. This model was also used by Honma and Takahashi (2001), however, using more precise calculation models. In this work, this same model has been improved by its conversion from an axial symmetrical system to an asymmetrical system, taking into consideration the calculations in axially asymmetrical loads. As a result of the studies carried out, it has been shown that the state of usable loads increases the value of stresses in soft tissues and contributes to the faster development of bedsores. It has also been demonstrated that the use of soft cushions as layers supporting the buttocks has a beneficial effect on reducing stresses, especially in the layers of soft tissue. The more flexible the base, the lower the value of stresses on the border of soft and hard tissues, as well as in the middle area of soft tissues.

Brosh and Arcan (2000) presented a methodology to develop a realistic numerical model, using the finite elements method, a mutual effect of the anthropotechnic system human body-seat of the chair. The built model constitutes a corset of the chest and the lower part of the spine, surrounded by soft tissue. The properties of soft tissue were determined by the in situ method. The primary purpose of the work was to determine the behaviour of the soft tissues under load caused by a sitting position of the user. Energy deformations function:

$$W = (G/2)(I_1 - 3), \qquad (8.195)$$

where

$I_1$    the first variable factor of the deformations matrix, which has been described in the works of Chow and Odell (1994), Reddy et al. (1982) and Candadai and Reddi (1992)

The results of studies of numerical calculations were compared with the results of experimental measurements carried out in vivo in the system human-seat. Two approaches have been presented in the work to determine the shear modulus of soft tissues in a seated position of the user:

- the transformation of the modulus determined during contact to the modulus of figural deformations, and
- test of identification with treatability.

To determine the stresses between the body of the user and the horizontal rigid seat board, the method of displaying contact pressures was used (*Contact Pressure Display*) (Brosh and Arcan 1994, Arcan 1990). The centre of the load constituted a smooth, stiff sphere situated in the middle part of the soft tissue of the buttock. By loading the sphere, pressures of soft tissue on the seat were forced. The observed movements were measured using the LVDT system (Linear Variable Differential Transducer). By substituting subsequent values of loads and movements $(P, \delta)$, a graph was obtained that showed the behaviour of soft tissue during compression. Hence, it was calculated $G_1 = 11.7$ kPa and $G_2 = 33.8$ kPa. Then Brosh and Arcan (2000) built the two-dimensional axially symmetrical numerical model of the

human-seat system, made of the femur ($E = 20$ GPa, $v = 0.3$) and soft tissue, using
the values of moduli $G_i$ calculated above. The seat was defined as a flat board, with
a thickness of 30 mm. For a hard seat, it was assumed $E = 10$ GPa and $v = 0.3$, for a
semi-rigid seat $E = 20$ MPa and $v = 0.2$, while for a soft seat, $E = 3$ MPa and $v = 0.1$.
The calculations carried out gave convincing results that a change in the hardness of
the seat from a hard to semi-stiff one and then to soft causes a reduction of contact
stresses, respectively, by 54 % and 80 %.

Based on these studies, it has been demonstrated that soft tissues are a bimodular
material. Tissues accepts low value of shear strains when the contact stresses are
low. On this basis, one can look for new ways to an optimal design of anthropotechnic systems human-seat and to the ergonomic modelling of seats, by selecting
better linings, cushions or materials reducing contact stresses between the user's
body and the seat.

A numerical analysis of pressures of the buttocks of a physically impaired
person and sitting on a wheelchair was also presented by Linder-Ganz et al. (2005).
In particular in these studies, the distribution of stresses was defined in the middle
layers of soft tissue far from the surface of pressure. In the cited work, on the basis
of MRJ studies (Nuclear Magnetic Resonance), a cross section of a woman's hips
was established (29 years old, weight 54 kg) in a seated position. Based on this
data, a two-dimensional mesh model was build for numerical analysis using the
finite elements method and silicon phantom buttocks were constructed. The
phantom (Fig. 8.53) contained a model of bones, made of a rigid material of
Young's modulus 12 MPa, and the soft tissue surrounding the bone, which was
modelled using silicone of Young's modulus 1.6 MPa. In order to determine the

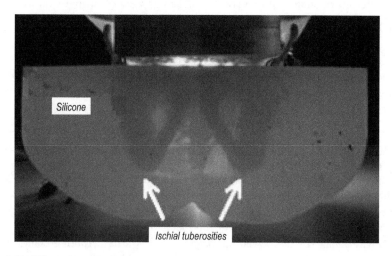

**Fig. 8.53**  Silicone buttock phantom (Linder-Ganz et al. 2005)

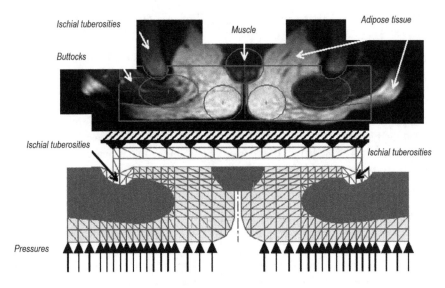

**Fig. 8.54** Mesh model: **a** actual cross section through the buttocks of a woman made using MRJ, **b** numerical model that includes the gluteal muscles, smooth muscles, adipose tissue and bone (Linder-Ganz et al. 2005)

internal pressure stresses, six ultra-thin pressure sensors were introduced to the model between IT (ischial tuberosities). Moreover, 14 sensors have been distributed on the seat surface in order to determine contact stresses. The silicone phantom was loaded with a weight from 50 to 90 kg. The measured stresses were compared with the results of numerical calculations.

The numerical model, necessary for calculations using the finite elements method, was developed on the basis of the analysis of the cross-sectional image of the buttocks, done using the MRJ method (Fig. 8.54). Individual materials in the numerical model were assigned the properties that a real silicone phantom had. Load, simulating the actual conditions of using an armchair for the disabled, was applied axially symmetrically, and also from the left and right side, bending the force vector by 15° to the left or right.

On the basis of the studies conducted, it was demonstrated that the value of stresses on the surface of the body is at the level of 0.4 to 63 kPa. A particularly high stress concentration is formed at the height of the ischiatic bone and amounts to 130 kPa. Inside the soft tissue—at contact of the ischiatic bone and soft tissue— compressive stress reaches a value of 160 to 200 kPa. The results of these studies enable to monitor in real time the actual pressure of stresses on the surface of the human body in a seated position. Furthermore, they enable more easily than before to plan exposure time of a paralysed person in a wheelchair.

## 8.10    Model of Interaction of the Human-Bed System

The development of mechanical models of interaction of the human-bed systems aims to determine the distribution of forces of mutual effect of the technical item and the human body. By using mesh phantoms, which are a reflection of the human body of the 5th, 50th and 95th centile (Fig. 8.55), the distribution of forces working in its support points can be determined. The calculation scheme of the mechanical human-bed system can be reduced to the form of a multi-joint beam with a length that corresponds to the height of the user's body. In the initial phase of building the model, the place of occurrence of reaction forces of the base has been established (like in the human-seat system) on the basis of the distribution of stresses of the body on the mattress, measured using a sensory mat of *Force Sensitive Applications®* system (Fig. 8.56) (Smardzewski et al. 2008).

Mass loads have been determined based on the coordinates of the position of the centres of gravity of individual parts of the body (Gedliczka 2001). The static scheme of the joint beam with points of application of active forces, corresponding to gravity forces of individual body parts, and reaction forces of the base have been illustrated in Fig. 8.57.

The calculations conducted show that the share percentage of pressure in a selected point of support in relation to the sum value of support forces of the tested body model is constant and does not depend on the anthropometric measure. Share of pressures of the smallest values were reported in the support point of the head (approx. 6 %) and feet (approx. 8 %). However, it should be noted that the contact of the body in these places takes place on a very small area, which in turn leads to the formation of very high stresses inside the tissues. The largest value of the base reaction on the user's body is observed at the contact site of the torso with the base. This is reflected in about a 44 % share of the total sum of all reaction forces.

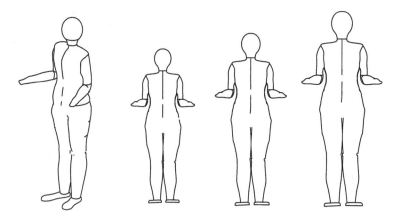

**Fig. 8.55** Models of the user's body corresponding to the measures of the 5th, 50th and 95th centile

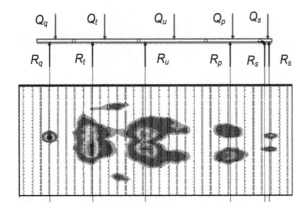

**Fig. 8.56** Determining the distribution of reaction forces of the base for the user in a lying down position

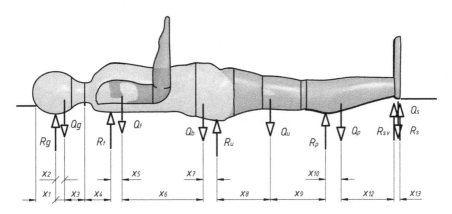

**Fig. 8.57** Static scheme of the human-bed system, where: $Rg$—base reaction at the contact site of the head, $Rt$—base reaction at the contact site of the torso, $Ru$—base reaction at the contact site of the thigh, $Rp$—base reaction at the contact site of the shank, $Rs$—base reaction at the contact site of the foot, $Rsv$—base reaction at the contact site of the additional support of the foot, $Qg$—centre of gravity of the head, $Qt$—centre of gravity of the torso, $Qu$—centre of gravity of the thigh, $Qp$—centre of gravity of the shank, $Qs$—centre of gravity of the foot

By increasing the number of supports (places of contact of the body with the base), not only the values of reaction forces can be reduced, but also the adverse high values of bending moments. In Fig. 8.58, the courses of bending moments have been compared, in the user's body, before and after introducing additional supports of the torso and thigh. Based on this, a clear reduction in the value of the bending moments can be observed, and as a result, an increase in the comfort of using the furniture piece.

**Fig. 8.58** Comparing the
course of variability of
bending moments in the
user's body, before and after
the introduction of additional
support at the contact site of
the torso and thigh with the
base

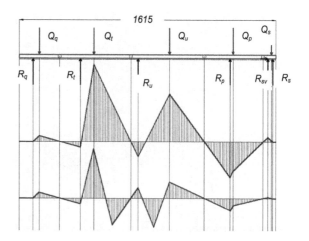

The analysis of the mechanical human-bed system provides information about the locations of concentrated forces occurring, being the reaction of the base's impact on the body. The analytical model also enables to specify the distribution of internal forces, especially bending moments, acting on a body that is resting on an upholstered base.

## 8.11  Numerical Modelling of Human-Bed Systems

Studies confirm that pressure in healthy blood vessels of the skin amounts to 32 mmHg (4.3 kPa) and enables proper blood circulation (Krutul 2004). By examining the stresses on muscle tissue, it was found that stresses up to 34.6 kPa lasting for 35 min cause its stiffening, which leads to pressure pain (Gefen et al. 2005). Adverse loads on soft tissues of the human body, caused by lying down on a base that is too hard, can be reduced by proper support of the user's torso using a soft and flexible material.

The conditions of the effect of the base on the human body in a lying down position can be illustrated using numerical calculations, using the algorithm of the finite elements method. Based on the transverse cross-sectional model of the human body, established at the height of the chest (Fig. 8.59), from the atlas of anthropometric characteristics measurements (Gedliczka 2001) the mass and dimensions of a person constituting actual load of the mattress have been determined. For the anatomical model selected in this way, using scanning, a two-dimensional mesh of finite elements was applied. During scanning, it was ensured that individual parts of the human body were covered by various grids, with varying degrees of density, depending on the type of tissue and skeletal system. In the static system, the symmetrical half of the analysed object was assumed for calculations (Fig. 8.60).

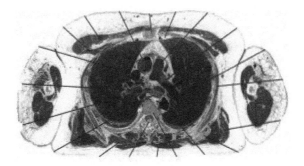

**Fig. 8.59** Transverse cross section of the chest at shoulder-height (http://www.meddean.luc.edu)

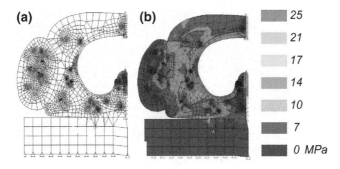

**Fig. 8.60** Mesh model reflecting: **a** the cross section of the human body on a flexible mattress, **b** the state of stresses in the human-bed system

The created cross-sectional model of the human body was propped up on a flexible mattress, and then it was assigned support bonds enabling vertical shifts. By defining the contact between the mattress and the human body, contact points were identified between the outer surface of the selected cross section of the body and the upper plane of the mattress. It was also assumed that the external load will be caused only by forces of gravity. For individual parts of the human body, the elastic properties provided by Gefen et al. (2005) were assumed and summarised in Table 8.6.

By analysing the contact between the human body and the mattress and the impact of flexibility of the mattress on the values of stresses in the human body,

**Table 8.6** Elastic properties of human body parts (Gefen et al. 2005)

| Type of tissue | Poisson's ratio | Young's modulus $E_t$ (kPa) | Shear modulus $G_t$ (kPa) |
|---|---|---|---|
| Bone tissue | 0.3 | $22.5 \times 10^6$ | $865 \times 10^6$ |
| Muscle tissue | 0.3 | 937 | 660 |
| Fat tissue | 0.3 | 100 | 38.4 |

a series of calculations were carried out, respectively, for a representative female, with anthropometric characteristics constituting the scale of the 50th centile (weight 65 kg), as well as the 95th centile (weight 87.8 kg). The results of the numerical calculations were presented in the system $\sigma_z = f(E_t/E_m)$, where $E_t$—module of tissue flexibility, $E_m$—module of mattress flexibility, for:

- point A located inside the mattress,
- point B on the contact surface of the body with the mattress, and
- point C inside the human body.

They were also illustrated in Figs. 8.61, 8.62 and 8.63.

It can be concluded from Fig. 8.61 that the stresses inside the mattress decrease together with the reduction of its stiffness. For a user from the 95th centile, stresses inside the mattress are about 22–25 % greater in relation to the stresses caused, in the same places, by a user of the 50th centile.

By analysing the contact stresses in point B (Fig. 8.62) it can be concluded that their value decreases, along with the reduction of the stiffness of the mattress to the value corresponding to the stiffness of human tissue. Then increases together with a clear decrease in the stiffness of the mattress in relation to the stiffness of the tissue.

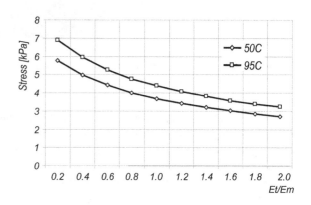

**Fig. 8.61** The stresses inside the mattress in point A, in the function of the coefficient $E_t/E_m$

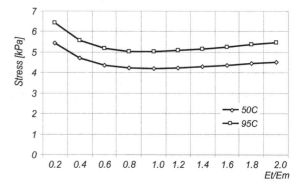

**Fig. 8.62** The stresses at the contact site of the mattress with the human body (point B), in the function of the coefficient $E_t/E_m$

**Fig. 8.63** Stresses inside the human body in point C, in the function of the coefficient $E_t/E_m$

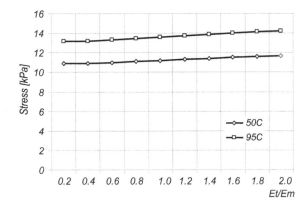

The smallest stress value, 4.2 kPa for the 50th centile and 5.0 kPa for the 95th centile, was obtained using the proportion $E_t/E_m$ equal to 1.

The changes of stresses shown in Fig. 8.63 in the function of the coefficient $E_t/E_m$ set out in point C have demonstrated that while lying down on the back, stresses inside the human body increase slightly along with a decrease in stiffness of the mattress. With a 20-fold decrease in the stiffness of the mattress, the stresses inside the human body increased linearly by 6.7 %, thus by 0.34 % for every 20 % in the reduction of stiffness of the mattress. The calculations carried out prove that we obtain the correct stiffness of the spring layer of the mattress if we use materials characteristic of Young's modulus similar or lower than the linear elastic modulus of the human body's soft tissue.

# References

Akimoto M, Oka T, Oki K, Hyakusoku H (2007) Finite element analysis of effect of softness of cushion pads on stress concentration due to an oblique load on pressure sores. J Nippon Med Sch 74(3):230–235

Anonim (2000a) ABAQUS Theory Manual, 6th edn. Hibbitt, Karlsson & Sorensen, Inc. USA

Anonim (2000b) ABAQUS User's Manual, 6th edn. Hibbitt, Karlsson & Sorensen, Inc. USA

Arcan M (1990) Non-invasive and sensor techniques in contact mechanics: a revolution in progress. In: Proceedings of the ninth international conference on experimental mechanics, vol 1. Copenhagen, p 1–18

Ashman R, Rho Y (1988) Elastic modulus of trabecular bone material. J Biomech 21(3):177–181

Brosh T, Arcan M (1994) Toward early detection of the tendency to stress fractures. Clin Biomech 9:111–116

Brosh T, Arcan M (2000) Modeling the body/chair interaction—an integrative experimental—numerical approach. Clin Biomech 15:217–219

Będziński R (1997) Biomechanika inżynierska. Zagadnienia wybrane. Oficyna Wydawnicza Politechniki Wrocławskiej, Wrocław

Candadai RS, Reddi NP (1992) Stress distribution in a physical buttock model: effect of simulated bone geometry. J Biomech 25(12):1403–1411

Chabanas M, Luboz V, Payan J (2002) Patient specific finite element model of the face soft tissues for computer—assisted maxillofacial surgery. Med Image Anal 7:131–151

Chow WW, Odell EI (1994) Deformations and stress in soft body tissues of sitting person. J Biomech Eng 100:79–87

Czysz HJ (1986) Experimentale Untersuchungen an Polyurethan-Iintegralweichschaum zur Bestimmung von Kennwerten zu Berechnungsgrundladen. Ph.D dissertation, Universitat der Bundeswehr Hamburg

Deuflhard P (2003) Biomechanical modeling of soft tissue and facial expressions for craniofacial surgery planning. Freien Universitat, Berlin, pp 60–63

DIN EN ISO 3386-1:2010-09 Polymeric materials, cellular flexible—Determination of stress-strain characteristics in compression—Part 1: Low-density materials

Douglas T (2000) Empirical relationships between acoustic parameters in human soft tissues. Acoust Res Lett Online 1(2):37–42

Enderle J, Blanchard S, Bronzino J (2000) Introduction to biomedical engineering. CA Academic, San Diego, pp 46–72

Fung Y (1993) Mechanical properties of living tissues. Springer Verlag, New York, pp 321–391

Gedliczka A (2001) Atlas miar człowieka. Dane do projektowania i oceny ergonomicznej. Centralny Instytut Ochrony Pracy, Warsaw

Gefen A, Chen J, Elad D (1999) Stresses in the normal and diabetic human penis following implantation of an inflatable prosthesis. Med Biol Eng Comput 37:625–631

Gefen A, Gefen N, Linder-Ganz E (2005) In vivo muscle stiffening under bone compression promotes deep pressure sores. J Biomech Eng 127:512–524

Gerard JM (2004) Indentation for estimating the human tongue soft tissues constitutive law: application to a 3d biomechanical model. In: Cotin S, Metaxas D (eds) ISMS 2004, LNCS 3078. Springer, Berlin Heidelberg, pp 77–83

Ghadiali S (2004) Finite element analysis of active eustachian tube function. J Appl Physiol 97:648–654

Golombeck M (1999) Magnetic resonance imaging with implanted neurostimulators: a first numerical approach using finite integration theory. Physical properties of human tissue. University of Karlsruhe, Germany

Guo X (2001) Mechanical properties of cortical bone and cancellous bone tissue. In: Cowin SC (ed) Bone mechanics handbook. CRC, Boca Raton, FL, pp 10–23

Guzik P (2001) Ocena adaptacji układu krążenia do zmiany kąta pochylenia w przebiegu próby pionizacji. Ph.D dissertation, Medical University, Poznań

Hazelwood S (1998) An adaptation simulation to predict bone remodeling around implant stems following hip replacement surgery. In: North American Congress on Biomechanics, Universty of Waterlo, Waterlo, Ontario, Canada, August 1998, p 14–18

Hill R (1978) Aspects of invariance in solid mechanics. Adv Appl Mech 18:1–75

Honma T, Takahashi M (2001) Stress analysis on the sacral model for pressure ulcers. Jpn J Press Ulcers 3:20–26

Hu T, Desai JP (2005) Characterization of soft-tissue material properties: large deformation analysis. In: Program for robotics, Intelligent Sensing, and Mechatronics (PRISM) Laboratory 3141 Chestnut Street, MEM Department, Room 2-115, Drexel University, Philadelphia, PA 19104

ISO 8307:2007 Determination of resilience by ball rebound

Kamińska J (2001) Opracowanie metody oceny obciążenia systemu mięśniowego pleców i kręgosłupa przy pracach siedzących w funkcji rodzaju siedziska. Centralny Instytut Ochrony Pracy Państwowy Instytut Badawczy, Warsaw

Krutul R (2004) Odleżyna, profilaktyka i terapia. Revita

Lees C, Vincent J, Hillerton J (1991) Poisson's ratio in skin. Biomed Mater Eng 1(1):19–23

Linder-Ganz E, Gefen A (2004) Mechanical compression—induced pressure sores in rat hindlimb: muscle stiffness, histology and comptational models. J Appl Physiol 96:2034–2049

Linder-Ganz E, Yarnitzky G, Portnoy S, Yizhar Z, Gefen A (2005) Real-time finite element monitoring of internal stresses in the buttock during wheelchair sitting to prevent pressure sores: verification and phantom results. In: Rodrigues H et al (ed) II International Conference on Omputational Bioengineering. Lisbon, Portugal, September 14–16

Mangurian L, Donaldson R (1990) Development of peroxisomal betaoxidation activities in brown fat of perinatal rabbits. Bid Neonate 57:349–357

Mills NJ, Gilchrist A (2000) Modeling the indentation of low density polymer foams. Cell Polym 19:389–412

Morcovescu V, Dragulescu D (2002) Reconstruction of the human femur based on the ct slices to perform the finite element analysis. Buletinul Stiintific Al Universitatii Politehnica Din Timisoara, Seria Mecanica 47(61):55–62

Ogden RW (1972) Large deformation isotropic elasticity: on the correlation of theory and experiment for compressible rubber like solids. Proceedings of the Royal Society of London, Series A. 326:565–584

Patil K, Braak L, Huson A (1996) Analysis of stresses in two—dimensional models of normal and neuropathic feet. Med Bid Eng Comput 34:280–284

PN-77/C-05012.03 Methods of testing flexible porous materials—determining apparent density

PN-77/C-05012.10 Methods of testing flexible porous materials—determining permanent deformation

Prior B (2001) Muscularity and the density of the fat—free mass in athletes. J Appl Physiol 90:1523–1531

Qunli S (2005) Finite element modeling of human buttock—thigh tissue in a seated posture. In: Summer Bioengineering conference, Vail Cascade Resort & Spa, Vail, Colorado, 22–26 June 2005

Reddy NP, Patel H, Cochran GVB, Brunski JB (1982) Model experiments to study the stress distribution in a seated buttocks. J Biomech 15(7):493–504

Renz R (1977) Zum zugigen und zyklischen Verformungsverhalten polimerer Hartschaumstoffe. Ph.D dissertation, TH Karlsruhe

Renz R (1978) Modellvorstellungen zur Berechnung des mechanisches Verhaltens von Hartschaumstoffen. In: Schaumkunstoffe, Fachferband Schaumkunstoffe e.V. Dusseldorf

Rho Y, Ashman R, Turner C (1993) Young's modulus of trabecular and cortical bone material: Ultrasonic and microtensile measurements. J Biomech 26(2):111–119

Schneck D (1995) An outline of cardiovascular structure and function. In: Bronzino J (ed) The biomechanical engineering handbook, vol 3. CRC/IEEE, Boca Raton FL 3

Schrodt M et al (2005) Hyperelastic description of polymer soft foams at finite deformations. Tech Mech 25:163–173

Smardzewski J, Grbac I, Prekrat S (2007) Nonlinear elastic of hyper elastic furniture foams. In: 18th Medunarodno Zanstveno Svjetovanje Ambienta'07, University of Zabreb, p 77–84

Smardzewski J, Kabała A, Matwiej Ł, Wiaderek K, Idzikowska W, Papież D (2008) Antropotechniczne projektowanie mebli do leżenia i siedzenia. In: Raport końcowy projektu badawczego MNiSzW nr 2 PO6L 013 30, umowa nr 0998/P01/2006/30

Srinivasan S (1999) 3-D global/local analysis of composite hip prostheses—a model for multiscale structural analysis. Compos Struct 45:163–170

Storakers B (1986) On material representation and constitutive branching in finite compressible elasticity. J Mech Phy Solids 34(2):125–145

Taylor M (1999) A combined finite element method and continuum damage mechanics approach to simulate the in vitro fatigue behavior of human cortical bone. J Mater Sci—Mater Med 10:841–846

Todd BA, Thacker JG (1994) Three-dimensional computer model of the human buttocks, in vivo. J Rehabil Res Dev 31:111–119

Wang Y, Lakes R (2002) Analytical parametric analysis of the contact problem of human buttocks and negative Poisson's ratio foam cushions. Int J Solids Struct 39:4825–4838

William M (1993) Modelling the mechanics of narrowly contained soft tissues: the effects of specification of Poisson's ratio. J Rehabil Res Dev 30(2):205–209

Zielnica J (1996) Wytrzymałość materiałów. Wydawnictwo Politechniki Poznańskiej, Poznań

CPSIA information can be obtained
at www.ICGtesting.com
Printed in the USA
LVHW080812100520
655298LV00001B/2